D1244120

IMPERIAL CHEMICAL INDUSTRIES: A HISTORY

VOLUME II The First Quarter-Century 1926–1952

IMPERIAL CHEMICAL INDUSTRIES

A HISTORY

VOLUME II The First Quarter-Century 1926–1952

W. J. READER

Research by
Astrid Baker   Pamela Jenkins
Elspeth N. D. Jervie   Rachel Lawrence

Technical Advisers
F. G. Lamont   J. D. Rose FRS

LONDON
OXFORD UNIVERSITY PRESS
NEW YORK   TORONTO
1975

Oxford University Press, Ely House, London W.1

GLASGOW NEW YORK TORONTO MELBOURNE WELLINGTON
CAPE TOWN IBADAN NAIROBI DAR ES SALAAM LUSAKA ADDIS ABABA
DELHI BOMBAY CALCUTTA MADRAS KARACHI LAHORE DACCA
KUALA LUMPUR SINGAPORE HONG KONG TOKYO

ISBN 0 19 215944 5

Printed in Great Britain at THE KYNOCH PRESS, *Birmingham*

Contents

PART III: THE GREAT DISTRACTION: REARMAMENT AND WAR 1935–1945

PART IV: ORGANIC SYNTHESIS: THE EMERGENCE OF THE NEW CHEMICAL INDUSTRY 1927–1952

List of Illustrations

viii

In Text

Figures

Tables

Abbreviations

THIS LIST covers the abbreviations used for sources. For abbreviations in the text, see index.

Volume I of this History (*The Forerunners, 1870–1926*) is cited as *ICI* I. Other published works, on first reference, are cited by author's name and full title; later, by author's name only. The full titles of all published works cited are listed under 'Published Sources' in Appendix v.

For other documents several times referred to (listed in full under 'Unpublished Sources'), the following abbreviations are used:

AD	Papers from ICI Agricultural Division Central File.
AGM	Reports of ICI General Meetings.
Bradley	Materials collected and some draft text for *History of ICI's War Effort* by F. Bradley.
Bunbury	Papers collected by H. M. Bunbury for his MS. 'History of ICI 1926–1939'.
Cab.	Cabinet Papers in Public Record Office.
CACM	ICI Central Administration Committee Minutes.
CCN	ICI Chairman's Conference Notes.
ChaP	Sir Paul Chambers's papers from ICI Head Office Records Centre.
CP	Sir William Coates's papers from ICI Head Office Records Centre.
CPCM	ICI Capital Programme Committee Minutes.
CR	Central Registry papers from ICI Head Office Records Centre.
DECP	ICI Development Executive Committee Papers.
ECM	ICI Executive Committee Minutes.
ECSP	ICI Executive Committee Supporting Papers.
FCM	ICI Finance Committee/Finance Executive Committee Minutes.
FCSP	ICI Finance Committee/Finance Executive Committee Supporting Papers.
FD	ICI Foreign Department Files from ICI Head Office Records Centre.
GPCM	ICI General Purposes Committee Minutes.
GPCSP	ICI General Purposes Committee Supporting Papers.
ICIANZLM	Minutes of Meetings of London Directors of ICIANZ Ltd. (at first sometimes called London Committee of ICIANZ or Australian Committee of ICI).

ICIBM   ICI Board Minutes.

ICIBR   Reports to ICI Board.

LDHC   Papers from Library Department Historical Collection, ICI Organics Division, Blackley.

MBM   ICI Management Board Minutes.

MBSP   ICI Management Board Supporting Papers.

MP   Papers of Henry Mond (2nd Lord Melchett from 27 Dec. 1930) from ICI Head Office Records Centre.

NLL   National Lending Library for Science and Technology papers.

OECP   ICI Overseas Executive Committee Papers.

PWWCP   ICI Post-War Works Committee Papers.

SCP   ICI Secretary's Confidential papers.

Stevens   F. P. Stevens's typescript 'History of ICI 1939–1945', 19 Feb. 1963.

TP   B. E. Todhunter's papers from ICI Head Office Records Centre.

USA   Documents and commentaries prepared for the defence to the action *USA v. ICI et al.*, from ICI Legal and Foreign Departments. For some of the sub-headed collections the following abbreviations are used:

     CSAE   ICI Legal Department 'Write-up'—'The Operations of CSAE [Compania Sud Americana de Explosivos] from 1925 down to the Present Day. Documentary Evidence for Outline Brief for the Defendant ICI', 6 volumes.

     DA   ICI Legal Department 'Write-up'—'Duperial Argentina, Phases I-VIII'.

     DB   ICI Legal Department 'Write-up'—'Duperial Brazil', in two parts with Appendices.

WR   War Records from ICI Head Office Records Centre.

WP   J. H. Wadsworth's papers from ICI Head Office Records Centre.

'Groups' in ICI Organization

From 1929 to the end of 1943 the main operating units of ICI in the United Kingdom were known as Groups, e.g. Alkali Group, General Chemical Group, Dyestuffs Group.

Under the reorganization scheme of 1944 these Groups were renamed Divisions, e.g. Alkali Division, General Chemicals Division, Dyestuffs Division.

These Divisions were formed into Groups, e.g. Group A, consisting of Alkali Division, General Chemicals Division, Lime Division, Salt Division; Group B, consisting of Dyestuffs Division, Pharmaceuticals Division.

Contemporary documents are not consistent in their references to Groups and Divisions, hence the text is not always consistent.

Author's Preface

WHAT I HAVE attempted to do in this volume is to provide an account of the way ICI was run in the quarter-century between 1926 and 1952. It is no part of my purpose to examine the workings of the business at the present day: that does not lie within the historian's field. During the period which I cover, those in charge of ICI's affairs had to face world-wide economic disaster in the thirties, world war in the forties, and the beginning, in the early fifties, of economic expansion in the industrial countries which immensely raised living standards but at the same time created tensions within nations and between nations which only began to make themselves fully apparent in the sixties and seventies. The chemical industry, during most of these twenty-five years, was undergoing a technological revolution more thoroughgoing than anything of that kind that had been seen since the late eighteen-sixties. It is against this background that the men and businesses discussed in the pages that follow, and their performance under conditions as testing as any that British business men have ever had to face, should be seen; and the standards applied should be those of their own day, not ours. It will fall to other hands than mine to present an account of ICI during the twenty-odd years since 1952.

My general statement of purpose and method appears in the Preface to Volume One and I need not repeat it here, but I should not like to finish my task without saying, once again, how extremely grateful I am to many people in ICI and elsewhere who have helped me at every stage of it.

For work specifically connected with this volume I should like to thank Mr. M. E. T. Boddy of ICI Records Centre; Miss U. M. Price and her staff in the Head Office Library; Mr. David Pratt of the Design and Production Section of Publicity Services; Miss J. H. Much and Miss Ann Smith in Secretary's Department; Mrs. M. Palmer, who has made the index; and Messrs. W. N. Cambridge, H. Gardner and M. W. Brothers who have provided financial data. Mr. S. D. Lyon and Mr. R. W. Pennock, Directors, have been kind enough to read the text, and I should like also to thank Mr. John K. Jenney and Mr. David Conklin, of du Pont, for a similar service. The names of those who have given me information, either verbally or through the loan of papers, are in Appendix v.

To those who have seen the entire work through to completion, many of them from start to finish and whose names appear in the Preface to

Volume One, I should like once again to offer my thanks for their goodwill and hard work over the years since 1964.

I hope the finished work will not disappoint any who have had a hand in it, for this is their book as well as mine.

October 1974   W. J. READER

PART I

HOPE AND HIGH PRESSURE 1926–1930

Figure 1. ICI Works in Britain, 1927

Source: Imperial Chemical Industries Record, Jan. 1927.

CHAPTER I

ICI at the Beginning:
an Introductory Survey

IN 1926 IMPERIAL CHEMICAL INDUSTRIES came into existence after a merger between the four largest companies in the British chemical industry:

Brunner, Mond & Co. Ltd.
Nobel Industries Ltd.
The United Alkali Co. Ltd.
The British Dyestuffs Corporation Ltd.

This merger, in fact though not in law, represented a takeover by two strong businesses—Brunner, Mond and Nobel Industries—of two others —United Alkali Company and British Dyestuffs Corporation—which were much weaker. United Alkali Company, indeed, was weak to the point of extinction, and although British Dyestuffs Corporation eventually proved to have plenty of life in it, it was not, at the moment of merger, in any condition to face Nobels or Brunner, Mond on a footing of equality. The truth of the situation was made brutally clear by the composition of the original Board of ICI, on which the Chairman, Sir Alfred Mond, the President, Sir Harry McGowan, and all six other full-time Directors came, in equal numbers, from Nobels and Brunner, Mond. The other partners had to be content with part-time representatives.

From the start, therefore, the nature of ICI was strongly influenced by the strengths and weaknesses of Nobels and Brunner, Mond which in important respects were complementary. Nobel Industries, already rationalized and organized under a holding company, had large resources in cash and securities but nothing very obvious to lay them out on. It was a group looking for a project. Brunner, Mond, rudimentary (by comparison) in organization, had a project—ammonia synthesis at Billingham—which had already outgrown its resources. It was natural, in consequence, that ICI should take over the principles and much of the structure of Nobel Industries' organization, as well as its resources, and apply them to the development of fertilizer plant, based on ammonia synthesis, at Billingham. Sir Harry McGowan, lately Chairman of Nobels, was well content, as President of ICI, to see

Nobels' wealth applied to what was acknowledged to be an outstanding technical achievement which seemed to be near the point of yielding very large profits indeed from supplying nitrogenous fertilizer to the farmers of the British Empire.

The technology of ICI at the outset, then, took its tone from Brunner, Mond and hence from the heavy chemical industry based on alkali in which for over fifty years they had been so successful. Success in alkali continued, and throughout the years between 1926 and 1939 the Alkali Group, selling comparatively simple products in huge quantities at prices yielding a moderate return on capital employed, earned profits without which, in the thirties, ICI as a whole would have been in danger of foundering.

With the concentration on large-scale operations at Billingham, where £20m. capital was invested during ICI's first four years, went a tendency at the top of ICI to undervalue the dyestuffs industry and the possibilities of the organic side of the chemical industry as a whole. Alfred Mond, indeed, would have been prepared to get rid of ICI's interest in dyestuffs to the Germans in return for their technical knowledge of the process for producing petrol from coal.[1] To the mind of one brought up in heavy chemicals, used to continuous processes delivering hundreds of thousands of tons a year, there was something piddling and inconsiderable about the 'pot and kettle' batch processes of the dyestuffs industry, in which a hundred tons of one product would be a large output, even though high scientific ability lay behind these tiny operations and a successful invention would yield a return on capital inconceivable in the alkali trade. Moreover the deplorable history of the dyestuffs industry in England generally,[2] and of BDC and its predecessors in particular, had produced on the dyestuffs side of ICI an air of penny-pinching gloom most uncongenial to anyone used to the expansive and cheerful affluence of Winnington Hall in Alkali Group.

Since the organic side of the chemical industry seemed of little consequence—in spite of the fact that it was the foundation of the activities of ICI's much feared and much respected rival, *IG Farbenindustrie AG*—very little provision was made for it by the founders of ICI. Dyestuffs Group, as we have remarked, was not represented by an executive Director on the main Board, which meant that the Group had no permanent 'friend at Court'. There was no attempt to bring together, for the benefit of ICI as a whole, the considerable body of knowledge of organic chemistry which lay scattered not only in Dyestuffs Group but in Explosives Group and elsewhere as well, so that when the potential importance of plastics began to be recognized, in the thirties, the vested interests of several groups clashed and it was difficult to get a united effort going. On the 'heavy organic' side, already important in 1926 and much more important later, ICI's control of the production process was incomplete. Ethylene and acetylene, for instance, were made respectively from alcohol and from carbide bought from outside suppliers. On the 'fine chemicals' side ICI, in strong contrast to IG, was not in pharmaceuticals or photographic chemicals at all, nor had the founders any plans for going into them.

The business which ICI took over from its predecessors was thus heavily biased towards the 'heavy'* chemical industry, and moreover the founders were disposed to keep it so. The decision to concentrate investment at Billingham pushed it still further in the same direction, and once that decision was taken there were not the resources, even if there had been the inclination and knowledge, to launch out on any other equally ambitious venture. Neglect and dispersion of effort on the organic side, being built into the original organization and purposes of ICI, had a very long-lasting effect, especially noticeable in a tardy and hesitant approach to plastics in the thirties.

When ICI was founded it was almost entirely a group of businesses supplying materials or finished products required, not by the final consumer, but by other producers, whether manufacturers, miners, civil engineers, or farmers. This was what the founders thought of as the proper function of ICI. They did not see ICI as the first Lord Leverhulme had seen Lever Brothers Ltd.—that is, as makers of finished goods to be sold under brand names to the public at large. Indeed in the general ICI attitude to such matters, for a good many years, it is possible to trace disdain. The technology of the consumer-goods industries seemed hardly worth a scientist's attention, and besides that, having come lately on the scene, they formed no part of the rugged structure of the 'basic industries' which seemed to be the indispensable foundation of British industrial strength. The leaders of ICI, like the shipbuilders and heavy engineers, were inclined to regard the manufacture and, still more, the advertising and marketing of consumer goods with contempt. In 1941, considering the possibility of some sort of merger between ICI and Distillers Company Ltd., McGowan wrote of 'the whiskey trade' (his spelling) as undesirable for ICI to be associated with, which seems unkind from so appreciative a consumer of its products.[3]

Since ICI supplied such a wide range of industrial materials, and in many of the most important, such as alkali and explosives, held what amounted to a monopoly, it followed that ICI had customers throughout British industry, and some of the most important were among the

---

* 'Heavy' refers to the scale of operations, not to the physical nature of the materials. In alkali and fertilizers, in the 1920s, it already implied continuous production with an annual output of hundreds of thousands of tons. 'Heavy organic chemicals' came nowhere near those figures until the 1950s, but when ICI was founded they were already being produced in quantities much larger than the organic side of the chemical industry had formerly been used to.

largest and most powerful firms in the country: Courtaulds Ltd., Lever Brothers Ltd., Distillers Company Ltd., Pilkington Brothers Ltd. Some of them, as well as buying supplies from ICI, also sold supplies to them, as the Distillers sold products based on industrial alcohol and Lever Brothers sold glycerine.

There was always the possibility that customers and suppliers might become competitors, either through the development of new products or in the course of industrial warfare. How disastrous the latter might be had been shown in the long struggle between Brunner, Mond and Lever Brothers between 1911 and 1925, when Levers had reacted against what they considered Brunners' unreasonable exercise of monopoly power in the supply of alkali by ostentatiously preparing to make alkali themselves, and Brunner, Mond, in response, had invaded the soap industry.[4] That episode had ended, just before the merger, with £1m. damages against Brunners and with murder and suicide. Clearly ICI would go to great lengths to avoid anything of the sort ever happening again.

ICI's settled policy, indeed, was to avoid competition with suppliers and customers, even if that meant forgoing attractive opportunities in new fields of the chemical industry. Thus between the wars ICI kept out of heavy organic chemicals to avoid competition with Distillers; out of rayon to avoid competition with Courtaulds; out of detergents to avoid competition with Unilever; and they accepted restrictions on their business in 'Perspex'—an ICI invention—to avoid competition with Pilkingtons. Any suggestion for a new venture would be met with cautious inquiries about its effect on undertakings, formal or informal, given to firms with which ICI had trading relations. Sometimes, as in the case of the 1939 agreement with Courtaulds on nylon, competition could be avoided by co-operation rather than abnegation, but the general effect of the disinclination to compete with trading partners was very restrictive. Inside ICI, where debate on policy was often outspoken, it was criticized, but it remained a ruling principle until a combination of circumstances, towards the end of the Second World War, caused it to begin to break down.

This outlook was consistent with the general principles of commercial policy inherited from the predecessors of ICI, especially Brunner, Mond and Nobels. There might be occasional aberrations, such as the fight between Brunners and Levers, but in general, for as long as the founders of ICI could remember—and longer—the heavy chemical industry and the explosives industry had been run, very successfully, within world-wide systems of diplomacy designed to regulate markets, to share trade, in all ways to seek agreement rather than conflict both at home and abroad. Brunner, Mond's affairs were governed by their agreements with Solvay et Cie; Nobels', by agreements with du Pont and (before

6

1914) with the Germans; and both had numerous subsidiary agreements covering specific products and markets.

McGowan's own reputation was founded on his skill in precisely this field of activity, not on technological or scientific eminence, which he never claimed. The stages of his career had been marked by a series of mergers, culminating in ICI itself, and the foundation of his foreign policy was Nobels' alliance with du Pont, which he was determined to carry over into ICI. Writing in 1941 to the second Lord Melchett, McGowan said he did not accept 'the theory . . . that competition is essential to efficiency'.[5] This was not a chance remark. It was the ultimate distillation, as he saw it, of his life's experience.

Sir Alfred Mond was of a like mind. Surveying the unpleasing industrial landscape of Great Britain in the early twenties, he saw the 'basic industries'—shipbuilding and heavy engineering; iron and steel; coal; cotton textiles—in the grip of a depression which had settled on them in 1921 and from which they have never since permanently emerged. The chemical industry, apart from dyestuffs, was not among the worst victims of the post-war depression, but Sir Alfred, on an all-embracing view, was convinced that radical reorganization and concentration, if necessary on an international scale, was needed to put matters right. He claimed to have invented the term 'rationalization' and in the South Wales coal industry, before the formation of ICI, he had already had a chance to try out his ideas by the formation of Amalgamated Anthracite Collieries, of which he was Chairman. Hence the attraction for him of ICI, in which he could operate in a more promising field on a much larger scale, and in an industry, moreover, to which he felt a strong sense of family loyalty.

In forming ICI both McGowan and Mond had the largest ideas. 'As becomes more apparent day by day,' wrote Mond in 1927, 'the trend of all modern industry is towards greater units, greater co-ordination for the more effective use of resources. . . . But this process . . . is leading towards a further series of economic consequences. One of the main consequences is the creation of inter-relations among industries which must seriously affect the economic policies of nations.'[6] McGowan was more succinct and, if possible, even more ambitious. The formation of ICI, he told a representative of du Pont in December 1926, 'is only the first step in a comprehensive scheme . . . to rationalise [the] chemical manufacture of the world.'[7]

This brings us to the immediate purpose of setting up ICI, in a great hurry, between the end of September 1926 and 1 January 1927. Probably no merger of such a size has ever been carried out so fast, and speed had its price. So urgent were Mond and McGowan that shares in ICI were issued in exchange for shares of the four predecessors at a rate so generous that the original ICI capitalization (£65m. authorized:

£56·8m. issued) had to allow for about £18m. goodwill—an item not revealed in the published accounts until 1941, by which time it had been somewhat reduced, and not finally written off until 1956.

The reason for haste was the formation, in the autumn of 1925, of IG Farbenindustrie AG. This was the culmination of a process of amalgamation which had been going on in Germany for half a century and it produced a concentration of power in the chemical industry massive even by American standards and by British standards overwhelming. It was all the more menacing because the Germans, like the British but unlike the Americans, traditionally directed a great deal of fearsomely efficient effort overseas. McGowan and Mond recognized that none of the four largest firms in the British chemical industry, on its own, was strong enough to face the new IG. An amalgamation of all four might be, and it could not be set up too soon.[8]

The merger was thus in the nature of a defensive alliance. Mond and McGowan made its purpose clear at once. 'We are Imperial in aspect and Imperial in name', they wrote grandly to the President of the Board of Trade in November 1926,[9] and they repeatedly proclaimed their strategic aim of keeping the development of the chemical industry in British territory in British hands, including, where necessary, those of local shareholders. The seal of the company, reproduced year after year on the cover of the Annual Report, shows Britannia looking confidently outward to a chemical works standing conveniently on the ocean shore (not too fanciful a representation of Winnington Works), while a steamer, presumably laden with exports, steers for the far horizon.

In return for unhindered dominance of Empire markets Mond and McGowan were prepared to leave markets outside the Empire partly or wholly to others. This was a policy fully in the traditions of the alkali industry and the explosives industry and it was one which they knew, from previous experience, the large German and American companies were prepared to accept, provided only that British hands were manifestly strong enough to keep their grip on what was claimed for them. ICI's tactical aim at the outset, therefore, was to use the newly united strength of the four founders in seeking agreement with other Great Powers in the world chemical industry—IG above all—which would guarantee the result ICI had been set up to achieve: the defence of 'natural' British markets in the United Kingdom and the Empire, with some desirable peripheral outposts, which might be shared, such as South America and China. ICI had no ambitions either in Europe or the USA, which they were content to regard as the 'natural markets' of other people.

ICI, as the choice of name was meant to show, was an expression of British imperialism: a sentiment which comparatively few people, in 1926, had yet learnt to be ashamed of. When ICI was formed, the

8

British Empire was at its largest, because after 1918 various former German and Turkish possessions had been brought under British control to be administered, under League of Nations' 'mandates', alongside the territories which had belonged to the Empire before the Great War. In ICI's early years it was possible to travel from Singapore through Malaya, Burma, India, and the Middle East, and from north to south through the length of Africa, by way of Egypt and the Anglo-Egyptian Sudan, without once passing beyond the protection of one manifestation or another of British power.

Mond, McGowan and their colleagues had been brought up, like others of their generation, to regard the British Empire as beneficent, permanent, unshakeable, and the guarantee of their country's position in the world. They saw nothing in the circumstances of the inter-war world to invalidate this view and much to reinforce it. It seemed therefore a matter of duty and commonsense for ICI to develop Empire markets in preference to others. Moreover it was widely hoped, in the twenties, that the Empire might be made into a self-sufficient economic system run by the co-operation of its members for the benefit of all. That hope was shaken by the results of the Ottawa Conference of 1932, but even from that the system of Imperial Preference emerged. In the world of the thirties, with economic nationalism and political instability widespread in traditional British markets outside the Empire, the attractiveness of markets within it grew greater rather than less. Until well after the end of the Second World War all major decisions taken in ICI had imperial considerations prominently in view.

ICI at the time of the merger was easily the largest exporter of alkali products in the world. The figures varied widely from year to year, but in 1927 about 380,000 tons* were sent abroad, representing about 40 per cent of ICI's total make of alkali products, against 86,000 tons exported by the large American makers.[10] That was much the largest item in ICI's export trade, which represented one aspect of ICI's commitments overseas.

The other main aspect was overseas manufacture. Brunner, Mond had never put up works overseas and the main manufacturing companies, overseas, which came into ICI were those founded by Nobels to make explosives in three of the main mining countries of the world: Canada, South Africa, and Australia. They represented a response to economic nationalism in the Dominions, each anxious to encourage the growth of home industries to displace imports, and there were local partners in each. Moreover, although each was in origin an explosives company, they were all developing into diversified enterprises roughly on the model of ICI at home, so as to take, in the growing chemical

* All alkali products expressed as soda ash.

9

industry of each Dominion, a place as dominating as that of ICI in the chemical industry of Great Britain. They were the main instruments of ICI's development policy overseas, because it was recognized that in the future, even in imperial territories, British imports would increasingly be kept out in favour of local manufactures and that the only way British interests could be developed overseas would be through manufacturing enterprises set up jointly with local investors to employ local labour.

Without foreign markets ICI could not live, as the Directors were well aware, and their outlook, inherited from the predecessor companies, was world-wide. Nevertheless ICI was not an international company in the same sense as Royal Dutch-Shell or Unilever. The great bulk of ICI's assets were in Great Britain; the management, at home and overseas, was overwhelmingly British; and the bedrock of policy was British imperialism, though not an aggressive imperialism. Like the nation generally between the wars, ICI would defend territory already held, but not attempt fresh conquest.

Defence was not all. Within the limits to policy-making already indicated, Mond and McGowan were determined that the strength of unity should be devoted not only to making the world safe for the British chemical industry but also to the modernization and growth of that industry itself. They set their hopes on Billingham, which was to become the heart of a great enterprise dedicated to the advancement of the Empire's agriculture on the basis of nitrogenous fertilizers. Then, when that was fairly going, they would develop a process of hydrogenation, technically closely similar to ammonia synthesis, which would produce petrol from British coal so that the country would no longer depend entirely on foreign oilfields, which in any case were thought to be rapidly running dry. Both enterprises blended public spirit with the profit motive in highly satisfactory proportions, so that Mond could feel that his political ambitions, particularly for the economic unity of the Empire and the salvation of the coal industry, might be realized in business even if they had been frustrated at Westminster.[11]

Round this central power-house of advanced technology, which was expected to support a complex of processes dependent on hydrogen and related in one way or another to ammonia synthesis and its products, there was to be erected a modern science-based industrial undertaking of the largest class, adequately supported by research. Mond was the son of an academically trained German scientist who, in the German manner, had applied his science to industrial problems. Probably, therefore, Alfred Mond was more keenly aware than most of his contemporaries in British business of the huge gap which still existed between the application of scientific knowledge to industry—especially chemical industry—in Germany and in Britain. He was determined to

10

build up ICI's research, on a basis of adequate expenditure and close collaboration, on the German model, with the universities, so that as soon as possible ICI might be able to deal with IG, technically, on something like equal terms, which in 1926 was out of the question. Mond himself was not scientifically trained (except in so far as he had failed the Natural Sciences Tripos at Cambridge), but he admired scientists, sought their company, and, with the resources of ICI at his disposal, found little difficulty in attracting young men to permanent positions and eminent Professors to well-paid consultancy. In the expansive years between 1926 and 1930 a recruiting drive for Alkali Group was set in motion, directed by F. A. Freeth, FRS, among able young men newly graduated or about to graduate. They were offered salaries above the going rates, attractive working conditions and rapid promotion, and the traditional mutual suspicion between British universities and British business became, so far as ICI was concerned, somewhat less inspissated.

The founders intended that ICI should lead not only in the modernization of British industry but also in industrial relations. Here again, in the early years, the imprint of Mond's personality is clear. From his father he inherited a tradition of liberalism combined with an autocratic temperament, a combination not uncommon among late Victorian paternalist employers, and in spirit he was closer to Sir Titus Salt, to William Lever, to the Cadburys than to the personnel managers of a modern corporation. Towards trade unions his attitude was ambivalent. In ICI he was insistent that the employees should look to the Company and not to the unions for the benefits he was determined they should enjoy, and to that end he made sure that pay and working conditions in ICI, closely controlled centrally from the start, should be better than unions had been able to get for their members elsewhere. At the same time, 'bloated plutocrat' though he was in appearance, manner and, very largely, opinions, he nevertheless inspired such men as Ben Turner and Ernest Bevin with sufficient trust to create a relationship through which co-operation was possible, even in the raw, resentful atmosphere after the General Strike and the even more disastrous strike of the miners. It was a remarkable achievement, and ICI's labour relations began well.

Thus between 1926 and 1929 the inherent caution of the chemical industry's traditions was balanced in ICI, perhaps even outweighed, by optimism and expansive planning. In the benign atmosphere of the mild boom of the late twenties—mild, that is, in Great Britain, though hectic across the Atlantic—ICI seemed, as F. A. Freeth put it, perhaps with a dash of acid, to be 'floating to prosperity on a wave of capital expenditure'. Capital employed in the business rose from £72·8m. in 1927 to £102·5m. in 1930, and it is safe to assume that between half and two-thirds of the increase went into Billingham, where by 1930 some £20m. capital was employed.

In 1929 the agricultural markets of the world, which Billingham had been built to serve, collapsed into catastrophic depression, just at the moment when Billingham's full capacity was becoming ready for production. At the same time the spreading financial disorders of the world destroyed the easy assurance with which money had been raised and laid out since 1926, and in the search for economies an obvious target was Alfred Mond's ambitious programme of research expenditure. At the end of 1930 Alfred Mond (Lord Melchett since 1928) died. How ICI, with the 'inherited' characteristics examined in this chapter, met and survived this crisis, or rather bundle of crises, which hit it so early in its history, and in what form it emerged, on the eve of the Second World War, will be the theme of Part I of this volume.

REFERENCES

[1] Lord Melchett, 'Dyes', 23 May 1929, CR 75/107/2.
[2] L. F. Haber, *The Chemical Industry during the Nineteenth Century*, Clarendon Press, Oxford, 1958 (reissued 1969).
[3] Lord McGowan to 2nd Lord Melchett, 4 May 1941, MP 'Post-War Problems'.
[4] *ICI* I, chapter 10 (ii), pp. 293–7, ch. 16 (iii); C. Wilson, *History of Unilever* I, Cassell, 1954, ch. 9.
[5] As (3).
[6] Sir Alfred Mond, *Industry and Politics*, Macmillan, 1927, p. 9.
[7] Wendell Swint to Jasper Crane, 17 Dec. 1926, USA 'ICI/DuPont Agreements 1929–39 (Craven)', p. 43.
[8] *ICI* I, ch. 19.
[9] Sir Alfred Mond and Sir Harry McGowan to the President of the Board of Trade, 5 Nov. 1926, CR 93/1/—.
[10] *ICI* I, p. 344.
[11] *ICI* I, p. 371.

The Making of ICI

(i) COMPOSITION AND CAPITALIZATION

WHEN SIR ALFRED MOND and Sir Harry McGowan landed at Southampton from RMS *Aquitania* on 12 October 1926 they brought with them plans, worked out during six days' voyage from New York, for carrying out the largest merger so far seen in Great Britain: a merger of Brunner, Mond & Company Ltd.; Nobel Industries Ltd.; United Alkali Company Ltd.; and British Dyestuffs Corporation Ltd. For the new corporation which they intended to bring into existence they had already chosen a name: Imperial Chemical Industries Limited.[1]

The merger companies and their subsidiaries could be divided, as the organizers of ICI soon recognized, into two main groups. Brunner, Mond, United Alkali, and British Dyestuffs operated in the chemical industry as it was usually defined; Nobel Industries were based mainly on explosives and related products but had important interests outside the chemical industry altogether. The chemical group we have already briefly discussed (ch. 1 above). Before we return to it in detail we must make a closer acquaintance with Nobels, if only because they brought into ICI wealth which the other partners stood urgently in need of.

The Nobel Industries group was heavily stamped with the personality and ideas of Sir Harry McGowan, who, having the instincts of a financier, had always regarded it as having been formed less to develop the explosives industry than to find profitable employment for funds set free by a deliberate policy of reducing the group's commitment to explosives manufacture.[2] On the explosives side of ICI, accordingly, the process of rationalization begun in Nobels before the merger was carried on into the early thirties, against a background of sluggish demand from the depressed coal industry and an almost total absence of military orders. Factories scattered about the country were closed and manufacture of explosives and accessories was concentrated at Ardeer. At the same time the technology of nitro-cellulose was developed, with help from du Pont, away from explosives towards leathercloth, paints, and lacquers.

All these products found a market in the motor industry, and it was the motor industry which particularly attracted McGowan. The 'metals side' or 'Birmingham end' of Nobels' business was built up, before the merger, largely with a view to making motor components, although zip fasteners were developed as well, and this policy, too, was carried over into ICI. Amac Limited, Nobels' carburettor business, was merged with Brown & Barlow, Birmingham and C. Binks (1920) Ltd., Eccles, to form Amalgamated Carburettors Ltd., incorporated on 24 September 1927. During 1928, in order to add the hot working of non-ferrous metals to the existing cold-worked processes, ICI took over British Copper Manufacturers of Swansea and Elliot's Metal Company Ltd. of Birmingham, each of which had subsidiaries. By the time they had taken over Allen Everitt & Sons Ltd., tube-makers of Smethwick, ICI had control of the largest group of non-ferrous metal producers in the British Empire and one of the largest in the world.

McGowan's enthusiasm for the motor trade thus carried ICI well beyond the conventional boundaries of the chemical industry. The Metals Group was not functionally or technologically related to any other Group in ICI, except in so far as propellants were needed for ammunition, and it operated more or less in isolation. But for McGowan, it is safe to say that ICI's metal interests would never have developed in the way they did, and the growth of the Group both before and after the merger shows how powerfully, against all probability, one dominant individual can influence the activities of a large corporation.

There was an even more unorthodox side to McGowan's belief in the future of the motor car. As well as directing Nobels' investment into gaining control of companies in and on the fringes of the motor industry, he directed large sums into companies which Nobels did not control, could not control, and had no intention of controlling. Just before the merger one such investment, in Rotax (Motor Accessories) Ltd., had led to McGowan being invited to become Chairman of Joseph Lucas when they took Rotax over.[3] But the boldest and largest investment of this kind was in General Motors Corporation of USA. At various times between May 1920 and the date of the merger Nobels bought 1,517,677 Common shares and sold 767,677, so that the holding they brought into ICI amounted to 750,000.[4]

Considered purely in stock market terms, there was no doubt of the success of these transactions up to the date of the merger and for some years after. General Motors, having been rescued from near disaster in the early twenties, entered a period of great prosperity just before ICI was formed. Net sales rose from $568m. in 1924 to $1,504m. in 1929, and earnings per share of common stock rose from $0·19 to $0·91, all of which made the market very happy.[5] The 750,000 shares brought into ICI at the merger were already worth more than the cost to Nobels of all the shares they had at one time or another held, and the market price was still rising.

Mond, in conversation with McGowan in the *Aquitania*, was much

14

The doors of Imperial Chemical House, London, by W. Fagan (page 24).
They are faced with nickel-copper alloy, weigh $2\frac{1}{2}$ tons each, and are opened
and closed electrically. They are 20 feet high and 10 feet wide.

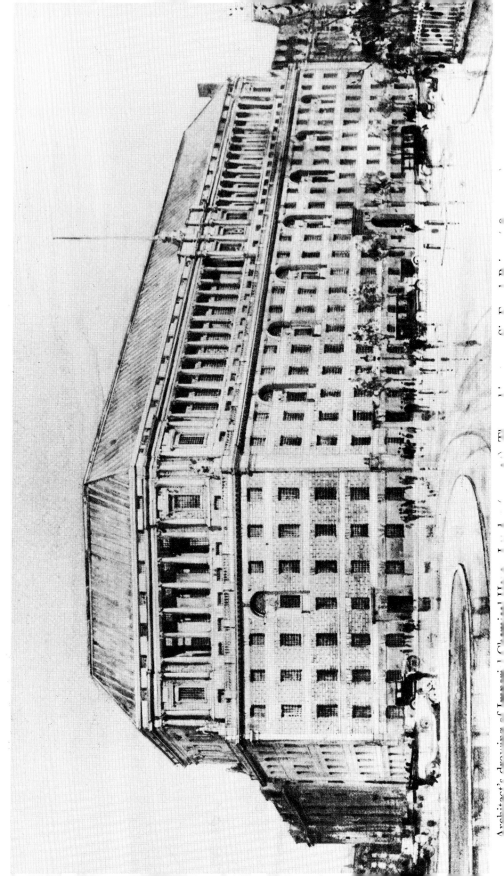

Architect's drawing of Imperial Hotel, London (*c. 1911*). The architect was Sir F. and Rickard's.

impressed with the hidden reserves which Nobels' interest in GM would make available to ICI.[6] During 1927 and 1928 375,000 shares—half the holding— were sold in three parcels on a rising market, bringing in £3,466,436, including a large element of capital appreciation both handsome and untaxed. Some £2·5m., to judge from guarded references in the 1928 Report, was at once laid out at Billingham, and the proceeds of the sales as a whole made a substantial contribution to the capital needs of ICI, being more than half as great as the sum of £6·5m. raised by a rights issue in 1928.

Nevertheless, the return on capital invested in GM was lower than might reasonably have been expected from investments in ICI's own business and, more important, the value of the investment fluctuated with the day-to-day movements of the stock market, and what went up might come down. The GM holding, which carried a seat on the GM Board for McGowan, no doubt induced, in McGowan at least, a pleasing sense of partnership with that mighty American corporation, but it begged the question, which as hard times set in after 1929 was increasingly often asked, within and outside ICI, whether the ordinary shares of companies over which ICI had no control were a proper home for substantial reserve funds. In the enthusiastic bustle of the early years, however, no such doubts seem to have arisen.

The 'explosives and general' side of ICI, then, brought into the merger a bulging portfolio of marketable investments. On the chemical side, likewise, there was a portfolio, in which much the largest item was Brunner, Mond's holding in Allied Chemical & Dye Corporation, which stood in the books in 1925 at nearly £1·6m. But Brunner, Mond had not invested in Allied, as Nobels had in GM, principally because they believed in Allied's future and wanted to share in it. They had taken the shares up because they could see no help for it, in exchange for their holding in the Solvay Process Corporation* at the time of the Allied merger, which they disliked.[7] Since then, Brunners had not seen fit to get out of Allied, but it was no part of their way of life, as it was of Nobels', to invest substantially in companies outside their control. Brunners' drive came from technology, not finance.

Brunner, Mond, together with the United Alkali Company, brought heavy chemical industry into ICI, and it is not far from the truth to say that the two companies owned the entire heavy chemical industry of Great Britain. The activities which they brought with them into the merger presented no such picture of tidy rationalization as Nobel Industries' did. Brunner, Mond, when the merger took place, was in the early stages of a thorough internal reorganization which would probably have produced a group of manufacturing subsidiaries in the

* Set up in 1881 by Solvay et Cie and Brunner, Mond jointly (*ICI* I, p. 98).

provinces, and of marketing subsidiaries overseas, under a holding company based in London, but the process had not gone very far. In the autumn of 1926 the function of Brunner, Mond as alkali manufacturers in Cheshire had not been separated from their function as the holding company of a diversified chemical group, with interests not only in alkali but in chlorine, acids, and, above all, ammonia synthesis as well.[8]

Brunners' business overlapped at many points with UAC's. In the central business of the ammonia-soda process, Brunners had works at and near Winnington and UAC had works at Fleetwood in Lancashire. Both companies produced soda ash and soda crystals, and both companies converted soda ash into caustic soda. Geographically, the main base of each company's alkali business was in Cheshire and Lancashire, but UAC had outlying works on the Tyne, in Glasgow, and in Bristol, and Brunner, Mond had soda crystal plant at Silvertown in London.

Brunner, Mond and UAC each controlled supplies of electrolytic caustic soda and chlorine. The most important plant was Castner-Kellners', at Runcorn, which belonged to Brunner, Mond who also owned Electro-Bleach's works in Widnes. UAC also had electrolytic plant in Widnes, and they had works at Newcastle as well as the Bay City plant in Michigan, USA.

In acids, again, there was duplication and triplication of effort. Chance & Hunt of Oldbury, one of Brunners' companies, was an important producer of sulphuric acid and Brunners also had a plant at Cwmbran in South Wales. UAC had a 50,000-ton/year plant at Widnes and another, smaller, in Glasgow. Nitric acid was made, at the time of the merger, by UAC, Nobels, and BDC, and Brunner, Mond were building plant at Billingham. Hydrochloric acid could be made by merger companies either electrolytically or *via* the operations of the Leblanc cycle.[9]

As a contribution to the manufacture of sulphuric acid the UAC could offer the new merger supplies of pyrites mined from its properties in Spain and carried on its own 3' 6" Buitron Railway. To this exotic acquisition there was perhaps only one parallel, namely, Brunner, Mond's Magadi Soda Company in the wilds of East Africa.[10]

Apart from the general run of the heavy chemical industry—and, its advocates would have said, above and beyond it in conception, in difficulty of execution, and in promise—stood Brunner, Mond's Billingham complex, which by the time of the merger was at last in sight of fairly large-scale operation. The heart of it was ammonia synthesis, and it was expected at first to make its living from nitrogenous fertilizers, chiefly ammonium sulphate. In time, a great deal more was expected, including nitric acid, sulphuric acid from anhydrite mined beneath the site, cement as a by-product, methanol, and, above all and crowning all, oil-from-coal.

Brunner, Mond's attitude towards UAC, for many years before the merger, had been amiable, polite, and slightly contemptuous, and the two firms had regulated their affairs by numerous agreements to share markets and fix prices—hence their duplicated facilities which otherwise would have been abraded by competition. Both firms were run by men of broadly similar technical and social background, and the general tone of Brunners' management, set by education at the ancient universities, by country life in Cheshire, and by the Winnington Hall Club, and justified by the company's efficiency and prosperity, was matched in a muted, melancholy way in UAC by the aura of faded grandeur which hung about the representatives of the once-great Leblanc tradition. The leading men of Brunner, Mond and UAC, generally speaking, understood one another and could fairly easily contemplate life in industrial wedlock.

British Dyestuffs Corporation, reconstructed but not prosperous, was the other member of the chemical group of the new merger, and it differed both in spirit and in function from either Brunners or UAC. The management of BDC neither shared nor approved of the Winnington style, derived, as it was, from many years of affluence. Instead, at Blackley and at Huddersfield they hugged their penury to them like a beloved hairshirt, and for many years after the penury had disappeared— well into the 1960s—the tradition of the hairshirt lingered, lovingly and even reverently preserved.

BDC represented within ICI the technology which had led the Germans to their mastery not only of dyestuffs but of fine chemicals generally and of most branches of the chemical industry, especially the most advanced. But BDC was quite as full of problems as of promise and its whole technical approach, directed as it was to the production of complex substances in relatively small quantities, was fundamentally different from the large and sometimes rather crude operations of the heavy chemical industry. Moreover Lord Ashfield, Chairman of BDC before the merger and after it a Director of ICI, was inclined to regard BDC as his personal fief. Therefore although BDC's difficulties had set both McGowan and Mond off on their different paths to the merger, they did not find it an easy partner to deal with, and in ICI's early days there were even occasional more or less serious thoughts of getting rid of BDC.

The merger decided upon in the *Aquitania*, then, would have to be made to hold four large companies representing not only three very different branches of the chemical industry but also large-scale activities outside the chemical industry altogether. The Nobel group, already for some years organized under a holding company and drastically rationalized, came nearest to presenting a model of what the merger might eventually look like. The sprawling mass of the heavy chemical industry, divided between Brunner, Mond and UAC, had not been tidied up at all. Its plants and processes, which frequently duplicated each other, were scattered widely about the country and outside it and in design they ran all the way from museum-worthy obsolescence in the deeper forest reaches of UAC to some of the most up-to-date and ambitious chemical engineering in the world, at Billingham. BDC, its affairs at last on a rational basis but far from prosperous, represented fine chemicals and the organic side of the chemical industry generally. However great the underlying need for unity against the foreigner, the practical difficulty of welding these groups together would nevertheless be very great.

Practical problems notwithstanding, once the principle of the merger had been agreed upon, the work of getting it into existence went on very fast indeed, and ICI was open for business by 1 January 1927. The speed of this final phase was greatly helped by the proceedings earlier in 1926, during which the principal parties to the merger, and the British Government, had accustomed themselves to the idea that an amalgamation of some sort in the British chemical industry had better take place, and quickly. Nevertheless, the rapidity of the whole episode, leading to a merger of unprecedented size and complexity, contrasts with the gentle pace of previous large amalgamations in the chemical industry—with the formation of Explosives Trades Ltd. (later Nobel Industries Ltd.), which had taken over two years to complete, and with the leisured acrimony which had gone before the setting up of British Dyestuffs Corporation. The foundation of ICI was the response of practical men to a practical problem—the formation of IG Farbenindustrie and its implications—which all of them felt to be urgent. It was therefore put through fast and without very much regard to anything but the immediate necessities of the situation. There were no elaborate theoretical studies of the organizational difficulties of the proposition, nor much in the way of economic forecasting. That was all left until later. The first thing to do was to get ICI established.

To do that, the promoters of the merger had to work out terms for the capitalization of their proposed holding company—ICI—which would be sufficiently attractive to the shareholders of the four merging companies to persuade them to exchange their shares, and the promoters were clear in their own minds, from the first, that they did not intend to pay out any cash. The basis of the exchange had been agreed by 21 October, and a document of that date survives. It is signed by Mond, McGowan, and Ashfield, the Chairmen, respectively, of Brunner, Mond, Nobel Industries, and BDC. It is not signed by Sir Max Muspratt, the Chairman of UAC. Whether he had been consulted we do not know, but we do know that amongst the other three it had already been 'made clear that even if they [UAC] stayed out the remaining three companies should proceed'.[11]

To arrive at the terms to be offered, there is no evidence that the promoters used calculating machinery much more sophisticated than pencil and paper or methods less elementary than simple arithmetic. No elaborate computations have survived, and since the promoters were anxious to keep matters secret until an announcement could be made it is probable that most of the work was done in the principals' heads, or the heads of their senior assistants, especially H. D. Butchart for Brunner, Mond and B. E. Todhunter and W. H. Coates for Nobel Industries.

Only one point seems to have presented any difficulty. This was the matter of the profits which Brunners expected to make at Billingham. On 16 October 1926 Mond opened his mind to McGowan: 'I don't like capitalising future profits on the same basis as those ascertained by more experience: on the other hand they are so large I cannot give them away.'[12] A forward estimate, made a little later, put them at £1,235,000 for 1928, a figure slightly greater than the estimate (£1,225,000) for Brunner, Mond's profits in the same year from the manufacture of alkali.[13] The actual Billingham profits for 1925 were given in the same paper as £105,000, so a tenfold increase was expected in three years. To accommodate these brilliant prospects within ICI's capital structure, Mond suggested creating deferred shares entitled to 'a return to be agreed' after the payment of Preferred and Ordinary dividends, which 'would enable the present B. M. & Co. shareholders to obtain their legitimate return for the . . . risk they have run and the capital they have stood out of if Billingham is the winner we anticipate without burdening the other members of the Group with a capitalisation that would be unjustified if the winner were not so complete'. McGowan raised no objection—Nobels had great expectations, too, especially from GM shares—and the creation of Deferred shares was agreed to.

With this point settled, the general scheme of ICI's opening capitalization could be worked out. Preference shares in all the founding companies were to be converted on a 7 per cent basis. To arrive at the Ordinary capital, the actual earnings of Brunners and Nobels for 1925 were to be capitalized at 12 per cent. Nothing was said, no doubt advisedly, about the capitalization of earnings for UAC and BDC. Their shareholders were simply presented with an offer for their shares, the exact basis of which is nowhere explained in surviving documents. The Deferred capital was arrived at by capitalizing potential profits at 16·6 per cent, and the holders were to have one-third of any balance remaining after a dividend of 7 per cent had been paid to Ordinary shareholders.[14] The terms of the share exchange offer were summarized as follows:

Ordinary Shares

Brunner, Mond
Nobel Industries   Three new £1 shares for two old shares
United Alkali
British Dyestuffs   Eight new £1 shares for twenty old shares

Preference Shares

Brunner, Mond   Five new 7% for four old 7½% shares
Nobel Industries   One new 7% share for each old 6% share
United Alkali   One new 7% share for each old 7% share
British Dyestuffs   Eight new 7% shares for each twenty old Ordinary shares

Deferred Shares

Brunner, Mond
Nobel Industries   One new 10s. [50p] share for each old Ordinary share
United Alkali   One new 10s. [50p] share for three old Ordinary shares
British Dyestuffs   One new 10s. [50p] share for forty old Ordinary shares

These terms, by offering UAC and BDC shareholders Deferred shares, gave them a right, to which they were not, strictly speaking, entitled, to a taste of honey still to come, and in general the terms of exchange were generous enough to introduce a considerable element of 'goodwill' or, as counsel advising BDC austerely put it, 'dilution', into ICI's balance sheet. Counsel (Sir Edward Clarke) put the figure at £18·4m., representing the difference between the combined capital of the four merger firms (£42·4m.) and the capital proposed, in October 1926, for ICI (£60·8m.).[15] Coates, in 1934, quoted a sum of £17·9m., arising in the same way.[16] No doubt this goodwill had to be created as an inducement to the shareholders to come in, and in they came, but it turned out to be a heavy burden for ICI to carry into the thirties.

The authorized capital of the new company was finally settled at £65,000,000.[17] £56,802,996 was issued straightaway, as follows:

16,219,306 7% Cumulative Preference Shares of £1 each   £16,219,306
31,095,555 Ordinary Shares of £1 each   31,095,555
18,976,270 Deferred Shares of 10s. [50p] each   9,488,135

                              £56,802,996

Before the new company could be registered, an unexpected snag appeared. The Registrar of Joint Stock Companies refused to register a title containing the word 'Imperial'. Mond and McGowan, united, rose in outraged majesty and wrote at once to the President of the Board of Trade, Sir Philip Cunliffe-Lister:

To our very great surprise we have to approach you as to a communication we received this morning [5th November 1926] from the Registrar of Joint Stock Companies, informing us that we may not register the name 'Imperial Chemical Industries Limited' for our new Merger. . . . We would like to state that this name was chosen after most careful consideration and study and for very special reasons. . . . We are 'Imperial' in aspect and 'Imperial' in name. I [sic: the letter is signed by both: evidently they were excited] want to emphasise a point that is not sufficiently appreciated. Dealing as we have to, with powerful and influential groups in foreign countries, we have in all our negotiations continually to fight for 'Imperial' rights and 'Imperial' markets, as contrasted merely with British rights and British markets. The difficulty of making those conducting business in various countries, especially on the Continent of Europe, regard the British Empire as an economic whole is one of the chief troubles of those who, like ourselves, are not satisfied in allowing ourselves to be swayed by purely national and not imperial relations. . . . we would urge that no such combination of the same importance for the safety of the country which largely depends on the products which are manufactured by our combination—we might even characterise it as a source of safety to the Empire—is in existence or probably ever will be in existence. . . . The developments which this Company has in view we may confidentially inform you, will be of enormous value, both from the point of view of national defence and of the economic position of the Empire. . . .[18]

Eloquence (or verbal diarrhoea) prevailed. The Registrar was over-ruled, and the name Imperial Chemical Industries Limited was registered on 7 December 1926.

(ii) THE CENTRAL DIRECTION

In the nascent organization of ICI one point was clear from the start. Executive authority would lie with Sir Alfred Mond and Sir Harry McGowan, respectively Chairman and President of the new company. These offices, in themselves, carried no formal powers, but Mond and McGowan were also appointed joint Managing Directors, each with a service agreement for five years. These agreements gave the Managing Directors, 'subject to the control of the Board', direction and control of the Company's powers and duties. The 'control of the Board' was remote and shadowy, for the same agreements delegated to the Managing Directors, presumably acting jointly, 'all the powers of the Board'.[19] Within ICI Mond and McGowan were supreme: joint auto-crats over all others, including their executive colleagues on the Board.

It was given out, when the Chairman and President were appointed, that their standing and authority would be equal. Seen from below, no doubt it was, but joint co-equal power is notoriously difficult to exercise.

How did Mond and McGowan manage their delicate relationship, which might so easily have degenerated into covert feuding or exploded into open conflict?

McGowan had conceived the plan of a fourfold All-British amalgamation and had steered it to success, in New York, when Mond's own plan for an international consortium broke down. McGowan might therefore have considered himself to have a clear claim to the Chairmanship. His political instinct, however, as he had often previously shown, never sought to extract the last drop of advantage out of a successful negotiation. Having gained his major point, the formation of ICI, he gracefully conceded the Chairmanship to Mond.

Mond, as a Privy Councillor and former Minister, would probably not have been content with less and it is likely that the rest of the Board, including McGowan, considered a politician of Cabinet rank an appropriate first Chairman for Great Britain's largest company. Certainly Mond's political career, though never more than moderately successful and by 1926 in ruins, gave him a public position which overshadowed McGowan, merely a highly successful business man. Moreover, public affairs apart, Mond took his family connection with the chemical industry very seriously. In 1927, writing to Bosch of IG, he described himself as 'the Head of the Chemical Industry in this country, and the inheritor of the position of promoter of the Chemical Industry from one of its greatest leaders, my late father'.[20]

McGowan, it has been suggested, went in some awe of Mond because his public position was one which McGowan deeply respected but never, for himself, aspired to. Certainly the central policy-making documents which have survived from the period of the joint Managing Directors all show Mond in the lead with McGowan a barely perceptible half-pace behind. With no one but Mond would McGowan agree willingly to divide authority, and even Mond, it appears, was careful not to interfere on the explosives and general side of the business which had been brought in from Nobels. McGowan for his part kept away from Mond's domain on the chemical side, and no doubt the partnership owed a good deal of its success to the partners' good sense in knowing how to keep out of each other's way.

Once ICI was in being Mond began to pour out ideas for its organization and development, and McGowan's influence at this stage is far less easy to detect. This is partly, no doubt, because he left far fewer written documents than Mond, but there is also less trace of his contribution to oral discussion, and it seems fair to suggest that the guiding mind behind the setting-up of ICI, once the indispensable first step had been taken, was Mond's rather than McGowan's.

Whatever their separate contributions may have been to the making of ICI, there seems never to have been any serious dissension on fundamentals between Mond and McGowan. They brought to their task that essential ingredient in the success of any merger: willpower. The ICI merger was not universally popular in its constituent companies. No merger ever is, because in every merger there are threats as well as promises. Therefore a merger requires to be pushed through with mingled persuasion and ruthlessness, so that the emerging whole can be seen as more than the sum of the disappearing parts. It was a task for which Mond and McGowan, united, were well qualified.

The Board, which had been at work for some weeks informally, met formally for the first time on 8 December 1926. There were eight full-time Directors, who must have been among the highest paid men in the country, for they drew not only a salary but a share in the profits too, which in the case of Mond and McGowan and probably of others lifted their total pay well into five figures. Ex-members of Brunner, Mond's Board had compensation for loss of office, as well. The eight were drawn four and four, from Brunner, Mond and Nobels:

Sir Alfred Mond	Sir Harry McGowan
Henry Mond	H. J. Mitchell
J. G. Nicholson	John Rogers (appointed 8 Dec. 1926)
G. P. Pollitt	B. E. Todhunter

Besides the executive Directors, 'lay Directors' were from the first appointed in roughly equal numbers. Some, in the early days, were put in for temporary reasons, particularly as guardians of the memory of United Alkali or, in one case, because he was Sir John Brunner, 2nd Baronet. Until the Second World War there was always a representative of Solvay et Cie, who by exchange of Brunner, Mond shares had become much the largest shareholders in ICI, with about 6 per cent of the Ordinary capital.

The general principle, however, in choosing lay Directors was that they should be men distinguished in public life or in business, or both, who could bring independent judgement to bear on ICI's affairs. In a sense, although the point was never formally made, they were trustees of the public interest, both to guard against the abuse of monopoly powers and to watch over strategic materials, such as nitrogen, dyestuffs, explosives and ammunition, of which ICI was the only or almost the only supplier in Great Britain. Inside ICI, they could advise on the choice, when the need arose, of a new Chairman. Moreover they formed a force in reserve to resolve any crisis that might arise in the high politics of the Company, and when such a crisis did arise, in 1937–8, the influence of the 'lay Directors', as we shall see, was decisive.

Public figures of great eminence including, among others, Lord

Birkenhead (F. E. Smith), Lord Weir, and Sir John Anderson (Lord Waverley) served on the ICI Board in the twenties and thirties. The standing of the company was high—it was almost a branch of the Constitution—the element of public service was plain, and the fees might be welcome to a distinguished man of no fortune. In the earliest days the outstanding public figure among the lay Directors, and the most active, was Rufus Isaacs, 1st Marquis of Reading. He had been Lord Chief Justice of England from 1913 to 1921 and immediately after that, for five years, Viceroy of India. His son had married Alfred Mond's daughter Eva in 1914, so that he was connected to the Mond family. He joined the Board after he came back from India in 1926 and applied himself so assiduously to ICI's affairs as to be close to becoming, in fact though not in name, an executive Director.

At the first Board meeting the Directors appointed ICI's first Secretary, J. H. Wadsworth (1887–1954), from Brunner, Mond, and the first Treasurer, W. H. Coates (1882–1963), who had come from the Inland Revenue to succeed Sir Josiah Stamp, himself an ex-Revenue man, as Secretary of Nobels. If the Directors of ICI may be said to hold a position in some ways similar to Ministers in Government, then the Secretary and the Treasurer, along with a third permanent official, the Solicitor, stand rather in the position of senior civil servants, running central departments for the service of the Board and offering advice not lightly to be disregarded.

The Board took one more decision very early on: to build a head office. Like everything else connected with the foundation of ICI, the decision was made and carried through at top speed. At the first Board meeting plans were discussed. Soon afterwards Sir Frank Baines (1877–1933), whom Mond had known at the Ministry of Works, was appointed architect. A site of just over one acre on Millbank, a little way upstream from the Houses of Parliament, was taken on a building lease from the London County Council, and a nine-storey block was designed to house 1,500 people. It was put up by John Mowlem in twenty-six months.

Ornate and stately, the building was in the style then considered appropriate to the head offices of companies which were pillars of the British economy. There is another example of the style, built three or four years later, in Unilever House at Blackfriars. At the main entrance was hung a pair of enormous metal doors elaborately decorated, and all faces of the building were provided with statuary, including portrait heads of eminent scientists; of the Founding Fathers (Ludwig Mond and Alfred Nobel); and of the reigning Chairman and President. At the back, in Smith Square, by a piquant juxtaposition, Imperial Chemical House abutted on to Transport House, opened in May 1928, where the Labour Party and the Transport and General Workers Union had their headquarters. Before the end of 1928 the first occupants of the new offices had moved in.

During the autumn of 1926, before ICI was legally constituted, technical and commercial committees were set up to consider the reorganization and rationalization of what was already called the Chemical Group. They met for the first time on 30 November 1926 'for the purpose of launching the new organization of Imperial Chemical Industries, Limited'.[21] Mond, supported much more briefly by Mc-Gowan and, for UAC, F. W. Bain, addressed the gathering.

Mond and McGowan each insisted on the paramount need to sink the identity of the four old companies in the identity of ICI. 'I want us all', said McGowan, ' . . . to forget our old Companies.' Both speakers insisted that the trend of the day, in Germany and America, was towards larger groups, and the British chemical industry must not be left behind. There were reasons of public policy, too. 'We are on trial', said Mond, 'before the eyes of the entire world, and especially the eyes of our fellow citizens and of the Empire. We are not merely a body of people carrying on industry in order to make dividends, we are much more: we are the object of universal envy, admiration and criticism, and the capacity of British industrialists, British commercialists and British technicians will be judged by the entire world from the success we make of this merger.' McGowan echoed him: 'Primarily of course the duties of a Board of Directors are to make dividends for their Share-holders, but in forming this combine we are doing something for the Empire.'

The message of unity cannot have been welcome to all who heard it. Some, even among those very close to the planning of the merger, had assumed that ICI would form a fairly loose federation within which the four main companies, at least, would continue in separate existence, and to some in Brunners and Nobels the idea of association with the other party was disagreeable.[22] These are normal sentiments in any merger, and equally normal was the anxiety, evidently widespread in the merging companies, about security of employment in the face of the rationalization that was certainly coming. 'Any man . . . in this concern', said Mond, 'who can prove his value need have no fear whatever that there is not room for him', but perhaps that promise was not universally reassuring.

Mond in his address moved quickly from general principles to practical proposals for making sure that ICI should gain the full benefit of the merger, particularly in the rationalization of production and the establishment of central commercial services. Nobels had already had six years experience 'as a Merger Company', but Brunners and the other companies had still to go through it. Hence the two committees, technical and commercial, to deal with the problems of the Chemical

Group. Each would have a Nobel representative—'a great advantage to us', said Mond. 'It is essential', he went on, 'that the personnel of these Committees should become acquainted without delay.'

In planning to rationalize production Nobels' procedure of the early twenties was closely followed. Nobels had set up technical committees to consider which factories were redundant. So, too, would ICI. 'The Technical Committee', said Mond, 'will deal with . . . the overlapping in the manufacture of . . . Chlorine and its derivatives, Nitric Acid, Solvents, Soda Ash, Caustic and other alkalies [sic], and Sulphuric Acid from all its sources, and indeed thousands of problems connected with not only to-day's products but many which we have not yet even discussed.' Research, too, would have to be looked at from the same point of view. 'I think in the near future a sub-committee on Research will have to be formed . . . to ensure that the investigation into problems is not being duplicated and that the best use is being made of the best plants and the best staff, in order to prevent waste of energy, time and money.'

The Commercial Committee was told to deal with the centralization of certain services required throughout the business, particularly sales, buying, traffic control, estate control, labour control, and the standardization of methods and materials. In raising these issues Mond, knowingly or not, was driving at a fundamental problem in large-scale industrial organization, namely, how far to centralize control of essential profit-making operations, thus reducing the responsibility of subordinate managements for the success of their activity. The crucial point was selling. If ICI sales policy, including prices, were to be controlled in London, then the men in charge of the production units would be left with no more than a factory manager's responsibility for keeping costs down, and their general commercial success or failure would be merged in the success or failure of ICI as a whole. Was this really what ICI wanted? Different answers were given at different times and by different people.

Mond's ideas on the development of ICI's business generally were never anything but the largest, and they were reinforced by the ideas of his colleagues. When he spoke on 30 November 1926 he must have had in mind a paper by Pollitt, dated about a fortnight earlier, which had discussed various products of the Chemical group, including all the most important, on the assumption, express or implied, that ICI would have a monopoly of production in the United Kingdom and probably in the British Empire.[23] 'My ideal', said Mond, speaking of the fertilizer business based on Billingham, 'is to see, in this *as in other chemical manufactures,** our great organisation controlling the production and sales of

* WJR's italics.

the whole chemical manufacture within the Empire.' The 'other chemical manufactures' must have included chlorine, alkali in its various forms, many of the products of Billingham, and presumably explosives and dyestuffs.

To establish their monopoly position, which depended essentially on the possession of very large resources and on economies of scale, ICI put up the capital, which was large, took the risks, which were heavy, and endured a degree of public suspicion and unpopularity, particularly where the harsher consequences of rationalization were felt. In return, Mond and his colleagues considered that ICI had something approaching a moral right to such benefits as monopoly conferred, especially since monopoly was a necessary condition of providing a secure base from which to fight for the national interest, which they considered was identical with the company's interest, in the markets of the world. At the same time, bearing the 'public service' aspect of ICI in mind, they recognized that their British monopoly imposed duties upon them, chiefly in treating customers fairly and in guaranteeing the supply of materials of national importance, such as dyestuffs, even if they were not a very attractive commercial proposition.

Mond rounded off his remarks with an appeal for *short* reports—'we want our information rapidly . . . concisely; we do not want to dot our i's and cross our t's, because the Board of ICI Ltd. have to make decisions for a Company with a capital of £65,000,000 and not for one with a capital of £500,000. The whole matter is one of proportion . . . .' A sense of proportion, however, was not to interfere with due economy and efficiency. 'We expect', Mond finished, ' . . . to produce results in such a manner that we shall justify the very drastic, courageous, or perhaps you have heard it expressed, dangerous, step we have taken in forming Imperial Chemical Industries, Limited.'

While the Technical and Commercial Committees, under the chairmanship of Pollitt (Technical) and Nicholson (Commercial), applied their minds to the problems of the Chemical group, the planning of a central organization for ICI as a whole went rapidly ahead. Mond relied heavily on the advice of Nobels' experienced mergermen, especially W. H. Coates and H. J. Mitchell. McGowan must have been close at hand all the time, but he did not write much down and his influence is difficult to detect. Mond, on the other hand, went on to paper frequently and copiously, and usually in the first person singular, though no doubt he would have refuted any suggestion that he was taking McGowan's agreement for granted.

A paper written early in 1927, unsigned but probably by Alfred Mond, gives first place, among the problems to be solved, to finding a means for giving the ICI Board 'absolute and rapid control over all the activities of the four constituent companies'. Coates and Mitchell had their answer ready, for they had faced exactly this problem in Nobel Industries and had solved it by requiring the Directors of companies which Nobels controlled to resign and then putting Nobel Industries in as sole legal Director and Manager. There were then no independent Directors with a statutory right, and probably a duty, to query policies which they disagreed with.[24]

This device immediately commended itself to Mond. He would listen to advice and even enter into argument, but when his mind was made up he expected the powers of an autocrat, and he had no intention of giving ICI's subsidiaries any right whatever to control their own destinies, which he regarded as essentially a matter for the main Board. Nobels' method of control, given legal force by resolutions of the ICI Board on 30 March 1927 and applied to all companies in which ICI owned upwards of 75 per cent of the capital, was therefore made the centrepiece of a scheme of organization (see Fig. 2) promulgated with notes by the Chairman on 6 April 1927.[25]

The Chairman and President, together, remained the Managing Directors with plenary powers, but the remaining six full-time Directors, called Executive Directors, jointly formed an Executive Committee and individually became departmental managers and Delegate Directors of subsidiary companies. Two—Henry Mond and H. J. Mitchell—joined the Chairman and President in a separate Finance Committee, with Coates as its Secretary. Mond, in his note, succinctly explained the relationship between the Managing Directors and the rest: 'The ICI have really entrusted the President . . . and myself with the powers of management of the concern; the Executive Committee operates under our supervision. Its functions are divided by its constitution into Technical and Commercial, Development, Finance, etc., and it is for these officers in their respective spheres to exercise supervision and control of the operations of the Subsidiary Companies.'

Under this first constitution of ICI the personal position of an Executive Director was powerful. He had under his hand one or more of the great Departments, which were expressly intended to *control* the subsidiaries' operations, and he would also be a Delegate Director of several of the subsidiaries themselves. The Delegate Boards consisted partly of ICI Directors and other London officials and partly of local men, and the affairs of a subsidiary company were run by a Management Committee consisting both of the London members and the locals. The Chairman of the Delegate Board might or might not be on the Management Committee.

In his notes accompanying the scheme of organization Mond explained bluntly why he had decided to destroy every shred of independence in the subsidiaries by adopting the Nobel method of control. 'The whole picture of ICI', he wrote, 'has to be kept in view by those

THEN—AND NOW

An extract from a little book on the art of correct letter writing, published about the middle of last century, with a modern example

FOR I.C.I. GIRLS

CONDUCTED BY "VERA"

directing its affairs and . . . they cannot be asked to discuss matters of ICI policy with Boards of Directors who are not themselves ICI Directors, firstly because a great deal too much of ICI policy . . . would become disclosed . . . and secondly, the natural view point of those who are Directors of ICI and those who are merely Directors of Subsidiary Companies might quite legitimately be widely separated.'

He went on to give examples of conflict between the interest of a subsidiary and the interest of ICI. BDC, he said, had been unwilling to go on selling indigo at a loss and had asked the Executive Committee for permission to give it up. Permission had been refused, partly because of profits made by other subsidiaries out of materials supplied to BDC, partly because new methanol plant at Billingham was likely soon to bring down the cost of indigo, and partly because of intricate negotiations likely to be put in train with IG and with Allied. Discussing matters of this sort with the Board of BDC, in Mond's view, would be 'quite an unnecessary duplication of work', and he did not intend to let himself in for it.

Again, referring to the matter of alkali manufacture, where it was obvious that a good deal of redundant plant would shortly be rationalized out of existence, Mond was quite plain. 'It is a matter of practical indifference to ICI', he wrote, 'whether BM & Co., the UA Co., or any other of its constituent companies manufactures any article. What is important to ICI is that the greatest maximum profit should be made.'

No one could have stated more frankly the primary object of industrial organization: to make the maximum profit. With that simple end in view, ICI, in the first months of its existence, had rapidly been provided with a directing organization which concentrated all power very narrowly at the centre. What remained, in the spring of 1927, was to put the new organization to work to bring the profits in.

REFERENCES

[1] *ICI* I, p. 464.
[2] *ICI* I, p. 388 and ch. 17 (i) generally.
[3] *ICI* I, p. 420.
[4] Treasurer to Finance Committee, 16 Dec. 1937, CP 'General Motors'.
[5] Alfred P. Sloan, Jr., *My Years with General Motors*, Pan Books Ltd., 1967, pp. 238–9.
[6] Notes of Interview between Sir Alfred Mond and Sir Harry McGowan, 7 Oct. 1926, Bunbury folder 7.
[7] *ICI* I, p. 293.
[8] Scheme of Reorganisation, Bunbury folder 13; Memo to Executive Committee [of Brunner, Mond], 28 Sept. 1926, CR 34/1/R; 'BM & Co. Ltd.—Group Office' (undated memo), CR 34/1/2; 'Notes on proposed Amalgamation . . . ' by D. Marsh, 5 Oct. 1926, TP 'Amalgamation'.
[9] On duplication of production facilities, see memo by G. P. Pollitt, 17 Nov. 1926, CR 215/–/2R.
[10] Buitron Properties—*ICI* I, p. 228; Magadi—ibid. p. 342.

[11] Exchange terms—see CP, 21 and 25 Oct. 1926, 'Merger Capitalisation'; for UAC—see 'Development of the Chemical Industry', Note of Meeting 14 Oct. 1926, TP 'Amalgamation'.

[12] Sir Alfred Mond to Sir Harry McGowan, 16 Oct. 1926, CP 'Merger Capitalisation'.

[13] 'Summary of Profits of the Group', CP 'Organisation 1926/27'.

[14] Capitalization of ICI, as submitted to NIL Board, 25 Oct. 1926, CP 'Organisation 1926/27'.

[15] Sir Edward Clarke's Notes, 22 Oct. 1926, CP 'Merger Capitalisation'.

[16] W. H. Coates to Chairman, 5 Feb. 1934, p. 5, CP 'Capital (4)'.

[17] Letter of Offer to Shareholders of Brunner, Mond, Nobel Industries, United Alkali Co., British Dyestuffs Corporation, 15 Dec. 1926, CP 'Organisation 1926/27'.

[18] Sir Alfred Mond and Sir Harry McGowan to President of the Board of Trade, 5 Nov. 1926, CR 93/1/–.

[19] Clause 2 of the Agreement as cited by J. E. James to Lord McGowan, 5 Jan. 1931, CR 93/8/125.

[20] A. Mond to C. Bosch, 24 Nov. 1927, CR 107/2/ .

[21] Privately printed *Addresses . . . [to] . . . the First Meeting of the Commercial and Technical Committees of Imperial Chemical Industries Limited (Chemical Group)*, 30 Nov. 1926.

[22] On separate existence, Marsh, 'Notes on proposed Amalgamation', 5 Oct. 1926, TP 'Amalgamation'; Todhunter's paper 'Amalgamation of A B C and D Companies', 10 Oct. 1926, drawn up in the *Aquitania*, CP 'Organisation 1926/27'. Evidence of mutual dislike from private information.

[23] Memo by G. P. Pollitt, 17 Nov. 1926, CR 215/–/2R.

[24] *ICI* I, p. 392, where Sir Josiah Stamp's comments are given.

[25] 'Organisation' (Scheme and Chairman's Note), 6 April 1927, from which all subsequent quotations in this section are taken, CR 93/2/C.

Naphthalene Black 4093
1 kilo nett

CHAPTER 3

The Foundations of Foreign Policy 1926–1929

THE FIRST OBJECTIVE of the newly installed Board of ICI was to establish ICI's position in the chemical industry of the world. On 16 December 1926, in one grand and simple sweep, McGowan outlined the aims of ICI's foreign policy:

> Sir Harry explained [to Wendell Swint, du Pont's European manager] that the formation of ICI is only the first step in a comprehensive scheme which he has in mind to rationalize the chemical manufacture of the world. The details . . . are not worked out, not even in Sir Harry's own mind, but the broad picture included working arrangements between three groups—the IG in Germany, Imperial Chemical Industries in the British Empire and Du Ponts and the Allied Chemical and Dye in America. The next step is an arrangement of some sort between the Germans and the British.[1]

Of the businesses mentioned by McGowan—the Great Powers of 'the chemical manufacture of the world'—much the most formidable, as he spoke, was IG Farbenindustrie AG: 'IG'. The magnitude of IG's resources, the range of its operations, the quality and scale of its research (never fewer than 1,000 chemists), all backed by the towering prestige of German chemical technology, gave IG strength unmatched, in 1926, by any other chemical business in the world, and in every chemical business in the world they knew it.

After the Second World War, when the Americans had examined oral and documentary evidence from the high officials of IG and from IG's archives, they described IG as 'the most powerful single industrial combine in Germany and indeed in Europe'. IG produced, they said, 'all of Germany's magnesium, nickel, synthetic rubber, and nearly all of its dyestuffs. It produced the bulk of Germany's nitrogen, synthetic gasoline, and numerous important chemicals. It produced half of Germany's pharmaceuticals and more than half its photographic supplies. It dominated the German explosives industry. It enjoyed close relations with the German Government long before Hitler came to power.'[2]

This passage refers to IG's activities at full stretch in the late thirties and during the Second World War, but by 1927 IG's potential was already sufficiently apparent to generate enormous bargaining power. Not ICI only but the Americans and everyone else in the chemical industry respected and feared IG's versatility. Accordingly each was anxious at least to gain protection from IG's competition, and at best to negotiate some share in the benefits that flowed from IG's technical omnicompetence.

IG's general principle of development was to concentrate on new, advanced products and to keep clear of what was rather patronizingly called 'the so-called old chemistry', so that sulphuric acid, soda ash, cyanide and many other products of the orthodox heavy chemical industry found no place in IG's activities except so far as might be necessary to keep up with new methods of production or to supply IG's own requirements. The main development effort, before the war, went into plastics, synthetic rubber, synthetic resins, solvents, the products of hydrogenation (including oil-from-coal), synthetic detergents and insecticides.

In 1926 IG's turnover was chiefly in two roughly equal product groups, nitrogen products and dyestuffs, including intermediates and pharmaceuticals. The turnover in each of these groups represented about 36 per cent of the whole, and the balance was made up with general chemicals (20 per cent) and photographic materials. The range of products rapidly broadened, especially in the later thirties, and by 1938, when the total turnover was much greater than in 1926, metals, synthetic rubber and plastics, petrol and cellulose fibres were all increasingly important.[3]

The growth points in IG were all in materials useful in war, particularly if blockade were to cut off supplies of natural materials such as rubber, wool, cotton, or oil. The point was noticed, and in England a good deal of ignorant scorn was poured on German *ersatz* materials, which were taken as evidence of strategic weakness in the German economy. What was less readily comprehended—even, perhaps, in some high quarters of ICI—was that these materials were characteristic of the most advanced practice of the contemporary chemical industry, especially in the organic field, and their development foreshadowed the lines along which the industry was to advance towards the world of the nineteen-fifties and sixties.

This broadening of IG's activities was founded, as the activities of the German chemical industry always had been founded, on a massive research effort. Never between 1927 and 1939 were fewer than 1,000 chemists employed, and when the depression set in they were not dismissed, although salaries were drastically cut:

IG Research 1927–38

	Expenditure RM millions	$£$[a]	Estimated no. of chemists employed
1927	154	7·5	1,000
1928	140·3	6·9	1,050
1929	137·2	6·7	1,100
1930	99·9	4·9	1,100
1931	70	4·4	1,050
1932	40·5	2·75	1,000
1933	42·8	3·3	1,000
1934	43·8	3·4	1,000
1935	57·1	4·7	1,020
1936	68	5·5	1,060
1937	82·3	6·7	1,100
1938	93·3	7·6	1,150

[a] Calculated at annual average rates given in *The British Economy, Key Statistics, 1900–1966* (Times Newspapers Ltd.), Table L.

Even at the bottom of the depression, it appears, IG's research expenditure did not fall below the equivalent of £2¾m. ICI's expenditure, before the war, never rose much above £1m., and in 1932 it was only £568,504[4] though how far these figures are directly comparable with IG's it is impossible to say.

The strength of IG was heavily concentrated on foreign trade, partly by way of direct export, partly through an expanding multiplicity of subsidiaries and associates throughout the world, particularly in Europe, the USA, and Latin America. Some of these connections were open, some concealed, and some ran outside the chemical industry. There were strong financial links with the Ford Motor Company in the USA and Germany (prompted, perhaps, by the example of du Pont and General Motors?) and there was a joint interest in hydrogenation, which we shall later examine, with Standard Oil.

As for exports, they generally represented, in the late twenties and early thirties, something between 60 and 66 per cent of the total value of IG's turnover. By 1938 the export proportion had fallen to 34 per cent but the total IG turnover had almost doubled since 1932. As a consequence IG's share in the German chemical export trade rose from just under 40 per cent in 1926 to over 53 per cent in 1938. IG's share in the chemical export trade of the entire world, reckoned at just under 11 per cent in 1926, reached nearly 14 per cent in 1938.[5]

The German chemical industry had been one of the wonders of the world since the eighteen-seventies, and everybody in IG knew it, just as they knew that German leadership in chemical science and technology was acknowledged, with gratifying reluctance, in the great chemical businesses of Britain and the USA. IG as a corporate body, therefore, was imbued with pride in its own technical and scientific excellence, inherited from the dyestuffs firms of Imperial Germany, and in its all-round commercial competence, which relied quite as much on fierce and efficient marketing as on the intensively organized research which the dyestuffs businesses had led the world in establishing.

That the Germans should have been able to organize such a group as IG within seven years of total defeat in the Great War was evidence of the inherent strength of the German chemical industry—indeed, of German industry as a whole—and of the tenacity of purpose of those at the head of German industrial affairs. The German Army might be overcome, the German Empire might collapse, the foundations of German society might crumble, but Carl Bosch and his colleagues in IG could reconstruct the German chemical industry and face their international rivals with unshaken self-assurance.

Across the Atlantic two businesses in particular engaged the attention of the founders of ICI: Allied Chemical and Dye Corporation and E. I. du Pont de Nemours & Co. Inc. ICI had close links with Allied through Brunner, Mond; with du Pont through Nobel Industries.

Allied had been brought into existence, in 1920, by a fivefold merger.[6] One of the merging companies was the Solvay Process Company in which Solvay et Cie and Brunner, Mond had a majority interest, and by exchange of shares they had become the largest shareholders, though far from holders of a majority interest, in Allied. Brunner, Mond's interest had passed to ICI, but relations between the two groups were not cordial. Orlando Weber, at the head of Allied, had wrecked Alfred Mond's plans in New York in September 1926, thereby precipitating the ICI merger, and he had every intention of continuing to go his own way.[7]

Allied in 1926 had greater capital employed ($324·2m.) than du Pont ($308·1m.), but there was no equivalent to du Pont's investment in General Motors. Moreover, Allied was growing more slowly, perhaps because Weber's ruthless measures of rationalization had ejected a number of able men who had promptly gone into competition with Allied. By 1929 Allied's capital employed was put at $374·3m. against du Pont's $520m.[8]

Allied was heavily committed to alkali, through the Solvay Process Company, and to ammonia synthesis. Neither was of much interest to du Pont. Both, however, were of great interest to ICI, particularly in the export trade. Hence the desire for agreement with Allied.

Du Pont, in contrast to Allied, IG, and ICI, was not the product of a recent merger. Its origins ran back to 1802, when the original E. I. du Pont de Nemours (1771–1834), a Frenchman, had set up powder mills by the Brandywine River in Delaware. In 124 years there had been many changes of form and ownership, but in 1926 and for many years afterwards the business was controlled, as it always had been, by members of the founder's family. The President was Irénée du Pont (1876–1963). His elder brother Pierre S. du Pont (1870–1954) was Chairman of the Board, having handed over the Presidency in 1919.[9] Among the Vice-Presidents was Lammot du Pont, younger brother of Pierre and Irénée, and there were other relations by blood or marriage. Control was exercised through Christiana Securities which, wrote Lammot du Pont in 1928, 'is almost entirely owned by a small [family] group that are intimately connected and whose interests are much the same'.[10] The capital employed in this family business, converted at $4·8 to the £, amounted to £64m. in 1926 and £108m. in 1929, against figures for ICI—the largest manufacturing business in Great Britain—of £73m. (1927) and £99m. in 1929.

The Great War had shown up the deficiencies of the American chemical industry, especially in comparison with the Germans. It had also opened the eyes of du Pont and others to attractive possibilities in filling the gaps. Not only that, but war contracts had been profitable enough to provide funds for research and development and for new investment. Du Pont's management, during the Great War and after, sought diversification outside the explosives industry in which up to 1914 almost all the resources of the company had been employed. By the mid-twenties, after a great deal of investigation, many experiments and some expensive failures, du Pont were fairly launched as a diversified chemical enterprise on the largest scale. They were in dyestuffs in a large way; in viscose rayon and cellophane; in paints and lacquers; in photographic film; in coated textiles (leathercloth); and they were looking at the range of possibilities in the organic field that was being explored by the world's other advanced chemical companies—such products, or groups of products, as synthetic rubber, plastics, and insecticides.[11]

Du Pont had no interest in alkali, which they regarded as insufficiently profitable, and their venture into heavy chemicals generally, through the purchase in 1928 of the Grasselli Chemical Company, was not a striking success. They had a synthetic ammonia plant at Belle, West Virginia, but it occupied by no means so central a place in du Pont's scheme of things as Billingham in ICI's, being chiefly valued as a source of urea, for urea-formaldehyde plastics, and of methanol and other industrial alcohols. They did not make fertilizers, largely because, as ICI also found, farmers had great difficulty in finding the money to pay for them at the time of purchase, in the spring, when the crops that were to be grown from them could not be harvested until the autumn. Oil-from-coal presented no attraction to du Pont at all.

Du Pont's development policy, like IG's, was directed chiefly to the possibilities of advanced research, of which one in a thousand, perhaps, might turn out profitable. Then it would be very profitable while the exclusive rights lasted. Up to about 1920 they usually expected a return of 15 per cent on capital invested in a project, but after that their expectations began to rise, perhaps because they were seeing 30 per cent return in General Motors.[12] By 1926 they looked for at least 20 per cent at home, though abroad, rather curiously, they were prepared to accept 15. They sometimes showed a faint surprise that ICI saw fit to engage heavily in low-return bulk products like fertilizers.

Du Pont and ICI stood very differently in their home markets, and that partly explained the different view they took of return on capital employed. ICI's position, on the face of it, was much the stronger, since ICI was very much the largest chemical business in Great Britain and in many fields a monopolist, whereas du Pont was merely the largest of a number of competitors. But ICI, by virtue of its dominant position, had the dual character of a private profit-making enterprise and a public service with an obligation to supply certain essential materials, whereas du Pont was wholly a private business.

The consequence was that du Pont had greater freedom of action than ICI, being free to move into or out of any field without reference to any standard of judgement but the return on capital employed. ICI, on the other hand, had a responsibility towards the public interest which, in fields where ICI had a monopoly—ammonia, for instance, or chlorine —might override strictly commercial considerations. Moreover a monopolist always lies open to political attack, as in ICI they were well aware.

Du Pont had one large commitment outside the chemical industry altogether—their holding in General Motors. It amounted to more than 25 per cent of GM Common Stock, the only class of stock normally carrying votes. Lammot du Pont was the non-executive Chairman of General Motors, and among GM's officers and directors there were several names prominent in du Pont's hierarchy also. The capital employed in GM's business, by 1939, amounted to $1,207·2m., against du Pont's $520m., and in 1929 ICI's Intelligence Department, without stating the source of their information, said that du Pont's income from GM was 'considerably in excess of that derived from [du Pont's] own operations'.[13]

Du Pont's links with Nobel Industries, many and close, were all taken over by ICI. The main agreement between the two was a successor to agreements between du Pont and European explosives firms running back to 1897 and even before. It was the Patents and Processes Agreement of 1920. It provided for the exchange of technical information and for cross-licensing of patent rights, and apart from its purely technical purposes it was designed to keep the two companies out of each other's way in what they regarded as their exclusive territories—that is, the British Empire for Nobels and the USA for du Pont. It referred only to explosives and related products and it had not been extended to the wider fields in which both du Pont and ICI, by the later twenties, were engaged.[14]

Outside the 'exclusive territories', du Pont and ICI had numerous special arrangements. In South America there were two joint companies, Compania Sud Americana de Explosivos, a manufacturing company in Chile, and Explosives Industries Limited, a selling company for other South American countries. In this last company, IG were also indirectly interested through shares held by Dynamit AG of Hamburg. In Canada, which was regarded as a special case within the British Empire, du Pont and ICI jointly owned Canadian Industries Limited, and they had also one joint company—Nobel Chemical Finishes Limited—in the United Kingdom itself. In Europe, just before the ICI merger, du Pont and Nobels had each invested the equivalent of £375,000 in Dynamit AG and another German company, Köln-Rottweil. K-R, soon afterwards, were absorbed into IG, and by exchange of shares du Pont and ICI had both become shareholders in IG, though their holdings were very small in relation to IG's total capital. Finally, ICI's controversial investment in General Motors, made originally at du Pont's suggestion, carried with it a seat for Sir Harry McGowan on General Motors' Board, which reinforced his already close personal links with the top men in du Ponts.[15]

(ii) IG WOOED BUT NOT WON 1926–1927

In spite of what McGowan said to Swint (p. 32 above) about the next step in ICI's diplomacy being 'an arrangement . . . between the Germans and the British', there is little doubt that his own instinct would have been to found ICI's foreign policy, from the start, on an understanding with du Pont, with whom he personally and Nobels corporately had so long a tradition of close and cordial friendship. Alfred Mond, however, and his son Henry were inclined to play down this established relationship and to seek first a triple alliance between ICI, Allied, and IG.

Allied loomed thus large on the Monds' horizon partly because of ICI's large holding of Allied stock and partly because of the importance of Allied in the alkali industry. That, besides being the bedrock of Brunner, Mond's prosperity before the merger, was the greatest contributor to ICI's profits after it, and to anyone with a Winnington background it represented the chemical industry as explosives and dyestuffs never could. In the way of an agreement with Allied, however, stood Orlando Weber, Allied's President. Already, in New York in September 1926, he had wrecked one attempt and had thereby driven Sir Alfred

Mond, against his earlier inclination, into the arms of Sir Harry McGowan. After ICI was formed, however, Sir Alfred, undeterred by a spiky personal interchange with Weber in October 1926, was determined to keep on trying, and so was his son. Du Pont, in their view, was a second-class power. In the summer of 1927 Henry Mond, discussing the proposed triple alliance, referred to 'these three great Companies'—ICI, IG, and Allied—'and in a smaller way Solvays and du Ponts'.[16]

McGowan did not greatly like the Monds' ideas. He told Swint that 'he did not propose to have the twenty years' friendship which the Nobel Company and he personally had enjoyed with the du Pont Company and du Pont officials terminated as a result of the formation of ICI' and he promised to use his influence on du Pont's behalf with 'the other members of the amalgamation'. He went further. 'He would see to it', Swint reported, 'that no hostile steps . . . against the du Pont Company were taken by ICI or any combination which they might arrange with the Germans *without telling us* [du Pont] *of it beforehand.*'[17] Nevertheless, in this as in other matters, McGowan seems to have been prepared to defer to Alfred Mond, who, early in 1927, began to apply himself to negotiations with IG.

From the first it was evident that relations between ICI and IG were likely to crystallize round three main topics: Nitrogen, Oil-from-Coal, and Dyestuffs. In other fields the two sides either did not overlap, or did not attach great importance to those where they did.[18] In any bargaining over markets, it could be assumed that ICI would claim the British Empire and IG, Europe. Claims to other territories, especially in Latin America, would be negotiable.

In nitrogen, ICI's productive capacity was only a fraction of IG's. Large extensions were planned at Billingham, but even after they had been made ICI would still only produce ammonia equivalent to 170,000 tons a year of nitrogen, against the Germans' 700,000. Moreover the Germans' home market for fertilizers was expected to take more than the whole of the British Empire, on which ICI's hopes were chiefly set. Nevertheless, ICI's bargaining position was stronger than might have been thought. They were well advanced with the technology of synthetic ammonia, and Pollitt was convinced, and was convinced that IG were convinced, that if necessary ICI could go much lower than IG's prices and still make a profit.[19]

In oil-from-coal and dyestuffs ICI were far less strongly placed. Broadly speaking, they sought German technical knowledge on both subjects, and the onus lay on them to show the Germans that they had valuable consideration to offer. The Germans, in oil-from-coal, might prefer the big oil companies as partners, and negotiations were already on foot with Standard Oil, while Royal Dutch-Shell had prudently interested themselves in the patent rights originated by the German inventor Bergius. In dyestuffs IG held the rest of the world in sovereign contempt, and British Dyestuffs Corporation, perhaps, especially so.

Having such considerations as these in mind, and being determined to show the Germans that ICI were not technically so negligible as they had been accustomed to think, Alfred Mond led a strong delegation to Paris, in January 1927, to meet an equally strong party from IG. Mond had with him McGowan and all but two of the ICI executive directors. For IG Carl Bosch himself appeared, supported by Wilhelm Gaus and Hermann Schmitz, both members of the Central Committee of the *Vorstand* (IG's Board of Management).

Bosch suggested a plan for joint working which was typical of the way the Germans liked to do these things, and to which, in Volume I, we have seen parallels in the organization of the Nobel–Dynamite Trust and its associates and in the negotiations of the early twenties between the first IG and BDC. The basis, Bosch suggested, should be a massive interchange of shares, so that ICI and IG should become interested in each other to the extent, perhaps, of 30 per cent. As well as that, there should be a profit pool, and Schmitz said helpfully that on the basis of existing earning power he thought IG would get two-thirds of the total. 'ICI', says Pollitt's note of the meeting, 'showed no interest in this proposal.'[20]

What ICI would have liked, according to a draft outline agreement drawn up by Mond, would have been a full and free exchange of patents and technical information, with a number of ICI shares allotted to IG in consideration of their superiority in dyestuffs. Markets should be shared in the accustomed way—British Empire for the British, Europe for the Germans; South America and the Far East divided—or perhaps sales quotas might be arranged. In oil-from-coal, ICI would invest in production facilities for which IG would supply whatever patent rights, knowledge and help might be needed to ensure commercial success. In this field, though not in the business generally, profits would be divided.[21]

The two sides were a long way apart. No negotiations seem to have been attempted in Paris to bring them closer together. After two days (21–22 January 1927) the meeting broke up. The eminent gentlemen went home again, presumably to think. ICI, for their part, promised to consider the idea of a profit pool. They did so, and in April accepted it, taking as a guiding principle 'the avoidance of economic wastage, of duplication of plant, of capital expenditure, and of competition between the parties'—a perfect short statement of the rationalizers' creed, to which Mond so heartily subscribed.[22] A meeting followed, in London on 21–22 April 1927, and afterwards Bosch, taking the pooling of profits for granted, wrote to press for an IG interest of

50 per cent in British Dyestuffs Corporation and in Synthetic Ammonia and Nitrates, ICI's Billingham subsidiary. He offered ICI a joint oil-from-coal company, presumably for trade within the British Empire, but he insisted that although ICI's holding would be equal to IG's, yet IG's technical contribution would be so much the greater that there ought to be compensation for it.[23]

Mond and McGowan, replying, retreated once again from the idea of a profit pool, saying: ' . . . it appears to us so liable to misconstruction, and so full of difficulties of a political, financial and legal character, that . . . we have come to the conclusion that at any rate for the present we had better proceed on simpler lines.' 'Simpler lines', no doubt to the Germans' surprise, turned out to be based on an exchange of shares, the very idea which ICI had so far rejected. The shares were to carry no voting rights, but the exchange was to be sufficiently massive to give ICI and IG 'a permanent interest in each other's welfare, and to give a return for any advantages to be mutually derived by closer co-operation, technical and commercial'.[24]

With some colour of indignation, Mond and McGowan rejected Bosch's demand for shares in BDC and SA & N. 'We do not understand why this proposal has been made', they said, ' . . . as it would entirely destroy the scheme of reorganisation we are building up.' They did not say so but they may have wondered whether this was precisely what Bosch intended, since dyestuffs and nitrogen were IG's two strongest lines. In any case, said Mond and McGowan, the British Government would never allow such a thing, since they regarded both dyestuffs and nitrogen as 'war industries'.[25]

Mond and McGowan then turned to what really interested them—oil-from-coal. They showed great impatience to get going, and to get IG's co-operation in 'the utilisation of your process for the liquefaction of coal in England and in the British Empire'. This was an entirely new development, they said, and 'the creation of a new Company would seem to be entirely suitable'. The proposition was so big, however, 'that it appears doubtful whether your Company, or ours, would care to establish the whole of this development'. No doubt they were thinking of the oil companies. They suggested a company in which ICI would have a 60 per cent interest and IG 25, remarking that the company 'must be under British control' and 'as we shall . . . have full responsibility for the finance, management, shares, patents and development of this vast problem in the British Empire, I [Mond ?] think you will agree that it is not unreasonable that we should claim this percentage of the profits.'

They suggested joint committee meetings with IG to get rid of 'unnecessary competition' in dyestuffs and nitrogen products. They ended with a gentle admonition:

. . . while we quite admit the inventive resource, technical abilities and research facilities of your Company, we can claim that . . . we on our side have produced and are producing technical improvements, inventions and research . . . which we trust will be of value to your Company. We would especially point out the . . . developments of Vat Dyes; working out of new Alizarine process . . . and also that we consider we have made very considerable progress in Urea practice, and the technique of the production of Ammonia products.

While, therefore, we are willing to try to find a basis of some compensation for the exchange of technical information we wish to make it quite clear that it must not be taken that we should be the only party which would benefit by an exchange of technical information.

This exchange of ideas, in which ICI had shown most of the flexibility' prepared the way for a serious attempt at agreement. On 12 May 1927 Mond, McGowan, Ashfield, Reading, and Stamp—the Chairman, the President, and three lay Directors, an interesting combination— formed a committee to consider German proposals, and on 21–22 June 1927 the two sides met in London, each company once again being represented at the highest level. The two Monds, McGowan, and Mitchell appeared for ICI: Bosch, Schmitz, Gaus, and three others for IG. In the German party there was a representative of the Supervisory Board (Dr. Carl Müller) and of the Associate Members of the *Vorstand* (Dr. August von Knierem), which suggests that the Germans expected to bring the negotiations to a decision, or very near one.

Bosch put forward a plan for an international company under (his phrase) a Super Board of Control. French, Swiss, and Norwegian interests were to come in, and profits were to be pooled. He rejected the idea of sharing markets on a territorial basis because 'this form of fusion is liable to attack and unlikely to survive', and instead he proposed a device which the Germans, as opposed to ICI, always greatly preferred —a production cartel. He said it would aim at reducing, not increasing, prices, 'the main objective being to regulate production in each country and to determine what each country can best produce, due consideration . . . being given to obligations which the various manufacturing units . . . may have towards their respective Governments'.[26]

What could be more reasonable, what closer to the rationalizers' habit of mind? Mond and McGowan, however, who do not seem to have been expecting a plan so lordly, reacted cautiously, Mond especially so. Companies joining the cartel, they said, would lose their identity, which ICI at least had no mind to do. Moreover, said Mond, the British Government certainly would not allow dyestuffs and synthetic ammonia, those two 'war industries', to come under any international Board of Control, however Super.

The ICI party seem to have been in some disarray. Mond persisted in his scheme for a share exchange with IG—which, after all, was close to what IG had originally said they wanted—but McGowan seems to have given some support to Bosch's idea for a profit pool. Mond, on that point, indicated that he didn't care for the possibility of having to explain to ICI's shareholders that they were going to be called upon to make good losses incurred by 'say, the French or Swiss' in uneconomic production. Bosch's reply was that the German and British companies would have to see to it that the French and/or Swiss companies 'became dividend-earning units of the Cartel'.

There is no doubt that what Bosch was aiming at was drastic rationalization of the entire European chemical industry. There is no doubt, also, that in nitrogen and dyestuffs, at least, there was good economic reason for it, because both industries were facing problems of over-production. On the other hand, and the point cannot have escaped the ICI delegation, any rationalization of the kind proposed, in the circumstances of 1927, was bound to result in even greater concentration of manufacturing power in Germany, particularly if the aim was to bring down prices. McGowan and Mitchell both showed anxiety about the threat to the autonomy of the ICI Board. They had good reason.

Bosch must have known that he was asking a great deal of ICI. He therefore offered what he doubtless hoped would be irresistible bait. Let ICI agree to join IG, along with the French and Swiss, in nitrogen and dyestuffs, and 'the IG would be prepared to extend the understanding to embrace Petroleum'. This above all was what ICI wanted from IG. The obstacle would be the British Government, but surely ICI could point out the vital national importance of oil fuel, and then announce that 'it could obtain a concession in regard to the production [of oil fuel] provided the Government were prepared to release ICI from its obligations regarding the Ammonia and the Dyestuffs Industry'? That would be a fair bargain for all parties, surely?

The London meeting turned out as inconclusive as the Paris one. Its main upshot was the appointment of four committees on each side, one each for dyestuffs, nitrogen, and oil-from-coal and one for accountancy, legal matters and taxation. They were to make proposals 'for the most economic exploitation in the joint interest, whether by selective manufacture, allocation of output, reservation of markets, apportionment of sales or otherwise, having regard to the governing principle that national requirements must be the prime factor'. That appears to have put the whole problem back in the melting pot again, and ICI left it on record that 'any Committee or Council . . . must be advisory only' and that 'share exchange . . . would involve such taxation charges as make it impracticable at present', so that it appears that the grand design for an international company, with its Super Board of Control, had been decisively turned down.

During the summer of 1927 IG's diplomats were very active, though not with ICI. ICI heard rumours, some accurate, some not, and none reassuring. In August IG were reported to have signed an agreement with Standard Oil over what H. A. Humphrey called 'a new process for the production of synthetic petrol'. This was true and turned out to be important (p. 167 below), but at the time ICI had no very precise information.[27] On 1 September Ernest Solvay told Alfred Mond that the Standard–IG agreement would not be signed until IG had settled with ICI and the French, and that great plans were afoot for the regulation of the French and Swiss chemical industries, in which, apparently, both IG and Solvay would be concerned. 'He said it was time that Solvays and ourselves definitely laid down our ideas and jointly presented them to IG. He thought this was the only way we could get anything settled with the IG.'

Since the spring ICI committees had been at work with just that end in view, and one of them had worked out proposals for dyestuffs based on the usual ICI formula of reserving Empire markets for themselves and bargaining about South America, the Far East and 'sundry other markets'.[28] On 5 and 6 September a party headed by H. J. Mitchell presented these views to the Germans at Frankfurt. S. A. Whetmore, a member of Mitchell's party, who had had experience of the Germans in the negotiations with BDC before the merger, had already prophesied hard going.

Mitchell found himself confronted with a much larger party than his own (four, including himself) and mounting much heavier metal. Dr. Bosch himself was there, with nine supporters of whom one (Carl von Weinberg) was a member of IG's Supervisory Board and five were Managing Directors (members of the *Vorstand*). Mitchell was the only Director on his side of the table.[29]

The Germans were aggressive from the start. Almost as soon as the meeting opened they announced that they 'held very strongly that ICI should efface itself altogether from the Export trade [in dyestuffs], and expressed a willingness to take over for resale any quantity produced in the UK (within an agreed manufacturing programme), which could not be sold there.' Mitchell naturally refused to agree.

Dr. Georg von Schnitzler, sometimes known within ICI as 'handsome George', then reviewed the British dyestuffs industry, 'and he desired that it should be clearly understood at the outset that the German Group considered that fusion of the existing British Dyestuff entities was an essential preliminary to the completion of any negotiations between the British and German Groups'. It might, perhaps, have been thought that the ICI merger would have been sufficient for the purpose, but apparently not, since there still remained firms 'who, although comparatively insignificant, could, if not embraced within a Dyestuffs

Merger, be a real nuisance'—to the Germans, that is. Mitchell, rather surprisingly, said ICI had expected this demand from the Germans and would do their best to carry it out.

By blunt questioning, the Germans cleared up one or two matters of detail that interested them in ICI's relations with other dyestuffs firms, and advised them to keep clear of the discussions which IG were having with the Swiss IG about Clayton Aniline, which the Swiss owned. They asked whether ICI would consider lobbying the Government for the suspension of the Dyestuffs Act ('Mr. Mitchell stated he personally did not feel that would be proper'), and then turned to BDC and the export trade. 'The accomplishments of BDC', von Schnitzler said, 'could not be regarded as brilliant', and when Whetmore challenged some of von Schnitzler's views he responded by saying that the only profitable British dyestuffs company was Clayton Aniline, and that was because they had Swiss technical advice. The reporter of the meeting, almost certainly Whetmore, thought that these tactics were intended to induce an inferiority complex in the British, and perhaps he was right.

When von Schnitzler came to review the export trade he said 'the British industry was offering no *real* competition in overseas markets' and IG 'felt that the British Group should abstain altogether' from the export trade, except possibly in Australia and Africa. In return, IG would be prepared to make concessions. What the 'concessions' turned out to be was a proposal that BDC should confine itself to making the United Kingdom's requirements of bulk colours, and IG would then agree not to compete with them in these lines in the United Kingdom. As to the country's national safety, the IG's view was that it would be effectively ensured if BDC kept to 'Indigo, Sulphur Colours, some Direct Colours, and in general the staple articles, together with a few of the intermediates required for these'. These 'concessions', in IG's view, 'would in effect give ICI a manufacturing production of approximately 70% of the UK requirements, whilst the balance composed of specialities should . . . be imported into the UK from the IG.'

These astonishing demands, which would certainly have ruined the British dyestuffs industry and prevented development of fine chemicals generally, were followed, at a meeting in The Hague in October 1927, by a demand that ICI should hold up extensions at Billingham until the nitrogen production of IG and their associate, Norsk Hydro, had been absorbed. Pollitt flatly refused, and added a threat to cut prices which, he believed, the Germans took seriously.[30]

The German show of force virtually ended the attempt to get a general agreement between ICI and IG, and it was never revived, although numerous agreements on various aspects of the two companies' activities, some very far-reaching, were later entered into. The Germans, at Frankfurt and The Hague, had repeated the overbearing tactics which they had found so nearly successful with BDC before the merger.[31] ICI had been founded to undertake international bargaining at its toughest, and they were far more formidable than BDC. Moreover in Alfred Mond they had a Chairman who was determined to be treated by the Germans as an equal. 'These gentlemen', he wrote to Armand Solvay, in October 1927, 'still fail to realise that the day of their supremacy in the chemical industry has passed away, and that when they deal with an organisation like ours they have to deal on terms of equality.'[32]

Correspondence went on through the autumn of 1927 and into 1928, but the end really came in November 1927 when Bosch, in a letter to Mond, stuck to the essence of IG's demands. He claimed that Mond had long ago agreed to restrictions on dyestuffs as the price of IG's co-operation in oil-from-coal. As for nitrogen, he said that ICI's production ought to be restricted or, as he put it, 'ought to conform to the fundamental principles laid down by us at the commencement of our various discussions and which, owing to a too extensive interpretation of "National Requirements", are naturally in danger of being broken down.'[33]

This letter stirred Mond to considerable flights of eloquence, particularly on the subject of dyestuffs:

We cannot carry on in this country a skeleton industry without hope of development or expansion, which is the inevitable outcome of your proposals, as no one will realise more clearly than yourself. Quite apart from the economic results, I cannot doom . . . energetic and able scientists to a life of sterility and stagnation. We cannot obviously retain their services under such conditions, and if such a course is adopted the probable result will be that their energy and information would be carried to . . . competitors. . . . Nor can I, as the Head of the Chemical Industry in this country, and the inheritor of the position of promoter of the Chemical Industry from one of its greatest leaders, my late father, accept such a position, and even if we wanted to we should not be allowed to from the point of view of consumers, public opinion or the Government of this country.[34]

He turned Bosch's demands down finally, and from that moment on began to look across the Atlantic, rather than across the North Sea, in his search for allies.

(iii) THE DU PONT ALLIANCE 1926–1929

The collapse of the IG negotiations completed the destruction of a foreign policy which Alfred Mond had been carrying on, with great persistence, since before the formation of ICI. He conceived the welfare of the British chemical industry to require an alliance with IG in

46

Lord McGowan
(1874–1961), by
Harold Knight, R.A.

Lord Melchett (Sir Alfred
Mond) (1868–1930). From
a photo by Logan,
Birmingham.

The Bridge, by 'Spike'. From the first issue of the *ICI Magazine*, January 1928.

Europe and with Allied in USA. The continuance, in some form, of the alliance between Brunners and Solvay was taken for granted, and relations with du Pont would be settled, from a position of strength, as a matter of secondary importance.

This was an alkali-maker's policy, based on a view of the world from Winnington. Orlando Weber dealt the first blow at it when he denounced Allied's agreement with the Solvay group in the summer of 1926 (*ICI* I, p. 462). A little over a year later the overbearing nature of IG's demands began gradually convincing Alfred Mond that the whole policy would have to be abandoned.

There was an alternative, associated not with Sir Alfred Mond and Brunner, Mond but with Nobel Industries and Sir Harry McGowan. This was the policy of close alliance with du Pont, founded on the Nobel Industries–du Pont agreement which had been carried into ICI with thirty years' experience of close association, by 1927, behind it. It required broadening to cover the wider interests of ICI and of du Pont themselves, but negotiators would at least start from a basis of mutual goodwill, which was more than could be said for those who negotiated with IG.

The formation of IG Farbenindustrie caused quite as much uneasiness to the management of du Pont as to the founders of ICI, because they held the Germans quite as much in awe and had an equally great respect for their technical versatility. 'The IG', wrote Jasper Crane to Irénée du Pont early in 1926, 'is very aggressive and is particularly seeking to establish itself in the United States.'[35]

During the summer of 1927 du Pont, at the same time as ICI but quite separately, were trying to bring IG to a general agreement. They were probably seeking IG's technical help generally, but in particular they discussed the possibility of help with ammonia plant in the USA. At first IG seemed disposed to agree, but in the autumn their attitude changed and they began to talk about the danger of over-production of nitrogen. It was at about this time that IG tried and failed to get ICI to agree to limit Billingham's capacity, and after a meeting in November, according to Crane, 'it became apparent . . . that the aim of IG was to control, restrict, limit or hamper the development of ammonia production in America rather than promote it'.[36]

This change of attitude Crane found surprising. Perhaps if he had been privy to the negotiations between IG and ICI it would have surprised him less. As it was, he put it down to the breakdown of their attempts to come to agreement with ICI.

The IG's fundamental purpose [Crane concluded] is to dominate the nitrogen industry throughout the world. . . . If the British had agreed to

47

C

restrict their production there would be a great advantage to IG having a large voice in the American business, but without being able to restrict the British activities they apparently see less advantage with an American association and may seek to maintain their supremacy by other methods, notably by lowering of prices. The most unfavourable side to this is that this monopolistic or dominating policy, if persisted in, would prevent IG from being an acceptable partner in an American enterprise. Instead of promoting its interest they seek to restrict its development.[37]

Du Pont's experience of IG was the same as ICI's. They found the Germans confident to the point of arrogance and determined to have their own way.

There were some in ICI and probably some in du Pont who had no great enthusiasm for closer ties between the two companies.[38] Equal disenchantment with IG, however, allied with equally healthy respect for IG as a competitor, was making for greater togetherness. McGowan had never ceased to work for it, even at the height of the IG negotiations. For years he had hoped for a share exchange between Nobel Industries and du Pont, and in March 1927 he saw, or thought he saw, a chance of carrying ICI part of the way along that road. He proposed to Pierre S. du Pont, a personal friend as well as Chairman of the du Pont Board, that ICI should sell 100,000 GM Common Stock and invest the proceeds in 45,000 du Pont. Pierre agreed, but then J. J. Raskob, a du Pont Vice-President, suggested letting ICI buy, not du Pont stock, but stock of Christiana Securities. That would have been almost like letting McGowan marry into the family, and, as Lammot du Pont observed, 'if we get another substantial stockholder [in Christiana], such as ICI, the chances are that its interest would not be wholly with the other stockholders'. No such thing could be allowed, and McGowan's whole plan wilted. Its reception by Raskob did show, however, that on the du Pont side as well as McGowan's there was one, at least, who hoped for closer rather than more distant relations with ICI, whatever might be the upshot of the talks with IG.[39]

At the beginning of December 1927 there were meetings in London between du Pont's European representatives and a strong party from ICI—Alfred Mond, McGowan, Rogers, Mitchell, and Pollitt. The idea of extending the Patents and Processes Agreement beyond explosives and related products came up. McGowan, and probably Rogers and Mitchell, would have been favourable, but Mond was not easily convinced. He still wanted an agreement with IG, or rather, at this time of day, he did not want to do anything which might block an agreement in the future. This, he thought, was what ICI might very easily do if they came to an agreement with du Pont covering nitrogen and dyestuffs, and he saw no likelihood of ICI learning anything from du Pont, on either of these matters, which would compensate for the enmity of IG. Besides that, he was very much against upsetting Allied, and again he feared the effect of co-operating with du Pont in nitrogen and dyestuffs. 'It has also to be remembered', he wrote, 'that any connection with Duponts that the Allied might take as hostile might involve us in very costly competition with them in Alkali, in which Duponts do not share.' He saw some possibility of a territorial agreement, provided ICI had the British Empire (except Canada, to be handled through CIL) and if possible the Far East, and he concluded with a lukewarm suggestion: 'Could we not have a general understanding that either party will offer to the other any new inventions, patents and processes before offering them to any other party on terms to be agreed?'[40]

During March 1928 McGowan, Mitchell, Nicholson, L. J. Barley, and R. B. Brown—a strongly Nobel delegation—discussed possibilities at length with du Pont in Wilmington. The meeting was inconclusive, and it became clear that the main obstacle to an agreement with du Pont was ICI's holding in Allied, which drove them to seek a triple alliance of themselves, Allied, and du Pont. The plan had no chance. It simply irritated du Pont's men. 'I think', wrote John Jenney, one of the younger of them, 'the feeling of all our people here is that the ICI is in the very delicate position of flirting with us on explosives and with the Allied Chemical in the Alkali business. . . . Taking a very narrow view, it strikes me that they feel that everything the du Pont Company has to offer should be given to them because of the old Nobel–du Pont understanding, to be used by any branch of ICI, but whenever the ICI has something in which du Pont would be interested it seems to belong to Brunner, Mond or one of the other companies, and therefore cannot be offered.' Jenney admitted he was not being entirely fair, but he insisted that 'in measuring actual results . . . this about describes the present situation'.[41]

In October 1928 McGowan went to the USA again. This time Sir Alfred Mond, newly created Lord Melchett, went with him. The meeting decided in principle that it would be desirable to extend the Patents and Processes Agreement to other products, and appointed representatives 'to seriously explore the subject'.[42] That decision in itself might have meant very little. What gave it force was Lord Melchett's change of mind. 'When in America', wrote McGowan, 'Lord Melchett and myself, after a good deal of deliberation, decided that, as a matter of high policy, it would be well that we should be "foot loose" as regards shareholdings in Companies cognate to our own. This carries with it the decision to sell our shares in the Allied Chemical Company. . . .'[43] Melchett, evidently, had at last come round to McGowan's view of foreign policy. There should be an alliance with du Pont, and, since that carried the possibility of conflict with Allied, the Allied shares should be sold.*

And sold the Allied shares were—to Solvays. They bought 105,600, which, for some reason now lost, left 10,316 still in ICI's hands. Solvays' purchase, added to their existing holding, gave them 457,195 Allied shares, or 20·9 per cent of Allied's issued capital. Solvays set up the Solvay American Investment Corporation especially to hold them.[45]

The sale of the Allied shares opened the path to an extended Patents and Processes Agreement between ICI and du Pont. The representative appointed to carry out the detailed negotiations on behalf of du Pont was Dr. Fin Sparre, head of du Pont's Development Department. ICI was represented by G. W. White of the New York office and Francis Walker of the Dyestuffs Group. Two series of meetings were held, one in the USA in March 1929 and another in London in May. At the March meetings the ICI men had the benefit of the presence of McGowan and Pollitt.

Both sides intended to have an agreement but neither displayed any great readiness to make concessions to the other, so that the negotiations were long and occasionally rather acrimonious. Sparre in particular, skilful and unyielding, was described by White as 'one of the "Die-Hards" of the du Pont organisation'[46] and, although genial enough socially, he came to be regarded with great circumspection by the ICI negotiators. Over a number of years they came to know him well—too well, sometimes, for their comfort.

ICI were obliged to look carefully at prior agreements, especially the Solvay agreement, which might conflict with the principle of free exchange of knowledge with du Pont, and each side, from time to time, cast a lingering glance at whatever possibility still remained of a joint or separate accommodation with IG. There were also differences of opinion about the stage at which information should be exchanged—whether before or after it had been protected by patent—and about whether, and if so how, information should be paid for.

What the argument chiefly turned on, however, was the division of markets, and especially the definition of 'exclusive territories'. At the first meeting of the first series, in March, White and Walker made an attempt, which Jenney described as 'ridiculous', to get du Pont to accept North America, except Canada, as 'exclusive territory', leaving

---

* The decision led to a minor comedy. At the same time, ICI decided to get rid of one or two other holdings, including 5,000 du Pont Common Stock, and McGowan politely told Lammot du Pont what was intended. Du Pont replied that McGowan must be mistaken—du Pont's Treasurer knew of no such holding. Perhaps McGowan had meant to say General Motors. But McGowan had meant exactly what he said, and he had to explain, somewhat embarrassed, that the shares had been held by a nominee, so of course du Pont's Treasurer wouldn't have been able to identify them. It is not clear whether they were sold, for in the 1930s ICI were still (again?) holding du Pont stock and wondering whether to sell.[44]

South America as common ground and the rest of the world as ICI's 'logical markets'. Sparre rejected the claim and referred to Lammot du Pont. Agitated telephoning between him and McGowan, in New York, produced agreement in principle that the USA should be regarded as du Pont's exclusive territory and the British Empire as ICI's, with the rest of the world as common ground and agreements to meet special cases as might be necessary.[47]

Territorial bargaining raised the question of the nature of the agreement being negotiated, and whether it could be legal for the Americans to enter into it. If the matter came into court, the question would be asked whether the agreement was genuinely directed at exchange of technical knowledge, or whether it was a device for restricting competition. An agreement to exchange technical knowledge, carefully drafted, would come within the limits of anti-trust legality, or so du Pont's lawyers thought (as they had thought, it may in passing be recalled, in 1907, only to reverse their opinion in panic-stricken haste six years later).[48] An agreement to share markets with a competitor, on the other hand, could never be legal.

An exchange of technical knowledge was certainly desired by both sides, especially since the IG negotiations had broken down. If they did not exchange knowledge, they would be obliged to duplicate each other's research, which neither side wanted. Du Pont, for their part, were so anxious to get at ICI's knowledge of ammonia synthesis that Jasper Crane managed to persuade ICI's Executive Committee to allow a party from du Pont into Billingham even before any agreement was signed.[49] Other du Pont divisions were in less of a hurry, and suspicions arose on the ICI side that, with Sparre's connivance, they were hanging back from divulging information to ICI. However, Lammot du Pont was persuaded to issue a directive instructing them to be more forthcoming, and in general there is no reason to doubt the genuineness of each side's desire to profit from the technical knowledge of the other, and to come to an agreement which would permit each to do so.

It is certain also that ICI, if not du Pont, had market-sharing in mind. ICI had no need to disguise the fact and made no attempt to do so. Du Pont's were obliged to be cautious. Sparre, at one of the London meetings in May, said there must be no pooling or profit-sharing in common territory,[50] and it was held that technical information which passed between the parties would have to be paid for. The general conclusions of the negotiators were summarized in the consolidated May minutes as follows:

. . . agreed that the agreement must conform to the laws of the United States and . . . must take the form of an agreement to purchase and sell exclusive and non-exclusive licenses to patents and secret processes, and that for this reason the phraseology of the agreement would be left to the Du Pont legal department.[51]

When du Pont's draft was received, in August 1929, H. J. Mitchell wrote to the other Executive Directors:

The document has been drawn up with special reference to the anti-trust legislation in force in the United States and it is a matter of vital importance to Duponts that no flaw should exist from this point of view and that no subsidiary documentary evidence such as correspondence between the two companies should indicate that the agreement is worked in a way substantially different from the intentions indicated in the legal document.[52]

Mitchell's point is made. We need labour it no further.

The general form of an agreement acceptable to both parties was in draft by the middle of April 1929, and the final document, on which Mitchell was commenting, was signed in November to run from the preceding 1 July. The preamble recited that ICI and du Pont were 'engaged in the development, manufacture and sale of a broad line of chemicals and chemical products, both in their respective home countries and in other countries, and maintain[ed] research and development organisations for . . . expanding their present activities as well as developing new industries.' Each of the parties, the preamble went on, desired 'the right to acquire licenses in respect of the patented and secret inventions of the other party'.[53] All of which was perfectly true, as far as it went.

Each party undertook to disclose to the other 'as soon as practicable, or in any event within nine months from the date of this agreement, or from the date of filing application for letters patent covering patented inventions, or from the time any secret invention becomes commercially established, information in respect of all patented or secret inventions . . . owned or controlled by it, relating to the products hereinafter specified, sufficient to enable the other party to determine whether it desires to negotiate for license covering any or all of such inventions.'

The list of products was comprehensive. The important exceptions were military explosives, 'products of the general alkali industry', and tetra-ethyl lead. Military explosives were usually subject to some kind of government regulation, which made them difficult subjects for private agreement. Disclosure of alkali information, by ICI, would have run foul of the Solvay agreement, and since du Pont had no interest in alkali there seemed to be little practical disadvantage in admitting the exception. Tetra-ethyl lead, as an anti-knock agent, was important to the development of motor fuel and was probably kept out of the agreement at the insistence of ICI, preoccupied as they were with the alluring possibilities of oil-from-coal. Other exceptions were made from time to time, and there was a clause allowing either party to remove a 'major invention' from the agreement altogether, so that they could make special terms. Both nylon and polythene were dealt with under this clause.

Since both parties still hoped to come to terms with IG a sub-clause was provided which permitted either to do so, separately, 'and upon the execution of such an agreement . . . this agreement in so far as it relates to the dyestuffs industry shall cease and terminate'. Each party undertook, however, 'that in negotiating or upon entering into such an agreement . . . with said IG Farbenindustrie AG, it shall use its best efforts to extend same to include the other party hereto'.

The grant of licences was based on the principle that ICI and du Pont should each have a right, within certain territories, to an exclusive licence from the other to operate any patented or secret invention belonging to it. Technical help would be provided, and each party would support the other in the maintenance of its rights. Existing trade by one party in the 'exclusive license territory' of the other, at the date of the agreement, was recognized and would be tolerated, but clearly the whole force of the agreement was to prevent the development of new trade by one party in competition with the other.

The world was divided into 'exclusive license territories' along lines familar for many years in the explosives industry. Du Pont were to have 'the countries of North America and Central America [defined as the region from Colombia to Mexico, inclusive, between the Caribbean and the Pacific], exclusive of Canada, Newfoundland and British possessions, but otherwise inclusive of the West Indies, and . . . all present and future colonies and possessions and mandated territories of the United States of America.' ICI were to have the British Empire, including Egypt and mandated territories, but not including Canada and Newfoundland. Canada and Newfoundland, excluded from the 'exclusive license territories' of each party, were the territory of the jointly-owned Canadian Industries Ltd. The agreement does not make it clear what the licensing arrangements were to be there. In all other countries, the parties were to grant each other, upon request, non-exclusive licences.

Knowledge exchanged would have to be paid for. ICI did not like this provision because they thought the exchanges would be roughly equal in value and the prospect of having to pay might inhibit frank co-operation.[54] Very similar objections had held up the 1907 agreement between Nobels and du Pont[55] but there was no escape then and there was to be no escape now. The lawyers insisted, and du Pont's Foreign Relations Committee agreed to tell McGowan 'that if he could not see his way clear to signing the agreement in conformity with [American] laws there could be no agreement at all'. The operative clause, no doubt to mollify Sir Harry, was soothingly drawn. 'Compensation', it said, 'will be determined under broad principles giving recognition to the mutual benefits secured or to be secured hereunder, without requiring detailed accounting or an involved system of compensation.' The Patents and Processes Agreement of 1929, entered into by ICI hesitantly and, so far as Melchett was concerned, very much as a second-best, became the bedrock of ICI foreign policy throughout the thirties and, in a revised form, for almost the whole of the rest of the period with which this book is concerned. It was an agreement simple in principle, simply drawn, and for smooth working it depended even more than most agreements on the mutual goodwill of the parties, which wore thin at times but never entirely disappeared.

The technical purpose of the du Pont Agreement—the exchange of knowledge—was perfectly genuine, but the licensing system had the effect, intentionally, of cutting out competition between ICI and du Pont in the 'exclusive' territories and inhibiting it elsewhere. It also, again intentionally, made it difficult for anyone else to enter the field. This may seem all very reprehensible to a later generation, but when the agreement was devised the chemical industry in the USA and Great Britain was still underdeveloped, relatively to the industry in Germany, and it seemed only sensible to avoid the expensive duplication of research and development which Anglo-American competition would have led to. Moreover, as Melchett and other rationalizers of the twenties often pointed out, supply and demand must be matched— excess productive capacity did no one any good—and the Patents and Processes Agreement, along with others which we shall examine, was directed partly to that end. No doubt it could also have been achieved by ruthless competition, but to whose benefit? Almost immediately the agreement was signed, the depression set in, and Melchett's point was driven home by circumstances in the crudest possible fashion.

REFERENCES

Note: Many of the documents on IG Farbenindustrie are held by the National Lending Library for Science and Technology (NLL).
1 Wendell Swint to Jasper Crane, 17 Dec. 1926, USA 'ICI/DuPont Agreements 1929–39 (Craven)', p. 43. F. J. P. Craven used documents later destroyed.
2 *Trials of War Criminals* VII—'The Farben Case', Military Tribunal VI, Case VI, The US v. Karl Krauch, Hermann Schmitz etc., NLL; see also L. F. Haber, *The Chemical Industry* 1900–1930, Clarendon Press, Oxford, 1971, pp. 281–2 and generally.
3 Fritz ter Meer, *Die IG Farben*, Econ. Verlag., Düsseldorf, 1953; Mil. Trib. VI, Case VI, defence document books 1–7 ter Meer (IWM Box 321), Doc. Bk. III, ter Meer, p. 61, NLL.
4 Mil. Trib. VI, Case VI, Doc, Bk. III, ter Meer, NLL; for ICI research budgets and expenditure see below, ch. 4 (iii), p. 87.
5 *Rationalisation of German Industry*, National Industrial Conference Board Inc., New York, 1931, pp. 124 ff.; Basic Information, Defense, Doc. 9 and Doc. 14, NLL.
6 *ICI* I, pp. 318–19.

[7] *ICI* I, p. 462.
[8] *Moody's Manual of Investments*, New York, 1930, p. 476 (Allied), p. 2263 (du Pont).
[9] A. D. Chandler and S. Salsbury, *Pierre S. du Pont and the Making of the Modern Corporation*, Harper & Row, New York, 1971, p. 429.
[10] Ibid. p. 571.
[11] W. S. Dutton, *Du Pont—One Hundred and Forty Years*, Scribner, New York, 1942, Book Four generally.
[12] A. D. Chandler, *Strategy and Structure*, MIT Press, 1962, Paperback Edition 1969, p. 95.
[13] 'Brief Notes on and Links between ICI and Du Pont, Solvay, etc.', ICI Intelligence Dept., 23 April 1929, p. 12, MP 'Links with Du Pont'.
[14] *ICI* I, pp. 395–6.
[15] *ICI* I generally, esp. ch. 17.
[16] H. Mond, 'Notes on General Considerations of ICI World Agreements', 3 Aug. 1927, CR 93/50/1.
[17] Wendell Swint to Jasper Crane, 17 Dec. 1926, USA 'ICI/DuPont Agreements 1929–39 (Craven)', p. 44.
[18] Nitrogen—report by Pollitt, 6 Jan. 1927, CR 107/2/17/5; Lynex, 14 Jan. 1927, CR 107/2/17/1.
Oil-from-coal—two undated reports (one by H. Mond), CR 107/2/17/2 and /3.
Dyestuffs—report by E. F. Armstrong, 17 Jan. 1927, CR 107/2/15.
[19] Pollitt to Executive Committee, 17 Oct. 1927, CR 84/7–1/3.
[20] CR 107/2/18.
[21] 'Synopsis of Draft Agreements', 13 Sept. 1927, CR 107/2/–.
[22] CR 107/2/18A; CR 107/2/–.
[23] Bosch to ICI, 29 April 1927, CR 107/2/20.
[24] Mond and McGowan to Bosch, 19 May 1927, CR 107/2/27.
[25] Ibid.; phrase 'war industries' from 'Notes taken at Meetings', 21–22 June 1927, CR 107/2/38.
[26] 'Notes taken at Meetings', CR 107/2/38.
[27] H. A. Humphrey to H. Mond, 5 Sept. 1927, CR 107/6/2; G. P. Pollitt to A. Mond, 8 Sept. 1927, CR 107/6/3.
[28] Report by Pollitt, 13 June 1927, CR 107/2/–X; report by S. A. Whetmore, 25 May 1927, CR 75/21/C.
[29] 'Notes of Meetings . . . 5th and 6th September 1927', CR 107/2/43.
[30] As (19).
[31] *ICI* I, pp. 445–6.
[32] A. Mond to A. Solvay, 26 Oct. 1927, CR 93/3/22, p. 4.
[33] C. Bosch to A. Mond, 15 Nov. 1927, CR 107/2/51.
[34] A. Mond to C. Bosch, 24 Nov. 1927, CR 107/2, p. 3.
[35] USA 'ICI/DuPont Agreements 1929–39 (Craven)', p. 7.
[36] Ibid. p. 24.
[37] Ibid.
[38] Ibid. pp. 3–4 quoting G. W. White's opinion.
[39] Chandler and Salsbury, p. 571.
[40] Notes by the Chairman, 5 Dec. 1927, CR 595/1/1.
[41] As (35), p. 68.
[42] Minutes of Meeting, 12 Oct. 1928, CR 595/4/2; I. du Pont to McGowan, 23 Nov. 1928, CR 595/4/3.
[43] McGowan to L. du Pont, 21 Nov. 1928, CR 595/7/1.
[44] Letters between McGowan and L. du Pont, 21 Nov. 1928, CR 595/7/1; 5 Dec. 1928, CR 595/7/2; 4 Jan. 1929, CR 595/7/2.
[45] ICI Intelligence Dept. Report on Allied, CR 138/–/R2, p. 3.
[46] White to Major, 12 Nov. 1929, CR 595/4–4/8.
[47] Minutes of first Meeting, 4 March 1929, CR 595/4/18, p. 4; as (35), Jenney to Swint, 8 March 1929; McGowan to L. du Pont, 6 March 1929, CR 595/4/20.
[48] *ICI* I, p. 213.
[49] Crane to Pollitt, 28 June 1929, CR 595/4/39; ECM 1427, 9 July 1929; Crane to Pollitt, 15 July 1929, CR 595/4–5/–; White to McGowan, 8 Nov. 1929, CR 595/4–4/7.
[50] Minutes, 6 May 1929, CR 595/4/29.

⁵¹ Consolidated minutes, 6–14 May 1929, CR 595/4/33.
⁵² H. J. Mitchell to Executive Directors, 23 Aug. 1929, CR 595/4/ .
⁵³ 'DuPont Agreement', 1 July 1929, CR 595/4/39, reprinted in full as Appendix IV.
⁵⁴ Du Pont Foreign Relations Committee, 23 Sept. 1929 (Jenney to L. du Pont, 24 Sept. 1929); as (35), pp. 176–7.
⁵⁵ ICI I, p. 200.

CHAPTER 4

Building for the Future 1926–1939
Labour, Management, Research

THE NEGOTIATIONS described in the last chapter were intended to meet the challenge which had brought ICI into being: the challenge of IG. Their purpose was to bring the strength of the new combine to bear on the balance of power in the world's chemical industry. That strength, as Mond and McGowan well knew, was potential, not realized, and there was no time to spare. Therefore while the bargaining went on, with IG first and then with du Pont, decisions of great importance were being taken with great rapidity on many and diverse subjects, and Mond's influence on all of them is powerfully apparent.

So much was going on, at such a furious pace, during the first three or four years of ICI's existence that comprehensive coverage would obliterate the narrative in a morass of detail. But even if only the most important matters are selected for detailed discussion, discussing them separately conveys an impression of orderly deliberation which must be false to the atmosphere of the time. Nevertheless for the sake of clarity the separation must be made, artificial though it is, and before we turn to the first great investment decision in ICI's history—the decision to put £20m. into fertilizer plant at Billingham—we shall discuss the formulation of policy in three other fields which the founders of ICI held to be essential to the success of any investment they might choose to make. These were labour relations; management selection and training; and research.

(i) LABOUR RELATIONS

ICI went into business, on 1 January 1927, with some 33,000 people employed at factories in the United Kingdom. In the early, hopeful years the numbers shot up, to 47,000 by the autumn of 1928 and to 57,000 in 1929. In the Depression they fell to 36,000 (1932), and they did not recover to the level of the late twenties until 1937.[1] From the start, therefore, ICI was a large employer—so large, indeed, as to be in a position, along with two or three other very large firms, to set the tone of policy towards labour, staff, and management in British business generally.

The founders were well aware of their responsibility, particularly in the uneasy atmosphere following the General Strike and the miners' strike of 1926. Between them they caused the loss of 162m. working days, a figure not remotely approached again until the early nineteen-seventies. Of the two strikes, there is not much doubt which was the more calamitous. The General Strike lasted for nine days. The coal strike, which set it off, outlasted it by six months, threw half a million out of work besides miners, disrupted the export trade, failed, and added to the folklore of the miners a resentful tale which played its part in em-bittering the next national coal strike more than forty-five years later. Sir Alfred Mond, in 1927, called the coal strike of 1926 'the longest and most devastating industrial dispute in the history of the country'.[2]

During ICI's early years the unions, in the aftermath of defeat, were weak and, if anything, growing weaker. Their members evidently had a poor opinion of the leaders, for they deserted in hundreds of thousands. There were about 5·2m. members of trade unions in 1926, but under 4·4m. in 1932. That was the bottom figure, but it took the unions ten years to recover, in terms of membership, from the disasters of 1926, for the 1926 total was not reached again until 1936.[3]

Among the union leaders the diehard Left, represented most vo-ciferously by A. J. Cook (1883–1931) of the Miners' Federation, were determined that no defeat should be allowed to moderate their un-relenting class warfare. Over against them, however, there was a group who accepted, in an undogmatic way, the existing structure of capital-ism and were prepared, without giving up socialist aims in the long run, to gain what they could by negotiation rather than conflict. The influence of this group was rising. At its head were the redoubtable figures of Walter Citrine (b. 1887) and Ernest Bevin (1881–1951).

During the early part of 1927 Bevin found himself approached with proposals for talks between employers and union leaders.[4] They came from Lord Weir (1877–1959), a Scottish engineer who was the head of his family firm. More important than that, he was a considerable public figure who, as Secretary of State for Air, had presided over the begin-nings of the RAF in 1918 and who, throughout the twenties and thirties, was never far from the centres of power in public affairs. He knew McGowan and Mond well and was on the Board of ICI from 1928 to 1953.[5]

Weir had no love at all for traditional union policies, especially in the craft unions, whose main aim, so far as he could see, was to preserve their members' jobs by rigidly obstructing any kind of technical innovation. He had fought the iron-moulders bitterly in his own business and he had gone in for factory-built housing in an effort—which was defeated—to break the restrictive practices of craftsmen in the building trades.[6] For these activities and others he was widely known among the working classes as 'bluidy Wullie Weir'.

Nevertheless, if Weir could break through the restrictive shell which encased so much union thinking he was anxious for co-operation between unions and employers. In Bevin he found the man he was looking for: an immensely powerful, combative personality, quite prepared to face the outrage of the Left if he thought that by co-operation he could do better for trade unionists generally and his own members in particular. As General Secretary of the Transport and General Workers' Union, in any case, he was probably not over-sympathetic to the craft unions.

Weir's proposals eventually blossomed into a conference, on 12 January 1928, between an influential group of employers, including Weir himself, Mond and McGowan, and the members of the General Council of the TUC. To the fury of A. J. Cook and the Left, this conference was followed, during 1928 and 1929, by committee work on both sides and more joint meetings. The leading figure for the unions was the General Council's Chairman for 1927–8, Ben Turner (1863–1942), and for the employers, Alfred Mond.

The practical upshot of the Mond–Turner talks was small, but they did produce a formal statement:

That it is definitely in the interests of all concerned in industry that full recognition should be given to affiliated Unions or other bona fide Trade Unions . . . as the appropriate and established machinery for the discussion and negotiation of all questions of working conditions, including wages and hours, and other matters of common interest in the trade or industry concerned.[7]

This was taking matters a great deal further than most employers had formerly been prepared to go, and it went too far for a good many in 1928. Even Mond evidently had reservations, for he was unwilling 'to commit ICI in any way to the resolutions or report of the Conference'.[8]

Mond's outlook on labour relations was essentially that of the late Victorian paternalist employer, in the same tradition as his father. Like the first Lord Leverhulme, he placed great emphasis on the idea of 'co-partnership'. 'We still read', he said in 1926, 'about employers and employed, about masters and men; whereas we all know that they are all employed. . . . The true phrase to-day is "co-workers in industry". They are co-workers in different capacities, and at different salaries, but are all dependent upon the prosperity of the industry for their remuneration or reward, whatever it may be.'[9]

This was a sentiment with which Bevin might have found himself in cautious agreement. He and Mond were like-minded enough to negotiate fruitfully with each other, and Bevin, to the scandal of his opponents within the trade unions, was prepared to contemplate industrial modernization, without socialism, if it would lead to greater efficiency, because in greater efficiency he saw better competitive prospects against the Germans and Americans and hence greater prosperity for his members. But he did expect employers to consult unions when their modernizing proposals were likely to lead to redundancy.[10]

Mond had no notion, any more than Lord Leverhulme, of extending co-partnership to matters of policy in the running of the firm. That would remain with the management. The core of co-partnership, for Mond, lay in profit-sharing, in which he had great faith both as a means of serving the self-interest of everyone in the firm and of binding the labour force to the firm's interest. 'The best answer to socialism', he wrote in 1927, 'is to make every man a capitalist.'[11]

As Chairman of ICI, Mond found himself at the head of a group in which, as he said, 'the history of most of the industries which have now been amalgamated . . . has been . . . happy and peaceful . . . and almost entirely free from industrial disputes'.[12] His own family firm, Brunner, Mond, had led the way towards the eight-hour day, holidays with pay, share-ownership schemes, housing, schools and recreation clubs, and in general towards an enlightened, if paternalist, labour policy based, in Mond's words, on 'foreseeing reasonable demands and . . . granting them even before they were asked'.[13]

Mond was justifiably proud of the Brunner, Mond labour policy. He had every intention of importing it wholesale into ICI, rather as Nobel methods of organization were imported. In 1927 Richard Lloyd Roberts (1885–1956) was appointed Chief Labour Officer in ICI. He had at first been a civil servant in the Post Office and in Labour Exchanges, but since 1916 he had been Brunner, Mond's Chief Labour Officer. He quickly set up a Central Labour Department with himself at the head, reporting to Henry Mond who sat in the next room to his father's.[14]

With the Brunner, Mond tradition intact and Brunner, Mond's Chief Labour Officer at hand to act on it in a wider setting, a labour policy for ICI could be worked out very fast. On 7 October 1927 Alfred Mond met the Press with a 'Complete Statement on the Labour Programme of Imperial Chemical Industries Ltd.'[15] In spirit and in general outline the programme announced on that day remained unaltered until national conditions began to alter radically during the Second World War.

It was important to get the policy announced as soon as possible. 'When ICI was formed', as Henry Mond put it in 1930, 'the many changes in the directorates of the component companies, the whole conception of a vast and, as described in the Press, "a soulless merger", and the concentration at headquarters of many influential personalities, gave rise—very rightly and properly—to the fear that the workingmen in the Company might feel that they were likely to be abandoned . . .

and that in this frame of mind they would easily come under the influence of the wrong type of trade union leader, and that we might see a change from the friendly feeling and confidence that existed in the past to an atmosphere of hostility and even rebellion.'[16]

Sir Alfred Mond, therefore, in public statements made as early as the autumn of 1926, was at pains to emphasize that the formation of ICI was not evidence of a capitalist plot to 'worsen the conditions of those employed in the respective companies'.[17] Nothing, he said, could be more untrue, and quite the reverse was in the minds of 'those directing this enterprise'. When he revealed the full programme, in October 1927, he insisted that the broad aim was to perpetuate 'the more than half a century of industrial peace which Brunner, Mond & Company, Ltd., has enjoyed' and to give a lead in the same direction to 'other companies and to other industries'.[18]

The purpose of the scheme, which Mond did not hesitate to announce, was to win the men's loyalty to ICI by giving them security and status directly derived from their position as ICI employees. The problem of communication, Mond realized, would be crucial to success. In Brunner, Mond he claimed that contact had been maintained 'from father to son between the heads of the firm and those working in the firm' for fifty years.[19] How could the feat be repeated in ICI, so much larger and so much more widely spread?

Mond's answer was to take a system of Works Councils already in operation at Brunner, Mond and implant it, suitably modified, in the body of ICI. There would be local Works Councils, with equal representation of management and democratically elected workers, meeting once a month in all factories. At the next level up there would be General Works Councils, appointed by the local councils from their own members, to discuss the affairs of groups of factories. At the peak of the pyramid the Central Works Council, again appointed from the level below, would represent the management and labour of ICI as a whole. 'Over this body', said Mond, 'I shall preside myself. As Chairman of the Company I shall thus be able to maintain that constant and close personal contact with the interests and operations of all the workers of the combine, and will be able to consult and discuss with them on all questions affecting their general well-being.'[20]

The councils were purely consultative, having no executive power, and the sort of subjects they would deal with, it was suggested, would be 'the comfort, safety, health and well-being of all employees . . . including matters affecting sport and recreation; the ways and means by which time, material or expense may be saved; the administration of the Sick Benefit, Benevolent and Hospital Fund schemes'.[21] In other words, a hostile critic might have remarked, any subject but the really important ones: wages, hours, conditions of work.

The councils were also designed to give the Management 'opportunities of giving information and discussing matters concerned with improvement in the plant; changes of process or organisation; new construction programmes, etc.,' and Mond even spoke of 'informing the workers of such questions as output . . . and general matters affecting the industry'.[22] But the same hostile critic might have observed that the council members had no *right* to any information whatsoever, and if he had had access to the deliberations of the Executive Committee of ICI he would have found a decision 'that any discussion of production costs should be limited to very general statements and it would be undesirable to give any figures of any kind'.[23]

Nevertheless, even with these limited functions the first council elections, in April 1929, attracted 1,132 candidates for 387 seats, and 93 per cent of the electorate turned out.[24] Moreover the system of Works Councils, substantially as Mond and Lloyd Roberts designed it, was still vigorous over forty years later. It did in fact provide the 'close personal contact' which Mond had hoped for. At council meetings there were no hierarchical barriers and at meetings of the Central Council men from the factories could talk even to those remote figures of power, ICI Directors, not merely in set debate but, more importantly, in the intervals of informality carefully provided between the sessions.

Alongside the Works Councils the other main channel of communication within ICI was to be a monthly sixty-four-page illustrated magazine dealing with 'all matters of interest to workers, particularly their social activities'. It would circulate among all ICI's employees and Mond asserted confidently that it would make them all feel they were 'a band of brothers'.[25] A degree of scepticism is no doubt permissible, but nevertheless in the first year—1928—of the magazine's life over 400,000 copies were sold, suggesting that the magazine was bought not once or twice, but regularly, by the majority of those who worked for ICI in Great Britain. The gross cost of each copy to the company was 9·54*d*. (4p) and the selling price (less than 1p) was heavily subsidized.[26]

Plans for giving ICI's workers a sense of status and security centred mainly on two devices: the Staff Grade Scheme and the Share Ownership Scheme. The Share Ownership scheme had precedents in Nobel Industries and in Brunner, Mond but the Staff Grade Scheme, Mond claimed, was 'quite new in any private concern'.[27]

Staff Grade was for hourly-paid workers of more than five years' service. The main benefits were a weekly wage instead of hourly rates (with elaborate arrangements for extra time); the right to a month's notice of discharge; payment for Bank Holidays whether worked or not; and payment of full wages (less National Health Insurance Benefit) for all certified sick absence up to six months in any year.[28]

These were valuable privileges and they were not indiscriminately granted. Promotion to Staff Grade lay entirely within the discretion of the Directors, and the original intention, in Mond's words, was that 'even up to 50 per cent of the men eligible may be promoted'. In fact the first promotions made when the scheme came into effect, on 1 June 1928, were on a much less lavish scale. Of 16,862 eligible, 4,319—25 per cent—were promoted.[29]

'I have always said you cannot make the world more prosperous by making the rich poorer. What you want to do is to make the poor richer.'[30] With these engaging remarks Mond introduced the ICI Workers' Shareholding Scheme. The basic provision was that a worker might buy ICI Ordinary shares at 2s. 6d. ($12\frac{1}{2}$p) under mean market price, and pay by instalments. The price was lowered still further by the grant of free shares on a sliding scale linked to the purchasers' pay. Anyone paid less than £200 a year—that is, the majority of hourly-paid and Staff Grade workers, as well as many office staff—would get one share free for every four bought. The scheme was open to everyone earning up to £2,000 a year—that is, to all except the most senior managers—and at that level one free share was granted for every eight bought.

Unlike some companies' employees' share-purchase schemes, the ICI scheme was not restricted to a special class of share. The shares issued under the scheme, and Mond dwelt on the point, were ICI Ordinary (and some Preference) shares freely transferable on the market, though speculation was discouraged. Therefore the worker who bought shares under the scheme, on the preferential terms outlined, stood to gain by market movements like any other shareholder. He also stood to lose, of course, but that was not prominent in anyone's mind in the optimistic atmosphere of 1928. During the year 1,821 monthly staff and 3,731 other employees bought 237,359 and 129,514 shares respectively, and the average holding for monthly staff was 130·3, for the rest, 34·7.[31] The market price of ICI shares during the year ranged between 29s. 3d. and 42s. 3d. (146p and 211p).

The administration of labour policy in ICI was designed to allow very little independence at the factories. The local managements had control over engagement and dismissal, over purely local administrative matters, over the maintenance of works councils, and over sports and social activities, 'except that when any question of principle arises, reference should be made to ICI Labour Department for guidance'.[32] Even this degree of freedom did not extend to 'Trade Union negotiations and all questions involving changes in the wages or working conditions of any group of employees'. These matters Lloyd Roberts kept firmly under his own hand, to the scandal of works managers, who told Henry Mond they were afraid that they were 'drifting into the position where no dispute of any importance at all will be able to be settled without long and delayed correspondence between the Head of the men on the one hand and some important functionary in London on the other'.[33] Henry seems to have been sympathetic, but it does not appear that his sympathy had any very far-reaching effect. The *Central* Labour Department had been set up and *Central* it was to stay.

ICI's labour policy, with its emphasis on loyalty to the Company in return for benefits granted, flew in the face of the unions, and that was recognized, not without satisfaction, from the start. 'An improvement in the official grade and status of the older and more reliable workmen', said Henry Mond, writing on his own behalf and Lloyd Roberts's in 1927, 'would be enormously appreciated by them and would bind them more closely to the Company's interests than to the interests of the "Working Classes".'[34]

Rather more than a year later Lloyd Roberts carried the same thought further. 'There is a fundamental antagonism', he wrote, 'between the Company's policy and that of the Unions . . . their whole effort is directed towards allying ICI Workers with Workers generally, whereas the Company's policy continually tends to ally the Workers with the Company. In the degree in which the Company's policy is successful, the Worker's growing inclination is to regard his natural contact as being with his Management, whereas the Unions consider it should be with his Union.'[35] And, indeed, by the autumn of 1929 Henry Mond was convinced, though he had no exact figures to prove his point, that men had been leaving the unions 'as they feel there is no particular point in paying their Union contribution and maintaining Union officials, many of whom they do not trust, when the Company provides them with a method of ventilating their grievances . . . through the Works Councils. The Company also provides them with the opportunity of reaching Staff Grade, which gives them greater security than any form of Trade Unionism could possibly give them.'[36]

The situation was clear enough to Ernest Bevin, who greeted the Staff Grade scheme with a blast of hostility. It was presented to the unions, he said, 'as a fait accompli . . . irrespective of collective agreements', and he was very disturbed by the proposal to promote only a proportion of the qualified men. 'It seems to us', he wrote, 'to be an attempt to drive a wedge between the workmen in the factory; assuming for a moment that 100 men apply to go on the staff and 25 are taken on, we can imagine what is going to be the effect on the feelings of the remainder.'[37] He took it, and we have just seen evidence that he was right, as a blow at working-class solidarity and the power of the unions, delivered within ICI at a moment when the Chairman of ICI, in his capacity as a leading figure at the Mond–Turner conferences, was on the point of agreeing to the principle (p. 59 above) of 'full recognition' of the unions 'as the appropriate and established machinery for the discussion and negotiation of all questions of working conditions'.

However disgruntled Bevin may have been, he had to recognize that he was in no position to start a fight, even if fighting had been his general policy. Apart from the weakness of the unions generally, he had to take account of the weakness of unions within ICI and of his own union especially. Union membership had never been universal on any side of ICI's business, and now it was falling. Central Labour Department had no exact figures of membership because it was settled policy not to inquire into it when taking men on, but such estimates as they could make they found, from their point of view, encouraging. 'So far as we can ascertain,' wrote Henry Mond in the autumn of 1929, 'in the Explosives side of the business, employing in all some 6,000 men, the number of Trade Unionists has fallen from 80% to 50%. In the Chemical side [including Billingham] . . . employing over 30,000 men, the number of Trade Unionists has fallen from 60% to under 20%. . . . In the Metal and Leathercloth side . . . [8,000 men] there are not more than 30% of Trade Unionists at all.'[38] Bevin's own union, Transport & General Workers, had practically no members outside Metal Group's works at Swansea. Of the total number of union members in ICI about 30–40 per cent belonged to J. R. Clynes's General and Municipal Workers which, said Henry Mond, 'can easily be called the senior Trade Union for ICI'. The situation allowed plenty of scope for inter-union jealousy and intrigue, a point not lost on those responsible for ICI's labour relations.

TABLE I

PRINCIPAL UNIONS WITH MEMBERS EMPLOYED BY ICI, 1931

National Union of General and Municipal Workers
Transport and General Workers' Union
Amalgamated Engineering Union
Boilermakers', Iron and Steel Shipbuilders' Society
National Federation of Building Trades Operatives

Source: R. Lloyd Roberts, 'ICI Labour Policy', 9 Jan. 1931, MP, 'ICI Labour Policy, Annual Cost of Various Items, 1931'.

In these circumstances it was beyond even Bevin, though he tried hard enough, to get ICI to recognize the unions as the men's only representatives for the purposes of collective bargaining. After all, as Lloyd Roberts pointed out in November 1928, they only represented a minority in ICI as a whole (he thought, at that time, about 30–40 per cent), and 'the value to the Company of . . . benefits, privileges and concessions depends on the Workers recognising them as voluntary acts by the Company and not as the products of negotiations with the Trades Unions'.[39]

Bevin had no more success in getting ICI to encourage men to join unions, still less to put any pressure on them to do so. 'On the whole', wrote Henry Mond, in a secret and confidential letter to his father, 'we find that where there is a good mixture of Trade Union and non-Trade Union labour matters proceed most smoothly.'[40] The younger Mond saw ICI's relations with the unions quite simply as a matter of power politics. 'The Trade Unions', he told his father, 'are extremely useful to us in bringing to our notice matters that we should not otherwise be aware of, and at the same time they are not in a sufficiently powerful position to make themselves aggressive or difficult to deal with. It seems to me clear that our main line of policy should be firm but friendly.'[41]

Henry Mond and Bevin, it seems clear from surviving correspondence, did not greatly like each other. Bevin preferred to deal with Sir Alfred if he could, which made Lloyd Roberts suspect, probably rightly, that Bevin was trying to cut out the Central Labour Department, which he regarded as usurping the unions' rightful place in contact between the Board and the workers.[42] With Sir Alfred, Bevin could work closely, and where the elder Mond's influence is detectable it can be seen as tending to bring ICI and unions closer together rather than further apart.

If Sir Alfred Mond's published writings are to be taken at anything like their face value, he diligently sought industrial peace, and he evidently convinced Ernest Bevin, probably not the most naïve of men, that he was prepared to take the unions into partnership to get it. He was evidently not prepared to push his son and Lloyd Roberts faster in that direction than they were willing to go. What McGowan's views were on these matters is uncertain, but it is not likely that they were more advanced than Alfred Mond's.

ICI's labour policy was therefore launched against the impotent hostility of the unions, and indeed one union leader, whose name is not revealed, told Lloyd Roberts that the Staff Grade Scheme was 'the most dastardly blow against the Trades Unions that any employer could devise'.[43] They disliked the Works Councils almost as much, as alternative channels to the unions for organized approach to the management, and it was partly to meet this complaint that in 1928 a Labour Advisory Council was set up. Its six ICI members, headed by the Chairman, were all Directors—and the union members represented the five main unions (p. 65 above) with members in ICI.

The policy thus critically received had at once to stand the strain of the rationalization which had been so widely feared from the moment the formation of ICI was announced. During 1927 and 1928 1,934 employees were displaced by closing down some works and improving others. They were dealt with as follows:[44]

Transferred to other ICI Works	590
Granted Gratuity or Pension	739
No action taken	362[a]
Still working in 1929	252[b]
	1,934

[a] includes persons with less than five years years' service, under 21 years of age, or 'who unreasonably refused other work'.
[b] this category presumably includes people kept on at a site where plant had been closed or modernized, but in different jobs: 106 were at Electro-Bleach and By-products Ltd. and 94 at the Muspratt Works of the old United Alkali Company.

It appears, therefore, that the much-dreaded merger only caused about 360 uncompensated redundancies among 40,000 or more employed, and over roughly the same period the net increase in employment at ICI factories was about 4,000 by takeover of other firms and about 10,000 by the creation of new jobs. It must have been very obvious, especially at Billingham in the depressed north-east, that ICI's activities were generating much more work than they destroyed.[45]

Another fear expressed at the time of the merger was that the power of the new combine would be used against wages: if not actually to reduce them, then at least to prevent them from rising. The power was certainly there, being inherent in ICI's position as the largest employer in many branches of the chemical industry and the only employer in some, and it was unrestrained by countervailing power on the union side. In 1931 Lloyd Roberts calculated, for McGowan's information, that ICI's budget for items of welfare expenditure 'over and above the employment conditions obtaining in industry generally' amounted to a sum equal to $8\frac{1}{3}$ per cent of the wages bill. He went on:

The policy adopted by ICI hitherto has been designed to get the confidence and active co-operation of the workers by methods that so far have been but little used in British industry. It is not an exaggeration to say that the Staff Grade Scheme and the Works Council system have achieved more in this direction than any practicable increase in money wages could possibly have achieved—and at a fraction of the cost. . . . It has been the aim of the Central Labour Department throughout to lay a greater stress on the attainment by the workers of a sense of status and security rather than on the mere rate of wages, and it is believed that ICI workers now fairly generally accept this view. Naturally, a fair minimum wage is pre-supposed, but given that, it is believed that our policy is certain to be more successful than the more materialistic outlook generally adopted.[46]

Lloyd Roberts was a man of subtle mind, devious methods, and great determination. He was justifying to McGowan, not noted for an enlightened outlook on labour relations, a policy which McGowan might have considered unacceptably expensive, especially in the circumstances of 1931. Hence, presumably, his insistence that the benefits conferred by ICI on their workers represented a bargain highly favourable to the company, in so far as they cost ICI less than the 'mere wages' which ICI might otherwise have had to pay.

'Status and security' may have been worth some sacrifice of pay. In the conditions of the early thirties, security would have been very valuable if it could have been guaranteed. The workers, however, were not asked their opinion of the bargain, which was one which the unions would certainly have rejected. ICI's early labour policy, however enlightened, was thoroughly in the Victorian paternalist tradition—for what the good employer saw fit to bestow, the good employee should be truly thankful to receive. And no doubt many of them were.

'The mere rate of wages', taking ICI as a whole, was not particularly high. The minimum rate at the beginning of 1931 was £2 9s. 6d. (£2·47½) on the Chemical side, £2 8s. 6d. (£2·42½) on the Explosives side, and in the Metal Group, where rates were determined by conditions in the depressed engineering industry, only £2 2s. 0d. (£2·10). The chemical and explosives rates, said Lloyd Roberts, were 'higher than those prevailing in the districts where the various Works are situated and, of course, considerably higher than the rates paid in the "heavy" industries'—but then the 'heavy' industries were those which since 1921 had been suffering from chronic depression.[47]

Minimum rates were supplemented by a variety of bonuses to make sure that ICI got the best men. It appears from another source that a labourer's average wage, in ICI, might be about £2 10s. 0d. (£2·50), that a semi-skilled man might get about £3 and a shiftman as much as £4, which would bring him up to the lowest level of salary earners.[48] But the weighted average male wage for the whole of ICI, at £2 17s. 0d. (£2·85), did not compare well with a national average, for men in manufacturing industry in 1931, of £2 19s. 0d. (£2·95).[49]

Since ICI labour policy was designed to foster a sense of security it might have been expected that a formal pension scheme for wage-earners, conferring rights against contributions, would very early have been set up to replace pensions granted at the discretion of the Directors. In fact, pensions policy developed slowly, perhaps because it was at first hoped to finance pensions, as well as certain other benefits, from profit-sharing based on cost-reduction in the factories, a favourite idea of Alfred Mond.[50] Lloyd Roberts, in 1927, prepared and costed a scheme in considerable detail, and it would have provided comparatively generous weekly pensions linked with the pensioners' former wage levels (after 40 years' service, 16s. 8d. (c. 83p) for a labourer; 20s.

68

(£1·00) for a semi-skilled man; 26s. 8d. (£1·33) for a shiftman, with related benefits for retirement through sickness and premature death), but the administrative complications were alarming, the cost was high, and the scheme was dropped.[51]

Foremen, like management, had been very generously provided for at Brunner, Mond under a bonus scheme. That was replaced in ICI, not very much to their liking, by a contributory pension scheme launched on 1 June 1928. There was no contributory scheme below the level of the foremen until 1937. Instead, retirement gratuities and pensions were granted at 65 on a scale drawn up in 1928, as follows:

Length of Service	Men	Women
Less than 15 years	Gratuity equal to two weeks' wages for each year of completed service	
	Per week	Per week
15 to 24 years	10s. (50p)	7s. (35p)
25 to 34 years	12s. 6d. (62½p)	9s. 6d. (47½p)
35 or more years	15s. (75p)	12s. (60p)[52]

The scale was modified in 1930 and 1931 and an attempt was made to relieve the acknowledged inadequacy of its provisions by means of the Workers' Voluntary Pension Fund, inaugurated in February 1930, which was financed by the workers and managed, free of cost, by ICI. Nevertheless, when every allowance is made for the falling cost of living, which meant that until the mid-thirties the purchasing power of a steady wage was rising, and when due account is taken of the company's welfare measures, ICI still does not look like a generous paymaster at the lower levels. Higher up the scale, among the foremen and, especially, the management, the story was different, but the wage figures for the majority show, as Lloyd Roberts was proud to point out to the Chairman, the cash price, to the worker, of the policy of granting benefits rather than pay increases.

Apart from the cash saving, the psychological dividend was handsome:

The response accorded to the Company's labour policy by the more intelligent workers [Lloyd Roberts told the Chairman] has increased as they have come to realize that they are not being made the subject of philanthropy, but that the principles involved in the labour policy truly represent the views of the Directors of the Company as to the manual workers' proper place in the scheme of industry.

That passage by itself might have been capable of more than one interpretation, but Lloyd Roberts went on:

The practice adopted by the Chairmen of Works Councils, of the Group Councils and of the Central Council in giving as complete information as possible with regard to the Company and its developments, and the consultations that regularly take place between the representatives of the Management and of the men on matters affecting the efficiency and prosperity of the various factories, are steadily building up in the minds of the workers the conviction that the workers of ICI *do really count.* . . . It cannot be overemphasised that the existence of this spirit, which is likely to prove of such advantage to ICI in the future, is a product entirely without relation to mere money wages, but is the result of treating each individual worker as a contributor, in his own sphere, to the prosperity of the Company.[53]

The tone of this passage, and of Lloyd Roberts's letter generally, suggests a cynical bargain by which 'benefits' were traded off against hard cash—'mere money wages'—so that ICI contrived to gain a labour force both contented and comparatively cheap. No hint of any such bargain reached the factories, where ICI's labour policy was administered and accepted at its face value: that is, as being fair and generous beyond the general standards of the day. It may be that Lloyd Roberts put his gloss on his late Chairman's policy as a temporary tactical device to gain his ends with the new Chairman, McGowan. Nevertheless his letter leaves an odd flavour behind, forty years on.

(ii) THE MANAGERS

Winnington Hall Club was unique in British business and probably in the world. The building which it occupied, a country house encapsulated within a heavy chemical works, was the mansion which Lord Stanley of Alderley had obliged Brunner and Mond to buy, much against their will, as a condition of letting them have the land which they wanted for their alkali works.[54] The club was founded in 1891, when Brunner and his family left the Hall, as an amenity for the management of Brunner, Mond.

But not for all the management. Membership was by election, and election was determined by social qualifications as well as by standing in the company. Broadly speaking, it might be said that the social structure of Brunner, Mond's management reproduced the split in the English middle classes between 'the professions' and 'trade', and membership of the Club was the badge of professional standing. A university degree was a sure passport to early election and so was a qualification in law or accountancy, but engineers, unless they were graduates, were less readily accepted, reflecting their uncertain standing in English society at large; and on the commercial side chances of election were even poorer. Commercial men and engineers would both be elected when their positions in the Company were thought to justify election, but they were likely to become members much later in their careers than chemists, lawyers, or accountants, who were not required to earn the privilege in the same way.

This remarkable institution passed, considerably liberalized, into ICI, in which Alkali Group was the reincarnation of Brunner, Mond. Within the Hall, in its bars and on its lawns, the 'fellows' moved in an agreeable, masculine atmosphere, very English in tone, of cordial informality. Dress, with a deprecatory glance at the black coats and hard collars of Dyestuffs Group in nearby Manchester, was kept carefully casual—sports jackets and flannel 'bags'. Conversation was on a basis of christian-name equality, regardless of standing in the company, food and drink were excellent, conviviality consorted genially with high ability, and in talk at the bar or over lunch ideas, information, and advice passed easily from mind to mind. The whole setting was reminiscent of an officers' mess or an Oxford college, which was not surprising, since by the time ICI was founded Oxford was the university which many of Alkali Group's scientifically-educated staff were coming from. There was also a Cambridge contingent, but other universities were sparsely represented. Irrespective of university background, there was a strong public-school element, reinforcing still further the social homogeneity of the Winnington technical Establishment. Once elected, however, a member with a different background had little cause to fear ill-will arising from snobbery.[55]

Billingham, as a Brunner, Mond colony, inherited something of the Winnington tradition, though considerably modified. In the first place, the demands of high-pressure technology made Billingham a better place for engineers. Secondly, there was perhaps rather less devotion to Oxford, possibly because Pollitt, the founding father, had been educated at Manchester and Zurich Polytechnic and his successor after the merger, R. E. Slade, was from Manchester too. Nevertheless there was a management club, Norton Hall, modelled fairly closely on the club at Winnington, and although there may have been less of a college atmosphere yet another side of traditional English life—field sports—was cultivated both by Pollitt and by Slade.

In the other dominant partner to the merger, Nobels, the management's background was quite different, being strongly Scottish. Here were no ex-public-school boys but the products of Scottish day schools, either Board schools of the type which John Rogers attended or, perhaps, the solidly established foundations in the cities of Scotland such as Glasgow Academy, Alan Glen's School, where McGowan was educated, Edinburgh Academy and others. Managers with degrees would have gained them, for the most part, in Scottish universities, and there was not likely to be the English prejudice against engineers. Indeed the whole English tradition, cherished at Winnington, was deeply suspect in high quarters at Nobel House:

Concerning Public School and University training, [wrote John Rogers in 1926] there is no doubt whatever in my mind that in the case of a man of natural ability such training is bound to be of the greatest advantage, but in many cases this training is given . . . not because of natural ability, but because of the accident of position, and when this is the case, the person is in my opinion rendered less useful . . . than he would have been had he not had the training. This is because he undoubtedly gets views and ideas as to his capacity and future which are not likely to be fulfilled on merit alone and, therefore, he becomes discontented.[56]

The Winnington world was the world of a gentleman's club, with a background of affluence, expensive schools and the ancient universities in the afterglow of their Edwardian glory, and with the hard cutting edge of professional efficiency always present but never indelicately exposed. Ex-officers of the Great War (Major Freeth, Major Hodgkin, Colonel Pollitt and many more) followed the general custom of the day in continuing to use their Army rank, and altogether it was a world in which John Buchan's heroes would have felt perfectly at home. There was a sufficient basis of shared assumptions and of mutual trust to make for easy and frank intercourse between different levels and different functions of management, and even under the autocracy of McGowan policy was discussed with great freedom. Moreover, in spite of official secrecy, no insuperable obstacles would be placed in the way of a technical manager, even a junior one, seeking information which he needed for his work.[57]

This was all very English, at the level of the educated, cultivated, upper middle class. There was also in ICI's management another English tradition, much closer to the Scottish way of looking at things and related to a social level with no assured background of affluence, less social grace, and a more puritanical cast of mind, finding expression in greater formality of dress and manners and in distrust of the whole extravagant way of life which Winnington Hall seemed to symbolize. 'Here', says Peter Allen, a Winnington man, speaking of BDC, later Dyestuffs Group, in 1926, 'were such solecisms as butterfly collars and high tea and in the canteen men who had known each other for thirty years would say "may I trouble you for the salt, Dr. Brown?" '

Dyestuffs Group, in 1936, was employing 397 chemists (31% of its total male staff)—more than in any other ICI Group and far more than in Alkali Group (77—11% of male staff). The next biggest employer of chemists was General Chemicals Group, with 255 (23%), followed by Fertilizers (157—17%) and Explosives (133—21%).[58] Probably it was the relatively small number of chemists needed in Alkali Group which made it possible for that Group to discriminate so heavily in favour of public school and Oxford men. Certainly the other Groups cast their nets more widely. The Dyestuffs or General Chemicals man was more likely to have gone to a grammar school than to a public school and to have taken his degree at Liverpool, Leeds, Manchester or, above all, London (perhaps externally) than at either Oxford or Cambridge.[59]

On the commercial side of the business, in any of the Groups, a university education of any sort would be exceedingly rare, particularly in the first few years of the merger. P. C. Dickens (1888–1964), briefly Secretary and later Treasurer of ICI, had been at Eton and Trinity Hall, Cambridge, before qualifying as an accountant. W. H. Coates (1882–1963), Treasurer and later on the Board, had taken a London degree externally, but he had been brought into Nobels chiefly on the strength of his experience at the Inland Revenue Department. These were the exceptions. Generally speaking a boy would join direct from school as a junior clerk, as both McGowan and Mitchell (Chairman and President in the thirties) had done, and work his way up if he could. The road to advancement through the commercial side was there, but it was by no means so broad nor so well trodden—in spite of the example of McGowan himself, Mitchell, J. G. Nicholson and others— as the technical route.

In the selection and training of managers the Groups were very much their own masters. Of the dominant partners to the merger, Nobels came into ICI with no systematic procedure,[60] but Brunner, Mond's connection with the universities, especially Oxford, was well established and their selection system, if 'system' is the right word for the highly idiosyncratic methods of one man, was in the hands of Major F. A. Freeth, FRS. For some years his influence in the same field was strong in ICI, both within and beyond Alkali Group, and of that man of wayward brilliance something must here be said.

We met Freeth in Volume One as an innovator in the manufacture of high explosive.[61] We shall meet him again as research manager in ICI. As a recruiter and, indeed, in all he did he conducted his activities with panache, genial hospitality, violent personal prejudice, snobbery, and witty, caustic talk often aimed at Authority. He was unmethodical, detested administration, and went about broadcasting ideas brilliant, perverse, crackpot by turns, but never orthodox or soothing. His later career was marred and almost wrecked by drunkenness, but by extraordinary will-power he rescued himself and for the last 32 years of his life (he died in 1970, at 86) he was a total abstainer. It says much for the perceptive tolerance of those at the top, first, of Brunner, Mond, and later of ICI, that they were able for so long to accommodate such a determined and articulate anti-organization man. He was most active in the twenties and early thirties, in the last days of Brunner, Mond and the first of ICI. He relied on personal contacts at the universities to put candidates forward and on his own intuition in choosing from amongst them: not at all, despite his own eminence as a scientist, on scientific selection procedures. His detractors said that with the starting salaries he was able to offer—Alkali Group outbid most other employers—it was not difficult to attract the ablest candidates, and that his judgement of personality was eccentric and by no means infallible. Moreover we have already observed that the number of vacancies to be filled in Alkali Group was nowhere near so great as in Dyestuffs or General Chemicals. Certainly there were Freeth failures as well as Freeth successes, but many of the leading figures in ICI, from the nineteen-thirties to the nineteen-fifties and later, owed their original offer of employment to Freeth and acknowledged his influence upon them in their later careers. He never, says Peter Allen, talked down to young men, for which they were willing to put up with a lot.

Selection for other Groups came under the control of men equally strong in character, if less flamboyant. At Billingham F. E. Smith concerned himself especially with the recruitment and training of graduate engineers.[62] In Dyestuffs Group C. J. T. Cronshaw struggled continually, in the face of a good deal of scepticism but with the goodwill of Freeth, to build up the level of research and inventiveness on which Dyestuffs' prosperity and indeed survival depended, and he could not do that without a continual—and rising—intake of able scientific graduates.

From the very early days of ICI there was a move towards bringing management recruiting under some measure of central control and for establishing management policy, like labour policy, as the responsibility of a head office department. This was consonant not only with the general centralizing tendency prevalent in ICI but also with a growing concern for management selection and training, as an important function of the highest central direction, which was beginning to be observable in some of the larger British companies of the late twenties and thirties.

Hence during 1927 Central Staff Department came into existence. At the head of it was Major-General F. J. Duncan, CB, CMG, DSO (1889–1960), who had retired from the Army in 1924 after commanding 61 Division during the Great War and serving as Military Attaché in Bucharest for five years. His reign was brief and he was succeeded in 1931 by Captain (S) Sir Frank Spickernell, KBE, CB, CVO, DSO, RN (Retd.) (1885–1956), who had joined Sir Alfred Mond's personal staff in 1928 after completing his final naval appointment—as Secretary to the First Sea Lord.

When a central scheme of management selection began to emerge, to be run by General Duncan's Department alongside recruiting by the Groups for their own requirements, it was on lines devised by P.C. Dickens and Henry Mond, an Etonian and a Wykehamist. With such parentage it is hardly surprising that the scheme was based firmly on

74

the public schools and universities, and as a mirror of the times it is worth examination.

In September 1927 every member of the Headmasters' Conference, representing about 160 schools, received a letter on ICI paper signed by the Headmasters of Eton, Harrow, and Winchester (C. A. Alington, Cyril Norwood, A. J. P. Williams). They said they had been approached by ICI 'to assist them in securing a larger proportion of public school boys on their staff' and that ICI would be looking for chemists, engineers, or physicists for research and works management, and for 'men to act as assistants to senior officials . . . or for work on the commercial side'.

ICI expected to take twenty to thirty boys a year, and the proposal was that Headmasters should put candidates to a selection committee consisting of H. J. Mitchell, Henry Mond, G. P. Pollitt, F. A. Freeth, P. C. Dickens, General Duncan and the Headmasters of Eton, Harrow, and Winchester. Boys accepted for technical jobs would be expected to read a scientific subject at the university and to do a year's research. The rest might read some arts subject, followed, it was recommended, by economics or an accountancy qualification. ICI gave an unconditional promise that boys selected at school who succeeded in the course recommended would be offered posts at £350 to £400 a year.

In 1927, therefore, there was sufficient confidence in the future to commit ICI, four years or so in advance, to taking 20–30 men a year for well-paid positions intended to lead to the highest levels in the business. Moreover the commitment was to be made while the men concerned were still schoolboys, and ICI would have little or no control over their development during the four years or more between selection and appointment.

It was still rather rare in 1927 for a boy educated at an English public school and a university, especially Oxford or Cambridge, to decide on a business career unless he had a family firm to go into. All the weight of tradition would press him towards the professions, teaching or Government service. Dickens himself, before 1914, had chosen business. So had Herbert Davis, who became a Vice-Chairman of Unilever, and in 1926 the comparatively small firm of Barclay & Fry, tin box-makers and printers, took on, from Cambridge, a future Chairman of the Metal Box Co. who had turned down the Indian Civil Service in favour of business. In general, however, there was a great weight of ignorance and prejudice, on the side of business as well as in the schools and universities, to be broken down, and Dickens said that one of the objects of his scheme was to enable the public school boy 'to appreciate the possibilities of obtaining good and well-paid work outside the so-called learned professions which are overcrowded'. Parents also, he added, 'should be enlightened to such possibilities'.[63]

75

The scheme was not very successful. In the first eighteen months there were disappointingly few applicants, perhaps because the proposals had not been properly presented to the headmasters and candidates, or perhaps because parents were unwilling to face the expense of the four years or more of higher education and/or professional training which would be required between selection and salary.[64] During 1929 the response was much better, but almost immediately the deepening depression made it difficult to see how jobs were to be found for the 45 or so who, by November 1931, were due to join between 1932 and 1935. No promises were broken, but one candidate at least—F. C. Bagnall, later Managing Director of British Nylon Spinners and later still a Director of ICI—was recommended, when his time for joining came after he left Oxford, to spend a year at the business school, partly financed by ICI, which had just been set up at the London School of Economics. At the end of 1931 the scheme was suspended and in March 1932 it was brought to an end.[65] 'The Scheme', wrote H. J. Mitchell to Dr. Alington of Eton, and to the Headmasters of Harrow and Winchester, 'has proved an interesting experiment.'[66] What the Headmasters said, or thought, does not seem to be on record.

Apart from the difficulty of providing jobs during the Depression, the policy of promising schoolboys jobs which they might or might not be fit for in four or more years' time had proved, in Spickernell's words, 'rather an embarrassment than otherwise'. After the promises had been given, better candidates often appeared, especially since ICI was in close contact with the universities through Freeth, Cronshaw and others.[67] There was, in any case, a slowly growing demand from British business at large for men (not, for many years, women) with degrees, and in the universities themselves appointments boards were beginning to see 'business' as a possible occupation for their clients, though they would still be unlikely to put forward the names of the most brilliant on their lists, just as the most brilliant would be unlikely to apply. The pull of the traditional occupations was still very strong, especially at Oxford and Cambridge, and it remained so until well after the Second World War.

Linked with the problem of finding potential managers there was the problem of training them. Other large companies had similar problems and in 1929 Dickens was exchanging letters with L. V. Fildes, the Secretary of Lever Brothers, who had just set up a scheme for recruiting half-a-dozen trainees a year from the universities for non-technical posts.[68] In training, as in selection, the Groups were left largely to make what arrangements suited them best, but the central authorities would back promising schemes with an application throughout ICI—as, for instance, a programme devised in 1933 by F. E. Smith, at Billingham, for selecting and training engineers—and they paid particular attention to the training of commercial staff. They also followed the careers of promising men and tried to arrange for broadening their experience within ICI.

By the time ICI had been in existence for ten or twelve years it had a very high concentration of scientific and technological manpower— much the highest in British industry, and even in the universities it may be doubted whether any single university would have been able to match ICI's staff, at any rate in point of numbers, in certain branches of chemistry and chemical engineering. The merger itself had brought together the staffs of the four largest companies in the British chemical industry, and then the recruiting policies we have been discussing had thickened up the concentration, particularly in the expanding Dyestuffs Group, where organic chemists were required in large numbers, and at Billingham, where they needed engineers.

The commercial side of ICI, by contrast, was by no means so strongly staffed, and this was an unintended consequence of the pursuit of technical excellence. A. J. Quig, an ICI Director with a background in paints and lacquers, where commercial shrewdness, not to say sharpness, was a condition of survival, put the matter thus: 'I find [in 1941] that the Group Chairmen are more interested in chemical and technical aspects of the Groups than in the commercial—a side which they leave almost entirely to the Group Commercial Director or Managing Director and I think this is a mistake.'[69] With selection, training, and promotion up to middle management effectively in the hands of the Groups, the prospects for a young manager joining ICI in the twenties and thirties varied greatly from one Group to another. Starting salaries varied, increments varied, the number of highly paid posts varied, and there was little movement between Groups or between Groups and the centre except at high levels, so that where a man joined he was likely to stay, and his prospects would be determined accordingly.

TABLE 2

EMPLOYMENT OF CHEMISTS IN ICI GROUPS, 1935–1938

	Alkali	Gen. Chem.	Dyes	Expl.	Fert.	Paints
September 1935[a]	77	222	378	n a	151	n a
June 1936[b]	78	255	397	133	157	n a
March 1938[c]	58	194	417	139	111	53

Sources: [a] 'Number of Chemists in Groups', 18 Sept. 1935, CR 157/9/10
[b] 'Chemists in the Dyestuffs Group', 2 June 1936, CR 157/9/10
[c] 'Chemists in ICI', March 1938, CR 157/9/10

The disparity, as it affected graduate chemists, was several times investigated in Central Staff Department between 1935 and 1938. The Department reached the disturbing conclusion that prospects were worst in the two Groups—General Chemicals (194 chemists) and Dyestuffs (417)—where the number of chemists employed was greatest. They were about equally bad in Paints and Lacquers, but the number of chemists was much smaller—53.[70]

The trouble arose partly from low starting salaries and small annual increments; partly from the extreme scarcity of posts above £600 a year in General Chemicals and £500 in Dyestuffs. In General Chemicals it was calculated that it took ten years from joining to reach £500 a year, and that since there were only 70 posts out of 194 assessed at more than £600, the chances were that a good many chemists would never get that far. In Dyestuffs Group the position was even gloomier. Starting pay in 1935 ranged from £250 for a man with four years' degree work to £275 for one with five years' and £300 for one with six. Of the Group's 417 chemists, 280 earned £500 a year or less, and to get even as far as that was likely to take ten years. In 1942 only one member of the Research Department was getting more than £1,000.

Prospects in Explosives Group and Fertilizer Group (Billingham) were slightly better. It was calculated that in both Groups chemists, starting at £300, would take about twelve years to reach £600, but that after that the majority might hope to reach posts worth between £725 and £800, with some prospect of going higher. In Fertilizer Group there were 14 posts out of 111 assessed at more than £1,000 a year, compared with 22 out of 417 in Dyestuffs and 13 out of 194 in General Chemicals.

TABLE 3

MAXIMUM SALARIES OF CHEMISTS IN ICI GROUPS, 1938

	Alkali	Gen. Chem.	Dyes	Expl.	Fert.	Paints
Over £1,000	11	13	22	6	14	2
1,000	5	26	23	13	9	5
850	9	10	14	10	9	3
725	15	21	30	22	11	2
600	18	124	48	88	68	3
500			280			38
	58	194	417	139	111	53

Source: As [c] in Table 2.

TABLE 4

ICI Groups: SALARY RANGES OF STAFF FROM £500 PER ANNUM UPWARDS 1937 DIVIDED INTO TECHNICAL AND COMMERCIAL

	£501–£800 Tech.	£501–£800 Comm.	£801–£999 Tech.	£801–£999 Comm.	£1,000–£1,999 Tech.	£1,000–£1,999 Comm.	£2,000–£2,999 Tech.	£2,000–£2,999 Comm.	£3,000 upwards Tech.	£3,000 upwards Comm.	Totals Tech.	Totals Comm.
Alkali Group	59	11	20	5	24	4	3	5	2	1	108	26
General Chemicals Group	98	46	15	7	20	9	3	3	1	2	137	67
Dyestuffs Group	108	18	21	6	18	12	4	1	2	1	153	38
Explosives Group	76	14	14	10	11	5	3	3	1	3	105	35
Fertilizer Group	106	28	18	5	10	7	2	4	4	1	140	45
Leathercloth Group	5	9	2	5	2	1	—	1	—	1	9	17
Lime Group	19	8	—	—	2	2	1	2	1	—	23	12
Metal Group	37	37	9	10	8	15	1	2	—	2	55	66
Paint and Lacquer Group	18	28	4	3	2	3	—	1	—	1	24	36
Plastics Group	6	4	1	1	—	—	2	1	—	—	9	6
Head Office	5	104	1	23	7	49	4	21	3	10	20	207
Divisional Sales	—	93	—	12	—	21	—	5	—	3	—	134
Totals	537	400	105	87	104	128	23	49	14	25	783	689

Source: Central Staff Department, 2 Sept. 1937, CR 157/9/9.

If this was the outlook for chemists in several of ICI's biggest groups it may be assumed, though there is little written evidence to go on, that the outlook for non-technical staff was worse. The commercial side, generally speaking, was not held in high esteem and the difference between the way technical staff, on the one hand, and commercial staff, on the other, were selected made it certain that the ablest men would usually be found on the technical side, so that posts in general management, which multiplied as ICI grew, would be filled from that source. Such, at any rate, was Pollitt's view. 'It was found impossible', he wrote in 1933, 'to fill the new administrative positions without calling on the technical staff.' He listed a baker's dozen (he could easily have found more), including five future Directors (W. A. Akers, W. F. Lutyens, H. Gaskell, D. R. Lawson, W. J. Worboys), who had been 'transferred from purely technical to commercial and administrative posts'.[71]

He might have pointed also to the preponderance of Alkali Group in his list. That Group, from every point of view, came best out of the 1938 investigation. Starting salaries were higher, the number of chemists was lower, and the prospects of reaching £850 to £1,000 a year were better than in any other Group. Allen reached £1,000 eleven years after joining, and J. L. S. Steel was being paid £1,400 at the age of 29, early in the 1930s. It was the most profitable Group in ICI, and managers coming into it found an atmosphere of hope and advancement, congenial companionship, and the means to live comfortably. Above all, if they were good they could expect advancement young. W. F. Lutyens, Group Chairman in the thirties, was appointed in 1931 at the age of forty, and he had at least one Director and two works managers in their early thirties. Other young men were sent on important missions abroad in their late twenties. 'We were good and we knew it', Peter Allen has said of Alkali Group in these days; '[we were] rather brash and not very popular; we didn't care.'[72]

At Billingham also there was an atmosphere very attractive to able young men, an atmosphere of great things to be done and every encouragement in the doing of them. The effects of the Depression, as we shall see (chapter 6), were much worse than at Winnington, but the hopeful atmosphere was not altogether destroyed; and at Billingham, too, promotion came early. At Billingham, as at Winnington, there were Directors on the Group Board in their early thirties and from Billingham, as from Winnington, Directors of ICI came much more numerously than from other Groups. Of the 29 Executive Directors appointed between 1926 and 1951, nine, if the two Monds are included, might be classed as Winnington men, and six had a Billingham background. Nobels supplied five, all in the earlier years, and no other Group more than two. One only—Cronshaw—was a Dyestuffs man.

In ICI's early years, then, there was little in the way of central control over the methods by which managers were recruited, trained, and developed. A structure grew up—it was not planned—on foundations laid by the merging companies, and the strongest foundations were those laid by Brunner, Mond. As a consequence a situation arose, during the latter part of the period covered by this volume and for some time afterwards, in which the higher direction of ICI was dominated by men whose early careers had been spent at Winnington or Billingham. Other sides of the business, especially Dyestuffs Group, which was close to much of the emergent technology of the forties and fifties, were under-represented. This is the background, so far as management is concerned, to the narrative which later chapters will unfold.

(iii) RESEARCH

Among the tasks that urgently needed to be done, when ICI was set up, probably none was more congenial to Sir Alfred Mond than the organization of research. He had always before him the example of his father, who founded Brunner, Mond's technical success on the German principle of applying academic science to the solution of industrial problems, and he was determined to repeat the process in ICI so as to create, eventually, a business at least equal in technology to IG. For this purpose he had at his disposal immensely greater resources than Brunner, Mond ever had, and it had always been in the minds of the founders of ICI that the new concern should go in for research on a scale much larger than any British firm, or group of firms, had formerly contemplated.

Mond set to work at once. By the end of November 1926 he had already induced Professor Sir Frederick Keeble, FRS (1870–1952), whom he described as 'a very old friend and colleague', to give up the Sherardian Chair of Botany at Oxford in order to 'devote himself to Research and Propaganda in fertilizers'.[73] This move shows the direction of Mond's thoughts. He meant to link research in ICI with research in the universities, directly after the German model, and he would be helped in doing so by Brunner, Mond's connection with the universities, which we examined in section (ii) of this chapter, as recruiting grounds for technical management.

In the autumn of 1927, building on another Brunner, Mond precedent, Mond set up a Research Council, in which eminent academic scientists were invited to deliberate with the leading figures of ICI research. The precise functions of the Council were kept vague, probably on purpose, but not the rewards of attendance. Members drawn from outside ICI were offered fifty guineas, plus expenses, for every meeting they attended, and the original intention was to meet every two months.[74] Moreover, membership of the Council was intended to provide opportunities for suggesting promising lines of work, not necessarily of a directly commercial nature, and the Council had a budget which could be used to subsidize work in university laboratories.[75]

Those who accepted ICI's invitation included Professor F. G. Donnan, FRS (1870–1956), already familiar with ICI's problems as a long-established consultant to Brunner, Mond; Professor F. A. Lindemann, FRS (1886–1957), a close friend of Winston Churchill and from 1932 onward his scientific adviser; and Professor Robert Robinson, FRS (b. 1886), later President of the Royal Society. Mond became President of the Council, with Pollitt as Chairman and Dr. Christopher Clayton as Vice-President. The other ICI members were Dr. E. F. Armstrong, FRS, F. A. Freeth, FRS, H. A. Humphrey, Sir Frederick Keeble, FRS, John Rogers, and R. E. Slade. It must have been as strong a scientific group for its size as could have been assembled in Great Britain at the time, and its purpose, Mond explained, was to help 'the object with which Imperial Chemical Industries was organised . . . [which is] to place the Chemical Industry of the Empire in a position second to none in the World'.[76]

No project is more characteristic of Alfred Mond's ideas and methods than the Research Council. Designed to serve both the country's cause and the interests of science, it was statesmanlike enough to satisfy Mond's political instinct and at the same time it served the equally important commercial purpose of strengthening ICI. The sums which it disposed of, chiefly in grants for the purchase of chemicals and apparatus and in subsidies for research in universities, were considerable by the standards of the day:[77]

Research Council Expenditure 1931–1935

1931 £29,487	1933 £33,000
1932 £28,482	1934 £50,343
	1935 £53,277

Neither Pollitt nor Slade liked the Council. Pollitt vigorously attacked Mond's original proposal, which he evidently thought would merely create an agreeable luncheon club for eminent scientists—an alternative, perhaps, to the Athenaeum—and cut across the established ICI organization.[78] Slade greatly resented 'the Professors'. 'When we look back', he wrote in 1936, 'upon what was decided as important fundamental work we find that it was always the work that they themselves were doing at the time.'[79] While Mond lived, however, the Research Council was safe. The Professors were diligent in attendance—much more so than the ICI members—and the minutes show that meetings settled down to a fairly regular rate of two a year in most years. The establishment of the Research Council, like other early decisions of policy, represents a determination to establish some measure of central control over activities already in progress when ICI was formed and to chart their future development with the interests of the whole business, rather than individual parts of it, in mind. This was a particularly delicate matter in research, because the needs and resources of the Groups varied so greatly and also because of the problem of finding a home for research projects which might eventually be of advantage to ICI as a whole but showed no immediate benefit to any one particular Group. The argument for central control, indeed for the establishment of a central research laboratory, was strong, but the argument for close attention to the Groups' individual needs was strong also, and we shall see that the conflict was not readily resolved.

Research in ICI, when the Research Council was set up, was going on in nine laboratories, mostly very small. The total expenditure for 1926/7 was given by Mond as £221,000. By the autumn of 1928 it was running at an annual rate of £350,000, of which all but £50,000 was going on work in five laboratories—Billingham, Winnington, Runcorn (Castner-Kellner), Widnes (UAC), and Ardeer. Among these five big spenders the Dyestuffs laboratories at Blackley and elsewhere do not appear, a sure indication of the low regard in which Dyestuffs were held in ICI at the time, for a dyestuffs business without extensive research amounts almost to a contradiction in terms.[80]

Alongside this expenditure on ICI research £16,000 was going to what Freeth called 'direct assistance of science in general', which meant £1,000 for the publication fund of the Royal Society and £15,000 divided more or less equally between physical chemistry, organic chemistry, and physics. 'Direct expenditure on the Research Council activities' Freeth put at £14,000, chiefly for sponsored work at Oxford, Imperial College, Liverpool, Manchester, Amsterdam, and University College London. The work at UCL, directed by Donnan and by Freeth himself, was taking £4,000 of the £14,000 in employing five young scientists, at £400–£500 a year each, on the investigation of the colloid properties of dyes, the effects of electrical discharge, and physical constants.[81] 'Direct financial result from this type of research cannot be counted on', wrote Pollitt; 'its chief object is the training of research men and the investigation of problems which should eventually result in additions to our general chemical knowledge. . . . The Research Council grant was not intended in any way to subsidise research having an immediate works value. It was ICI's main contribution to the raising of the standard of scientific work in the country.'[82] With this judgement Alfred Mond would not have disagreed. It sums up one of his main purposes in founding the Research Council.

ICI's main research effort was summarized by Mond, towards the end of 1927, in a paper sent to members of the Research Council. He played down ammonia-soda ('it is the refinements of the process only to which attention is now, continually, being directed') and heavy inorganic chemicals generally, but under Heavy Chemicals (Organic) he mentioned work at Billingham on fusel oil and other solvents, on the production from coke-oven gas of acetylene, acetic acid and ethylene (for glycol), and work by Nobels and BDC on synthetic resins, both 'the hardening type for the manufacture of pressed goods' and 'the soft type' for varnishes. He passed over dyestuffs lightly, mentioned that in synthetic ammonia, 'as with the ammonia-soda process', only refinements remained, and turned to fertilizers, again at Billingham, and the work being done on urea, ammonium phosphates, nitro-chalk and compound fertilizers based on potash, phosphoric acid and nitrogen. Other products that he mentioned briefly included paints and varnishes, activated charcoal, carbon black, rubber vulcanizing accelerators, anti-knock compounds and wetting agents, but he placed his main emphasis on fuel research. 'This', he said, 'is by far the most important problem', and he divided it under three main heads—the hydrogenation of coal by Bergius's process (p. 162) to yield oil; the hydrogenation of crude oil to yield petrol; and low-temperature carbonization of coal with super-heated steam to yield both oil and 'a smokeless solid fuel'.[83] 'Great importance', Mond concluded, 'is attached to the general question of oil from coal and the work is being prosecuted as rapidly as circumstances permit. It is considered that not only oil, but the whole field of organic products will be based upon coal as a raw material in the near future.' Coal, then, was to be the rock upon which the architects of ICI's future would chiefly build. Here was another argument, and in Mond's view evidently the main one, for concentrating research, as well as capital expenditure, in the general area of Billingham.

With Mond's influence behind it, the enlargement of ICI's research effort went rapidly ahead. We have no complete record of the increase in scientific staff, the key element in the process, but it was certainly on a scale hitherto unprecedented in British industry. In February 1928 the Executive Committee, following a recommendation by Pollitt based on a report by the Research Staff at Winnington, agreed to the engagement of twenty-four additional chemists and engineers and the necessary expenditure on equipment.[84] During the spring of 1928, accordingly, Major Freeth was very busy. He interviewed, reported, recommended, and some of his remarks have survived. He was recommending salaries varying upwards from £400 for a graduate of 21 with research experience to £600 for an Associate of the Royal College of Science of Ireland, aged 30, with a doctoral thesis in preparation, and he wanted permission to revise the salary—upwards, naturally—in about twelve months.

The burgeoning of research in ICI, in theory universally approved of, was in practice watched, in some quarters, with narrowly critical eyes, for to the tidy commercial mind it had grave disadvantages. First, everything about it—the genesis of ideas, the method of carrying them out, above all, the results—were irritatingly unpredictable and not at all susceptible to methodical planning according to well-known rules of procedure. Secondly, it was freely admitted to be very expensive, but nobody, least of all the scientists themselves, had devised a method for calculating the return—if any—on the heavy investment that was demanded.

While times were reasonably good, in 1927 and 1928, little was heard from the critics, but in November 1929, after the economic weather had changed decisively for the worse, McGowan pounced on a request for £20,000 from the Research Manager at Winnington, H. E. Cocksedge, Freeth's wartime collaborator and brother-in-law. Cocksedge wanted to extend the Winnington laboratories to meet 'the increase in scope and urgency of "Oil from Coal" '.[85] 'Generally speaking', wrote McGowan, commenting on Cocksedge's request, 'the financial situation in the world is very obscure. . . . Cannot the necessary work be done with our present appliances?' He went further, asking the Executive Directors to look narrowly at all projected extensions, not Cocksedge's only. 'I should like it to be understood', he finished, 'that we cannot authorise expenditure which . . . can be postponed till the financial situation is clearer.'[86]

McGowan had Mitchell's enthusiastic support. Mitchell, before the end of the year, mounted an attack of his own, aimed at the difficulty of forecasting the return on research expenditure. 'I submit', Mitchell wrote, 'that it should be possible and that it is desirable to get at the commercial aim behind many items of our research programmes and to examine that aim from both economic and policy standpoints before serious expenditure is engaged in.'[87] He called for a general inquiry into projects on hand and proposed, calling for the reasons behind them, the probable cost, and the 'possible commercial advantage'.

Figures produced by Pollitt, several months later, showed a startling rise in ICI research expenditure. Freeth, in 1928, quoted £350,000 as the going annual rate. Pollitt, in 1930, estimated £1,224,355, laid out as follows:[88]

Capital Expenditure on Extensions and Equipment	£ 57,300
Work on existing processes	345,375
allied processes	140,360
Work on new processes—Coal Oil	333,100
others	110,945
Propaganda and Sales Development	204,475
Grants and subscriptions to Universities etc.	32,800
Total	£1,224,355

No direct comparison between Freeth's figures and Pollitt's is possible, though it seems certain that Pollitt allowed for some items which Freeth did not, notably the £204,475 for 'Propaganda and Sales Development' (chiefly for fertilizers), of which Pollitt observed: 'Without a first-class system of this kind it would be waste of money to carry on . . . research . . . directed towards the manufacture of new products.' Even without that, however, the rising trend was so steep that it was bound to attract hostile attention as the threat of hard times set in.

Pollitt, in the reports accompanying the figures, pointed out that for ICI to show a good return on capital it was 'necessary that new and temporarily very profitable processes should be discovered and put into operation at regular intervals' and that the research needed would cost 'a considerable sum of money'. This his critics did not deny, but according to Pollitt they did suggest that 'an investigation of the existing market or commercial prospects should invariably be the prelude to the expenditure of any but the smallest sum on a particular line of research'. That prompted him to assert: 'The really lucrative processes are likely to be those manufacturing products which are entirely novel, and it would be unfortunate indeed if research . . . were barred because an investigation indicated that there was no market.' An unhappy turn of phrase: what presumably he meant was that promising ideas should not be given up too early on hypothetical grounds: let them at least be carried to the stage of a testable prototype.

But how to identify promising ideas? Pollitt was very anxious that tidiness should not strangle creativity by imposing, in the interests of economy, too rigid a system. Ideas for research, he pointed out, came from several directions—some from the research departments themselves, some from the Commercial Department at headquarters, some from his own Technical Department, and 'this apparently rather haphazard system . . . works excellently and should not be hampered', although he was willing to admit that money might be saved by investigating commercial prospects before moving from the comparatively inexpensive laboratory stage to much more costly 'semi-technical' operations. That had not always been done in the past: it should be done as a matter of routine.

Pollitt, essentially, was defending the general principle of research as a necessary industrial function, but not one to be judged by criteria valid for more predictable operations. His case was a strong one, but it had to be disentangled from a sprawling mass of verbiage. How many of his colleagues read the 57-page report? One at least did—H. J. Mitchell. 'The report', he said, writing a week after it had been issued, ' . . . does not to my mind pay sufficient consideration to the financial aspects of the question. In fact it appears to proceed on the assumption that finance is available to any extent.'[89] This would have been an unfortunate impression to create at any time. In 1930 it was quickly fatal to any hopes Pollitt may have had of pushing his estimates through. On 16 July there was a Chairman's Conference, Mond being supported by Lord Reading. McGowan was not present, but Coates and Mitchell were, as well as Pollitt, Henry Mond, Todhunter, Wadsworth, and a number of senior officials.[90]

The occasion must have been a melancholy one. Mond, speaking of research expenditure, but evidently with other matters in mind, opened by saying that the Company was passing through difficult times. 'It was necessary, therefore, to revise the views which had formerly been held. The Capital, Running, and Overhead expenses must be reduced, even at the expense of cherished ideas.' They talked inconclusively for a while, Pollitt remarking that the moment was favourable for reducing expenditure, and then the Chairman 'suggested that better results might be obtained by concentration on fewer subjects. He thought the allocation should be reduced by £200,000.' It was only too plain what the result of that would be. 'It was explained that this would mean dismissing a number of chemists, salaries being roughly half the cost of Research.' Mond, who no doubt knew this as well as anyone and disliked it quite as much, pressed on. 'He said the amount susceptible to reduction appeared to be £700,000, and the Technical Department was to prepare a scheme for reducing this £700,000 to £500,000.'

In judging the result of this decision we are handicapped by having no series of expenditure figures for research for the years before 1931. It is certain, however, that the rising trend of the late twenties was put into reverse until 1932/3, and even after that it was not until 1937 that a figure as high as £1m. was budgeted for again, although from 1933 onward the sums spent, from year to year, consistently ran ahead of the budget figures:[91]

Research Expenditure 1931–1937

	Budget	Actual
1931	£ 771,031	£ 737,305
1932	585,995	568,504
1933	565,601	623,767
1934	698,308	821,333
1935	897,073	925,029
1936	944,381	982,283
1937	1,034,339	1,066,710

Pollitt's estimate for research expenditure in 1930 was probably not exceeded, either as a budget or as a sum actually spent, until 1939 or later.* The Depression, therefore, may be said to have had the effect of

* See note on Research Expenditure, p. 96.

postponing the expansion of ICI's research effort for some ten years. Whether the world's other great chemical businesses were equally badly affected is impossible to say with any precision, though in the case of IG there are indications. Figures collected after the war by the Americans (see p. 34) suggest that a decline in expenditure set in earlier (1927–8), continued, as in ICI, until 1932, and was not fully made good until 1938, but detailed comparison with ICI figures would almost certainly be misleading.

TABLE 5

ICI GROUPS—RESEARCH BUDGETS, 1931 AND 1937

Group	1931 Budget £	1937 Budget £
Coal-Oil		
(a) Fertilizer Group £211,850		
(b) Alkali Group £ 35,000		
	246,850	
Fertilizers		
(including Jealotts Hill)	166,235	212,695
Dyestuffs	26,960	262,064
Alkali	85,000 (excluding (b) above)	135,000
Metals	12,466	37,480
General Chemicals	50,320	178,100
Explosives and Leathercloth	95,000 (and including 'Miscellaneous Groups')	106,400
Lime		15,020
Paints and Lacquers		46,350
Plastics		24,000
Research Council	28,200	
G. P. C. Research Committee		17,230
	£771,031	£1,034,339

Source: GPCM 94(c), Annex II, 22 April 1931, and GPCSP, 9 Feb 1938, 'Notes on the Research & Development Budget, 1938'

By the mid-thirties, when once again it became possible to think of spending more, rather than less, on research, great changes were beginning to come over the chemical industry, and by the time war broke out the nature and direction of ICI's research were not at all the same as before the Depression. The point is illustrated in Table 5, above, which compares the research budgets for the various groups in two years, 1931 and 1937, for which we have comparable figures. The features which stand out are the disappearance of 'coal-oil' as a separate item (by 1937 it was presumably included under 'Fertilizers', which was the Group responsible) and the rising importance of Dyestuffs. These shifts in research emphasis signalize the swing away from operations based essentially on high-pressure technology, whose natural home in ICI was at Billingham, towards operations based on organic chemistry whose natural home was with Dyestuffs. That swing will form a major theme of this volume, and detailed discussion of it would be out of place here, although certain preliminary remarks will help to set the scene for the narrative which is set out in succeeding chapters.

All ICI research was carried out in Group laboratories, which meant that work on any new projects that arose, as they were continually arising during the thirties, had to be fitted somehow into the existing organization, and if more than one Group could lay claim to a new activity, then inter-Group rivalries might seriously hamper develop-ment, as was notoriously the case with plastics. On the other hand there was always the fear that if some promising possibility did not attract any of the Groups, it might die of neglect.

Pesticides, pharmaceuticals, rubber chemicals, fibres and plastics were among the most important of the new developments which were coming to the fore during the thirties. Dyestuffs Group could and did make a strong claim to work on all of them, but other Groups' interests were very strong. The Fertilizer Group, representing ICI's rising agricultural business, had an obvious concern with pesticides. Chlorine, needed for many products including plastics, was an important product of the General Chemicals Group. Explosives Group had long been interested in fibres as an alternative use for cellulose, particularly with the explosives industry no longer expanding. Plastics, in one form or another, attracted the attention not only of Dyestuffs Group but of Fertilizers (urea), Explosives, and General Chemicals, and what turned out to be the most important ICI plastic of all—polythene—was dis-covered in the laboratories of Alkali Group. That Group then pro-ceeded, very naturally, to hold on to the technical and commercial development of polythene instead of handing it over to Plastics Group when that was formed (pp. 348-9 below).

Towards the end of 1933, when there was again some prospect of renewed growth in research, Pollitt began to campaign for a Central Research Laboratory, for a Research Controller, and for a generally more imaginative and expansive approach to the whole matter. He did not want to hinder the Groups in pursuing their own interests—far from it—but he did feel, with others, that preoccupation with day-to-day commercial and technical problems, and with their own profitability, stood in the way of free-ranging speculative research. Moreover who was to pay for it, individual Groups or ICI? 'If', Pollitt argued, ' . . . we want to take the position of leadership in chemical development which the chief British Chemical company should hold, to enlarge our range of products, to develop to the full all the ideas that come to our staff . . . to sell new processes to others instead of always buying them, something further than our present commitments will have to be incurred.'[92]

'I am with you, heart and soul,' wrote Professor Donnan of the Research Council, 'in your patriotic effort to secure that good British brains and honest British hearts should, at least in the chemical industry, obtain the results and achieve the aims which are worthy of them.'[93] Expressions of support within ICI were more restrained. The Group Chairmen, consulted by McGowan, were chary of the proposal for a Central Laboratory, which no doubt they suspected would overshadow their own laboratories, and preferred the idea of grants in aid of centrally sponsored research carried out in Group laboratories. W. H. Coates, too, came down against the proposal for a Central Laboratory, though he favoured the appointment of a Controller.[94] McGowan was disposed to listen to all sides. The upshot, perhaps predictably, was that a Research Committee was formed, a Research Controller was appointed, and a decision on a Central Laboratory was put off.

Rogers took the Chair of the Research Committee, set up in November 1934, with W. H. Coates and Holbrook Gaskell as the other members.[95] The Research Controller, appointed early in 1935, was R. E. Slade, formerly Managing Director at Billingham, who during 1934 had investigated the matters raised by Pollitt and reported on them to McGowan. He was anxious to see the Groups take a broad and enterprising view of their responsibilities. 'The group Boards', he told McGowan, 'should consider themselves entrusted by ICI with an industry, not just with a works for the production of a certain number of products.'[96]

Thus, for the first time, ICI was provided with a central organization for the direction of research. The powers of the Committee were substantial. It was required to examine the whole field of research open to ICI, to allocate activities to the Groups, to keep the Groups working properly with each other and with institutions outside ICI, and 'to initiate, continue or terminate research activity in the light of commercial, financial or policy considerations'. To back up these functions the committee had power to review the Groups' research budgets and to decide how much should be spent on general ICI research.[97] On the other hand the committee had no laboratory directly at its own disposal and therefore depended ultimately on the Groups' goodwill for getting its policy carried out.

In fact the Committee, under Rogers, was on principle against interference with Group autonomy, and Slade was very much of the same mind. The staff of Slade's department was therefore kept small (five qualified staff, including Slade himself and F. A. Freeth, in 1937), and the control it exercised was deliberately light.[98]

One demolition job Slade undertook with relish. He destroyed the Research Council. He was convinced, and so was Pollitt, that the academic members used it simply for their own ends, and one at least of 'the Professors'—Donnan—took the same view. 'I agree with you', he wrote to Pollitt in 1934, 'that the Research Council is an unnecessary luxury . . . simply another method of enabling Professor X to spend money on some pet scheme.'[99] Slade, with McGowan's goodwill ('I am not concerned with the vested interests of the Research Council'),[100] stopped subsidizing research through the Council and took away from it all control of expenditure, so that it became purely advisory and gradually faded away. Fees for attendance were still budgeted for, on a diminishing scale, as late as 1939, but no minutes of meetings have survived later than 1936.

Thus, reviled and unlamented, except by those who lost subsidies, passed one of Alfred Mond's grand designs. Like some of his other ideas it had more breadth than depth, and he never seems to have realized that the intrusion of eminent scientists into ICI at the policy-making level gave great offence among ICI's own scientific staff. They were quite prepared, indeed eager, to work with 'the Professors' in other ways, and individual members of the Research Council served on the Dyestuffs Committee, the Dyestuffs Technological Committee, and the Pest Control Research Committee, but the function of these bodies was purely consultative and closely bound to ICI's immediate interests. There was no question of committees like these sponsoring purely academic work, although money was forthcoming if it was needed for university research of direct interest to ICI.

Spending of the kind formerly controlled by the Research Council, essentially a branch of high-level public relations, was taken over by the Research Committee. The 1937 Budget allowed £17,230, but by 1939 the figure had reached £26,900. The Committee, unlike the Council, no longer undertook to finance research projects for a term of years or indefinitely, which in Slade's view only tied men up in blind-alley jobs.[101] Instead they made grants for specific purposes, including contributions to publication funds of learned journals, and they provided scholarships, for which £1,000 was allocated in the 1939 Budget. Among 'Prestige Payments giving only an indirect benefit to ICI' one item stands out: 'Grants or salaries paid to German refugees' (£2,792 in 1937, £1,925 in 1938, £1,100 budgeted for in 1939—the sums fell as those to whom they were paid found permanent employment).[102] These refugees were scientists, chiefly Jewish, driven away by Hitler, and in arranging to help them McGowan worked closely with Lindemann.[103]

Slade, as time went on, became increasingly uneasy about the practical benefits of ICI research. In 1938 he was saying that £400,000 to £500,000 a year was steadily being spent on evolving new processes and introducing new products, but that outside the Dyestuffs Group, of which he made a specific exception, 'this expenditure has led to little profitable investment of capital'.[104] Slade put forward various proposals for the better analysis and control of the Groups' research projects and expenditure, but the fundamental problem which confronted him was what it always had been: how to harmonize the Groups' perfectly proper concern for their own short-term interests with ICI's need for long-term investigations, whether of basic theoretical problems or of possible new products, which might have little or no relation to the existing interests of the Groups and would certainly be speculative and expensive. These were obvious candidates for central control and financing.

Slade gave two examples of the kind of general problems he had in mind. One was the question of organic synthesis under pressure, especially reactions with carbon monoxide, ethylene and acetylene, which he thought would be 'very suitable for the Coal Oil research section if . . . it was decided to slow down on this research'. The other was the whole subject of polymerization, which overlapped all Group boundaries, and which Slade thought should be taken in hand by 'a good research man, possibly straight from academic life'.[105]

As 'existing research which ought to be taken over' centrally, he mentioned two projects of the first importance. They were 'the protein work at Stevenston' which eventually produced the fibre 'Ardil' and 'the polymerisation research based on ethylene at Winnington' which had already produced polythene, although its potentialities, when Slade wrote, were still only very dimly perceived. Thought along these lines led naturally to the revival of Pollitt's proposal for a Central Research Laboratory, and the example was quoted of du Pont's 'Experimental Station' at Wilmington, which Slade visited, along with the rest of du Pont's research organization, about the end of 1937.[106] This time the proposal went through, and in 1939 land was bought at Butterwick near St. Albans.

Slade told a meeting of Group Research Managers, in March 1939, that the intention was to start the Central Laboratory on quite a small scale, with eight to twelve senior staff. He and other representatives of the Research Executive Committee (as the Research Committee had by then become) went out of their way to assure the Groups that the central laboratory would not undertake any work 'which a Group was really anxious to do itself', but their suspicions were not easily dispelled. As Baddiley of Dyestuffs remarked, 'the Central Laboratory must not monopolise all the fundamental or "highbrow" research of the

Company, nor must it be the only part of the Company to possess expensive scientific equipment'.[107] War broke out before the laboratory could be built, and the argument was broken off, to be revived, with considerable acrimony, when post-war plans were under discussion in 1944.

By the time ICI had been in existence for a dozen years, then, the central research organization was still far from fully developed. Indeed its fundamental principles had hardly been laid down, apart from the somewhat negative one of leaving the Groups alone as much as possible. Probably the damage done by the onset of depression was compounded, late in 1930, by the death of Alfred Mond, by far the most powerful advocate of research in ICI. Pollitt was equally, if differently, enthusiastic but he had not the same authority, and in any case from 1934 onward he withdrew more and more from day-to-day executive duties. Rogers, presiding over the Research Committee, was by temperament and conviction a non-interferer, and Slade, the Committee's chief executive officer, had not the advantage of a seat on the Board.

Nevertheless, research in the Groups, where alone practical work was carried on, expanded greatly in the mid-thirties. Between 1932 and 1936, both inclusive, 361 research chemists were taken on from twenty-four universities and colleges in the British Isles and six from universities in Canada, Australia and New Zealand.[108] In 1938 the total number of qualified research staff in ICI was 615, of whom six were at Head Office and the remainder distributed as follows:[109]

Dyestuffs	195
General Chemicals	102
Fertilizers	88
Explosives	84
Alkali	59
	528
Other Groups	81
Head Office	6
	615

The pattern of this distribution emphasizes the direction ICI's development was taking during the late thirties. Its significance will emerge as the narrative unfolds.

Slade, visiting du Pont in 1937, observed: 'The most striking difference between du Pont's business and ours arises from the existence of free competition in America.'[110] By that he meant that research was not impeded by agreements not to compete with other businesses, or

93

indeed by any agreement among du Pont's divisions not to compete among themselves. He thought competition led to a more enterprising spirit in du Pont, but in the actual management of research he did not think they were better than ICI. Their research expenditure over all their fields, at 2·81 per cent of gross sales, was very little greater, proportionately, than ICI's (2·40 per cent), but in both businesses it varied greatly from one activity to another.

Research in IG, the great object of ICI's emulation, was still no doubt on a considerably greater scale than ICI's own. IG were said to employ about 1,000 chemists and expenditure figures running into millions of pounds have been quoted, though without knowing what items they cover it is impossible to make a fair comparison with ICI's expenditure. What will emerge, in the course of the narrative, is that the technical gap between ICI and IG, acknowledged to be very wide indeed in 1926, was very much narrower in the late thirties, and in IG, no doubt somewhat to their surprise, they were beginning to have some respect for ICI's competence. To that extent, at least, the aims of the founders of ICI, and especially of Alfred Mond, were beginning to be realized, though a great deal more remained to be done.

REFERENCES

[1] Labour Department Report 49, 'Employment in ICI', 26 Oct. 1928, CR 209/66/R; Appendix II.
[2] Sir Alfred Mond, *Industry and Politics*, pp. 2–3.
[3] Henry Pelling, *A History of British Trade Unionism*, 2nd edn., Macmillan, 1972, pp. 288–9.
[4] Alan Bullock, *The Life and Times of Ernest Bevin*, Vol. I—*Trade Union Leader 1881–1940*. Heinemann, 1960, p. 392.
[5] W. J. Reader, *Architect of Air Power*, Collins, 1968, ch. 6 and 7 and generally.
[6] Ibid. ch. 6 (II).
[7] E. Bevin to Lord Melchett, 20 Sept. 1928, MP 'Men's Unions 1928–29'.
[8] H. Mond to Chairman (only), Secret and Confidential 'Memorandum on the coming Conference with Trade Union Representatives', 28 Oct. 1929, MP 'Men's Unions 1928–29'.
[9] As (2), p. 110.
[10] Bullock, p. 403.
[11] As (2), p. 4.
[12] Sir Alfred Mond, 'Complete Statement on the Labour Programme of Imperial Chemical Industries Limited', 7 Oct. 1927, MP 'ICI Labour Policy 1927 & 1930–31'.
[13] As (2), p. 3.
[14] Hector Bolitho, *Alfred Mond, First Lord Melchett*, Secker, 1933, pp. 312–13.
[15] As (12).
[16] Henry Mond, 7 Jan. 1930, MP 'ICI Labour Policy 1927 & 1930–31'.
[17] Statement of December 1926, quoted in (12), p. 2.
[18] As (12), p. 3.
[19] Ibid. p. 5.
[20] Ibid. p. 6.

21 'Suggested Address in connection with Works Council Propaganda', about March 1929, CR 209/70/18A.
22 Ibid.; as (12), p. 6.
23 ECM 1319, 9 April 1929.
24 'Comments on Works Councils Election Results', 17 April 1929, CR 209/70/18A.
25 'Details of Labour Programme to supplement General Statement', 7 Oct. 1927, p. 4, MP 'ICI Labour Policy 1927 & 1930–31'.
26 '1928, Report of the Central Labour Department', 25 June 1929, p. 4, MP 'Labour General, 1927–31'.
27 As (12), p. 7.
28 As (25), p. 4.
29 As (26), p. 3.
30 As (12), p. 9.
31 As (26), p. 3.
32 'Labour Department Organisation', 30 June 1927, ECSP; ECM 245, 12 July 1927.
33 Henry Mond, 'Memorandum on the Organisation of the Labour Department', 3 Oct. 1927, MP 'Labour Dept. Organisation'.
34 'ICI Partnership Scheme', Preface by H. Mond, 5 July 1927, CR 209/2/R8.
35 Labour Department Report 51, 6 Nov. 1928, p. 3, MP 'Men's Unions 1928–29'.
36 As (8), pp. 5–6.
37 E. Bevin to H. Mond, 22 May 1928, covering copies of letters from Bevin to Lord Melchett and TU Colleagues, MP 'Men's Unions 1928–29'.
38 As (8), p. 6.
39 As (35), p. 2.
40 As (8), p. 7.
41 Ibid.
42 As (35), p. 4.
43 Lloyd Roberts, 'ICI Labour Policy', 9 Jan. 1931, para XII, MP 'ICI Labour Policy, Annual Cost of Various Items, 1931'.
44 As (26), p. 6.
45 As (1).
46 As (43), p. 3, paras VI, VII, VIII.
47 Ibid. pp. 3–4, para IX.
48 Labour Department Report 5003, 'ICI Partnership Scheme', 5 July 1927, ECSP.
49 The British Economy Key Statistics 1900–1966, Times Newspapers, n.d., Table E.
50 Sir Alfred Mond, 'Co-Partnership and Profit-Sharing', in (2), pp. 109–27.
51 As (48), pp. 8–12 and pp. 16–22.
52 As (26), p. 5.
53 As (43), pp. 7–8, para XVII.
54 ICI I, pp. 91–2.
55 I am indebted to Sir Peter Allen for much information about Winnington Hall Club in the thirties, but the interpretation put upon it is entirely my own.
56 J. Rogers, 'Notes on Mr. Coates's Memorandum on Personnel', 10 Dec. 1926, CP 'Staff Recruitment and Training'.
57 Private information.
58 'Chemists in the Dyestuffs Group', 2 June 1936, CR 157/9/10.
59 Central Staff Department to Dr. A. Proven, 'Research Chemists Entered . . . ', 19 March 1937, CR 157/9/10.
60 W. H. Coates, 'Personnel, its Recruitment and Organisation', 8 Nov. 1926, pp. 4–5, MP 'Public School Boys 1926–32'.
61 ICI I, p. 285.
62 Sir F. Spickernell, 'Engineers Training Scheme', 8 June 1933, Central Administration Committee Supporting Papers; CACM 674, 12 June 1933.
63 P. C. Dickens, 'Scheme for encouraging the Employment of Public School Boys in ICI', 4 May 1927, CR 215/7/–.
64 'Public School Boys Scheme', 13 June 1929, MP 'Public School Boys 1926–32'.
65 GPCM 283 (b), 16 March 1932.
66 H. J. Mitchell to C. A. Alington, 'Scheme for the Employment of Public School Boys', 18 March 1932, MP 'Public School Boys 1926–32'.

⁶⁷ Sir F. Spickernell, 'Scheme for the Employment of Public School Boys', 18 March 1932, filed with (66).
⁶⁸ P. C. Dickens, 'Internal Staff Training', 4 Oct. 1929, and attachments, CP 'Staff Recruitment and Training'.
⁶⁹ A. J. Quig, 'Groups Central Committee', 21 March 1941, CR 93/395/400.
⁷⁰ Central Staff Department, 'Chemists in ICI', March 1938, CR 157/9/10.
⁷¹ G. P. Pollitt, 'Recruitment and Training of Staff', 27 June 1933, CP 'Staff Recruitment and Training'.
⁷² Sir Peter Allen to author.

Sect. (iii)

Research Expenditure: Figures compiled in Research Department in 1960 and reproduced at Appendix II give expenditure on Research & Development, excluding Technical Service, from 1927 to 1952. The figures from 1927 to 1937 are estimates and run at a much lower level than figures obtainable from contemporary sources (General Purposes Committee minutes, etc., 1931–1939), which show in detail how the totals are arrived at. In the body of the text, therefore, I have used the contemporary figures, but the other figures are reproduced as representing a longer series which is internally consistent and showing the same general trends.

⁷³ *Addresses . . . [to] . . . the First Meeting of the Commercial and Technical Committees of Imperial Chemical Industries Limited (Chemical Group)*, 30 Nov. 1926, p. 8; 'Industrial Research' in Sir A. Mond, *Industry and Politics.*
⁷⁴ Letter of Invitation to Professors, 27 Oct. 1927, CR 215/12/8.
⁷⁵ 'Functions of the Research Council', 6 Oct. 1927, CR 215/12/8.
⁷⁶ As (74).
⁷⁷ Gpcsps.
⁷⁸ G. P. Pollitt to Sir A. Mond, 7 Sept. 1927, CR 215/3/13 and 15 Sept. 1927, CR 215/12/–; Minutes of Technical Committee; J. Rogers, 'Research Council', 15 Sept. 1927, CR 215/12/–; as (75).
⁷⁹ R. E. Slade to H. G. Littler, 6 Aug. 1936, CR 50/12/54.
⁸⁰ F. A. Freeth, 'Research and the Research Council', 19 Oct. 1928, CR 215/12/R1.
⁸¹ Ibid.
⁸² G. P. Pollitt, 'Research Council Expenditure', undated (probably Sept. 1930), CR 215/2/23.
⁸³ 'Short Summary of Lines of Research', 23 Nov. 1927, CR 215/3/R1.
⁸⁴ G. P. Pollitt to H. Glendinning, 21 Feb. 1928, CR 31/8/C.
⁸⁵ 'Proposed Extension to new Laboratories, Winnington', 7 Oct. 1929, CR 31/8/13; ECM 1529.
⁸⁶ President to Chairman and Executive Committee, 7 Nov. 1929, CR 31/8/14.
⁸⁷ H. J. Mitchell to Executive Directors, 'Research', 23 Dec. 1929, CR 215/3–13/1.
⁸⁸ G. P. Pollitt, 'Research Expenditure for 1930', 8 July 1930, p. 14, CR 215/2/23.
⁸⁹ H. J. Mitchell, 'Research Report', 16 July 1930, CR 215/2/23R.
⁹⁰ CCN 16 July 1930.
⁹¹ GPCM 1931–1939; Research Budgets in CR 93/5/342 and CR 93/5/539.
⁹² G. P. Pollitt, 'ICI Research', 3 Jan. 1934, CR 50/51/51.
⁹³ F. G. Donnan, 'Notes written for Col. G. P. Pollitt DSO after perusal of his draft report on ICI centralised research', 1 Jan. 1934, CR 50/51/51.
⁹⁴ GPCM 633, 17 Jan. 1934; W. H. Coates, 'ICI Research', 16 Jan. 1934, GPCSP.
⁹⁵ Chairman to GPC, 1 Nov. 1934, 'Research Expenditure', GPCSP; GPCM 815, 7 Nov. 1934.
⁹⁶ R. E. Slade to Sir H. McGowan, 'Development and Research', 14 May 1934, CR 50/51/51.
⁹⁷ As (95).
⁹⁸ Organisation Chart, Aug. 1937, CR 93/34/192; R. M. Winter, reported by Bunbury, 3 May 1950, Bunbury folder 44A.
⁹⁹ As (93), p. 12.
¹⁰⁰ Slade to the President, 10 July 1936, MS. annotation by McGowan, Slade Papers; see also 'Chairman's Remarks' to the 17th meeting of the Research Council, 14 July 1936, CR 50/46/51.

[101] As (79), p. 3; as (100), 2nd refce.
[102] As (91).
[103] Lord Birkenhead, *The Prof. in two Worlds*, Collins, 1961, pp. 101–4.
[104] R. E. Slade, 'Reorganisation of ICI Research', 25 Jan. 1938, CR 50/110/110.
[105] Ibid.
[106] R. E. Slade, 'The Dupont Research Organisation', 6 Jan. 1938, GPCSP.
[107] 'Notes of a Meeting held . . . on 9th March 1939 to discuss ICI Research', 21 March 1939, CR 50/110/110.
[108] Memo from Central Staff Department, 19 March 1937, CR 157/9/10.
[109] 'ICI Research Department Establishment as at 1st March 1938', 17 March 1938, CR 50/115/139.
[110] As (106), p. 7.

CHAPTER 5

The Crucial Investment: Fertilizers at Billingham 1926–1930

(i) THE DECISION TO EXPAND

FOREIGN POLICY, labour relations, management selection and training, research—all the matters discussed in previous chapters—were all secondary to the development of market power adequate to sustain ICI's position in the world.

Where was that power to come from?

Sir Alfred Mond had had his answer ready since the merger was first thought of. It was to come from the advanced high-pressure technology of ammonia synthesis and related processes which Brunner, Mond had been developing at Billingham since 1919. Mond had little difficulty in carrying McGowan with him, and it soon became evident that McGowan would be perfectly willing to see Nobels' cash brought to the aid of Brunner, Mond's technology.

We turn now, therefore, to the first major investment decision taken in ICI: the decision to concentrate capital expenditure, on a massive scale and over a period of four or five years, on expansion at Billingham. Not until we come to the decision, taken over twenty years later, to develop petro-chemicals at Wilton shall we find another investment decision of comparable importance to ICI.

Billingham's main marketable product was nitrogen, and the intention was to sell it to the farmers of Great Britain and the British Empire as fertilizer. It could be applied to the soil through the medium of ammonium sulphate, ammonium nitrate, urea and various other materials. Billingham's staple product was ammonium sulphate, which contains one ton of nitrogen in every 4·7 tons of its own weight.* The market for fertilizer nitrogen, in the twenties, was large and growing fast. J. G. Nicholson, in the autumn of 1927, took an optimistic view of the future founded on figures which showed an annual average increase of consumption, throughout the world, of nearly 12 per cent, which had raised the tonnage from 775,000 in 1922 to nearly 1,300,000 in 1926.[1] M. T. Sampson, about the same time, drew curves showing the world's

---

* It is a convention of the fertilizer industry to express tonnages of nitrogenous materials in terms of the tons of nitrogen contained ('tons N'). One ton N is contained in 4·7 tons of ammonium sulphate, 6·1 tons sodium nitrate, 2·9 tons ammonium nitrate, 2·1 tons urea.

consumption of fertilizer nitrogen rising at a cumulative annual rate of 6 to 7 per cent, and A. E. Hodgkin regarded these rates as 'minima', to be increased by vigorous 'propaganda'—that is, sales promotion.[2]

Hodgkin's temperament was cautious and sceptical. If he could take such a view, there was likely to be no limit to the optimism of his ebullient superior, George Pollitt. As early as June 1926 Pollitt was urging Brunners' Board to think in terms of £15m.–£20m. capital expenditure between 1927 and 1932. At that time Billingham was within sight of bulk production, with two ammonia units working and a third building. Pollitt wanted the capital chiefly to put two more units in hand as soon as the third was in production, and then go on to three more. Each unit was to be designed for a daily output of 165 metric tons of ammonia, representing 230,000 tons a year of ammonium sulphate (48,900 tons N), so that Pollitt's plan was aimed at an increase of 230,000 tons of sulphate a year for five years, which he considered would provide one-third of the world's increased demand. Demand, he thought, would be stimulated by a drop in price from the prevailing level of £10 to £8 per ton, allowing Billingham a margin of £3 10s. 0d. (£3·50) 'which is ample'.[3]

When ICI succeeded Brunner, Mond as owners of Billingham there was a great increase in resources and no dampening of enthusiasm. Sir Alfred Mond, confident of prosperity close ahead, demanded special provision for Billingham's profits in ICI's capitalization (p. 19) and Sir Harry McGowan raised no objection. Studies directed at the extension of Billingham's fertilizer capacity were pushed energetically ahead. The home market, it had to be recognized, was already plentifully supplied with ammonium sulphate made from ammonia which was a by-product of coke ovens. Billingham's output, added to the output of 400 works belonging to the British Sulphate of Ammonia Federation, would bring the total far above any figure that farmers at home were likely to be persuaded to take up. Pollitt, in his 1926 proposals, put home consumption at the equivalent of 160,000 tons of sulphate a year, adding characteristically: 'It [the home market] is capable of absorbing economically up to 1,000,000 tons per year if the application of Nitrogen to grassland as well as to arable is developed.' The large gap, at home, between demand and supply seems to have caused no one any uneasiness because, as Hodgkin put it, 'in talking of fertilisers we are envisaging the requirements of the whole world, but of the British Empire in particular'.[4] IG was known to be a very much larger producer and an energetic exporter, but everyone seems to have assumed that a market-sharing agreement on conventional lines—the British Empire for ICI: other markets for IG—would be arrived at.

Nitrogenous fertilizers did not cover the whole market, and although they had been available for thirty or forty years British farmers were

99

inclined to regard them as new-fangled. A much longer-established fertilizer was superphosphate, which had been used in Great Britain since the eighteen-forties, and there was a trade in that amounting to 600,000 tons a year, partly in the hands of small, unprofitable makers but much of it imported. Pollitt, on the principle that 'the policy of the merger should be to obtain in its own hands the whole of the fertiliser industry in this country',[5] suggested that ICI should baffle the foreigner by putting up 'a number of suitably situated plants . . . round the coast in England and Scotland, each plant having a capacity of 100,000 tons a year or upwards'. Hodgkin, less ambitious, thought ICI should aim only at 100,000 tons of superphosphate a year, but the whole idea was dropped when studies by the engineering department showed that works of this size would return only 5–6 per cent on the capital employed.[6]

There remained the question of 'complete' fertilizers, 'containing', as Hodgkin put it, 'N[itrogen], P[hosphorus] and K [potassium] in one bag'. They might take the form of mixtures or of products such as mono- and di-ammonium phosphate ($21 \cdot 2\%$ N, $53 \cdot 8\%$ P_2O_5) and 'nitro-phoska' (17% N, $12 \cdot 7\%$ P_2O_5, $21 \cdot 1\%$ K_2O). These latter, as compared with mixtures, showed great advantages in transport costs, being much more concentrated. IG was making them. Should ICI, or would it be better to get a really large output of nitrogenous fertilizers going first? F. C. O. Speyer, of the British Sulphate of Ammonia Federation, thought, in September 1927, that for the next few years 'the great part of the additions to world nitrogen consumption . . . will be as sulphate'. Sir Frederick Keeble, ICI's Agricultural Adviser, on the other hand, pointed out 'that unfortunately most of the British Empire thinks it needs phosphorus much more than nitrogen'.[7]

It will by now be apparent that beneath the surface of the general optimism about Billingham's future there were soggy areas of uncertainty, doubt, and downright ignorance. Very little attention, for instance, seems to have been given to the question of competition from Chile nitrate, the natural product for which synthetic nitrogenous fertilizers were substitutes. 'Bueb [of IG] and Speyer', wrote Alfred Mond in November 1927, 'think Sulphate of Ammonia is overdone, and cannot replace more Nitrate of Soda than has already been achieved . . . should we provide such large capital for Sulphate of Ammonia rather than more plant for Nitro-Lime and the more direct competitors of Chili Nitrate?'[8] There is no evidence that this very searching question was ever satisfactorily answered, and Mond does not seem to have pressed it.

Then again, was there a possibility of danger from what Hodgkin called 'the various small plants throughout the world'—ammonia plants, that is—which were being supported by their Governments for reasons of economic nationalism and defence? The question was raised but not, apparently, pursued, although they were an obvious (and recognized) obstacle to the export trade on which an enlarged Billingham would so heavily depend.

Had the forecasts of Billingham's increased output been properly worked out and thoroughly checked? Alfred Mond, as late as November 1927, thought they might be exaggerated.

Above all, how much was known within ICI of the real fertilizer needs of farmers in ICI's chosen market, the British Empire? Not, it would appear, very much. 'We require', wrote Hodgkin, 'a proper survey of our prospective markets from the agricultural and physicochemical point of view. . . . Clearly this is a huge task and will take years to accomplish.'[9]

Nevertheless, in the autumn of 1927 the decision was taken to go ahead with plans for the future of Billingham which, it was realized, would eventually represent an investment of some £20m. The matter lay in the hands of a Fertilizer Committee, which at a meeting on 6 September 1927, attended by Mond, McGowan, Pollitt, Nicholson, and others, 'decided . . . that a market could be found for the whole of the increased production from Billingham units 4 and 5 provided that a reasonable proportion of this production was in the form of nitro-chalk [a mixture of chalk and ammonium nitrate], di-ammonium phosphate and possibly nitrophoska'. From this preamble the Committee passed to their substantive decision:

. . . in view of the fact that ICI must be in a position to supply complete fertilisers and forms of nitrogen other than ammonium sulphate in order to satisfy the needs of the United Kingdom and the Empire, to recommend the Board to make the necessary financial provision so that orders for the plant required for units 4 and 5 at Billingham may be placed immediately, on the understanding that

(i) the capacity of the sulphate of ammonia plant of units 4 and 5 is substantially equal to the total output of ammonia,

(ii) that work be started simultaneously on nitro-chalk and di-ammonium phosphate plants in economic units; the total capacity of these subsidiary plants to be equal to the output of one ammonia unit.[10]

Finance does not seem to have presented any very grave problem. Sales of General Motors shares, on a rising market, provided a ready source of cash, and Mond told the first Ordinary General Meeting, on 31 May 1928, that by the sale of investments some £5m. had been raised for 'the extensive programme of construction of the various companies admitted at the time of the merger'.[11] During 1928 the Ordinary shareholders took up rights which provided £6·5m., but the biggest fund-raising operation was in 1929. Coates advised, no doubt quite properly, that it would be better to raise fresh capital than to realize more investments, and by the issue of 7 per cent Preference shares at 23s. (115p) and Ordinary shares at 33s. 6d. (167½p) some £15·25m. was brought in, most of which was required for Billingham.[12] By the time the money was received the price of ICI Ordinary shares had fallen below the price of issue, but to the sinister implications of that fall we shall return.

(ii) EXPANSION: THE AGRICULTURAL MARKET

Technically, the decision to expand at Billingham was sound. There was no doubt that the ammonia process developed under Pollitt's leadership would work on a large scale at reasonable cost. The technology of ammonia synthesis, no longer a problem, was as well understood within ICI as anywhere in the world, including Germany. The marketing problem, on the other hand, still remained, and it was far less thoroughly grasped. Judgement, from Sir Alfred Mond downward, was clouded by optimism, and the decision to go for a large output of fertilizers thrust ICI into a market which none of their leading men knew much about—the market for agricultural supplies.

The moment chosen for entry into this market, in spite of the optimistic forecasts which supported the decision to expand, was scarcely fortunate, especially at home. The British farmer, never a very enthusiastic buyer of ammonium sulphate (consumption was at a rate of 7·1 lb. per acre, against 13·8 lb. per acre in Germany),[13] could get all he wanted before the enlarged Billingham supplies came to hand, and in the nineteen-twenties he was not very receptive to suggestions that he should buy more. Farmers had endured a generation of depression before the Great War and then, after brief prosperity, they had been thrust back into depression by the refusal of peacetime governments either to guarantee minimum prices for cereals or to break with free-trade principles by charging foreign food with import duties.

The outlook overseas, in the mid-twenties, was rather less bleak, although there were already signs of the over-production of food crops which was to blight world trade in the depression of the thirties. Nevertheless there was encouraging evidence of rising demand for nitrogenous fertilizers. That in itself, however, would not guarantee sales from Billingham. There would have to be 'propaganda'—the term was consistently employed within ICI—to convince sceptical farmers of the value of ammonium sulphate and other nitrogen products. Propaganda would not be convincing unless it was based on detailed knowledge of the farmers' problems, and they might be growing cotton in the Sudan, rice in China, tea in Ceylon, wheat in Australia, or anything else that was growable in the markets which ICI liked to think of as their own.

Here was the germ of a project much to the taste of Sir Alfred Mond. Brunner, Mond had long engaged in agricultural propaganda in

China, but they had never had a research centre designed specifically to back their propaganda with scientific knowledge. ICI clearly required such a centre, and the work done there must command respect at Rothamsted Experimental Station, at other similar institutions, and among agricultural scientists generally. It might not be easy to convince the scientific world of the impartiality of ICI's results, especially since research and propaganda were lumped together and paid for out of the same budget. Nevertheless the attempt must be made.

There can be little doubt that this design, like the design for ICI research generally (p. 81 above), was substantially Mond's creation, for it bears all the marks of his personality and habits of mind. It was based on the kind of scientific endeavour which he admired and sought to promote. It was directed closely enough at the national and imperial interest to satisfy his political instinct. It would—he hoped—help to make a lot of money for ICI. Before ICI was even formed, Mond had clearly in mind the grand objects he was aiming at and had chosen his man to carry out his scheme.

Great efforts [said Mond in November 1926] will be required in organising this business, which will not only be of great service to this country but to the whole of the Empire by developing and extending the area capable of producing valuable food crops. My ideal is to see, in this as in other chemical manufactures, our great organisation controlling the production and sales of the whole chemical manufacture within the Empire. For this purpose we must take into our services . . . the best brains available. I was, therefore, very pleased indeed that I was able to induce a very old friend and colleague of mine, Professor Sir Frederick Keeble, to relinquish his important position as Professor of Botany at Oxford and to devote himself to Research and Propaganda in fertilisers.[14]

Sir Frederick Keeble (1870–1952) had been a Fellow of the Royal Society since 1913 and he had held important positions, including the Directorship of the Royal Horticultural Society's garden at Wisley, before he was elected to the Sherardian Chair in 1920. His wife was Lillah McCarthy (1875–1960), a leading actress-manager who had previously been married to Harley Granville-Barker. Of Sir Frederick's scientific distinction and general standing there could be no doubt, and probably Mond could have made no shrewder choice to surround ICI's nascent agricultural research with the desired aura. At the same time it might not prove easy to fit so eminent but so uncommercial a figure, already middle-aged, into ICI's senior management, itself not lacking men of comparable scientific attainments.

In July 1927 ICI bought two run-down farms, six miles or so from Maidenhead and about forty from London, of which the larger was called Jealott's Hill, and here they proceeded to set up their agricultural research station. With 90 acres added in 1929, they had by 1930 160 acres of arable land, 350 acres of grass, and 20 acres of gardens and orchards. They put up a laboratory (£15,000) and other new buildings, improved others, and put in electricity, gas, water, and drainage, as well as making good farm roads, hedges and ditches. By May 1930 the capital outlay was put at £103,785.[15]

The new research centre opened in June 1929. Speaking soon after the opening, Keeble said that since he had taken charge of 'the new organisation' the number of agricultural advisers (that is, propagandists) in Great Britain had been almost doubled. There were now more than twenty 'and we can claim to be covering the country much better'. He wanted all advisers 'to realise that Jealott's Hill is their spiritual home' and he hoped every adviser would spend some part of each year there 'refreshing his knowledge'. He said that all advisers must 'be prepared to go abroad', and he stressed the imperial nature of the service he was providing: 'We must regard ourselves as an advisory service not only for the British Isles, but also for the Empire. We regard the service

TABLE 6

APPROXIMATE DIVISION OF £212,975 TO BE SPENT ON FERTILIZER RESEARCH EXPERIMENTS, PROPAGANDA AND SELLING CHARGES IN 1930

	Research	Experimental	Propaganda	Selling Charges	Total
Incurred in UK on account of	£	£	£	£	£
World Organization	21,000	21,500	19,600	6,400	68,500
Incurred directly in and for:					
UK				40,775	40,775
Italy			700		700
Portugal		1,000	1,500		2,500
Spain		500		9,000	9,500
China		5,000		40,000	45,000
Japan				10,000	10,000
Palestine/Syria		500		5,500	6,000
Java, Straits &c		1,500		2,000	3,500
Africa		1,000		3,000	4,000
Central & South America		2,000		500	2,500
Australasia		4,000		6,000	10,000
India/Ceylon		5,000		5,000	10,000
Total	21,000	42,000	21,800	128,175	212,975

Source: H. J. Mitchell, G. P. Pollitt, J. Rogers, 'ICI Agricultural Research Station, Jealott's Hill', 5 May 1930, ECSP.

which we have built up at Jealott's Hill as an imperial general staff for agricultural research and we regard the advisory service in the same way as the nucleus of an imperial service.'[16] By May 1930 99 people, including 36 'research and experimental' staff and eight trainees, were based at Jealott's Hill, and more research and experimental staff were expected during the year. There was a wide range of research in hand, and the geographical spread of activities is illustrated in Table 6, opposite, showing how money was to be spent in 1930.

This table, reproduced from a report on the research station at Jealott's Hill by Mitchell, Pollitt, and Rogers, shows expenditure on research proper coming to rather less than 10 per cent of the total sum to be spent. About twice as much is shown for 'experimental' work, 'half-way between research and propaganda . . . to obtain propaganda material or to test under field conditions results obtained by the research staff'. It was roughly comparable with development work on other ICI activities, and research and development together may be said to have accounted for about 30 per cent of the total. The rest was to go in selling charges and in propaganda—'expenditure on account of sales, but not attributable directly to the selling organisation'.

If this was how three Directors of ICI thought of Jealott's Hill, its activities and its expenses, it is apparent why Sir Frederick Keeble, about nine months earlier, had felt the necessity of saying to his assembled staff:

It is a most difficult task to decide what is the proper relation between the advisory and sales services. I feel that there can ultimately be only one opinion. This is that there must be the closest relation between sales and propaganda. There must be ultimate direction by the head of the Sales Department and there must also be a complete divorce of the chief advisory officers from direct selling work.[17]

But the mingling of research expenditure with selling charges strongly suggested that in some quarters of ICI, at least, nothing like a 'complete divorce' would be recognized. On the contrary, it was a direct invitation to consider research expenditure as part of the selling costs of the product and to reduce the amount spent on research if selling costs were being reduced.

While plans for research and advisory services based on Jealott's Hill were being thus boldly pressed ahead, J. G. Nicholson and others were exploring the depressing reality of the agricultural market in Great Britain. Being used to the contemporary power politics of the chemical industry, they had taken it for granted that ICI would dominate the supply of fertilizers to the point of monopoly, but they do not seem to have looked much further. Certainly they do not seem to have taken the point, until Speyer brought it to their attention towards the end of

1926, that ICI, the national monopolists, would find themselves at the mercy of numerous much smaller local monopolists, the corn merchants in the country towns.[18]

These merchants bought the farmer's produce, and to him they sold essential supplies including, as one item among many, fertilizers. The farmer, in England as in many other countries, was commonly in debt to the merchant. The merchant, drawing his profit from many different lines of stock in trade, was under no sort of pressure to promote one kind or brand of fertilizer rather than another, and his natural inclination, said Speyer, would be to seek the highest profit, regardless of quality:

> If any competitive form of nitrogen turns up, which is not greatly inferior to Sulphate of Ammonia in effect, and which can be sold at a lower price per ton (though not necessarily per fertiliser unit) to yield the merchant a better profit, he will push [it]. With the increase in numbers of the better educated farmer, this danger is to some extent offset. . . . But many farmers are so tied up to the merchant that they are forced to buy materials which they know to be inferior.[19]

This is a partisan statement, and Speyer never underplayed dramatic effect. Nicholson, however, came to agree with him, observing in 1928: ' . . . the [corn] chandler . . . exploits the farmers' difficulties for his own profit'.[20] Both pointed to the merchant's control of credit as the source of his strength against the farmer, and to the relative unimportance of fertilizers in his business as a source of strength against ICI. The farmer was in no position to demand ICI fertilizer from the merchant, even if he could be induced to wish to, and ICI could give the merchant no very weighty reason for stocking Billingham products if something else paid better.* This was the more unfortunate since although fertilizers were a matter of such indifference to the merchants, ICI depended on fertilizers for about 90 per cent of Billingham's profits.

Then why not sell directly to the farmer, cutting out the merchant? Speyer said it would never do 'unless you are prepared to render him the same services [presumably including credit] as the corn chandler'. Nicholson, once again, found himself bound to agree. 'The farmer . . . refuses to regard fertilizers as a special class of his supplies, but expects to buy them along with seed and machinery. ICI has therefore been in no case able to go direct to the farmer.' Instead, ICI had depended on the chandlers, and as a consequence, said Nicholson, had 'no effective selling organization for fertilizers in the Home Markets'.

This was in July 1928. It was a position which Nicholson was no

---

* The makers of compound feedingstuffs, and no doubt of other branded supplies also, were very similarly placed.

longer prepared to tolerate. 'I do insist that ICI . . . must take prompt steps to substitute a more healthy condition. We must overcome the peculiar hold which the corn chandler has obtained over the agriculture of this country, and we can only do this by giving the farmer the same service (and better) as that to which he has been accustomed.' In other words ICI—as Speyer had come close to suggesting in 1926—should go into the merchant business themselves.

Nicholson proposed an operation in two stages. First, ICI should amalgamate 'some of the large fertilizer mixers which are at present combining various "straight fertilizers" to form "special" or combined" fertilizers'. It was from these firms that the merchants bought products which competed with ICI's output. ICI would regulate quality, rationalize production, and organize supplies from Billingham to the amalgamated mixers. The next stage in the operation would be for ICI to acquire merchants' businesses of their own. Some would be bought in with mixing firms which owned them, particularly in Scotland, but in England and Ireland 'it will be necessary . . . to absorb gradually sufficient corn chandling businesses to cover the entire country'.

Suspicion would certainly arise among farmers, who would probably conclude that ICI, too, was against them and was simply seeking to raise prices. Nicholson therefore proposed to limit ordinary dividends from the newly amalgamated companies to 10 per cent, and to distribute any surplus among customers in proportion to their purchases. This was the co-operative principle, which British farmers were well known to be averse to. Nicholson, however, was optimistic, though with what precise justification it is hard to see.

The amalgamations were to be formed under holding companies controlled by ICI, one each for England, Scotland, and Ireland, each called Agricultural Industries Limited with the appropriate national prefix. Nicholson predicted the hoped-for consummation:

. . . the final position will, I think, place ICI and the agricultural industry in a position to co-operate usefully for some time to come without serious readjustment. It is obvious that those chandlers not included in Agricultural Industries Limited, must quickly conform to its methods or go out of business; and as the proposed system demands capital not at the disposal of small concerns, the probable result will be the survival of a few sound concerns with which co-operation might ultimately be arranged. Meantime the farmers will be obtaining on advantageous terms all their requirements including fertilizers of standard types, suited to varying needs and scientifically constituted, together with sound and expert advice on their methods from the technical salesmen of Agricultural Industries Limited. I do not think I need elaborate the advantage to ICI of such a position.

This bland forecast of universal rustic harmony, under the benevolent despotism of Millbank, was not fulfilled. The only one of the projected holding companies to be formed was Scottish Agricultural Industries. In Scotland Nicholson had prepared the ground before proposing his scheme for an amalgamation to the ICI Board, and during the autumn of 1928 half a dozen of the leading Scottish mixers and chandlers were fairly easily brought together. The authorized capital of SAI, in Preference, Ordinary and Deferred classes, was £1,750,000, and ICI took up 62·5 per cent of the Ordinary capital, leaving the rest with the former owners of the merging firms.

The fertilizer mixers elsewhere in the British Isles saw no attraction in ICI's merger proposals. Their position as matters stood was very strong. Banded together in the Fertilizer Manufacturers' Association and buying jointly, the major firms of mixers were ICI's largest customers for ammonium sulphate in the home market, taking about 50 per cent of Billingham's output. Moreover they were not captive customers, for there were alternative suppliers of ammonium sulphate abroad and they could get ammonia liquor at home. Their product was a compound of ammonium sulphate with superphosphate and potash which merchants—many of them controlled by mixers—were very happy to sell. There was no reason, in 1928, why the mixers should listen to ICI, and they did not.

In the early thirties the position began to change as ICI produced concentrated complete fertilizers of their own which, according to Speyer, gave the farmers better value for money and could be produced at costs competitive with those of the fertilizer mixers. The mixers, in 1934, apprehensive of competition, approached ICI for an agreement, and terms were worked out which were intended 'to provide ICI with a reasonable probability of selling concentrated fertilizers to capacity in the shortest possible time but at the same time safeguard FMA members in their purchase of sulphate of ammonia for mixing while their trade in lower-grade compounds continues and make it possible for them to participate in the sale of concentrates as the demand for their compounds declines'.[21]

The mixers, that is to say, not only fended off ICI's attempt to control them but blunted the edge of the competitive weapon brought by ICI against them. Early in 1935 an agreement was concluded under which the FMA agreed to buy sulphate of ammonia, nitro-chalk, concentrated fertilizers and synthetic nitrate of soda from ICI only. ICI in return undertook to sell concentrated fertilizers only through existing agents and not to appoint any more unless the FMA failed to buy a certain quantity of concentrated fertilizers. There were also provisions protecting prices, regulating terms, and sharing markets. The agreement was to run for five years. It was not at all what Nicholson had in mind when he conceived his scheme for Agricultural Industries Limited, and the whole episode showed that ICI's size, contrary to what the Board had been inclined to assume, did not automatically enable them to dominate any industry they chose to enter, especially if they did not properly understand it.

In Scotland, where the plan for a merger succeeded, the outcome, from ICI's point of view, was even more perverse. ICI in theory controlled the Scottish mixers, but the mixers refused to be controlled. They had capital invested in superphosphate plant and mixing plant which they had no intention of scrapping as long as it gave them a product which the merchants could profitably sell. They therefore went on doing exactly what the merger had been intended to stop—selling their own products, which ICI said were inferior, instead of ICI's. ICI, up to 1934 at least, found no way of preventing them, and so the Scottish mixers were in the happy position of enjoying the backing of ICI capital while they competed profitably with ICI goods. 'There is in SAI', Pollitt sadly concluded, 'a conflict between ICI and the management, which we have failed to overcome and ICI have consistently been on the losing side in this conflict.'[22]

Long before Pollitt found occasion to bemoan SAI's selfishness, the development of ICI's entire agricultural enterprise, and through it the development of the Billingham investment, had been overshadowed by events in the world at large culminating in the onset of the depression of the early thirties. For convenience, we have followed events in the home market well beyond that point. We must return to matters as they stood in 1929, when Jealott's Hill had just opened and Scottish Agricultural Industries had just been formed.

(iii) COLLAPSE: THE ONSET OF THE DEPRESSION

First the test-tube,
      Then the pail,
Then the semi-working scale,
Then the Plant,
And then Disaster,
        Faster!
          Faster!!
            Faster!!!
              Faster!!!!
F. A. Freeth, about 1929.

By the summer of 1929 it had been apparent for some time that the optimism about the development of the world market for nitrogen, on which the Billingham investment decision had been based, was false. This, originally, had nothing to do with world depression, the signs of which were not obvious, at any rate to the generality of business men, in the latter part of 1928 and the early months of 1929. What was obvious, and daily becoming more so, was that far too much ammonia plant was being put up in the world. Even without a depression, it was unlikely that demand would keep up with the increase in productive capacity.

This was a new development. In the early twenties, as we saw in Volume One, ammonia synthesis was one of the most difficult problems in chemical engineering, but by 1926, when ICI was founded, the process was well understood and it was becoming fairly easy to get plant built. As time went on, in the late twenties and the thirties, ammonia technology became widespread, partly through independent development work, particularly in Italy, and partly through the dispersion of American engineers who had worked on plant controlled by Allied Chemical and Dye Corporation. When they were sacked, in the course of Orlando Weber's drastic rationalization, they took their knowledge into the market. There were plenty of customers. In the rising mood of economic nationalism, through the late twenties into the thirties, an ammonia plant seemed desirable, even essential, to governments all over the world, and an investor prepared to finance one could be reasonably sure of protection or other helpful measures. Moreover it would often pay the existing large operators, IG or ICI, to build or assist in building plant abroad rather than let someone else do it. By 1936 plant had been designed by Billingham engineers for Czecho-slovakia and South Africa, and schemes had been examined for plant in China, Spain, Egypt, the Irish Free State and Mysore, though some-times the examination had been sufficiently discouraging—to ICI's relief—to prevent the plant going any further.[23]

The threat of over-capacity seems first to have been taken seriously in ICI in the latter part of 1928. In November of that year the statistical department produced figures suggesting that between 1927 and 1931 the quantity of nitrogen that could be produced from the world's synthetic ammonia plant was likely to rise from 670,000 metric tons a year to 1,800,000. As well as this, of course, the coke-ovens and gas-works would be producing ammonia and there would be ample supplies of nitrate from Chile. Much the largest synthetic producer, throughout this period, would be IG, whose capacity at Oppau and Merseburg, where very large additions were in progress, would rise from 450,000 metric tons of nitrogen in 1927 to about 750,000 in the early thirties. The next largest producer, increasing fast but a long way behind IG, would be ICI. By the time Nos. 3, 4 and 5 units were fully in action at Billingham, about the beginning of 1931, ICI would be able to produce 210,000 tons of nitrogen a year.[24] All these figures would represent much larger tonnages of fertilizers, whether as ammonium sulphate, ammonium phosphate, nitro-chalk, or in other forms.

Part of the Billingham works of **Synthetic Ammonia & Nitrates Ltd.** Artist's impression by Keesey, 1927 (see chapter 5).

Parabolic silo for ammonium sulphate, Billingham, 1932 (see chapter 5).

Below: Hydrogen plant at Billingham, 1932 (see chapter 5)

With the increase in productive capacity and in output went a steady fall in prices. The price of ammonium sulphate, which had been as high as £14 10s. (£14·50) a ton in the spring of 1924, was at £10 6s. (£10·30) in the spring of 1929,[25] and the net naked realization* per ton at Billingham, taking account of seasonal fluctuations in prices, was much lower, though up to 1929 rising sales kept the figures of total realization rising also:

Ammonium Sulphate—Sales and Realization, 1927–1929[26]

	Tons sold	Net naked Realization	
		Per ton	Total
		£	£
1927	76,834	8·672	666,258
1928	204,300	8·485	1,733,465
1929	455,294	7·1889	3,273,057

Despite the rising trend of these figures, Mond, in the spring of 1929 was sufficiently disturbed by the continuing price fall to write a note asking whether it would be better to restrict output and maintain prices or 'to try and get full output at low prices'. He was concerned, as well he might be, with the return on capital employed at Billingham, which in 1929 was below 5 per cent. 'We have invested', he wrote, 'about £20,000,000 in Fertiliser works, and every £100,000 loss [of profit] is equivalent to one half per cent on that capital.'[27]

In the autumn of 1929 the troubles of the fertilizer makers were compounded by the collapse of the boom in the USA, with its catastrophic consequences for the commodity markets on which the sales of fertilizer depended. In these circumstances it became clear that the choice of tactics proposed by Mond in the spring did not exist. Stocks of unsold fertilizer were mounting so alarmingly at Billingham that there could be no prospect of trying to get full output at any conceivable price level. Moreover, calculations of costs and profits based on the full capacity of the plant, where No. 5 unit was just coming to completion, would be futile and misleading. In December 1929 Mitchell put his view of the situation bluntly before the Finance Committee of the Board:

The facts with regard to the Billingham position should be faced at once. They are:

1. The capacity of the Plant will substantially exceed the demand for the material produced, at any rate for some time to come.
2. A fundamental rule of sound commercial management is that stocks of finished products should not exceed a figure reasonably calculated to cover the normal development of the trade. . . .

* The ultimate net realization, excluding packing, delivery, and other distribution charges.

E

3. Costs based upon 'capacity' and profits earnable based upon 'capacity' are of no practical use unless there is a market for the total output. . . .[28]

The inescapable conclusion from Mitchell's observations was that plant at Billingham would have to be shut down. This would be a reversal of much the most important investment policy followed by ICI since the merger, an admission that the assumptions underlying it were untenable, and a cruel blow at the self-respect of those members of the Board, especially Mond, Nicholson, and Pollitt, who had been foremost in pressing for the development of Billingham and of the fertilizer business. But perhaps that did not greatly perturb Mitchell, himself a Nobel man and usually at feud with Nicholson. He took it for granted that plant would have to be closed, but he only referred to the matter obliquely and left the substantive recommendation to come from others.

It came from Nicholson the next day, in a rather muddled memorandum which perhaps reflects agitation and hurry. Its first paragraph, however, is clear enough:

The object of this memorandum is to recommend that production of nitrogen at Billingham should be reduced to a figure which will supply the demand and provide the stocks mentioned in this note and no more.

The deliveries he expected from Billingham were 100,000 tons of nitrogen in 1930, 110,000 in 1931, and 120,000 in 1932, and he recommended 'that the plant should be run at the rate . . . which will give the above quantities and that the storage capacity for ammonium sulphate shall not exceed three months' estimated requirements (including byeproduct sulphate [made by members of the British Sulphate of Ammonia Federation, not by ICI]) viz: about 160,000 tons of sulphate'.

The capacity of the plant, when No. 5 unit was complete, would be about 987,000 tons of ammonium sulphate a year, representing 210,000 tons N, and as well as that ICI was responsible for selling by-product sulphate from the British Sulphate of Ammonia Federation which in 1929 represented nearly 68,000 tons N. The Federation makers would no doubt be required to reduce their output also, but even so it appears that what Nicholson was suggesting, for an indefinite period of years, was output from Billingham at little more than half the rated capacity, perhaps less.[29] The repercussion on the labour force, to say nothing of the cost of idle resources, was bound to be very serious indeed, especially in an area like Tees-side, worse depressed than most. Nevertheless Nicholson's recommendation was endorsed by the Chairman's Conference on 23 December—they lost no time: they must have been very worried—and the Commercial Department was given 'discretion to reduce the price for Sulphate of Ammonia'.[30]

As the depression set in, ICI and IG were engaged in trying to negotiate their way out of some of the difficulties which surrounded them by means of an agreement to regulate the world market for nitrogen. The negotiation had followed on, almost without a break, after the collapse of the attempt to make a general agreement between ICI and IG (pp. 45–6). At the beginning of 1928 it was evident that no general agreement was going to be made, and the parties transferred their attention to the narrower field of nitrogen, where IG, at least, were already apprehensive of over-production, particularly from the new makers coming into the industry. After a brief, reluctant glance at the possibility of fighting each other, the two groups, in May 1928, met for a conference. The meeting place was SS *Luetzow*, cruising in the Mediterranean. She was known in ICI as The Good Ship Nitrogen.[31] The meetings in the *Luetzow*—there were two—produced a disagreement, common in ICI/IG relations, in which ICI asked for a territorial division of markets whereas IG insisted on sales quotas based on sales in all markets taken together. The two parties failed to resolve their difference, but they went on working together, as they or their predecessors had done since before the formation of ICI, to control fertilizer prices as well as they could, and during 1928 the price of ammonium sulphate varied, according to the season, a few shillings either side of £10, whereas in 1927 it had at one time been nearly up to £12.

During the early part of 1929 the need for agreement became more urgent as the danger of over-production grew. Both ICI and IG would have liked a comprehensive scheme covering the new producers of synthetic nitrogen and the Chile nitrate producers as well, and suggestions were drafted. First, however, ICI and IG had to agree with each other. During the summer and autumn drafting and re-drafting went on and the final document was signed, as the world's farmers slid into depression, on 25 February 1930.[32]

In this agreement ICI and IG divided the markets of the world along conventional lines, in so far as ICI had 'the sole right to maintain organizations for the sale of nitrogenous fertilizers' in the British Empire and IG had similar rights in Europe, except the Iberian Peninsula and Canary Islands, and in Central America. ICI also had sole rights in the Netherlands Indies. North America was specifically excluded from the agreement and South America was not mentioned, but in the Iberian territories separate selling organizations were to continue 'until a different arrangement is agreed', and in Japan and China arrangements were set up for joint working, although separate selling organizations were to be maintained.

These were the marketing arrangements, but how much was to be marketed? As a basis, ICI and IG undertook not to extend their existing plant, complete or under construction, so long as demand was insufficient to absorb their total joint output, working at full capacity. IG's full capacity was put at 750,000 tons of nitrogen a year plus 90,000 from Norsk Hydro, ICI's at 180,000 tons nitrogen in 1929/30, rising thereafter to 210,000, plus 70,000 from the British Sulphate of Ammonia Federation The two groups agreed to divide joint sales throughout the world according to a formula based on the ratio between the figures for the total capacity of each, which worked out at $19\frac{1}{2}$ ICI to $80\frac{1}{2}$ IG, and elaborate arrangements were made for such matters as the sale of fertilizers bought in from third parties and for under- or over-selling quotas. Prices were to be set in Germany by IG, in the UK by ICI, and elsewhere by agreement, except that the price of a product made solely by one party (IG had a much larger range than ICI) might be fixed by the party making it, after giving 'consideration' to the views of the other party.

A clause to which ICI attached great importance, and over which they bargained hard, allowed them an unrestricted right to put up or extend works in the British Empire, provided they controlled them and brought 100 per cent of their deliveries into the pool, without paying compensation to IG. A similar right was granted to IG in their territories. As ICI pointed out, 'It is certain that as soon as the Empire Markets show a demand sufficiently large to justify the creation of a home nitrogen industry, that home nitrogen industry will be created, if not by us then by others. It is to the joint interest of both parties that we should participate in such plants, rather than that they should be put up by third parties'.[33]

With this agreement the world's largest makers of fertilizers battened down their hatches to face the economic storm. The details evidently represent a compromise between ICI's demand for exclusive marketing territories and IG's insistence on sales quotas. In its broader aspect, the whole tone of the agreement, which was for ten years, makes it clear that neither side had any hope of demand catching up with supply in the foreseeable future. Both groups, not ICI only, had over-invested very seriously in fertilizer plant. They had misjudged the market, and they had to live with the consequences.

REFERENCES

1 Memo by J. G. Nicholson, 28 Oct. 1927, CR 84/3/R.
2 Memo by A. E. Hodgkin for Pollitt on general Fertilizer Position, 2 Sept. 1927, CR 5/48–2/1.
3 Pollitt, 'Rate of Development at Billingham', 21 June 1926, CR 36/17/–.
4 As (2), p. 1.
5 'Problems for Technical Committee', 17 Nov. 1926, CR 215/–/2R.
6 Departmental Annual Reports, 1927, 'Fertiliser Investigations'.
7 'Advantages of Fertilisers of the Type of Diammonphos or Nitrophoska as compared with Mixtures . . .', M. T. Sampson, 2 Sept. 1927, CR 5/9/R2.
8 'Nitrogen Position—Notes by the Chairman', 8 Nov. 1927, CR 5/61–1/1.

⁹ As (2).
¹⁰ Minutes of third meeting of the Fertilizer Committee.
¹¹ AGM, 31 May 1928, Chairman's Speech.
¹² FCM 315, 8 Jan. 1929; AGM, 18 April 1929; FCM 459, 12 Nov. 1929.
¹³ Sir F. Keeble, 'Nitrogen Position', 14 Nov. 1927, CR 5/61–1/1.
¹⁴ Addresses . . . [to] . . . the First Meeting of the Commercial and Technical Committees of Imperial Chemical Industries Limited (Chemical Group), 30 Nov. 1926, p. 8.
¹⁵ H. J. Mitchell, G. P. Pollitt, J. Rogers, 'ICI Agricultural Research Station, Jealott's Hill', 5 May 1930, ECSP.
¹⁶ 'Proceedings of a Conference . . . June 25th and 26th 1929', CR 153/6–5M/–.
¹⁷ Ibid.
¹⁸ F. C. O. Speyer, 2 Dec. 1926, CR 179/1/8a; Fertilizer Sales Conference, 3 Sept. 1929, 153/6–9/R; J. G. Nicholson, 'ICI and Agriculture', 6 July 1928, CR 153/6/R.
¹⁹ Ibid. 1st refce.
²⁰ As (18), 3rd refce.
²¹ F. C. O. Speyer, 'Proposed Agreement between ICI and the Fertiliser Manufacturers Association Limited', 22 Sept. 1934, CR 3/43/146; GPCM 794c.
²² G. P. Pollitt, 'ICI—Scottish Agricultural Industries . . .', 23 Feb. 1934, GPCSP.
²³ 'Survey for Policy Meeting . . . Fertiliser Group', 25 Sept. 1936, GPCSP.
²⁴ ICI Statistical Department, 'World Nitrogen Building Programmes, by Processes', 2 Nov. 1928, CR 84/6/R4A; ICI/IG Nitrogen Agreement, IG Draft Proposals, 28 Oct. 1929, CR 84/10–1/3.
²⁵ J. Bueb, 'Nitrogen Economics, Retrospect and Prospect', CR 84/6/R6.
²⁶ Table with 'Fertiliser Budget Summary', 16 Nov. 1934, CP 'Fertilisers, Billingham Commercial Reorganisation'.
²⁷ 'The Chairman's own Note', 2 May 1929, CR 84/12/ .
²⁸ H. J. Mitchell to Finance Committee, 16 Dec. 1929, CR 36A/10–1/1.
²⁹ J. G. Nicholson to Executive Committee, 'Nitrogen Position', 17 Dec. 1929, ECSP.
³⁰ ECM 1629, 13 Jan. 1930.
³¹ F. C. O. Speyer, 'Probable immediate Effect of a Nitrogen War with IG', 13 Jan. 1928, CR 84/7–1/R1; W. H. Coates to H. Mond, 17 Feb. 1928, CR 84/7–1/4; ship's nickname from R. A. Lynex.
³² Heads of Agreement 18 June 1929, CR 84/10–1/ ; as (24), 2nd refce.; Final Text, ECSP 10 March 1930.
³³ ICI to IG (Draft), Dec. 1929, CR 84/10–1/4.

CHAPTER 6

The Impact of the Slump 1929–1931

THE ONSET IN 1929 of the world economic crisis of the early thirties brings us to one of the central events of twentieth-century British history. The material results of the slump, in Great Britain, were neither so severe nor so prolonged as in Germany or the USA, but the psychological damage, in the long run, seems to have been worse. The basic industries of the country, closely identified with Victorian power and prosperity—coal, cotton, heavy engineering and shipbuilding—had never recovered from the collapse of 1921 and were still depressed when the new blow fell. In the older industrial districts, where 20 or 30 per cent of the wage-earners had been out of work for years at a time, self-respect had been undermined, and the events of 1929–31 reinforced long-enduring misery and bitterness. Even outside the depressed areas, in regions where normally 90 per cent or more of the wage-earners would be employed, knowledge of the crumbling towns of Wales, Scotland, and Northern England, with their increasingly demoralized inhabitants, bit deeply into the national consciousness. Legends of 'the thirties' grew up and passed into folklore, to influence politics and industrial relations far into the future.

Outside the basic industries the events of the early thirties were not superimposed on so grim a background nor were their effects so catastrophic. In the chemical industry there was never anything to match the 60 per cent rate of unemployment which ruled in shipbuilding in 1932–3. Nevertheless the immediate impact of the slump was heavy enough to shatter the promise of ICI's early years and to turn expansion into standstill or contraction for three or four years.

The effect in the long run, working in the same direction as profound technical changes in the chemical industry, was to alter entirely the nature and direction of ICI's development. With that we shall increasingly be concerned in later chapters. Here we shall examine the direct consequences of world events upon ICI between 1929 and 1931.

The most spectacular victim of these events, in ICI, was the fertilizer plant at Billingham. The decision to invest £20m. there, taken in 1927, collapsed into ruin at the end of 1929. By that time most of the money raised for the project had been committed to it, and plant had been put up far in excess of any likely demand. ICI, within three years of its foundation, was flung into crisis: crisis so grave that it might have been mortal.

Perhaps the crisis was partly of ICI's own making. Perhaps the grandiose plan to monopolize the fertilizer trade of the British Empire was so ill-conceived that it never could have succeeded. We have seen indications in plenty that the market was never thoroughly investigated and that the effect of new ammonia plant coming into production throughout the world was not taken seriously until too late. But these weaknesses, fatal though they might have been eventually, would have been slow-acting. The direct cause of the crash at the end of 1929 was the collapse of trade in farm products throughout the world. 'We are the servants of agricultural and other industries,' said Lord Melchett in the summer of 1930, 'and their depression must naturally affect their power of consumption, and consequently our production.'[1]

Shutting down production at Billingham and, still more, stopping construction there, led at once to laying off several thousand men. In May 1930 the Annual Report for 1929 said that 6,741 men—not all, but probably most, at Billingham—had by that time been laid off, and many more were to follow as time went on. Accurate figures are hard to come by, but estimates set out in Appendix II suggest that between 1929 and 1932 the total number of ICI employees, in Great Britain, fell by about 20,000, and it is a fair assumption that a high proportion of those discharged, probably on the high side of one-third, came from Billingham. Many were taken on again later, but that did nothing to cushion the shock at the time.

Disaster on such a scale could not be hidden. Rumours got about, and eventually had to be publicly denied, that the fertilizer plant was to be closed entirely.[2] Confidence in ICI had been shaken as early as October 1929, when General Motors shares, caught in the crash of the New York Stock Exchange, fell from a level of about $72 (31 August) to $47·5, and went on falling. The price of ICI shares followed them down, because it was well known that ICI still had a large holding in GM (the holding had in fact been increased, though this was not publicly known, when GM fell to $47·5).[3] News from Billingham, as well as the general state of the market, accelerated the fall, and the price of ICI Ordinary shares, which had been issued at 33s. 6d. (167½p) in May 1929, to raise capital for Billingham, stood at barely half that a year later. At the bottom, in 1931, they fell to 9s. 10½d. (about 49p).

The worst sufferers were small investors, especially, as Melchett put it in a distressed letter of August 1930, 'a number of our smaller employees who took up shares of our last issue to some extent, I am afraid, on borrowed money and are now being compelled to sacrifice these shares at a quite unwarranted low figure'. He had evidently been buying for himself to support the market but, as he said, he could not go on indefinitely adding to his 'already large holding', and he was writing to ask Ernest Solvay if Solvay et Cie would come to the rescue.[4]

Stock Exchange prices are always at the mercy of rumour and rumour may have exaggerated the damage done to ICI by the slump. Melchett and McGowan, in public, maintained that it did, but in private they had to face a reality which was very bad. Capital expenditure on physical assets, representing plans for the future development of the business, was cut, cut, and cut again, to a point from which recovery was bound to be long-drawn and slow. From a peak of £11,023,720 in 1929 the figure fell to £945,491 in 1932, a fall of almost 92 per cent in three years.[5]

Capital spending in the lavish years had been very unequally distributed among the Groups. The two biggest spenders had been Alkali Group and Fertilizers, so that when adversity set in they were the most conspicuous sufferers.

In Alkali Group an ambitious programme of expansion reached a peak in 1928, when £1,229,307 went into 'physical assets'. In 1932 the corresponding figure was £109,707. In September 1930 the number of staff in the drawing office at Winnington was cut from 115 to 91. Work still grew less and in the summer of 1931 Lutyens, the Group Chairman, had to bring himself to the point of ordering still more to go, since by then there was plant available in the works for 40 per cent more output than was required. Nothing of the sort had ever happened at Winnington before, and Lutyens was very distressed, particularly because he could see no alternative to getting rid of middle-aged men, past their best, who would find it difficult to get other work, rather than younger and more efficient men who would be a growing asset in better times. 'Brunners', he wrote unhappily to Nicholson, 'have established a tradition for continuity of employment, and we should never recommend abandoning this tradition if it were possible to find work for the whole of the present staff. We should not have suggested that any of our senior and long service men (even though they are comparatively inefficient) should go in order to replace them with younger and more capable men.' In happier times, as Lutyens said, the policy had been 'to allow them to be carried along in the stream of younger men coming in below, the more capable of whom passed by their less efficient seniors'.[6] But in those days the volume of work had always been expanding.

In Fertilizer Group matters were much worse. At the height of euphoria, in 1929, £8·3m. of ICI's total outlay of £11·0m. for physical assets had gone into Billingham. In 1930 the figure was cut to £1·7m., in 1931 to £0·3m., and in 1932 to £93,677, scarcely more than 1 per cent of the 1929 expenditure. Capital spending at Billingham was virtually wiped out by the slump, carrying with it the most prolific source of employment on the site, since completed plant did not need so many men to run it, especially when half of it was closed down, as it had needed for putting up. No wonder people began to talk about the fertilizer factory being closed altogether and for good.

In 1931, with the value of world trade perhaps as much as 40 per cent below the 1929 figure, and volume badly down as well, ICI's net profits fell to the lowest figure ever recorded: less than £4m. on capital employed that still stood at nearly £100m. Running expenses as well as capital expenditure had already been cut drastically, stocks of raw materials had been run down,[7] and men had been dismissed in large numbers. There remained one other possibility of saving money: to cut wages and salaries.

It was a measure much in vogue at the time and, in the prevailing state of weakness of the trade unions, fairly easy to enforce. In September 1931 the new National Government, acting on proposals framed by the Labour Government before it fell, cut the salaries of ministers, judges, MPs and teachers, as well as unemployment allowances and the pay of the armed services, thereby provoking in the Navy the 'Invergordon Mutiny' which played some part in setting off the sterling crisis which drove Great Britain off the gold standard.

The trouble with the sailors was caused largely by the way news of the cuts in their pay reached, or rather failed to reach, the Fleet. In ICI, months earlier, they had handled matters more sensitively. The first proposal for a wage cut came in the spring of 1931 from the Yorkshire branch of the Chemical and Allied Employers' Federation. They suggested 10 per cent. McGowan was much in favour, but the Group Chairmen were horrified by so high a figure and Henry Mond supported them. McGowan, for once, gave way and agreed that ICI should go for 5 per cent, carrying the other employers, if they could, with them.

The proposal, once agreed upon between McGowan and the Group Chairmen, was put to the men with the utmost care. It was revealed first to ICI's Trade Union Advisory Council in London. Then, early in May, Group Council meetings were called at the factories. Mond insisted that a Director or some other senior representative of Millbank should attend and himself went to two, at Alkali Group and Dyestuffs Group. 'Both [meetings]', he reported to McGowan, 'have been most unqualified successes and the men really appeared to be extremely affected by the fact that they were taken into our confidence and that a Director had come down to explain the situation to them.'[8] There was no question of consultation or bargaining, but the gesture of public explanation seems to have been sufficient, in the atmosphere of national emergency then prevailing, to disarm any opposition there may have been. There is no record of any kind of resistance either by the unions or by the men themselves.

Only when ICI's men had been informed in this way was the proposal for wage cuts put formally to the Chemical Trade Joint Industrial Council for the chemical industry. A resolution was passed on 22 May 1931 which had the effect of reducing wages by about 5 per cent from the beginning of the second complete works week in June 1931.[9] For day labourers over 21 it produced a wage of 1s. (5p) an hour, amounting to £2 7s. 0d. (£2·35) a week.

Staff salaries were cut on 1 July and 1 October 1931.[10] By precisely how much they were cut is obscure. McGowan let it be known that he wanted savings of £190,000 to £200,000 a year,[11] and a cut of 10 per cent was suggested in April, at the same time as the proposal for a 10 per cent cut in wages. It probably aroused quite as much opposition among the Group Chairmen. Lutyens, for one, protested at length to Henry Mond, saying that staff at Winnington were already aggrieved by alterations in the bonus system since the merger. Moreover their Group's trade had not suffered so badly as the trade of some other groups and they felt they were being asked to suffer a reduction in salaries through having been incorporated in ICI.[12] No doubt other Chairmen protested also, but this time they did not win. A 10 per cent cut was certainly enforced in the higher ranges of the salary scale, though there is some reason for thinking that very low salaries may not have been cut by so much. P. C. Allen's salary of £400 was reduced to £371.

The cuts in wages and salaries did not last long, being restored in April 1933.[13] The saving to ICI was trivial, amounting perhaps to as much as £350,000 in a year of normal working, but probably less with conditions as they were.[14] Even in so bad a year as 1931–2, that sum could hardly make or break ICI. The main effect of the cuts was not practical but psychological. In the minds of those who suffered them—the lucky ones, who still had jobs—they rammed home the reality of depression, and the impression they made was permanent.

In 1931, according to figures prepared by the Treasurer in October 1932 (see Table 8, opposite), ICI had 'Capital and Surplus' amounting to £107·7m., of which £74·6m. was invested in manufacturing Groups in the United Kingdom. £22·4m. was invested in Fertilizers and Synthetic Products at Billingham—far more than anywhere else. On the F & SP capital there were gross sales, in 1931, of £3·6m., yielding a net trading profit of no more than £45,506, or 0·20 per cent.

Figures for the other large Groups, except Alkali, were dismal also, but it was in Billingham that ICI's central problem lay. 'The total value', says a footnote to the table, 'of the Property and Plant not in production during 1931 was £11,045,206.' Half the plant so ambitiously provided for fertilizer manufacture was out of action. Depression in world agriculture and over-capacity in the world's nitrogen industry lay with this crippling weight over Billingham, so that Billingham, far from

TABLE 8

TOTAL CAPITAL INVESTED IN ICI AND
EFFECTIVE PRODUCTIVE CAPITAL DURING 1931

Groups	Total capital invested	Non-productive capital	Gross sales	Net trading profit	Return on total capital
	£	£	£	£	%
Alkali	16,761,238	1,093,564	5,721,617	1,833,376	10·94
General Chemicals	12,532,354	809,941	4,888,535	345,143	2·75
Dyestuffs[a]	5,682,119	316,982	2,724,667	(195,355)	—3·44
Lime	838,775	244,680	496,104	84,272	10·05
Fertilizers and Synthetic Products	22,382,955	2,546,942[b]	3,657,363	45,506	0·20
Leather Cloth	1,896,057	189,766	936,392	200,398	10·57
Metals	5,562,987	406,918	4,304,835	221,384	3·98
Explosives	8,938,346	40,461	3,972,512	839,088	9·39
Misc. and Foreign Merchanting	7,650,194		9,674,562	153,800	2·01
	82,245,025	5,649,254	36,376,587	3,527,612	4·29
Investments	16,641,758			1,096,178	
Loans	1,126,771			150,375	
House Property, &c.	6,068,986			160,262	
Cash, Debtors, &c.	1,658,096		*Less* External Dividends, Debenture Interest, Income Tax, &c.	265,742	
Total Capital and Surplus	107,740,636		Profit as Published (Gross)	4,668,685	

Notes: a Figures in brackets show a loss.

b 'This figure represents the estimated value of the *permanently idle* Property and Plant. The total value of the Property and Plant not in production during 1931 was £11,045,206.'

Source: Treasurer to Finance Committee, 'Distribution of Capital by Groups & Products and the Return thereon 1930–1931', Statement A, 13th October 1932, FCSP p. 1836.

carrying ICI as had been recently intended, was a burden for the rest of ICI to carry.

The immediate effects of the Depression, on a business for the most part healthy, would pass and did pass. By the end of 1931 the worst, for most Groups in ICI, was over, but not at Billingham. At Billingham that enormous fertilizer plant remained, and the one thing that was

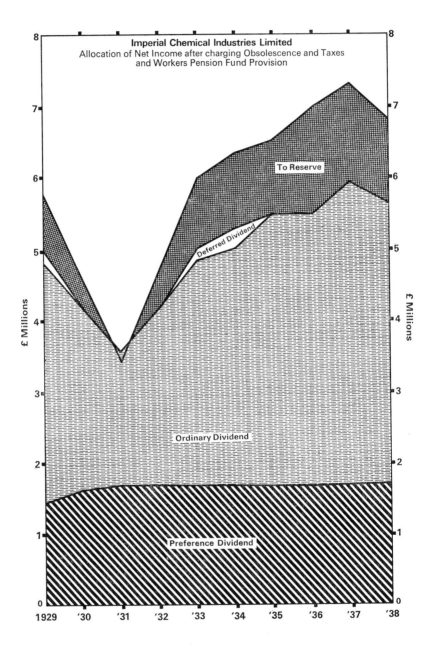

Figure 3. Slump and Recovery

Source: ICI Annual Report 1938.

becoming depressingly certain about it was that it would never operate at anything like its full capacity. That problem, though perhaps inherent in the planning of the enterprise, had been precipitated by the slump. When the slump's other direct results were little more than a horrible memory, it would still be there for ICI to solve.

REFERENCES

[1] Lord Melchett to Group Executive Boards, 25 July 1930, CR 93/44/107.
[2] McGowan's Speech, AGM, 1931.
[3] FCM 450, 29 Oct. 1929.
[4] Lord Melchett to E. J. Solvay, 16 Aug. 1930, MP 'Finance 1929-1937'.
[5] Treasurer to W. H. Coates, 23 June 1937, CP 'Capital and Financial Position'.
[6] W. F. Lutyens, 'Alkali Group—Staff Reductions', 8 June 1931, CR 31/12/61.
[7] Melchett as (1); McGowan on reduction of stocks, 18 Aug. 1930, CR 31/53/5; see also other correspondence in CR 31/53/53; 'Economies effected by Groups in 1930 and 1931', 2 Sept. 1932, CP 'Group Economies'.
[8] Lord Melchett to Sir H. McGowan, 6 May 1931, MP 'Wages 1931'; see also other correspondence in same file.
[9] Secretary, Chemical and Allied Employers' Federation (Lloyd Roberts) to Members, 'Wages Reductions', 22 May 1931, MP 'Wages 1931'.
[10] ICIBM 2207, 11 June 1931.
[11] Morris and Dickens to Spickernell, 12 May 1931, MP 'Wages 1931'.
[12] W. F. Lutyens, by telephone to Lord Melchett, 29 April 1931, MP 'Wages 1931'.
[13] ICIBM 3065, 15 March 1933; AGM, 1933.
[14] As (10); as (11); Central Labour Department, 'Estimated Saving by Recent Wage Reductions', 14 Sept. 1931, MP 'Wages 1931'.

PART II

THE CARTEL-MAKERS' WORLD 1930-1939

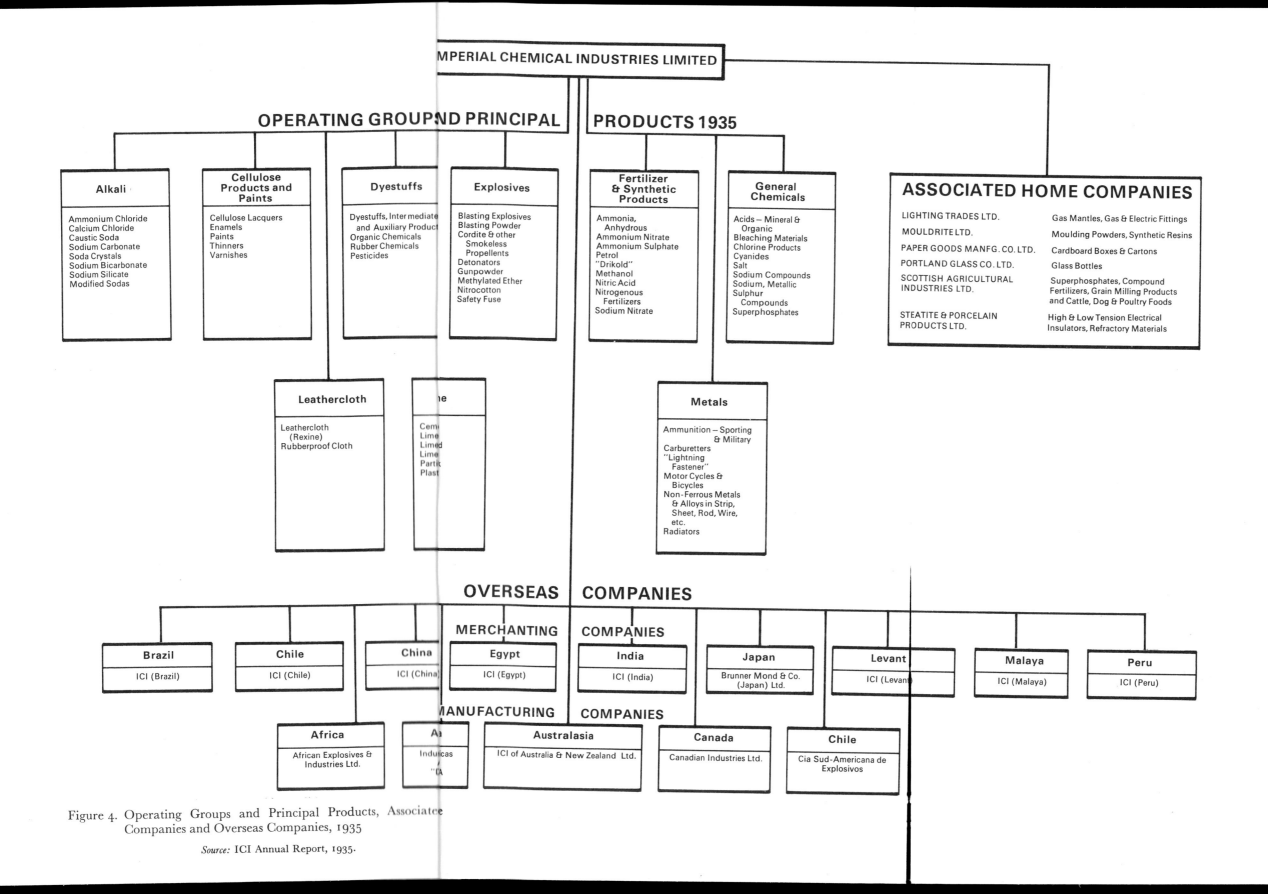

Figure 4. Operating Groups and Principal Products, Associated
Companies and Overseas Companies, 1935

Source: ICI Annual Report, 1935.

CHAPTER 7

Changing Course: ICI in the Thirties

THE CRISIS OF 1929–31 brought decisively to an end the first phase of development of ICI's industrial strength. That phase had lasted three years—no more—and it had been concentrated, almost to the point of monomania, on building up a world-wide fertilizer business based on Billingham. That enterprise was shattered at the end of 1929. Although the fertilizer side of ICI survived, there was never again any question of expanding it on the imperial scale originally planned. On the contrary it had to contract, with all the misery of idle plant and redundant labour which shrinking businesses have to face.

Figures in Appendix II give a shorthand account of the disaster. From 1930 to 1936 capital employed in ICI fell from £102·5m. to £91·6m. The greater part of the fall—certainly something like 80 per cent of it—can be directly attributed to the closing of fertilizer plant.[1] Capital employed did not rise higher than the 1930 figure until 1945.

Quite how serious an effect the fertilizer collapse had on ICI's renewed development, after the worst of the depression was over, is difficult to judge. The loss, in six years, of some 10 per cent of capital employed is a long step down the road to ruin, and if other major Groups had been so dependent on chronically depressed industries as the Fertilizer Group was on farming it seems unlikely that ICI would have survived in anything like its original form. As things were, the Company had to go to the High Court, in 1935, for a reduction of the nominal issued capital by £5,434,141.[2] Fortunately other Groups, including the main profit-earner, Alkali Group, had enough business outside the main disaster area of the depression to carry them through it to a fairly rapid recovery. From 1931 onward ICI's sales and profits both began to improve, and as early as 1933 they had passed the highest levels reached before the depression began.

The rate of return on capital employed, however, which had sunk to 4 per cent in 1931, did not reach double figures until 1939. Doubtless this miserable record reflects partly the weight of dead capital at Billingham, which was not finally dealt with until 1937. There was also in the balance sheet the very high figure for goodwill produced by the share purchase terms of the merger and another very high figure, brought in when Nobel Industries was liquidated in 1928, for the General Motors shareholding, which had been valued at the price ruling on the New York Stock Exchange on 28 September 1928, while the boom was running very strongly.[3]

All this made it very difficult to earn profits that looked respectable in relation to capital employed. But there is more to ICI's post-depression plight than that. It had been assumed since the earliest days of the merger that ICI's main profit-centre would be in the fertilizer plant at Billingham. With that crippled, ICI in the thirties was like a ship with the engines severely damaged. There was barely enough motive power to keep steerage way on so vast a vessel. Moreover it was no longer by any means clear where the vessel ought to be steering to.

The first answer to emerge was 'oil-from-coal'. It had always been assumed that the hydrogenation of coal to produce petrol would be put in hand as soon as possible, and great things were expected of it, especially by Alfred Mond. For ICI it had the advantage of relying on much the same high-pressure technology as ammonia synthesis and on the same raw material—hydrogen—which Billingham was equipped to produce in large quantities. From a wider point of view it had attractions, also. It would supply petrol from Great Britain's own coal and it would bring unemployed miners back to work by so doing. It was intended to be a project on the same scale as the fertilizer business, with operations throughout the British Empire, but its base was to be in the United Kingdom.

The oil-from-coal process had one flaw. It would not pay. There was never any real prospect that the costs of hydrogenation would be brought low enough to compete with the extremely low production costs of the oil companies, working from petroleum. Alfred Mond was always inclined to make light of technical problems, and until he died he continued to believe that the process would be developed into a commercial proposition. Elsewhere in ICI they had no such illusions or, if they had, the progress of development work dispelled them.

Nevertheless ICI held on to the process with great determination and backed it with investment which eventually matched the sums put behind the fertilizer plant.

Why?

Principally because oil-from-coal held out a prospect of saving something from the wreckage of the fertilizer project. Billingham was 'a factory for the manufacture and working up of hydrogen',[4] and if hydrogen plant no longer needed for fertilizers could be used in making petrol it would not have to be scrapped. The value of plant saved in this way reduced the new money required for petrol by about £1·7m., and surviving papers show conclusively that this capital saving was among ICI's strongest motives for undertaking a programme unacceptable on normal commercial grounds.[5]

Not that commercial justification was entirely lacking. A case could be made out, on grounds of public policy, for help from the Government with the economics of the process. It would help to relieve the problems of the coal industry and at the same time it would contribute to the defence of the realm and of the Empire. This case was presented to successive Governments with a tenacity which neither indifference nor outright rejection could destroy. ICI's negotiators could not afford to give up, and in the end they were rewarded. The British Hydrocarbon Oils Production Act, 1934, had the effect of guaranteeing a preference of 8d. (just under 3$\frac{1}{2}$p) a gallon for 4$\frac{1}{2}$ years in favour of petrol produced in the United Kingdom from British coal. In return, ICI agreed to keep the profits of the new venture to 5 per cent, before tax, on the capital employed in it and to apply them to writing off the new capital employed.[6]

The petrol plant opened early in 1935 in the presence of the Prime Minister, conveyed to Billingham by special train. It was presented as a fresh start after the miseries of the slump, and so it was, in a way. As time goes on, however, it looks much more like an end than a beginning. The terms of the Government's bargain were so unattractive, commercially, that they could only have been acceptable on the assumption that at some time in the future petrol from coal would become genuinely competitive, with or without a preference, against the oil companies' products. It never did become so, and even with the stimulus given to the petrol project by the approach of war it became steadily more evident that ICI's early hopes of prosperity through hydrogen and high pressure had been crushed beneath the wreckage of the fertilizer industry and were not going to be revived on a diet of petrol.

Even as Ramsay MacDonald's train steamed towards Billingham, ICI's true future was beginning to emerge, but in quite another place. The Billingham investment, originally, had been based on faith in the development of the kind of heavy chemical industry which Brunner, Mond were used to—an industry relying on one outstanding feat of technical development which would provide, for a long time to come, a limited range of fairly simple products (at Billingham, nitrogenous fertilizers; at Winnington, soda ash and its relations) which would sell cheaply in great bulk, without much call for innovation. This had been the great hope of the twenties, but in the early and middle thirties the most promising lines of technical development were beginning to point in quite a different direction, towards the organic side of the chemical industry, of which the natural home in ICI was neither Billingham nor Winnington but the hitherto despised and neglected Dyestuffs Group. There were outlying cells of it, too, in Explosives Group and General Chemicals Group.

In this kind of chemical industry, which was also attracting the attention of du Pont and IG (with whose ancestors, indeed, it had originated), the products were far from simple, value in relation to bulk was high, and profits arose less from an enormous volume of sales over an indefinite period than from the vigorous exploitation of patent rights while they lasted, both by direct sales at high prices and by the sale to other producers of licences and technical knowledge. Innovation was essential to continued prosperity, for without a continuing succession of new products profits would drop dramatically as exclusive rights to existing processes expired.

During the thirties a chemical industry based on the traditions of the late nineteenth century, and to a large extent on products and processes of that period, was beginning to give way to the chemical industry of the mid-twentieth century. A technical revolution was in its early stages which within twenty years or so would make plastics, man-made fibres, antibiotic drugs and agricultural chemicals, along with many other products of organic chemistry, as much the characteristic products of ICI as alkali and nitrated explosives had been characteristic of Brunner, Mond and Nobels. And in the course of the transformation a new heavy chemical industry would arise, to produce in huge quantities the materials from which the new products would be synthesized.

Very little of this was apparent until the late thirties. In ICI, in IG, in du Pont, in the oil companies, they could see that new possibilities were appearing, and their general nature, but no more. In particular they had no inkling of the scale on which the new industry would develop and, hence, the quantities of heavy organic chemicals that would be required.

During the thirties—the period we are concerned with in Part II of this volume—ICI was seeking new directions, rather like a super-tanker manœuvring in a fog: vast in size, slow to turn, uncertain of the right course to steer. By the time war broke out the fog was clearing a little: enough at least to enable the possibilities of nylon and polythene to be faintly and inadequately perceived. It was not until the late forties, however, as we shall see in Part IV, that the final decisions were taken which committed ICI to a specific programme of large-scale development in the newer branches of the chemical industry.

ICI could hardly have entered this uneasy phase of incipient transformation at a worse moment. The events which culminated in the world financial crisis of 1931 crippled international trade by destroying international co-operation. Every government sought to protect its own interests against the effects of depression and monetary chaos. Currencies were manipulated; markets were closed against the importer; home industries, whether economically viable or not, were built up; the transfer of profits and the repatriation of foreign capital was made as difficult as possible. Economic nationalism, that is to say, was greatly intensified and there was no government unaffected by it. In Great

Britain Free Trade was abandoned—for true Liberals the ultimate apostasy—and an attempt was made, at Ottawa in 1932, to turn the British Empire into a closed and self-sufficient economic system, but it was wrecked by the economic nationalism of the member states, especially Canada. In Germany, after the rise of Hitler, economic policy did not stop at self-defence. It was part of Hitler's foreign policy and correspondingly aggressive, as ICI increasingly found in dealing with IG from 1933 onward.

In these circumstances ICI's interests had to be defended. The only sensible thing to do seemed to be to share whatever trade might be available among existing producers and keep surplus capacity out of action, if necessary by paying producers not to export, as the Russian alkali makers were paid. Methods of management were therefore applied to the chemical industry of the nineteen-thirties which had been very successful in running the chemical industry of the late nineteenth century, and which had never been entirely abandoned—methods, that is, aimed at regulating competition, sharing markets, controlling output, matching supply to demand. The world was criss-crossed with international agreements covering nitrogen, dyestuffs, 'oil-from-coal', alkali, and every product of any importance in the chemical industry that entered into international trade. It was a cartel-maker's world.

In 1930 Alfred Mond, first Lord Melchett, sickened and died, almost as if the sickness at Billingham had been transmitted to Billingham's chief sponsor. In the summer, already ill, Melchett went to the USA. In September, when he came back, they carried him off the ship. He was active in the business, so far as he could be, throughout the autumn, but he was getting worse all the time, and on 27 December 1930 he died. He was 62.

Melchett's ambitions, throughout his life, were political. His achievements in public affairs were what he wished to be remembered by, and they are his biographer's central theme. He left Brunner, Mond when he went into Parliament in 1906 and he did not return willingly. He was thrust back, partly by the collapse of his political career and then, more powerfully, by the crisis in Brunner, Mond's affairs brought on by the threat of scandal over Lever Brothers' action for breach of contract. This grotesque twist in his affairs, compounded of two misfortunes, launched him into the outstanding success of his career, which he did not seek and which came right at the end of it—the founding of ICI. Life was never simple for Alfred Mond, nor did it ever go as it was planned, from the days of his terrifying father, when he failed the Natural Sciences Tripos at Cambridge, up to the moment when Roscoe Brunner confessed everything to Hulme Lever, thus forcing Mond into the Chair of Brunner, Mond, where he had no great desire to be. Even at this point, events took charge of him, rather than he of them. He saw the necessity for rationalizing the British chemical industry, but the conception of ICI was McGowan's, not Mond's, and at first Mond would have none of it. It was forced on him by the individualism of Orlando Weber and the appearance of McGowan in New York, at exactly the right moment, in September 1926.

Once committed to ICI, Mond brought the full power of his personality into play. While he was Chairman his influence was supreme, overbearing even the formidable McGowan. Of the two, once the original idea of the merger had been realized, there was no doubt that his was the more fertile and creative mind. He pushed policy along at a great pace in the direction he wanted it to go, and in the central and hardest task of all in a merger—making the whole more than the sum of the parts—he achieved enduring results in a surprisingly short time. Still, however, a strange malignity seemed to pursue him. The centrepiece of his grand design, at Billingham, collapsed with terrifying suddenness at the first onset of depression. In the shadow of that collapse Lord Melchett died. Surely it must have hastened his end.

REFERENCES

1 ICIBM 5296, 31 March 1937; J. G. Nicholson and A. R. Young, 'Report to Sir Harry McGowan', 22 March 1935, pp. 41–2, Nicholson Papers 'Fertiliser and Synthetic Products Ltd.'; 'Survey for Policy Meeting . . . Fertiliser Group', 25 Sept. 1936, Section I, p. 5, GPCSP.
2 ICI Annual Report, 1935.
3 W. H. Coates, 'Nobel Industries Limited. Liquidation', 4 July 1928, p. 4, FCSP p. 355; 'Liquidation of Nobel Industries Limited', 15 Nov. 1928, FCSP p. 571.
4 'Survey for Policy Meeting . . . Fertiliser Group', 25 Sept. 1936, Section I, pp. 1–2, GPCSP.
5 Chairman to Board, 'Manufacture of Petrol by Hydrogenation', 6 July 1933, CR 59/8/124 vol. 2.
6 Ibid.

CHAPTER 8

Dictatorship in the Thirties: McGowan in Control

(i) SOLE MANAGING DIRECTOR

MELCHETT'S DEATH had been foreseen and there was no interregnum. On 31 December 1930 the Board unanimously elected Sir Harry McGowan 'Chairman and Managing Director of the Company and of the Board'.[1] The depression was getting worse and the position called for strong nerves and an autocratic temperament. McGowan had both and now he had no colleague. He was the unquestioned head on earth of ICI, and the terms of his authority were set out in his Service Agreement:

It shall be the duty of the said Sir Harry McGowan to undertake (subject to the supervision and control of the Board) the general direction of the business of the Company, as far as possible in co-operation with the General Purposes Committee in regard to commercial matters and the Finance Committee in regard to financial matters. In particular he shall study and promote the efficient and economical management of the Company and its Subsidiaries and harmonious working co-operation and mutual assistance between their respective businesses.[2]

In making this appointment, as in much else, the Board of ICI followed Nobel precedent. Before the merger McGowan had been Chairman and Managing Director of Nobel Industries.[3] Now he took up the same combination of offices, with the same concentration of power, in a very much larger group. His agreement was for three years. It was renewed for another three years in 1934 and again in 1937. The effect, although the second renewal was cut short, was that for over six years in the thirties ICI was ultimately controlled, in far more than a purely formal sense, by the will of one man.

As President and joint Managing Director McGowan had been paid £20,000 and managing director's commission calculated at one-half of 1 per cent of ICI's profits over £2 million and a further one-half of 1 per cent of profits over £4 million. In his new splendour his salary went up to £30,000 for one year only—1931—and was continued at £25,000 from 1932. There was an expense allowance of £5,000 and commission was paid at the rate of one-third of 1 per cent of ICI's profits. The results were as follows:

Emoluments paid to the Chairman, 1927–1937[*][4]

£		£	
1927	37,015	1933	55,546
1928	47,912	1934	56,550
1929	50,620	1935	57,841
1930	28,896	1936	60,628
1931	50,562	1937	65,410
1932	53,136		

McGowan was fond of money. In 1931 he considered that the extra burdens thrown upon him, in that very difficult year, by the death of Melchett entitled him to a substantial tip, and he pressed his case so energetically that his service agreement provided for an extra £5,000 in that year. It did not pass without remark, among the two or three who knew the facts, that in the same year McGowan was pressing Group chairmen, already agonized by having to throw hundreds out of work, to enforce cuts of 5 per cent in wages and 10 per cent in salaries.

McGowan's claim no doubt lacked delicacy. It is fair to observe, however, that he was a salary-earner moving in circles in which his income would look moderate and his wealth inconsiderable. He would compare himself with such men as Lord Melchett, Sir Ernest Oppenheimer and Lord Weir, above all with the du Ponts, and they were all part-owners of large businesses, giving them capital resources many times greater than his own. Even among salary-earners he was a long way from the top, for in 1936 Lord Hirst, Chairman of GEC, drew £100,000.[5] At about the same time the Prime Minister was paid £10,000, a judge of the High Court £5,000, and the head of the Civil Service £3,500, but what they lost in salary was supposed to be made up in prestige of a kind which McGowan, as a mere businessman, could never hope to attain.

What kind of chief executive were ICI getting for their money? At 56, in January 1931, Sir Harry McGowan stood at the peak of his career, for he could climb no higher and his power could be no greater. True, there was the phrase in his agreement about 'the supervision and control of the Board', but what Board was likely to supervise and control Harry McGowan? He had not risen from Glasgow obscurity by being a dutiful subordinate. He had the supreme power in Britain's largest— almost Britain's only—science-based business. What was he likely to do with it?

* The comparable figure for the Chairman of ICI in 1972/3 was £65,695.

On the technical side McGowan had never in his life been creative. Neither his talents nor his training lay that way. He had been ready enough to fall in with Mond's grand designs at Billingham, both in fertilizers and in coal-oil, but they owed nothing to his initiative. He had not impeded Mond's energetic promotion of research, though he had shown no marked enthusiasm for it and made no effort to defend it—indeed rather the reverse—when the axe began to swing in 1930. During his period of sole power he was no quicker than anyone else to grasp the significance of developments on the organic side of the business, and although he was shrewd enough to applaud the renaissance of the British dyestuffs industry, largely for its short-term benefits, he gave no encouragement to the development of plastics. He did not, that is to say, see that a technical revolution was in the making in the chemical industry. Nor did anyone else on the Board, with the possible exception of John Rogers, but surely McGowan was not paid all that money to be like anyone else.

McGowan's reputation rested on far-sighted industrial diplomacy, culminating in the formation of ICI and continuing after it in the world-wide negotiations which we have examined in earlier chapters. He dealt in power, and his favourite instruments were commercial and financial, not technical or scientific. The enabling clause in his service agreement, defining his powers of 'general direction', refers specifically to 'commercial matters' and 'financial matters'. It makes no mention of technology—nor, indeed, of labour relations. Melchett had paid great attention to both, but McGowan's imagination was never captured by technology, and labour relations, though important, could be left to the management.

What was central, what fascinated him, as it has fascinated other great exponents of private capitalism, was the manipulation of financial resources to control industrial power. Much earlier than most British business men, McGowan grasped the principle of the holding company —not a manufacturing unit but a centre of financial power—and he put it into practice first in Nobel Industries and then in ICI. When he found these companies with cash to spare, his instinct was not to stack it away, as Courtaulds and others did, in boring Government securities, but to put it into real live industrial shares which might show a magnificent capital gain. Equally, they might not, but McGowan trusted his own judgement and as long as his gambles came off he found it easy enough to carry his Board with him.

He was nearly 6 ft. tall, bald, with a florid complexion and firm, self-confident features: very much the model of a successful capitalist, especially with one of his frequent cigars in hand. He had charm, being neither loud-voiced nor rude to his subordinates, though he made sure they knew their place. For many years he had been accustomed to living high; telling outrageous, sometimes childish, stories, often in dialect, very well; and travelling—he loved travel—like a prince. He had few interests outside the business. His wife was charming and capable, on occasion, of deflating his tendency towards pomposity. His relations with his children were tempestuous.

His conduct of company business, shrewd and far-sighted, contrasted sharply with his conduct of his own, which was disastrous, but that was not apparent when he took the Chair. He had no great intellectual attainments, though an excellent memory and a quick, probing mind, yet he ruled a business with a very high concentration of scientific talent. He was a natural autocrat without, so far as it is possible to judge, the smallest trace of introspection or self-distrust, which no doubt gave him an advantage over minds more analytical and less resolute.

With Melchett gone, McGowan dominated his Board. Indeed he dominated ICI. Those who knew him regarded him with great respect, though few with affection, and it seems doubtful whether amongst all his wide acquaintance he had many close friends. Among the executive Directors there cannot have been one who did not stand in awe of him: nor was there one, probably, whom he did not at some time bully. For an incoming Chairman of ICI, however, in the oppressive circumstances of 1931, very large allowances might be made.

(ii) THE DIRECTORS OVER-RIDDEN: THE DICTATORSHIP ESTABLISHED

King Henry VII would no doubt have understood McGowan's position very well. Henry had barons, McGowan had directors, and both monarchs had to decide what to do about the power of the magnates. McGowan might act as the strict letter of his agreement required, under the magnates' 'supervision and control', or he might make himself absolute. There is no doubt which way his inclinations lay.

The directors themselves, who must have known McGowan's autocratic tendencies very well, seem to have intended to provide some check upon them, for they made Lord Reading President. Reading, ex-Lord Chief Justice, ex-Viceroy of India, briefly (August-November 1931) Foreign Secretary, had a public position which towered above McGowan's. Moreover, with his son married to Alfred Mond's daughter, he reinforced the Mond influence on the Board, otherwise only represented by Henry Mond, second Lord Melchett, a much younger man than McGowan.

The powers of the President were indeterminate, but Reading made full use of his position, working almost full-time with an office and staff at Millbank. He was virtually an executive Director, serving—unlike other 'lay Directors'—on the Finance Committee and the General Purposes Committee, which brought him to the heart of ICI's policy-making, especially in finance. This was important, because ICI's financial policy was McGowan's policy, unorthodox in relying on large holdings of industrial shares. After Reading died, in December 1935, there was no independent public figure on the Finance Committee (p. 240 below): a point which was not missed, later on, when critics of McGowan began to make themselves heard.

The Finance Committee—McGowan, Reading, W. H. Coates, Lord Melchett (Henry Mond), H. J. Mitchell, with the Treasurer, P. C. Dickens, as Secretary—was McGowan's inner Cabinet. The most influential figure, after McGowan, was Mitchell. In the days of the Nobel-Dynamite Trust Mitchell may even have considered himself a rival to McGowan, but he had long been put in his place as a most efficient Chief of Staff, working out in detail plans which McGowan had approved in outline.

Mitchell, in Alfred Mond's time, had had to fight for his primacy against J. G. Nicholson. The two men disliked each other heartily, and their rivalry provides the sub-plot to the story of McGowan's rise to supreme power.

Nicholson's strength lay on the sales side and overseas. He wished to build up a commercial department at the centre, under himself, which by direction of sales policy would control the manufacturing groups. At the same time he did his best to dominate the foreign merchant companies. If his plans had entirely succeeded he would have become much the most powerful figure on the Board, certainly much too powerful for Mitchell's taste and probably for McGowan's, too. In fact his plan was defeated, and although he continued to inspire dread among those beneath him he was not, during the thirties, in the inmost circle at the top, if that was represented by the Finance Committee. But Jack Nicholson was hard-wearing: his time was still to come.

The strength of McGowan's Board was commercial and financial, not technical. McGowan himself and Mitchell were experts in industrial diplomacy and financial engineering, and McGowan's force found a complement in Mitchell's grasp of intricate detail, administrative as well as financial. Coates had a precise, pernickety mind and considerable powers of lucid expression which he loosed on his fellow-Directors in long, donnish papers on The World Economic Situation and other wide-ranging topics. It is to be hoped his close-knit arguments were read. Lord Melchett, like his father, had a well-informed grasp of company policy and applied himself particularly, again in the family tradition, to labour relations. He was young (b. 1898), popular, respected, and already beginning to appear as a possible heir-presumptive.

The full-time Directors not on the Finance Committee, apart from Nicholson, were B. E. Todhunter, J. H. Wadsworth, John Rogers, and G. P. Pollitt. Todhunter's interests were chiefly Australasian. Wadsworth, an amiable, mild-mannered lawyer, linked the Board with the

Central Administration Committee. That leaves the only two Directors with a scientific background: Rogers and Pollitt. 'Johnny' Rogers, three or four years younger than McGowan, had come up with him, a pace or two behind, from Nobel's Explosives Company, which he joined at Ardeer in 1898. His influence with McGowan he used chiefly as a conciliator and pacifier: the still centre of any storm that might be blowing, but unlikely to oppose the force of the storm itself. Pollitt, bluff, tactless, masterful, had been elevated from Billingham to the ICI Board on its formation. He was never entirely happy or effective there— his talents were suited rather to general management than to the duties of a Director—but it is a reasonable speculation that of all the executive Directors in 1931 he would be most likely, or least unlikely, to speak his mind to the Chairman without inhibition.

Before McGowan came to the Chair, events were already running his way: that is, towards the exaltation of the Chairman at the expense of the Directors. The 1927 organization, highly centralized and entrusting considerable executive power to individual Directors, may have been necessary to get the merger going, but as a permanent system it turned out to be unworkable. 'The existing system', wrote Wadsworth in June 1929, ' . . . has the very great disadvantage of absorbing by far the greater part of the energy and time of the Executive Directors in matters of detail affecting each of the many individual Companies. . . . The result . . . has been that the Executive Directors simply have not had the time to give as much assistance to the Chairman and the President in the study of major questions of policy as they should be able to give.'[6]

The remedy, as everyone could see, was to relieve the Directors of 'matters of detail' so that they could become advisers on 'major questions of policy'. In becoming advisers rather than managers they might lose their grip on the levers of power. That was not likely to be an unwelcome prospect to McGowan and it certainly was not unwelcome to H. J. Mitchell. His theory of the function of Directors he stated explicitly:

So long as the adjective 'Executive' precedes 'Director', there is a natural tendency for Directors . . . to be Executives rather than Directors. . . . Surely in plain English the true executive part of the organisation should be the senior officers of the Company, who should carry out the instructions of the Directors.[7]

The idea that the king should rely on royal officials, rather than on the baronage, would not have been unfamiliar to Henry VII either.

Movement in this direction began as early as 1928, with a reorganization of the Metal Group, which had a personality quite distinct from the rest of ICI and a tradition of local independence running back to

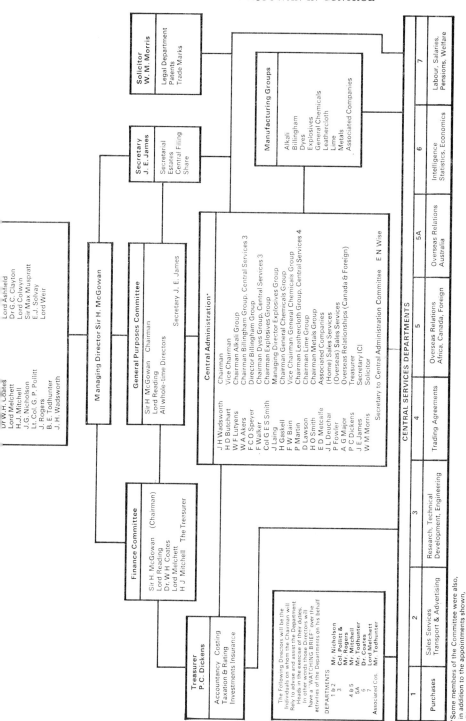

the time of Nobel Industries.[8] It had been enlarged by purchase soon after the merger (pp. 13–14 above) and drastically tidied up, leaving only four 'major companies' in the Group. In 1928 one Delegate Board and one Management Committee were appointed for all companies in the Group instead of one for each, which greatly reduced the duplication of work and the amount of detail which arose for the ICI Directors who belonged to both bodies.

This idea was seized upon by Pollitt. He suggested that the chlorine and acid producers in ICI should be similarly grouped.[9] Matters did not stop there. In June 1929 a scheme was put into force by which all the home companies of ICI were formed into eight groups (see Chart, Fig. 5) on the basis of similarity of products, and in this way the organization of Metals Group became a pattern for the organization of ICI as a whole.[10]

As the members of the main Board withdrew from close acquaintance with day-to-day management, officials, according to Mitchell's theory, began to move in. To run each new group an Executive Board was set up, including local management and 'senior . . . Headquarters staff'. All Executive Directors were members *ex officio* of every Executive Board, but the mere fact that they were members of *every* board made their membership a token and it was made clear that neither they nor their alternates were expected very often at meetings. 'The present purely managerial functions of the Executive Directors of ICI', said the document announcing the Group System, 'should almost immediately lessen by virtue of the scheme generally and particularly by the intro-duction of their senior officers formally to responsibility.'[11]

In the summer of 1930 Melchett and McGowan went to the USA and Canada. As they left they announced another move, worked out for them by Mitchell, with the declared object of freeing the executive Directors from 'some of their detailed duties' and of placing 'consider-ably greater responsibility' on 'Senior Officials of the Company'. They set up an Administrative Committee of fifteen, including the Secretary, Treasurer and Solicitor, but no Directors, 'to consider such matters as may be remitted to it by us or by the [newly proposed] Board of Management'. The Committee was to have 'power to take decisions on all matters of routine which do not involve the Company in contractual obligations, or which do not raise any questions of policy', and it was allowed to pass expenditure up to £5,000 and to fix salaries up to £1,000 a year.[12] These were large figures. The limits allowed to Group Executive Boards at the time were £2,000 for capital expenditure and £400–£600 for different classes of salaries.[13]

'Board of Management' notwithstanding, the lines of communication of this powerful Administrative Committee went straight to the two Managing Directors, leaving the rest of the Board in an ill-defined limbo. 'Owing to our early departure . . . ', said Melchett and McGowan with regal vagueness, 'the method by which we hope to derive the greatest assistance from our colleagues cannot be stated in full at the moment.'

The Group System and the Administrative Committee became the central features of an organization designed to concentrate executive authority in the office of Managing Director. When Melchett died the new organization was still being discussed, but it was brought very speedily to full development after McGowan became Chairman and sole Managing Director, giving him all the apparatus he needed for the exercise of supreme power.

The document which brought the new organization into effect was drafted as a memo to the Chairman by H. J. Mitchell late in January 1931, scarcely a month after McGowan had taken office. On 19 February Mitchell's memo, almost unchanged but this time over McGowan's signature, issued with other papers from the Chairman's office covered by a letter from McGowan addressed individually to the Chairmen and Managing Directors of Groups and to senior Head Office officials, twenty-four in all. The covering letter was uncompromisingly autocratic. McGowan made no mention of the Board and, having announced 'draft memoranda on the future organisation of ICI', went on in the first person singular throughout:

I propose that under this scheme you should occupy the position of . . .

It is my intention to issue these documents [the 'draft memoranda'] as instructions for the future conduct of the affairs of the Company, but before doing so I shall be glad to know whether they are quite clear. . . .

I do not want the documents to become public property; therefore you will restrict any discussion with your colleagues to those in whom you have complete confidence.[14]

The main document, drafted by Mitchell but edited in an egocentric direction by McGowan, rehearsed the disadvantages of committee rule which were said to have emerged from the system of Executive Boards, set up in 1929, and went on:

After consideration of many schemes I have come to the conclusion that any new organisation . . . should delegate all possible authority and responsibility to the Group Delegate Boards and should liberate the whole-time Directors of ICI from direct executive control of individual Groups, with the corollary that the various Departments at Millbank shall function as Service Departments and not as Executive Departments as hitherto.

This basic principle was put into effect, both in the Groups and at Head Office, by doing what Mitchell had suggested: that is, by removing all Directors except the Managing Director from the executive function, except by way of committees of the Board, to which we shall return.

The new model Group Boards, all ICI Directors finally and un-equivocally got rid of, were made up chiefly of local managers, and had each a Chairman who sat in London as a link with Head Office and a Managing Director 'located in the provinces or London, as may be most convenient for the direction of the day to day affairs of the Group'. At Head Office, managers, some of whom were also Group Chairmen, were appointed to run the service departments. Individual Directors remained associated with specified departments, but a note on an organization chart of 31 March 1931 (p. 139) calls them 'individuals on whom the Chairman will rely to advise and assist the Departmental Heads in the exercise of their duties. In other words those Directors will have a "WATCHING BRIEF" over the activities of the Departments on his [the Chairman's] behalf.' Advice and assistance can, of course, be very forcefully tendered, but the retreat of the Directors from management could hardly have been more plainly signalled.

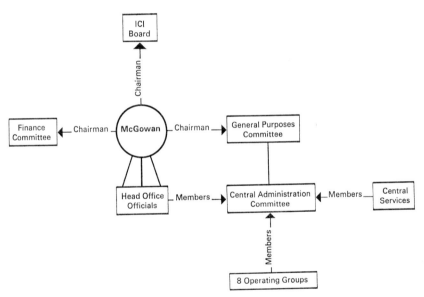

Figure 6. The Mechanics of Dictatorship

The organization was tied together at the top by a system of com-mittees. The Group Chairmen, the central officials (Secretary, Treasurer, and Solicitor), and some of the departmental managers had seats on the Central Administration Committee (formerly the Adminis-trative Committee) with a Director—J. H. Wadsworth, newly pro-moted from Secretary—as Committee Chairman. Above the CAC were

Pipe-bridge at Billingham, 1932.

Some products of Billingham, 1932.

two Committees of the Board: the General Purposes Committee, of all the full-time Directors, and the Finance Committee, much more select. McGowan was Chairman of each, as he was Chairman of the Board as a whole.

McGowan, as sole Managing Director, emerged as the only Director with personal executive powers. The others could only exercise power as a body, either through the Board as a whole—a cumbersome instrument of last resort—or through the Finance Committee and the General Purposes Committee, over each of which McGowan presided. Executive power had been delegated to the members of the CAC, but strictly in subordination to the General Purposes Committee which meant, in effect, to McGowan. From wherever you looked—from the Board Room, from the Treasurer's or the Secretary's office, from the factories, even from overseas—all the lines of power converged on one man: Sir Harry McGowan. The king was supreme over his barons, or, to shift the parallel, the dictatorship had begun.

The dictatorship was most oppressive at the top, in the dictator's own circle. For the Groups, the 1931 organization was something in the nature of a charter of liberty. True, buying, labour policy, finance, and selling were all controlled from Millbank and the Groups were adjured to use 'the Central Services available in Head Office [and] no other . . . unless by authority of the GPC'.[15] On the other hand the document setting up the system[16] insisted that the central service departments were to be advisory rather than, as formerly, dictatorial, and McGowan, speaking in 1935, said that 'the period of rigid centralisation' had only lasted for the first four years of ICI's existence.[17]

In one field of central interest to them, technology, the Groups had very considerable freedom. In theory, with their independent power of capital expenditure limited to £2,000, it might seem that the central control was very tight. In practice, since technology was not in McGowan's main field of interest and he was not qualified to criticize proposals on technical grounds, control was nominal. It was essential, of course, to keep within the limits of general ICI policy, and during the depression years, as we have seen (p. 118), capital spending almost ceased. Apart from that no Group Chairman with the most elementary political skill—and he was not likely to be a Group Chairman if he lacked that—need fear that a proposal which he considered sound would be rejected centrally. We shall see that the technical progress of ICI in the thirties, at the height of McGowan's dictatorship, depended far more on ideas sent up from below than on those handed down from above.

'The plain fact', as one Alkali Group director of the thirties wrote in 1971, 'was that no individual, whether he wielded supreme power or not, could conceivably have enough time and energy to decide on the vast array of problems that had to be dealt with.' It is a measure of McGowan's skill as an absolute ruler that he succeeded in freeing himself from an immense weight of managerial detail while keeping in his own hands those matters, particularly in major questions of manufacturing policy, in the regulation of competition, and in finance, which he considered of supreme importance to the welfare of ICI as a whole. In these matters, so far as one man could, he ran the business: to what effect, the following chapters will discuss.

REFERENCES

[1] ICIBM 1969, 31 Dec. 1930.
[2] J. E. James to the President and W. H. Coates, 30 Dec. 1937, SCP.
[3] *ICI* I, p. 390.
[4] SCP.
[5] Robert Jones and Oliver Marriott, *Anatomy of a Merger*, Cape, 1970, p. 87.
[6] J. H. Wadsworth, 'Notes on Reorganisation of Management of Subsidiary Companies', 6 June 1929, CR 93/5–11/R.
[7] H. J. Mitchell, 'Organisation', 1930, CP 'Organisation (6)'.
[8] *ICI* I, p. 393.
[9] G. P. Pollitt to Executive Committee, 'ICI Chlorine and Acid Group', 1 Oct. 1928, ECSP.
[10] As (6); see also 'Management and Control on a "Products" Basis', 31 Dec. 1928, p. A3, CR 93/2–2/ .
[11] As (6), pp. 2–3.
[12] Chairman and President, 'Organisation', 24 July 1930, CP 'Organisation (6)'. Based on (7).
[13] Brunner, Mond's 'Charter', CP 'Organisation (3)'.
[14] H. J. Mitchell to W. H. Coates, 'Organisation', 19 Feb. 1931, covering letter quoted and documents, CP 'Organisation (6)'.
[15] Ibid.
[16] Ibid.
[17] 'Ashridge—Address by Sir Harry McGowan . . . on the Organisation of a modern large-scale Unit in Industry', 7 Dec. 1935, Bunbury folder 14.

CHAPTER 9

The Nitrogen Cartel and the Transformation of Billingham

AT THE BEGINNING OF 1931, when McGowan became Chairman, ICI, along with IG, was facing the fact that the world-wide slump in agriculture had undermined the whole basis of the decision which both had taken to invest heavily in fertilizer plant designed to meet a rising export demand. Two interdependent tasks were urgent. One was to regulate the international trade in nitrogenous fertilizers to cope with the consequences of over-capacity. The other was to save as much as possible from the idle plant at Billingham and get it going again on a profitable basis.

The situation was bleakly sketched, in November 1930, by Dr. Hermann Schmitz of the Central Committee of the Board of IG. He told F. C. O. Speyer, of ICI, that he thought both IG and ICI 'would have to recognize that a fundamental mistake had been made in increasing our respective [fertilizer] plants to the extent to which we had increased them'.[1] The truth of his observation may readily be demonstrated from the statistics of the world nitrogen industry. Productive capacity grew, between 1930 and 1933, from 2·8m. tons (metric) of nitrogen to 3·3m., but over the same period the quantity actually produced dropped from 1·7m. tons to 1·5m.* Works based on synthetic ammonia were estimated to be working at 42·5 per cent capacity in 1930/1, and 37·5 per cent in 1931/2. There could hardly be a more dramatic illustration, in statistical form, of the impact of depression on industry.

IG had far more nitrogen to sell than anyone else in the world. Along with their own, they marketed the output of the *Stickstoff Syndikat*, representing most of the German makers of by-product ammonia, and the combined productive capacity, by 1933, was almost 1·3m. metric tons of nitrogen a year. ICI, very much larger producers than anyone else in Europe but very much smaller than IG, came next. They, like IG, marketed the output of by-product makers, and the combined capacity, by 1933, was about 275,000 metric tons of nitrogen.

---

* These figures take account of synthetic and by-product ammonia (2·4m. tons capacity 1930/31; 3·0m. 1932/33—production figures 1·2m. in each year), and of nitrogen produced by the cyanamide process and in Chile nitrate.[2]

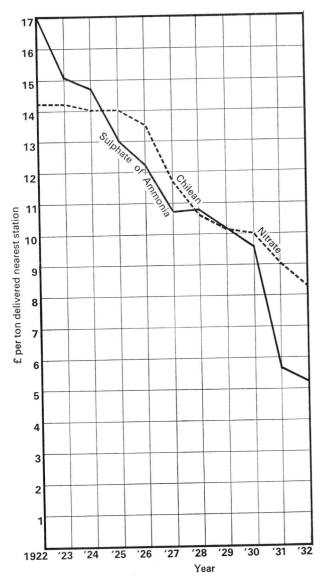

Figure 7. Selling Prices of Chile Nitrate and Sulphate of Ammonia, 1922–1932

Source: F. M. Irvine, 'Notes on the Present Position of the Chile Nitrate Industry', MP 'Nitrogen, 1932–33'.

Norsk-Hydro, with a capacity of 98,000 tons, was controlled by IG, and the German, English, and Norwegian makers, known (presumably from the German form of the national adjectives) as the DEN Group, usually acted together in the affairs of the world's nitrogen industry, being far and away the largest group in it.[3]

In the severely depressed nitrogen industry of the early thirties, size meant misery. The enormous productive capacity of the DEN Group was not a source of strength: indeed, quite the reverse. DEN's works had all been planned to serve a flourishing export trade, but after the slump set in export trade was not only not flourishing, it was very hard to come by at all. Producers of synthetic ammonia had come into being in every country of consequence in the world, and usually their Governments could be persuaded to protect them against foreigners. Consequently, as Schmitz put it to Speyer, '[IG] are pessimistic about their and our [ICI's] ability to maintain an export trade in nitrogen owing to the facilities which exist now in practically every country for putting up nitrogen plant and to the manifest desire of every country with a large population to be self-sufficing as regards nitrogen.'[4]

Nor was this all. On top of the very large potential output of synthetic nitrogen compounds there was the natural production of Chile nitrate, almost all of which had to be sold on world markets if it was to be sold at all. Nitrate represented a large part of the wealth and earning power of Chile. A good deal of the capital was held by foreigners, particularly the American Guggenheim group, who were no better liked in Chile than foreign capitalists usually are anywhere, especially in bad times. The pressure of hostile opinion, added to their own self-interest, made them extremely tough negotiators, and the Chile nitrate interests were unanimously regarded by the world's producers of synthetic nitrogen as the common enemy. They posed a sizeable threat, for the Chile industry's capacity, though much reduced over a period of years, was still put at 100,000 tons of nitrogen in 1932/3, and it was generally assumed that as much as could be produced would in fact come to market.

The situation described in the last two paragraphs meant that in the early thirties those producers of nitrogen were happiest, or least miserable, whose capacity did not much exceed the demand from the home market. By this standard the Dutch producers, among whom Royal Dutch-Shell was represented, were very well off, for even in an extremely bad year, 1931/2, they could keep their works nearly 90 per cent occupied. The Russians, the Swiss, the Czechs, and the Japanese were fairly well placed also. IG's works, on the other hand, were down to 30 per cent of capacity in 1931/2, and although Billingham, at the same time, was running at about 55 per cent, ICI's position was fundamentally weak. Along with the Belgians and the Consolidated Mining and Smelting Company at Trail, British Columbia, ICI were said by Speyer to share 'the unenviable distinction of possessing practically no market which [they] can call [their] own'.[5] The by-product makers alone could supply about twice as much nitrogen as British farmers needed, before Billingham's output was considered at all.

Whether or not continental producers had a fairly secure home base, they saw no reason to forgo export trade if there was any to be had. 'Belgium wants to export', Mitchell told Melchett in July 1930, and went on: 'Italy and Czechoslovakia all say if one has export why should we not have some. Poland wants to export more than we want them to.' All these people, said Mitchell rather bitterly, had 'nice remunerative home markets'.[6] The overriding interest of the DEN Group, in these circumstances, was to form a cartel, or cartels, in the export trade so that it would provide a living for everybody, especially themselves. The interest of the smaller producers in such an arrangement was strong but rather less compelling, because of the security offered by protected home markets. The Chilean producers, great though their need was for export trade, were extremely difficult to bring to terms. The difficulties in the way of the would-be cartel-makers were thus considerable. So important, however, to ICI and IG was a controlled market in nitrogen that throughout the thirties they never gave up striving to establish it. Their success varied. Their tenacity did not.

The Anglo-German allies turned their attention to the problem as soon as their own agreement was settled in February 1930. During the spring and early summer there was intense diplomatic activity until in August, in Speyer's words, 'after five weeks of arduous negotiations at Ostend [in the Hotel Splendide] and in Paris complete agreement was arrived at between the synthetic producers in France, Belgium, Holland, Poland, Italy, Czechoslovakia and Great Britain'. The Chilean Government and the Nitrate Producers' Association also came in.[7]

The settlement was in three parts. First, there were separate agreements between DEN and each of the producers in the countries listed by Speyer. Secondly, all these parties joined in a general agreement known as the Convention de l'Industrie de l'Azote (CIA) which provided for the creation of a common fund, to which all the signatories contributed, and which also laid down that exports to countries outside the CIA were to be made by ICI-IG or with their consent. Thirdly, ICI-IG came to an agreement with the Chilean producers under which the Chileans agreed to contribute to the CIA fund one-third of the amount contributed by the signatories to the Convention, with a maximum of 15m. marks (£750,000).[8]

The central purpose of all these agreements was to get producers to reduce their output 'in order to relate production to consumption', and to make it worth their while to do so. The underlying principle was 'that each country which is a party to the agreement has the right to supply its full home consumption out of its own production'. IG and ICI stood to lose far more than anyone else by accepting the principle of national self-sufficiency, but there was nothing else they could do. As Schmitz put it in 1931, 'it was the lot of the old established nitrogen industries to have to surrender part of their share to the newcomers'.[9]

Each synthetic producer who came into the agreement was required to accept a sales quota and to contribute to the common fund at a rate of 35s. ($£1·75$) per ton of nitrogen. By-product makers were politely asked for 25s. ($£1·25$), 'although Imperial Chemical Industries Limited recognise that there is no legal liability whatsoever . . . to contribute'. By these means, and with the Chilean contribution, it was hoped to raise £3m. which could be used to compensate makers who had to cut their output below an agreed level (70 per cent of capacity), to buy surplus stocks, and 'to compensate parties to the agreement who suffer from invasion of their home markets by free lances'.[10]

The reward for keeping to the terms of the agreement was to be a guaranteed price level for 1930/1 'which is not to differ greatly from last season's level'. Just before the agreement was signed prices were plunging downward. Continental makers had been offering sulphate of ammonia at prices which left them with 'a bare £4 per ton at works', and it looked as if, in the absence of an agreement, prices would have gone much lower than that. 'The agreement', said Speyer when it was concluded, 'has undoubtedly saved the nitrogen market from a complete collapse.'

The agreement of 1930 was limited to the year ending 30 June 1931, with the intention that while it lasted a new agreement should be worked out to cover 'a long term of years'. It was one thing, however, for ICI and IG to agree among themselves, as they had done, to regulate their affairs for ten years. It was quite a different matter to gain agreement between all the members of the CIA and the Chileans as well, and it is a mark of the direness of the times that the 1930 agreement itself could be negotiated. When it ran out, there was nothing remotely ready to take its place. Free competition broke out, the market collapsed, and the price of sulphate of ammonia in the United Kingdom fell from £9 10s. ($£9·50$) a ton in the spring of 1931 to £5 10s. ($£5·50$) a year later.[11] The British market, in spite of an appeal from McGowan to the President of the Board of Trade in the summer of 1931,[12] lay wide open to 'dumping' until a 20 per cent duty was imposed in April 1932.

This dose of classical economics was very unpleasant for producers heavily dependent on the export market, such as IG, ICI, and the Chileans. Sales of Chile nitrate, in particular, fell from the equivalent of 244,300 tons of nitrogen in 1930/1 to 114,000 in 1931/2.[13] For those who had a 'nice, remunerative home market', well protected, however, the situation had its attractions, because they could threaten the large exporters quite credibly with very cheap sales abroad. The smaller European producers, therefore, were in no great hurry to settle until they could get considerable concessions, in tonnage and cash compensation, from the DEN Group.[14]

Nevertheless, by July 1932 the Convention de l'Industrie de l'Azote was active again, and up to the outbreak of war it remained in precarious control of the nitrogen industry, which was regulated partly by the ten-year agreement between ICI and IG and partly by a network of successive short-term agreements between the various groups in the business. No long-term comprehensive agreement, of the kind envisaged in 1931, was ever negotiated. Probably it was impossible unless virtually the entire nitrogen industry of the world, or at least of Europe, were to be brought under common ownership and central control. Some such idea, which fitted well with the general gigantism of IG's outlook, was certainly in Schmitz's mind in 1931,[15] but it was never practical politics. Apart from anything else, it would almost certainly have led to the closing of Billingham, which IG more than once suggested. ICI's response was 'Why not close Merseburg?'[16] and it is difficult to believe that the British Government, or any other, would have allowed measures of this kind in the conditions of the thirties.

CIA was run by a committee who usually met in Paris or agreeable continental watering-places (also favoured by the CIA for its annual conferences). ICI's principal representative in these matters was Ferdinand Speyer who, as a cultivated Jew, is said to have derived considerable satisfaction from indulging his taste for opera at IG's expense. Since the committee was not a corporate body, but had nevertheless to deal with large sums of money, it was thought advisable to set up a Swiss company with capital of six million francs, jointly held by the members of CIA. This was the Compagnie Internationale de l'Industrie de l'Azote SA of Basle, managed by Dr. Jacobi of IG. Jacobi was a Jew, and as the outlook grew grimmer in the late thirties means were found to have him transferred to England. The Committee and the Company were both established by the terms of the agreement of 1930, and both survived the 'fighting', as it came to be called, of 1931/2. They became the permanent regulating bodies of the world nitrogen industry of the thirties.

By 1935, then, ICI were operating in the world nitrogen industry as a member of the DEN Group within the CIA. The foundation of ICI's policy was the ten-year agreement with IG, which was the foundation charter of DEN. 'The main object of the Agreement', as it was recorded in 1932, '[was] to secure the largest possible sales for the German-English-Norwegian group.'[17] Sales quotas for the two major parties, ICI and IG, were worked out on the basis of total productive capacity within the group, including the by-product capacity of the British Sulphate of Ammonia Federation with ICI's own capacity, and the capacity of the *Stickstoff Syndikat*, as well as of Norsk-Hydro, with IG's. The tonnage ratio between ICI's sales quota and IG's, as settled for three years when the agreement was revised in 1932, was $18\frac{1}{4}$ ICI to $81\frac{3}{4}$ IG.[18] The penalty for overselling was an obligation either to buy goods from the other party or to make a cash payment.[19]

A tonnage quota of this kind, backed by penalties but without territorial restrictions, except the reservation of the home market, was the method always preferred by IG and by German firms generally for regulating competition. ICI, on the other hand, always strove for agreements with a territorial basis, under which, in particular, their exclusive right to markets in the British Empire would be recognized, and this was the principle of their agreements both with Solvay and with du Pont.

These methods of avoiding competition were not easy to reconcile with each other, but the ICI-IG nitrogen agreement succeeded in embracing both. As well as the tonnage quotas, ICI and IG divided world markets between them and maintained selling organizations as follows:

ICI:
British Empire and mandated
   territories
Dutch East Indies [Indonesia]
Spain, Portugal, Canary Isles

IG:
Europe, except UK, Spain,
   Portugal, Canaries
Mexico, Guatemala, San Salvador,
   Costa Rica
Brazil
Asiatic Russia
Philippines

Certain markets they shared:

China	51% ICI,	49% IG
Egypt	20	80
Japan	40	60

but in 1935 it was noted 'in view of ICI's present oversold position the Japan quota is not adhered to'. Spain, Portugal, and the Canaries were granted to ICI largely by virtue of old Nobel connections in the ex-plosives industry, and were valuable markets, largely because of fruit and olive growing in Spain. In countries not specifically mentioned, each party had the right to maintain a selling organization. They kept clear of the market in the USA, partly because of Allied Chemical's power, if provoked, to strike back in Europe, and partly because of ICI's alliance with du Pont, although Speyer once reported 'almost tearful protestations' by Jasper Crane, of du Pont, that ICI 'entirely fail to take seriously du Pont's intention to enter the fertiliser field'.[20]

ICI and IG, acting together as the DEN Group, made separate agreements with the other members of CIA, who were 'Belgian, French, Dutch, Italian, Polish, Swiss and Czecho-Slovakian producers, and treated as a separate unit Sluiskill works'.[21] The works at Sluiskill, in Holland, belonged to the joint Belgo-Italian Coppée-Montecatini Group, and caused grave perturbation to DEN in export markets. 'The bases of the Cartel' (J. G. Nicholson's words) were said to be:

(a) The reservation of home consumption to national production.
(b) The allocation of export tonnage to open markets [i.e. those without 'national production'].
(c) The establishment of a common fund.

These objects were only achievable by hard bargaining within CIA and between CIA and outsiders, particularly the Americans and Japanese. Moreover the bargaining was almost continuous since most of the agreements, unlike the ten-year agreement between ICI and IG, covered one 'fertilizer year' (starting in the spring) at a time. The general basis for calculating export quotas depended partly on DEN export sales of ammonium sulphate and partly on DEN total export sales, but this was purely nominal since the manner in which each national group's affairs were conducted depended on the terms individually negotiated with DEN, which reflected the relative bargaining strength of the parties.

Sluiskill, strongly placed, were outside the ordinary quota arrangements altogether. They had a fixed quota of 15,000 tons of nitrogen and a right to buy 'additional quota' (Nicholson's phrase) from the Italian, Belgian, and Czech groups. The Belgians also, under certain conditions, had a right to buy 'additional quota' and the Italian and French groups were granted extra export quotas which they ceded to DEN in return for DEN paying the Italian and French contributions to the Common Fund. DEN acted as selling agents for the Belgian, Czech, Polish, and Swiss groups, but the Dutch, French, Italian, and Sluiskill groups sold their own, though they occasionally asked for DEN help. Some groups had fixed quotas in Spain and Portugal. In all export markets prices were 'agreed'.

CIA negotiated agreements with the Chileans, granting them a percentage in export markets and also in the home markets of CIA producers, and with the Japanese group also they negotiated agreements which for 1935 limited the Japanese to an export tonnage of 12,000 tons of nitrogen and regulated prices and markets. CIA as a whole, it appears, bought part of the output of Allied Chemical's

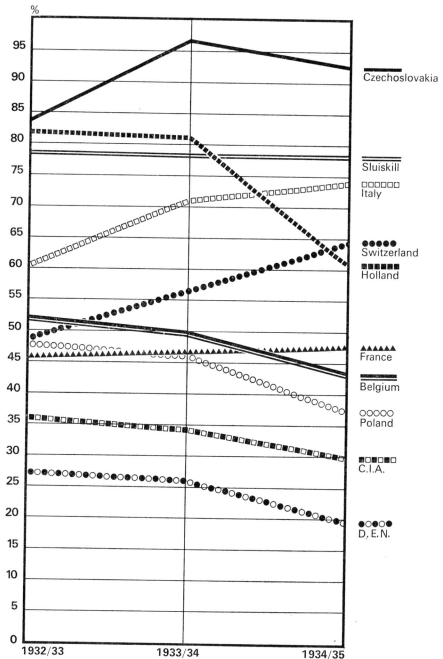

Figure 8. Underemployment of Nitrogen Factories. Percentage capacity working of CIA synthetic nitrogen factories

Source: GPCSP 25 March 1935.

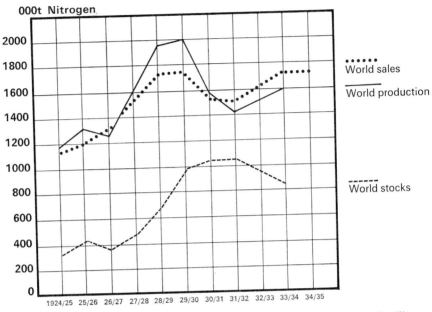

Figure 9. World production, sales and stocks of all nitrogenous fertilizers 1924/25–1934/35

Source: GPCSP 25 March 1935.

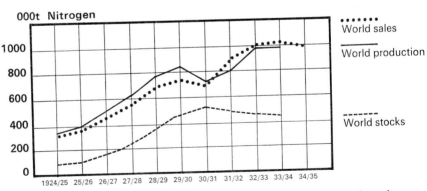

Figure 10. World production, sales and stocks of synthetic nitrogen 1924/25–1934/35

Source: GPCSP 25 March 1935.

'Arcadian Nitrate'—presumably enough to prevent Orlando Weber invading Europe.

All these operations required money. It was provided by the Common Fund, without which CIA would have been impotent. Members paid into it at the rate of one gold pfennig per kg. of nitrogen sold at home or for export, and the Chileans, in 1935, were contributing a million marks. The sums which the CIA Committee thus came to have at its disposal were large, running well into seven figures of sterling, which alone is sufficient to indicate the value the members of CIA placed on the elaborate, if sometimes creaky, machinery described above.

And what exactly was the CIA doing for its members, up to the mid-thirties, to earn its expensive keep? Briefly, it was regulating their export trade in nitrogen so that it ran down in an orderly way instead of collapsing uncontrolled. At the same time, it was asserting the interests of the smaller national producers at the expense of ICI and IG. Both points are illustrated in the following figures:[22]

CIA Export Trade in Tons of Nitrogen

	CIA Total Tonnage	DEN Total Tonnage	DEN Percentage
1930/1	399,300	340,200	85
1931/2	432,100	344,900	78
1932/3	374,200	284,700	76
1933/4	357,600	262,950	73
1934/5	270,900	190,700	70

NB: Figures for 1930/1 and 1931/2 are from a different source from the rest and may not be on exactly the same basis.

As Nicholson said in commenting on the figures from 1932/3 onwards, they 'show the effect of the concessions which have been forced on the DEN year after year in order to maintain international nitrogen relations'.

These concessions had the effect, by 1934/5, of bringing the working of 'synthetic' factories in the DEN Group below 20 per cent of capacity, at a time when the Czechs were working at over 90 per cent and in Holland, Italy, and Switzerland they were working between 60 and 80 per cent (Figure 8). The effect was worst of all in the great plants belonging to IG, but in ICI also they were desperate to get redundant plant going again, even, if necessary, at very low profitability in petrol production (pp. 171–2 below).

The reason why these deplorable conditions were in any way toler-able is revealed by the figures for production, sales, and stocks, from year to year, of nitrogenous fertilizers. They show (Figures 9 and 10) that after the formation of CIA sales of nitrogenous fertilizers overtook production so that, although both were rising in the early thirties, they

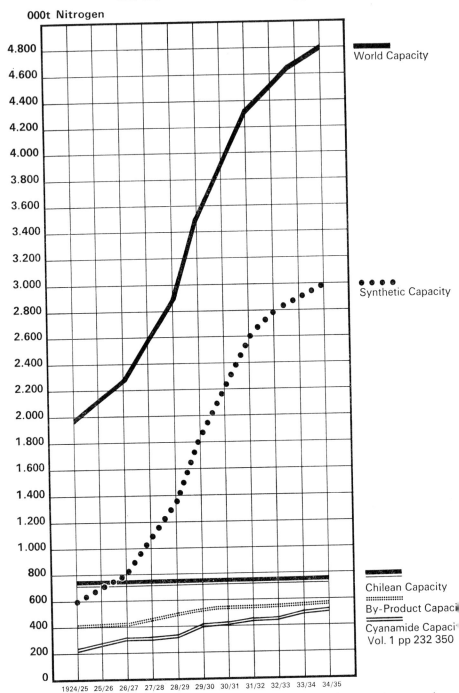

Figure 11. Nitrogenous fertilizers, world production capacity 1924/25–1934/35
Source: GPCSP 25 March 1935.

were not adding to stocks. Stocks, indeed, were falling, which meant that although over-capacity in the world's nitrogen industry had not been got rid of (capacity, indeed, incredible though it may seem, was still being added to (Figure 11)), yet over-production had been brought under control. It was for this relief, in broad terms, that ICI and other CIA members were willing to pay their contributions to CIA funds. ICI's figures for 1932/3 and 1933/4 were:[23]

	Contribution to CIA Fund*	Special Payment as Compensation	Total ICI Payment	Cost per ton N sold ex Billingham
1932/3	£63,603	£87,758	£151,361	£1 14s. 0d. (£1·70)
1933/4	£64,153	£74,392	£138,545	£1 19s. 0d. (£1·95)

On these figures Nicholson commented: 'The benefits ICI have obtained in freedom from competition and consequent higher price levels have fully justified these apparently heavy cash payments.' The special payment, he added, was ICI's share of 'export quota' purchased by the DEN Group from Belgium, Italy, etc. These sums were additional to payments for non-production from the CIA fund.

Nicholson meant that DEN, if pressed, would buy unwanted fertilizer rather than let it be thrown on the export market at an inconvenient price and that CIA would use its funds to pay producers not to produce. These manœuvres are so similar to those which many governments, with general approbation, had recourse to in the thirties, especially in support of agriculture, that it is hardly surprising to find representatives of IG, in 1932, pointing out 'that so far all Governments had respected international trade Agreements'.[24] CIA was one of the big international cartels of the depression years. In its day, its fitness for purpose was not widely doubted.

The existence of the cartel, nevertheless, was a permanent reminder that the nitrogen industry of the thirties was growing less and less likely to be what the men of the twenties had hoped. Plant at Billingham stood idle from year to year, other activities on the site grew while fertilizers declined, and it became more and more evident that a permanent change had come over world markets which would have to be reflected in permanent change at Billingham.

In the fertilizer business heavy and continuous pressure was brought

---

* The by-product makers (British Sulphate of Ammonia Federation) also contributed, their payments going through ICI, as follows:

1932/3 Federation payment £21,918 out of total British payment to CIA £85,521.
1933/4   £28,525   £92,678.

to bear to get costs down, especially in selling and especially overseas, where selling costs per ton showed a tendency to rise alarmingly. Sums spent abroad on propaganda and publicity dropped from £119,006 in 1930 to £64,014 in 1934. Publicity expenditure at home over the same period fell from £31,450 to £15,920.[25]

The agricultural research centre at Jealott's Hill paid a heavy penalty for being associated with sales propaganda (p. 105 above). It always had enemies, especially H. J. Mitchell, and Alfred Mond's death at the end of 1930 removed its most powerful protector at the worst possible moment. Sir Frederick Keeble's imperial dreams (pp. 104–5 above) faded in the bleak daylight of 1931. He found himself faced by McGowan, prompted by Mitchell, with a demand to bring his budget down from £65,000 to £30,000. That could only mean, as McGowan said, 'drastic depletion of staff'.[26] Keeble retired, no doubt bitterly disappointed, at the end of 1932. Expenditure at Jealott's Hill fell from nearly £50,000 in 1930 to £28,700 in 1934, and the work done there was severely restricted.[27]

All this severity did no more than bring down the cost of sales from plant which was chronically underemployed. It did nothing—it could do nothing—to cure the basic problem of over-capacity itself. That showed up in the books, year after year, as unproductive, over-valued assets, and in the spring of 1931 Sir William McLintock, as ICI's auditor, caused a crisis over the publication of the annual accounts by objecting to the value at which the Board proposed to take Billingham assets into the Balance Sheet.[28]

From that year onward, partly at the prompting of the auditors, ICI began a massive programme of writing off. In 1931 the capital invested at Billingham was put at £22,358,823. From 1931 to 1934, both inclusive, £4,549,587 was written off, and it is safe to assume that nearly the whole of this sum will have represented plant required in one way or another for fertilizer production.[29]

Over the same period £4,292,653 was invested in activities having nothing to do with the fertilizer business but related to the technology associated with it, especially the production of hydrogen. It was steadily becoming clear that the damage done by the slump was final; that the plans for the fertilizer industry laid in 1926–7 would never be realized; and that if the great complex of chemical plant at Billingham were to survive at all its functions would have to be radically altered.

Early in 1935 J. G. Nicholson, accompanied by A. R. Young of the Treasurer's Department, carried out a visitation. 'It seems impossible', they reported, 'ever to expect an adequate return on much of the capital involved', and they recommended writing off £4·7m. of the £7·5m. still employed in fertilizer plant. They reported on the management, too. The organization, they said, was weak, 'due largely to lack of direction', concluded that it lacked leadership and advised 'an infusion of new blood'.[30]

A management revolution was already going on, heralded, perhaps, by a portent. In 1934 George Pollitt, virtually the creator of Billingham, ceased to be an Executive Director of ICI (he remained on the Board) and took to farming. He was 56, full of health and energy. Why then did he go? He never confided in anyone, so far as we know, but it is permissible to speculate. All his forecasts for Billingham had gone wrong, all his plans had been overthrown. He could be regarded as the author of a gigantic blunder. It seems overwhelmingly likely that disappointment may have been one of his strongest motives, if not his strongest, for leaving a position in which, in any case, he had never been entirely happy. If it was so, then one of the side results of the slump was to drive George Pollitt from the centre of ICI's affairs.

The reconstruction of the Billingham management centred on the transfer of the Managing Director, R. E. Slade. He went temporarily to London in 1934 to report to McGowan on ICI Research (p. 90), and he stayed there as Research Controller. He was succeeded at Billingham by A. T. S. Zealley (1893–1970) who remained there, latterly as Chairman, until he became an ICI Director in 1951. There were other changes, on the Board and below, in 1935 and 1936; and finally, in 1937, the Chairman, W. A. Akers, who had been in office since 1931, was succeeded by Alexander Fleck.

Well before that, the remodelled management had asserted their independence. 'Much of the evidence', Young complained, 'points to their holding the opinion that little change in organization is necessary and that they are fully capable of controlling their own affairs. This no doubt explains their withdrawal, after the completion of the financial statement, from active contact with Mr. Nicholson and myself once we got going on the organization problem. Where we expected spontaneous interest, co-operation and active assistance, we were faced with an apparent self-sufficiency which completely frustrated any attempts to enter into any agreed line of investigation.'[31] To dispose in this way, as Akers succeeded in doing, of the formidable Nicholson was a notable victory for the circumference over the centre.

In October 1935 the new Board of Fertilizers & Synthetic Products presented to the General Purposes Committee of the ICI Board their own prescription for the cure of Billingham. They produced a chart (reproduced as Figure 12) of Billingham's activities and pronounced: 'Billingham is essentially a factory for the manufacture and working up of *Hydrogen*.' They then proceeded to show how, in their opinion, the interdependence of the various processes carried on at Billingham was a source of strength, not weakness. 'It is cheaper', they said, 'to make Drikold [dry ice—solid carbon dioxide] from waste gas from the

Hydrogen plant, than to make it from CO_2 prepared by burning coke. It is cheaper to liquefy the CO_2 for Drikold by means of liquid ammonia from the ammonia plant, than by special refrigeration. It is cheaper to make Nitrochalk [a fertilizer] with by-product chalk from the sulphate plant than with natural chalk, and so on.'[32]

They went on to show that the factory had changed rapidly from 'a Fertilizer factory, with Industrial side-lines, to an Industrial Chemical factory, with a small Fertilizer interest'. They also pointed out that the return on capital employed in fertilizers was miserable, but the return from industrial products (though not from petrol) was good. Their figures, summarized, are:[33]

	Capital applicable to 1935 Sales £m.	Gross Trading Profit 1935 £	1935 GTP as % on Capital
Fertilizers	5·8	72,101	1·2
Industrial Products	3·3	442,840	13·3
Petrol	4·3	62,598	1·5
Plant written down, under construction, 'temporarily not in production'	4·9		
Other capital employed	3·0		
Total	21·3	760,695	3·6
Deduct P/L Items		131,896	
Total Net Profit		628,799	2·9

The general tenor of these figures, with minute returns on so much of the capital employed, and so much of the capital not employed at all, may be thought sufficiently disastrous to daunt most managements. Nevertheless they represent the basis on which a new management was preparing to make a new start, and the figures indicate where the main effort was to lie: in the petrol plant.

In March 1937 the ICI Board decided to write off £5,110,000, 'representing excess values of plant, buildings and services at Billingham'.[34] That was the last great write-off of fertilizer plant and the ignominious end of ICI's first major investment decision. The financial burden of the past was lifted and at Billingham they were already busy with what they saw as their best hope for the future: oil-from-coal.

REFERENCES

1 F. C. O. Speyer, 'International Nitrogen Position', 20 Nov. 1930, Mitchell Papers 78.
2 'World Nitrogen Position Tables', 17 June 1932, CR 84/8/8.
3 Ibid.

[4] As (1).
[5] F. C. O. Speyer, 'Billingham Fertiliser Nitrogen, General Survey', March 1935, CR 84/18/121.
[6] H. J. Mitchell to Melchett, 7 July 1930, CR 84/12/12.
[7] F. C. O. Speyer, 'Report . . . of the International Nitrogen Conference', 26 Aug. 1930, CR 84/12/12.
[8] Ibid.; Speyer to Mitchell, 14 Aug. 1930, CR 84/12/12.
[9] Speech by H. Schmitz, 15 July 1931, CR 84/12/35.
[10] As (7).
[11] F. M. Irvine, 'Notes on the Present Position . . .', 20 June 1933, MP 'Nitrogen 1932–3'.
[12] McGowan to President, Board of Trade, 22 July 1931, Mitchell Papers 87.
[13] As (2).
[14] W. A. Akers, 11 July 1932, MP 'Nitrogen 1932–3'.
[15] As (9), p. 11.
[16] As (1); G. P. Pollitt, 19 Jan. 1931, CR 84/12/46.
[17] 'ICI–IG Nitrogen Agreement—Memorandum re Discussions in London, March 2nd and 3rd 1932', CR 84/4/4.
[18] To Speyer, 23 April 1932, CR 84/4/4 (2); Supplemental Agreement, 10 March 1933, CR 84/4/4 (2).
[19] Epitome of Agreement in J. G. Nicholson and A. R. Young, 'Report to Sir Harry McGowan', 22 March 1935, pp. 50–1, Nicholson Papers 'Fertilisers and Synthetic Products Ltd'.
[20] Speyer, 'Conversations in New York', 24 April–2 May 1933, CR 84/12/73.
[21] As (19).
[22] As (2) for 1930–1 and 1931–2; remaining figures from (19), p. 65 and Statement E.
[23] As (19), Statement F.
[24] As (17), p. 6.
[25] As (19), Statement H.
[26] Letters between McGowan and Keeble, 8, 16, 20 Jan, 1931, Mitchell Papers 'Capital Budget and Expenditure, Agriculture'; Mitchell to McGowan, 'Jealott's Hill', 25 May 1935, Mitchell Papers 'Jealott's Hill, Agriculture'.
[27] As (19), Statement H; B. E. Todhunter to General Purposes Committee, 21 March 1935, GPCSP; H. Gaskell, 'Jealott's Hill and Fertiliser Propaganda', 23 Aug. 1935, CR 153A/140/189; R. E. Slade to W. J. Worboys and S. P. Leigh, 21 Nov. 1935, CR 153A/140/189; H. Gaskell, 'ICI Farms', 29 Aug. 1935, CR 153A/190/190.
[28] ICIBM 2086, 5 March 1931.
[29] 'Survey for Policy Meeting . . . Fertiliser Group', 25 Sept. 1936, Sec. I, p. 5, GPCSP.
[30] As (19), pp. 42, 68.
[31] A. R. Young, 11 Feb. 1935, enclosed with (19).
[32] As (29), Section I, pp. 1–2.
[33] As (29), Table: 'Allocation of Capital . . . and actual Gross Trading Profits for the Year 1935'.
[34] ICIBM 5296, 31 March 1937; ICI Annual Report, 1936, pp. 25–6.

CHAPTER 10

The Hydrogenation Cartel and the
Billingham Petrol Plant

(i) THE POLITICS OF OIL AND CHEMICALS

THE POSSIBILITY of hydrogenating coal to produce petrol began to emerge before the Great War. In 1913 Dr. Friedrich Bergius (1884–1949) took out patents covering a process for producing light oils from coal or heavy oils by hydrogenating them at high pressure in temperatures between 300° and 500°C. Petrol, as exemplified by iso-octane, contains 85 per cent of its weight as carbon and 15 as hydrogen, whereas in coal the proportion of carbon is the same but only 5 per cent is hydrogen. The essence of Dr. Bergius's invention was the recognition that complex molecules in coal could be split into simpler hydrocarbons by heat, and that at the same time these newly formed molecules could be treated with hydrogen, under high pressure, to produce stable saturated hydro-carbons of low boiling point which would make very good motor spirit.[1]

Bergius's project had obvious technical links with ammonia synthesis. It needed large supplies of hydrogen, and the essential reaction was to be carried out at high pressure and temperature. Like ammonia synthesis, it was an operation of the heavy chemical industry, intended eventually to produce hundreds of thousands of tons a year. For these reasons it had always been considered, both in Germany and in Great Britain, as a natural companion process to ammonia synthesis; but when ICI was formed Bergius, retained at a handsome fee by IG, was still a long way from technical exploitation of his patents, although they were close to expiry.

The difficulties which had held Bergius up for thirteen years did not deter Alfred Mond, George Pollitt, and others of an optimistic and ambitious turn of mind. They had seen similar difficulties overcome in developing the Haber–Bosch ammonia process and any effort was worth while if it promised a great enough reward. In the view of the enthusiasts, the reward from the oil-from-coal process promised to be very great indeed, because it would provide the United Kingdom and other countries of the British Empire with petrol made from home-dug coal, not from imported petroleum.

162

This was just the kind of project to fire Alfred Mond's politico-commercial imagination. It would strengthen the British economy by saving foreign exchange. It would help the problems of the coal industry, with which he was closely identified through Amalgamated Anthracite Collieries Limited. It would contribute to imperial defence. And in doing all these things, so desirable politically, it would earn large profits. Everything depended, of course, on getting the technology right and the cost competitive with conventionally produced petrol, but once that had been done the outlook seemed brilliant. So thought Alfred Mond at the time of the ICI merger, when he would have been willing to bargain away the future of the British dyestuffs industry against access to IG's knowledge of the oil-from-coal process.[2]

In Royal Dutch-Shell and Standard Oil, naturally, they were aware of what was going on in IG and ICI. The threat to the oil companies, in the late twenties and early thirties, was distant, because a great deal of work had to be done in getting a process to work on a commercial scale and even more in getting the production costs down to a competitive level. Nevertheless it was a threat that had to be taken seriously. The best authorities thought the world's oil reserves were likely to run out quite quickly—perhaps in twenty years, it was said in 1933—whereas there was enough coal to last for centuries.[3] Moreover, even if petrol made from coal was more expensive than the conventional product, it might well be given preference by governments of countries like Great Britain and Germany which had ample coal but no petroleum at all.

By producing petrol, therefore, or even by preparing to do so, ICI and IG—not necessarily in alliance—would run a risk of colliding with two of the world's most powerful industrial enterprises. Royal Dutch-Shell and Standard Oil might be expected to react strongly, perhaps in unity, against any attempt to invade their territory, and they would be very strongly placed to do so. In the matter of costs, central to the whole issue, the oil companies in 1928 could sell petrol, retail, in Great Britain for less than 1s. 2d. (about 6p) a gallon, and at that figure petrol made from coal, to be competitive, would have to be produced for 4d.–5d. (about 2p) a gallon, whereas the estimated cost for large-scale production, in 1929, was 11·8d. (a little under 5p) at the works.[4] And besides that technical advantage, the oil companies had a distributive organization which would be impossibly expensive for ICI or even IG to copy.

Moreover, there was always the possibility that the oil companies would invade the chemical industry. Union Carbide & Carbon Corporation, in the USA, were already marketing materials worked up from refinery gases bought from the oil companies,[5] and what a chemical company could do with the oil companies' by-products (as they then were) the oil companies could, if they chose, do for themselves. Even nearer ICI's interests, Royal Dutch-Shell had a project at IJmuiden in Holland for making ammonia from coke-oven gas.[6] The oil industry and the chemical industry were beginning to have a common frontier, easy and tempting to cross in either direction.

In this kind of situation, where there was a risk of provoking opposition from powerful vested interests, ICI's almost invariable policy, between the wars, was to keep clear, as they kept clear, for instance, of the profitable artificial fibre industry in order to avoid trouble with Courtaulds (p. 366 below). It is a measure of the confidence felt by Mond and his colleagues and subordinates, especially those at Billingham, in the potentialities of oil-from-coal that they were prepared to steer ICI directly into the path of the Great Powers of oil. To avoid collision ICI would need extreme diplomatic ingenuity, but that was a talent they were not short of.

In February 1927 ICI bought control of the British Bergius Syndicate which owned Bergius's rights for the British Empire. To the rights themselves, which had only a couple of years to run, ICI attached little importance, but by owning them they prevented anyone else from challenging the development work they were carrying on at Billingham. Moreover they gained an unassailable position 'in the colonies', by which charmingly Victorian phrase Wadsworth meant, chiefly, Australia.[7]

Even with Bergius's rights, ICI were still well behind IG, who had Bergius. In 1927 IG began building full-scale plant at Leuna to work on brown coal (lignite) which seemed to offer better prospects than bituminous coal such as was plentiful in England. In ICI, meanwhile, by spending the large sum of £120,000 in two years, they found themselves ready, at the beginning of 1929, to put up plant large enough to hydrogenate ten metric tons of coal a day. Plant of this size, it was thought, would enable them 'to find out, once and for all, whether the production of oil from coal can under any circumstances be carried out economically'.[8]

To those of a less sanguine temperament than Mond and Pollitt it always seemed likely that the difficulty of the cost problem was being underrated. One ICI enthusiast, in October 1927, foresaw 'colossal profits to be made out of oil from coal of the order of hundreds of millions of pounds a year'—in 'ten or fifty years' time'. He went on to say that oil-from-coal had 'a great future'. 'I think', a critic observed, writing neatly in the margin, 'this wants to be demonstrated. Has it not become a catchword the truth of which is taken for granted?'[9] But negative thinking of this kind was disregarded between 1927 and 1929, just as it was on the subject of fertilizers, and in this heady atmosphere the expensive development work on oil-from-coal was pressed ahead.

Even Mond, optimist though he was, came to recognize that with petrol prices falling as they were in 1928 oil-from-coal, to be attractive commercially, would need Government help. He was hopeful of getting it. The strategic value of a home-based supply of petrol, and possibly of other oils, was obvious, and the Committee of Imperial Defence had been interested in Bergius's process since 1925.[10] They had even contributed £25,000, through the Department of Scientific and Industrial Research, to the cost of the German inventor's activities, presumably before he took up his appointment with IG. Therefore when, in the early part of 1928, Mond was approached by the Secretary of the Committee of Imperial Defence and of the Cabinet, Sir Maurice Hankey, for confidential information on ICI's development of the Bergius process, he seized the opportunity to put his case for Government help, which he thought was a good one.[11]

Mond told Hankey there was no technical difficulty in producing oil from coal, but the fall in the price of petrol had made the process unattractive commercially. Nevertheless the problem, he said, was 'vital from the point of view of national defence', and he wanted Government help—he was vague about the nature of it—with putting up plant for, say, 200 tons of petrol a day. 'He believed', Hankey reported to the Cabinet, 'the Germans within a few years would make themselves independent of imported liquid fuel. If we wished to do the same there was no technical difficulty: it was merely a matter of money.'

Mond had chosen a bad moment for his approach. Nobody in the Cabinet in 1928, least of all the Chancellor of the Exchequer, Winston Churchill, was going to take kindly to the idea of spending money for warlike purposes, except possibly the Service ministers. But when Mond in 1929 turned to the First Lord of the Admiralty for support, he found none. What Their Lordships were interested in was not petrol but fuel oil. A conversation between A. E. Hodgkin and the Fourth Sea Lord produced the interesting information that the USA, from which most of the Navy's oil came, was 'to be regarded as unfriendly' and might even be actively hostile, but motor fuel produced from coal would do nothing to solve that problem.[12]

In the autumn of 1929, not long after Mond had been rebuffed at the Admiralty, the Depression set in. The oil-from-coal project, far from offering prospects of enormous profits, began to show alarming signs of turning sour, for it quickly became apparent, about the turn of the year 1929–30, that in oil-from-coal ICI was confronted with the same problem as in fertilizers: the bottom was dropping out of the market. Productive capacity in the USA was far ahead of demand and the oil companies, greatly hampered by the anti-trust laws, were trying to persuade each other to cut down the flow from everybody's wells.[13] It was scarcely the time for ICI to come out with proposals for adding to the world's supply of petrol by hydrogenation, especially when their product was uncompetitively costly.

Yet to abandon the project would be very expensive. By the beginning of 1930 some £325,000, including the cost of the ten-ton plant, had been spent on developing oil-from-coal. That was a very large sum by the standards of the day. But full-scale plant was nowhere in sight, nor were costs remotely competitive with the oil companies' costs. More money would have to be spent if the sums already spent were not to be lost.

It was the familiar dilemma of research and development expenditure, and in January 1930 McGowan came down on the side of continued spending. He told the Chairman's Conference that £300,000 more was being asked for and they ought to authorize it. This was not because he saw great profits ahead. On the contrary, 'he was not convinced that we were on a very profitable line' and he feared 'an oil war ... in which we might become involved'. The Conference discussed the problem and 'agreed that if we did not go on with the development of the process we should have wasted what we had already spent'.[14] This argument, based on steeply mounting sums, was to be repeated in the months and years to come. In oil-from-coal, ICI had passed the point of no return, but in very bad times, and the outlook was anything but promising.

The expenditure was authorized, a large staff was collected, Kenneth Gordon was put in charge, and in February 1930 an Oil Division was created, within the Billingham Group, 'with a view to later separation from the main Billingham body'. By July estimates for large-scale plant had been worked out. For an investment of £9m. it would produce 210,000 tons of petrol a year, on a fresh site, at a works cost of 8d. (about 3p) a gallon, including obsolescence at 5 per cent.[15] This was a great improvement on previous cost estimates, but it was still far above the oil companies' figures.

In the circumstances of the oil industry in 1929–30, as in the closely similar circumstances of the nitrogen industry, ICI needed an agreement with other large producers to regulate the market. Moreover ICI needed all the help they could get with the technology of the process, implying some form of patent-pooling agreement. These lines of thought pointed towards negotiations with IG and with the two oil companies most closely interested in hydrogenation: Royal Dutch Shell and Standard Oil.

But in ICI's path there was an obstacle: the Patents and Processes Agreement with du Pont. Du Pont were not interested in oil-from-coal, so they had nothing to gain either from regulation of the market or from the pooling of knowledge. On the other hand they would strongly object to any possibility that information from them, passed to ICI under the terms of the Patents and Processes Agreement, might find its way to IG or to Standard Oil, whom they regarded with apprehension and mistrust.

IG, ahead of ICI in oil-from-coal technology, were ahead also in diplomacy. In 1927 they came to an agreement with Standard Oil, who in 1928, with IG's help, set up experimental oil-from-coal plant at Baton Rouge, Louisiana. In 1929 the two partners set up a joint American company, Standard–IG, to promote the development of oil-from-coal throughout the world. The basis of the bargain, presumably, was the pooling of IG's technical skill and Standard Oil's distributive and marketing organization.[16]

ICI's corporate peace of mind was seriously disturbed. Apart from what got into the papers, little was known about the details of the arrangements between IG and Standard, but any suggestion of an alliance filled ICI with apprehension, as well it might, considering the strength of the parties separately, let alone combined. The obvious counter-ploy was an alliance between ICI and Royal Dutch-Shell.

Royal Dutch were not on very good terms with IG. On the other hand they had already put a good deal of money—some £250,000— behind Bergius and presumably knew the difficulties and disadvantages of his process.[17] When ICI approached them, in 1928, trying to sell the cracking of heavy oils by hydrogenation, they replied that the costs were far too high.[18] By the following year, however, they had a process of their own to sell—the IJmuiden ammonia process—and that seemed to offer the basis for an alliance which ICI were looking for. 'It should be possible', wrote Pollitt in April 1929, 'to form a Company jointly between ourselves, Royal Dutch (and possibly Dupont for USA) which would have the task of developing all those products which are or will be based on the bye-products of the Oil Industry. This would cover Ammonia and also Heavy Organics. . . .'[19]

The plan, which seems to have wandered a long way from hydrogenation, contemplated selling ICI's ammonia plant at Billingham and Royal Dutch-Shell's at IJmuiden to a company financed 60:40 by ICI and Royal Dutch. Royal Dutch intended to put up ammonia plant in California, and this American venture—'a vast outlet for fertilizers'— was a large part of the attraction to Royal Dutch.[20] Other oil companies were thinking along the same lines—this was before the really disastrous collapse in the fertilizer market—and there seemed to be every prospect that they would shortly invade the nitrogen industry, overcrowded though it already was.

The plan was torpedoed by du Pont. ICI could not operate in the USA against du Pont, and had no intention of trying. Instead, they hopefully visualized an alliance between du Pont, themselves, and Royal Dutch-Shell. Du Pont would have none of it. Together with their prospective allies they examined the California plan and the findings of the joint committee 'indicated its extreme undesirability'.[21] That destroyed whatever faith ICI had ever had in the proposition, which was never much, and the matter was closed about the end of 1929 when du Pont went on to say that they did not want to risk offending other oil companies by joining Royal Dutch-Shell in hydrogenation, which Royal Dutch-Shell had brought into the negotiation after at first excluding it.[22]

Du Pont put ICI in a weak bargaining position against IG and Standard, but a bargain ICI must have, otherwise the competition, both technical and commercial, would be too strong. In March 1930 F. A. Howard, President of Standard-IG, offered an agreement, but on terms so harsh that Hodgkin did not think he seriously expected ICI to accept.[23] Some weeks of bargaining later a draft was agreed on which allowed ICI licences under Standard-IG patents for the whole of the British Empire, which was broadly what they wanted.[24] They were to be limited to 25 per cent of each country's market, but since that would represent, if taken up in full, an investment of some £210m., it was not regarded as an obstacle. There were reciprocal licensing provisions for ICI's patents.

Du Pont once again intervened. When the proposals were shown to them, with Standard-IG's agreement, they shouted angrily across the Atlantic: 'Do you realize that your action has injured du Pont interests because information we may transmit to you is likely to be transmitted to Standard Oil in this country?'[25] No information had at that time been transmitted, because no agreement had been signed, but du Pont seem to have been too excited to notice.

What chiefly upset du Pont, with reason, was ICI's rather too easy assumption that they would never be interested in hydrogenation; and, moreover, they were already very sensitive to the threat of competition from the oil companies in the heavy organic branch of the chemical industry. Du Pont were determined not to allow the oil companies access to high-pressure technology. In ICI, it was privately thought that du Pont were 'consciously or unconsciously' influenced by their dislike and distrust of IG, and that they were strongly against ICI being interested in the US market, however indirectly.[26]

ICI, in the summer of 1930, were not in a position to let du Pont or anyone else stand in the way of their policy. Their difficulties, as we have observed, were closing in. Melchett, far gone in his last illness, crossed the Atlantic with McGowan, and at a meeting in New York on 27 August 1930 Lammot du Pont 'stated he wanted the present spirit of agreement and good feeling to continue' and offered a solution to the dispute, in which du Pont explicitly recognized 'the importance of the [Standard-IG] deal to ICI'. Briefly, hydrogenation was to be kept within the scope of the Patents and Processes Agreement but, 'in view of du Pont's willingness to allow ICI to consummate its arrangement with Standard-IG', ICI gave du Pont a contingent promise to allow du Pont to remove 'subjects coming under the High Pressure field . . . from the [Patents and Processes] Agreement at any time they observe the exchange of information to be working to the detriment of their interests';[27] and in fact a whole group of subjects, known as 'the hydrocarbon field', was removed bodily from the provisions for exchange of information under the Agreement.

Thus for the second time within less than a year du Pont blocked negotiations, of great importance to ICI, between ICI and other companies in the oil-from-coal field. Under great pressure they relented and allowed matters, this once, to go forward. It was observed in 1940, however, that 'in view of the du Pont attitude towards the Oil Industry a sense of bad-feeling was created at the time', and du Pont came to regard the petrol plant at Billingham 'with a trace of suspicion . . . only . . . mitigated by very careful watchfulness in the matter of sending reports to du Pont'.[28]

With du Pont's opposition withdrawn, the Chairman and President of ICI, together in New York, could consider the agreement offered by Standard–IG. It had been revised to provide for a company, based on Liechtenstein, called International Hydrogenation Patents, to hold all hydrogenation patents belonging to parties to the agreement. Royal Dutch-Shell had come into the negotiations, and it was proposed that the shareholders in IHP would be Standard–IG and Royal Dutch-Shell, but not ICI. They would merely be granted licences for the British Empire, still on the basis of a 25 per cent share of the market.

McGowan would have initialled the agreement on the spot. His colleagues in London, horrified at his precipitancy, managed to hold him back. Negotiations with the Government, to which we shall return (p. 170 below), were in train once more, and they saw no reason to hurry. During the autumn of 1930, by a process of debate within the company marked by a stream of papers by Pollitt, Henry Mond, the dying Melchett, and others, ICI's policy towards Standard–IG and the oil companies, on the one hand, and towards the Government, on the other, was clarified and elaborated. ICI did not like the terms offered by Standard–IG, and indeed there is plenty of evidence that they were deeply resented. 'We are not prepared', said Hodgkin in one document, 'to subordinate ourselves to the extent now proposed and . . . we are therefore quite ready to proceed alone unless their terms are very radically modified.'[29]

That, however, was what in the end ICI were not prepared to do. The prospect of an alliance between Royal Dutch-Shell, Standard Oil, and IG, from which ICI were excluded, was something they could not face. 'We are not strong enough to oppose them', Henry Mond wrote in November 1930. 'We cannot contemplate the entry of such formidable antagonists into the general chemical field, nor does the commercial value of the oil process appear to be so attractive as to risk such a combination being formed against us.'[30] McGowan was terser: 'If . . . the Royal Dutch-Shell were to join with the IG–Standard Group on their own, we should be left out altogether; a situation which must not be allowed to arise.'[31]

Negotiations with the other powers of oil and chemicals went on, parallel with negotiations with the British Government, during the early months of 1931. On 10 April 1931, McGowan being in Monte Carlo at the time, four interrelated agreements were sealed which, taken together, became known as the IHP Agreement or, sometimes, the Hydrogenation Cartel.

They provided for ICI to share in the ownership of International Hydrogenation Patents and to have a seat on its Board, to which Pollitt was nominated, which was recognized on all sides as being a gesture towards ICI's prestige and not at all a concession of substance. There were cross-licensing arrangements between IHP and ICI, but ICI's licence rights were limited to the countries of the British Empire and to 25 per cent of the consumption of petroleum products in each. ICI's purchase of any petroleum they might need for hydrogenation and the marketing of ICI's hydrogenation products (with certain exceptions in favour of Government) were to go through nominees of IHP, which meant Standard Oil or Royal Dutch-Shell, and the marketing commission, on top of various charges for expenses, was to be 15 per cent 'on the . . . value of the marketing investment used within the market itself by the Oil Company . . . in distribution of products of the same grade and quality, before deduction of income or profit taxes'. Commission at this rate, S–IG maintained, was 'not particularly attractive' to the oil companies, and they resisted any suggestion of reducing it, no doubt because they knew that ICI on their own could not hope to market more cheaply. 'The Agreement in its final form', said Hodgkin, 'was considered to be highly satisfactory from ICI's standpoint and will undoubtedly enable us to reach the stage of commercial production much sooner than would otherwise be possible.'[32]

Thus another private international treaty was born out of the economic turmoil of the slump, which the official diplomacy of Governments was quite unable to control. To have any practical effect for ICI, it would have to be linked with whatever arrangements, within the United Kingdom, ICI could make with the Government of the day, because the costs of the oil-from-coal process were still too high to make it a normal commercial proposition.

(ii) THE LAME DUCK AT BILLINGHAM

In May 1930 McGowan told Sir Maurice Hankey, who told the Cabinet, that ICI had two problems in oil-from-coal—to reach some arrangement with the oil companies to avoid being driven out of the market and to get some form of Government support.[33] The IHP Agreement dealt with the first problem, though not on particularly good terms (it was, said Wadsworth, 'the best we could get',[34] but his tone was not enthusiastic). It did nothing to touch the second, and throughout the period of negotiation with Standard–IG parallel negotiations were going on with the Government, though on different lines from the abortive approaches by Alfred Mond in 1928 and 1929.

At the heart of the problem lay the cost of production of petrol by hydrogenation. In July 1930 the best figure the Billingham experts could offer was 8d. (about 3p) a gallon, but that was far too high for Pollitt to recommend putting large-scale plant in hand. Research, it was decided, would go on, on the basis of production from the Billingham ten-ton plant, to try to get down to 6d. ($2\frac{1}{2}$p) a gallon.[35] Even that would be too high for unaided competition, especially with oil prices falling as fast as they were in the early thirties.

Help could only come from the Government, but the form it should take was uncertain. Mond, in 1928–9, had sought supply contracts for the armed services, but the imposition in 1928 of an 8d. import duty on each gallon brought into the country raised the possibility of tariff protection instead or as well. Hankey, in May 1930, told the Cabinet he thought ICI might ask for both.[36]

The Cabinet asked the President of the Board of Trade, William Graham, and the Secretary for Air, Lord Thomson, 'to go into the whole question of extracting motor spirit from British coal'. The inquiry, which began with official visits to Billingham in August and September 1930, gave ICI a reason, which was not unwelcome to them, for drawing out the negotiations with Standard–IG, and it also gave them time to prepare a case for the Government.[37]

By this time, late in 1930, another element had entered the situation which was to become more and more important as time went on. With Government help, ICI might be able to bring back into use some of the plant at Billingham which had been seriously underemployed since the collapse of the nitrogen market towards the end of 1929. This applied particularly to hydrogen plant, but Pollitt encouraged hopes, also, that ammonia convertors might be turned over to the production of methanol for use as motor fuel.[38] 146,000 tons of liquid fuel, said Pollitt, could be produced that way on existing Billingham plant, although he admitted 'difficulties in using Methanol as a fuel of universal application'—chiefly, as Gordon pointed out, that it could not be mixed with petrol without a suitable blending agent, which would have to be developed.[39] Any prospect, however distant, of getting some return, however small, on the capital invested in redundant plant at Billingham became a leading motive, frequently mentioned, in ICI's plans for developing oil-from-coal. 'As we have raised capital to build this plant, estimated on a low basis of 5%,' wrote Melchett in the summer of 1930, 'if we got no profit at all beyond this we should be getting 5% where we are receiving none now. But if we only make 5% additional on this capital we shall be really earning 10%; that is to say, instead of paying 5% for nothing we are receiving 5%.'[40]

ICI had their case ready for the Government towards the end of November 1930. They based it on economic nationalism. 'Petrol', they said, 'is a foreign product for which we have and shall continue to have an urgent demand. Manufacture at home would therefore relieve the growing pressure tending to modify the equation of exchange between Great Britain and foreign countries.'[41] But oil-from-coal was 'at too early a stage of its life to be a good commercial risk', especially since it would run into heavy opposition from the oil companies, harassed by the problem of over-production and falling prices. 'The oil companies', it was pointed out, 'constitute probably the wealthiest industrial groups in the world. Economic warfare would without difficulty crush a new and developing synthetic industry. Security against this is essential.'

It was security against the oil companies during the early years of oil-from-coal, then, which ICI wanted the Government to provide. Their negotiations with Standard–IG, not yet concluded, could not be expected to achieve more than an undertaking that if petrol could be produced from coal at a competitive price, the oil companies would let it pass through their marketing organization. In getting a competitive product on to the market, naturally, they would not give ICI any help or encouragement, and it was this which ICI were looking to the Government to provide. The argument would have sounded very familiar to an Australian Government, to a South African Government, to any Government in South America, or indeed to a Government in any country where efforts were being made to develop home industries against imports.

ICI said the security they wanted could be achieved 'either by agreement or by compulsion', but since the oil companies in the United Kingdom were 'largely in a monopolistic position through their agreements with one another' and controlled a distributive system which it would be expensive and wasteful to compete with, agreement would permit them 'more or less to make their own terms'. 'The better method', ICI concluded, 'is compulsion.' Let the Government bring in legislation to compel importers and refiners of petroleum to buy a quota, proportionate to their imports, of home-produced fuel.

Given this security, ICI would undertake a scheme, worked out by Hodgkin, which would provide 27,800 tons a year of petrol, 9,800 of middle oil, and 38,000 of heavy oil, using 272,000 tons of coal a year and giving permanent employment to 1,420 men, including miners. The products could go into the quota system or the Government could contract to buy them. ICI would expect no financial help except the guarantee of a Debenture issue of not more than £1,500,000.

ICI's own hope from these proposals, discussed by McGowan and other Directors (Melchett was by now too ill to leave his bed) on 19 November 1930, was that a quota system would 'enable ICI to make large profits (between £500,000 and £1,000,000) by turning certain units at Billingham to the manufacture of methanol', which would be mixed with petrol made at Billingham and elsewhere. 'Billingham', it was observed, 'was capable of producing all the methanol required to mix with the present quantity of motor spirit consumed in this country.'[42] In their proposal for oil-from-coal plant, ICI deliberately kept well within their own borrowing powers, without seeking public money. They expected much larger schemes to follow if the large-scale pilot, which was what Hodgkin's scheme amounted to, succeeded, but an instinct of self-preservation deterred them from proposing really large expenditure immediately:

. . . it was not felt safe to embark on any larger scheme . . . owing to the novelty of the process, and it was therefore highly unadvisable to embark upon an expenditure of £8,000,000 even if this money were Govt. money. It would be extremely unpleasant for ICI, and especially those most responsible for directing its affairs, to have to face the Govt. with the fact that this expenditure of £8,000,000 had been unsuccessful and to have to face the parliamentary, national and commercial repercussions that would be bound to follow. . . .

ICI's proposals, though not the Board's private thoughts upon them, were presented to the Government by McGowan on 26 November 1930 at the offices of the Board of Trade. McGowan brought with him H. J. Mitchell, W. H. Coates, and W. A. Akers. The Government showed a certain sense of occasion—their leading representatives were two Ministers, the President of the Board of Trade, William Graham, and Lord Amulree, Secretary of State for Air since the death of Lord Thomson, who had been killed on 5 October when the airship R101 crashed on its way to India. But perhaps the civil servants present indicated even more plainly how seriously the Government were taking oil-from-coal. There were four, including Sir Alfred Hurst of the Treasury, and they were headed by the formidable, not to say forbidding, figure of Sir John Anderson, KCB.

Anderson, at the age of forty-eight, had a distinguished official career behind him and much greater things yet to come, including the Governorship of Bengal in times of trouble and high ministerial office, culminating at the Exchequer, during the Second World War. During the rest of his life he was to have dealings with ICI in many different capacities: as a negotiator in 1930–1; as their wartime overlord when he was Lord President of the Council (p. 292 below); and as a 'lay Director' for a few months in 1938 and again from 1948 to 1958.

In November 1930 Anderson was Permanent Under-Secretary at the Home Office and not perhaps the most obvious choice to lead a Government team in technical discussions with ICI. He had been educated in science, however, at Edinburgh and Leipzig, and he had a grasp of chemical technology rare at his level in the Civil Service. An austere and devastating critic of ICI's proposals he turned out to be.

In fact, he killed ICI's hope of getting help from the Labour Government. He was not at first hostile. Turning aside, rather than turning down, the idea of a quota, and following a recommendation of the Cabinet Sub-Committee on Coal Hydrogenation, he asked ICI to make proposals for a scheme intended to deal, eventually, with 1,000 tons of coal a day and requiring some £7m. to £8m. capital investment—a scheme, that is, of the size which had frightened the ICI Directors—and he 'indicated that the Government might consider taking a large financial interest'.[43]

Following Anderson's invitation, a scheme was worked out in ICI for a plant at Billingham to produce 214,000 tons of petrol a year at a works cost of 8d. (about 3p) a gallon, requiring an investment of £8m., to be run by a company capitalized at £10m. ICI were to have £2½m. in Ordinary shares, evidently partly in consideration for development work and partly for assets transferred. £7½m. was to be raised in cash 'as the Government wished'. A realization of 1s. 6d. (7½p) a gallon was to be guaranteed for four years, presumably by the Government, and after that, 1s. 0d. (5p).[44]

On 5 March 1931 the President of the Board of Trade, the Air Minister, and a distinguished band of civil servants met McGowan and a party from ICI to discuss the company's proposals. They had been asked for by Anderson, and presumably McGowan's party expected, not perhaps outright acceptance, but at least constructive criticism from which alternative suggestions might develop. What they got, instead, was a very thorough demolition job by Sir John Anderson.

The effect on unemployment, he said, would be so small and so far off as to be negligible. ICI's development expenditure so far had produced a process of no commercial value, and if they did any more work, with Government help, the benefit would go to the patent-holders. The IHP Agreement seemed to Sir John a bad one for ICI, 'though they, no doubt, knew their own business best'. At the prices ICI were asking for, Anderson calculated that the Government, in guaranteeing them, would have to bridge a gap of 8½d. or 9d.* a gallon, of which 4d.

* In the discussion of costs and prices in the remainder of this chapter, it should be borne in mind that 1p = 2·40d.

174

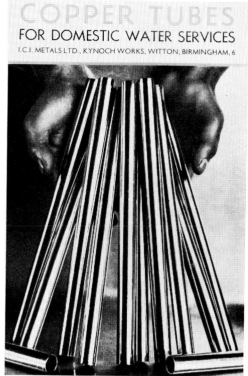

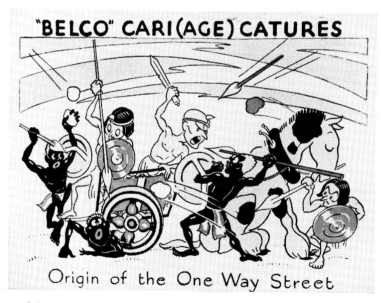

products of the thirties: paints, copper tubes, 'Belco' cellulose finishes.

Left to right: the Prime Minister (Ramsay MacDonald), Lord McGowan, and
Kenneth Gordon at the opening of the Billingham petrol plant, 15 October 1935
(page 180).

A general view of the petrol plant at Billingham, March 1935.

represented 'the element of profit' on the £1¼m. which ICI considered they had already spent on oil-from-coal. 'He asked how far this element of profit could be taken out.'[45]

This assault seems to have left the ICI delegation, including even McGowan, speechless—a considerable achievement—though whether from rage or mortification does not appear from Pollitt's note of the proceedings. All he says is that it was finally agreed that Sir John should submit a scheme of his own to ICI. After that the Cabinet would decide.

In ICI they were no doubt taken by surprise by Anderson's turn-about from apparently friendly helpfulness to this onslaught on 'the element of profit'. Yet the position of the Government—after all, a left-wing Government—was quite plain. They were being asked, as they saw it, to come to the rescue of a lame duck: a private venture which could not otherwise survive. They were unwilling to see ICI make more than a very small profit indeed as a result, and in particular they were unwilling to ensure a return on the money which ICI had already spent.

In ICI there was deep resentment. 'Industry', Mitchell wrote to McGowan in a long, indignant letter, 'is constantly being exhorted to research in order to keep abreast of or advance on foreign developments. We have spent this large sum'—he put it at '£1m. or so' rather than £1¼m.—'and can hardly be expected to waive all claims to credit for research, on the grounds that we have lost the money anyhow. This is scarcely a helpful attitude to the advancement of industrial research.'[46]

Proposal and counter-proposal followed, but the gap between the two sides was never bridged. ICI would not give up their claim to what they considered a fair return, especially on their development expenditure, and the Government would not concede it. Finally, in July 1931, ICI's corporate state of mind was summed up, probably by Henry Mond, in a document[47] marked 'DRAFT of letter it is *NOT* proposed to send to the President of the Board of Trade:'

Dear President,
We have your proposals as to the development of the coal oil manufacture in Gt. Britain.

These proposals give ICI no adequate return on the capital already spent on Research, no adequate payment for the services still required of them, and no advantage as investors in a new industry over the Government who have added nothing to the development of the process.

The proposals appear to be so grossly unfair to my Company that I can only conclude that they have been dictated solely from the point of view of party politics as distinct from national interests.

Under the circumstances it would appear that our course now is to close down our experimental plant, disband our trained staff and let the matter stand over until a Government less imbued with the determination to discourage British industrial development is in power.

We propose to make a full public statement of your attitude in the matter and to draw attention to the effect on employment in the coalfields and other industries of your decision.

This was the end, for the time being, of negotiations between ICI and the Government. Whether, with that particular Government, they could ever have been successful seems doubtful, for the tone of some surviving documents, particularly Henry Mond's unsent draft and Mitchell's private fulminations to McGowan, make it clear that some, at least, of the principal figures on the ICI side regarded Labour ministers as their natural enemies, and no doubt the feeling was reciprocated. Whatever the feelings on either side may have been, for several months after negotiations broke down the official side was in no case to revive them. At the end of August 1931 the Labour Government fell, overwhelmed by a sterling crisis, and it was not until after a General Election in October that a stable government could be constructed. Then, with a majority of 425 behind him, MacDonald came back to power at the head of what was known as the National Government. It was strongly Conservative in sentiment and more likely than its predecessor to find itself *en rapport* with McGowan and his colleagues. McGowan, moreover, was on good terms with Walter Runciman, the President of the Board of Trade.

Nothing, however, could be arranged quickly. Meanwhile so much oil was being produced in the world that petrol could be landed in the United Kingdom in March 1932 at about 3d. a gallon. Selling expenses and other costs of about 5½d., plus customs duty at 8d., gave a retail price for No. 1 motor spirit, directly competitive with the proposed product of hydrogenation, of 1s. 4½d. 'This price', observed Hodgkin, ' . . . is hardly sufficient to attract capital; and it is even less so when it is remembered how large a proportion of it is due to the tax, the incidence and magnitude of which is solely controlled by the Government.'[48]

Development work, under these circumstances, almost ceased. A meeting early in March 1932, presided over by Rogers and Wadsworth and attended by Akers, Gordon, and others of the Billingham management, 'unanimously agreed that at the present time, and in the absence of any support from the Government, expenditure on coal oil should be gradually reduced until a skeleton staff only remains, sufficient to enable ICI to keep in touch with hydrogenation developments elsewhere'.[49] Research expenditure, which had been running at over £200,000 a year in 1931, fell to little more than £55,000 in 1932 and a request for special expenditure of £1,600 went straight up to the General Purposes Committee of the main Board.[50]

When the gloom was at its deepest, on 23 May 1932, McGowan,

Mitchell, Rogers, and Wadsworth met Akers, Gordon and others concerned with oil-from-coal. Everybody agreed with McGowan that ICI could do nothing without Government help, that the Government's difficulty lay with the Treasury, and that 'at the present stage ICI would be well pleased with even 5% return on the million pounds expended on oil-from-coal development'.[51]

Then, in the early autumn of 1932, the oil companies succeeded in coming to terms with the Romanian producers and Standard Oil bought out Cleveland Oil, which had been behaving with distressing independence, in Great Britain. These measures put an end to price-cutting and the price of petrol in ship at London rose to what Hodgkin called 'the normal and proper price' of 5d.–6d. a gallon.[52] At that price, with the new cost figures which were becoming possible, and assuming a tax of 8d. on imported petrol (the prevailing rate), there began to be some prospect of making oil-from-coal pay.

In these slightly less depressing conditions a plan began to emerge for putting up, at Billingham, a plant to produce 100,000 tons of petrol a year. As originally conceived, it was expected to require about £3·9m. capital, and the total manufacturing costs, including 3·26d. for obsolescence, would be 7·93d. a gallon—a great improvement on 11·8d., the 1929 estimate (p. 163 above).[53] The scheme would need about £2·3m. new capital, but it had the great attraction that it would employ existing Billingham plant, roughly the equivalent of one ammonia unit, with a capital value of about £1·6m. (a figure later raised to about £2m.). There was no prospect of using the plant for its original purpose for many years, if ever, and if it was not used for hydrogenation it would either have to be more or less expensively maintained or pulled down and sold for scrap. In these circumstances, it was considered justifiable to look solely at the return on the new money invested, which naturally made the whole project much more attractive.[54] The employment of this ammonia plant, otherwise redundant, became a leading consideration in the minds of everyone concerned.

In the winter of 1932–3, with between 2½ and 3 million unemployed, another aspect of the matter presented itself:

It enables us [wrote Henry Mond] to give considerable employment to highly skilled men in the Middlesbrough area, who would otherwise suffer extreme privation during this winter, and also tend to lose their skill, a point which is not only of national interest but of local interest to us as potential employers on a large scale in the Middlesbrough district.[55]

It was expected that the plan would increase employment at Billingham, eventually, by some 1,000 men, and in the meantime men who would otherwise have been laid off would be kept on.[56]

ICI did not, this time, ask for Government capital. There was consequently no repetition of the embittered wrangle with Sir John Anderson over 'the element of profit' on ICI's development expenditure. Nevertheless this plan, like all others concerned with oil-from-coal, was based on the assumption that Government help in some form would be forthcoming. It was laid before Runciman on 14 November 1932, and the request made to the Government, through him, was for a guarantee to ICI, for ten years, of a works realization figure of 13·5d. per gallon, which was the figure currently assured to ICI by the oil companies, to whom they were obliged to sell their entire output, and from which they hoped to make a return of 7 per cent on their venture.

The request implied a contingent promise from the Government to support petrol prices if they fell below the level required for a realization of 13·5d. Not unnaturally, they refused. ICI, without very much conviction, looked into the possibility of proceeding without any Government help at all, but the prospect does not seem to have attracted them, in spite of the insistent urge to use the redundant Billingham plant and optimistic forecasts of the profits which, in certain circumstances, might be made.[57] By April 1933, with revised estimates, McGowan was back on the Board of Trade's doorstep, bargaining.

A basis at last began to emerge, apparently from discussions between W. H. Coates and Sir Horace Wilson, on which an agreement might be founded. The crux of it was the 8d. duty imposed on every gallon of imported petrol, which automatically gave home-produced petrol an advantage of 8d., in a retail price of 1s. 4d. or thereabouts, unless the Government imposed an excise tax. What ICI now sought was an assurance from the Government that they would resist the temptation to do so, or that if that was too much to ask, they would at least preserve a substantial preference, and that if they saw fit to reduce the customs duty, a preference would still be preserved.

This idea, essentially tariff protection for a new industry, had been present throughout the various phases of the discussions between ICI and the Government ever since the import duty had first been imposed, in 1928. It was an eminently reasonable scheme and one which had the advantage of costing the Government nothing—nothing, that is, except the purely notional 'loss of revenue' from petrol on which 8d. a gallon would have been charged if it had been imported instead of being made at home. This 'loss', it was supposed, no doubt rightly, the Government would be most unwilling to put up with, and it was discussed a great deal within ICI. To set off against it, Hodgkin and others made elaborate calculations to show the gains that might be expected, if petrol production started in England, from unemployment pay no longer required, from taxation on profits from the oil-from-coal business, and generally from renewed prosperity in depressed areas.[58] The Government seems to have remained quite unimpressed until

Coates and Sir Horace Wilson came together. Wilson, a senior civil servant like Sir John Anderson, was Chief Industrial Adviser to the Government. He was particularly influential with Neville Chamberlain, the Chancellor of the Exchequer, and when Chamberlain became Prime Minister, Wilson, in A. J. P. Taylor's words, 'became Chamberlain's closest adviser on practically everything, especially foreign affairs'.[59]

On 23 June 1933 McGowan called on Chamberlain. The bargaining was over and this was a ceremonial occasion: an occasion, moreover, when the 'public service' aspect of ICI would be prominent. Sir Harry had with him not only Coates and Akers but one of ICI's Directors who had been Viceroy of India (Lord Reading) and another (Lord Weir) who had been Secretary of State for Air. To this distinguished delegation the Chancellor 'intimated the willingness of H.M. Government to provide a margin of preference in relation to duties of Customs and Excise in respect of light hydro-carbon oils produced directly or indirectly from British coal'.[60]

The method of granting the preference was chosen to leave the Government, having made the basic undertaking, with as much freedom of action as possible. It was described by McGowan as 'a preference of 36 [old] pence-years . . . in the fiscal years following the 1st April 1935'.[61] 1935, being the earliest year in which plant was likely to be in production, was chosen as the base year for operating the preference scheme. By this formula the Government was left free to manipulate the rate of customs duty, or to introduce an excise duty, or to do both. ICI assumed that the period and rate of protection would lie between $4\frac{1}{2}$ years at 8d. and 9 years at 4d. On this assumption they prepared to go ahead with the erection of plant.

The Government's decision was announced by the Prime Minister in the House of Commons on 17 July 1933. Ramsay MacDonald evidently had some difficulty in understanding the method chosen for implementing the preference (he was not the only one), but he assured the House 'that, although the statement is rather long and complicated, when it is down in print it will be very simple to follow'.[62]

Legislation was needed, and a Bill was brought in on 21 December 1933. It was attacked from the Left with a demand for nationalization to replace ICI's monopoly of the hydrogenation process (other processes were covered by the Bill but hydrogenation attracted most attention) and also on the ground that the Government's 'loss of revenue', put at £4½m. during the life of the preference, was too high a price for the social benefits which the venture was likely to bring. In the circumstances of 1933, with over two million unemployed, with great distress in the coalfields, and with examples of very similar preferences granted by foreign governments, the attacks were not very damaging, the less so since the Government could show that no public money, in hard cash, was being asked for. The Bill passed into law, in the spring of 1934, as the British Hydrocarbon Oils Production Act 1934.

ICI did not wait for the Act. As soon as they were sure of the Government's intentions they went ahead with plans for plant at Billingham to produce 100,000 tons of petrol a year by hydrogenating coal, with a possibility of extending the plant to work on high- or low-temperature tars. In 1934 a 50,000-ton extension was agreed upon, and a contract was signed with Tar Residuals for creosote oil. The capital required was put by McGowan, in his formal recommendation to the Board in July 1933, at £4·4m. Of this £1·7m. was represented by assets and working capital already in existence, chiefly in the form of plant at Billingham. The £1m. or so already spent on development, which had caused so much trouble with Sir John Anderson, was not included in these figures, and preparations were being made to write it off. McGowan calculated that £2·7m. of new money would be required.[63]

The profits of the enterprise, in its early years, were to be limited. In negotiations with the Government, it was assumed that ICI would keep the profit to be taken out of the enterprise to a rate of 5 per cent, before tax, on the capital employed; that is, to £220,000. Taxation at the prevailing rate of 5s. (25 per cent) would bring the net profit to £165,000, and the whole of that was to be applied to writing off the new expenditure of £2·7m.* Quite why these restrictions were imposed is nowhere made clear. The reasons may well have been political, to avoid critics of the scheme saying that Government preference had made too large a present of profits to ICI. In any case, in the light of these figures it is clear that the attraction of oil-from-coal as a commercial proposition lay wholly in the future, when the industry would be established and much larger operations could be undertaken, at home and in the Empire. Meanwhile, some of Billingham's idle hydrogen plant could be started up again, which it never would be otherwise.

By the end of January 1934 10,400 men were at work, directly or indirectly, on Billingham hydrogenation plant, and ICI expected the number to rise to 12,000, who would be employed for about eighteen months. After that, there would be permanent work for about 2,500 (including miners) directly and for about 1,250 indirectly. Moreover, the money would start to flow. At the beginning of 1934 ICI had placed orders worth about £1·1m. and about £500,000 had already been spent by themselves or their suppliers. The Government was delighted with it all, and considered that they had made a very good bargain for the hypothetical loss of Exchequer revenue. On 15 October 1935 a

* Raised to £3m. by the creosote extension, and presumably the other figures were adjusted accordingly.

special train carried the Prime Minister to the opening ceremonies at the oil-from-coal plant, and on 7 February 1936 the first petrol began to flow. A very dangerous corner had been turned, in the early history of ICI, and the second major venture at Billingham was launched.

REFERENCES

[1] ICI Instructions to Counsel 1927, CR 59/1–1F/–.

[2] *ICI* I, p. 458 and ch. 19 generally; Melchett, 'Dyes', 23 May 1929, CR 75/107/2.

[3] L. Patrick, 'Oil from Coal at Billingham as ICI's own Enterprise', 2 March 1933, CR 59/8/279; M. T. Sampson to G. P. Pollitt, 'Oil from Coal', 11 Oct. 1927, CR 59/1/29B.

[4] A. E. Hodgkin, 'The Present Situation regarding the Production of Synthetic Petrol', 10 Jan. 1929, AD 102,164.

[5] L. J. Barley, 'The Development of Heavy Organic Chemicals including Plastics by ICI', 16 Feb. 1939, DECP.

[6] 'Relations with Royal Dutch-Shell', 22 May 1929, CR 84/1–37/3A; Sir R. Waley-Cohen to Lord Melchett, 25 Feb. 1929, CR 93/30/–.

[7] J. H. Wadsworth, 'Bergius', 31 Dec. 1927, CR 59/1–1L/– and other papers filed under 59/1. . . . Terms of purchase in A. Mond to H. Spence, 23 Feb. 1927, CR 59/1–1T/G2; Wadsworth to Executive Directors, 7 Nov. 1928, CR 59/1–1D/25.

[8] As (4); G. P. Pollitt, 'The Present Position as to Oil from Coal', 25 Jan. 1929, AD 102,274.

[9] As (3), 2nd refce.

[10] Cab. 21/322, 18 March 1925; Cab. 21/324.

[11] Sir A. Mond, 'Notes of Interview with Sir Maurice Hankey', 17 April 1928, CR 59/1–1S/1; Cab. 21/322.

[12] A. E. Hodgkin to G. P. Pollitt, 'Oil from Coal', 20 March 1929, MP 'Oil from Coal, Government Negotiations 1929–31'.

[13] W. H. Coates, 'Security during the Period of Development', 18 Dec. 1930, CR 59/8/124.

[14] CCN 6 Jan. 1930.

[15] L. Patrick, 'Oil from Coal, Government Negotiations 1927–33, a Résumé', 31 Aug. 1933, WP 'Hydrogenation/Oil from Coal'.

[16] ICI Intelligence Report on Standard Oil, 22 Nov. 1929, CP 'Oil from Coal, negotiations with Royal Dutch-Shell'.

[17] As (7), 1st refce.

[18] A. E. Hodgkin to Dr. Rule, 11 Feb. 1929, CR 41/13/18, and other papers.

[19] As (6); G. P. Pollitt, 30 April 1929, CP 'Oil from Coal, negotiations with R D-S'.

[20] W. H. Coates, 29 May 1929, CP 'Oil from Coal, negotiations with R D-S'; CCN 26 June 1929.

[21] G. P. Pollitt, 15 Nov. 1929, CP 'Oil from Coal, negotiations with R D-S'.

[22] CCN 16 Oct. 1929.

[23] A. E. Hodgkin, 28 March 1930, CR 59/34/34.

[24] Memo of 25 April 1930, CR 59/34/82.

[25] Du Pont to ICI, 27 May 1930, CR 59/34/117.

[26] 'ICI/SIG/Du Pont position', n.d., CR 59/34/117.

[27] 'Meeting of . . . ICI and Du Pont . . . in New York', 27 Aug. 1930, CR 59/34/117.

[28] L. Patrick, 'Du Pont-ICI-Oil Companies, Draft Summary for the Guidance of Mr. Francis Walker', 7 May 1940, CR 595/7/7.

[29] A. E. Hodgkin, 'General Notes on the Coal Oil Position . . .', 14 Nov. 1930, CR 59/8/124.

[30] H. Mond, 'The Hydrogenation Process', 19 Nov. 1930, MP 'Oil from Coal, Government Negotiations 1929–31'.

[31] As (22).

[32] A. E. Hodgkin to G. Ormsby Pearce, 10 April 1931, CR 59/34/82.

[33] 20 May 1930, Cab. 21/324.

34 J. H. Wadsworth, 'Notes on Discussion . . .', 30 Jan. 1931, CR 59/34/152.
35 L. Patrick, 'Oil from Coal, Government Negotiations 1927-33, a Résumé', 31 Aug. 1933, WP 'Hydrogenation/Oil from Coal'; K. Gordon, 'Position of Work on Oil from Coal in July 1930', 18 July 1930, CR 59/14/116.
36 As (33).
37 Lord Thomson to Lord Melchett, 31 July 1930, MP 'Oil from Coal, Government Negotiations 1929-31'; H. J. Mitchell to McGowan, 1 Sept. 1930, CR 59/34/82.
38 As (35), 2nd refce.; G. P. Pollitt to Board of Management, 'Coal Oil and Methanol', 18 Sept. 1930, CR 59/8/124.
39 K. Gordon, Memo 302, 20 Sept. 1930, CR 59/14/116.
40 'Note by the Chairman', 17 July 1930, MP 'Oil from Coal, Government Negotiations 1929-31'.
41 'Oil from Coal—Outline of Memorandum to H.M. Government', 25 Sept. 1930, CR 59/8/124.
42 'Memo on Discussion . . . , 19 Nov. 1930', MP 'Oil from Coal, Government Negotiations 1929-31'.
43 'Oil from Coal, Meeting at Board of Trade', 26 Nov. 1930, CR 59/8/124; Cab. 27/442.
44 As (35), 1st refce.
45 G. P. Pollitt, 'Memorandum on Interview with the Board of Trade', 5 March 1931, CR 59/8/124.
46 H. J. Mitchell to Sir H. McGowan, 'Oil from Coal, Etc.', 15 April 1931, CR 59/8/124.
47 16 July 1931, MP 'Oil from Coal, Government Negotiations 1929-31'.
48 A. E. Hodgkin, 'Oil from Coal—Suggested Memo for the Government', 17 March 1932, MP 'Oil from Coal, Government Negotiations 1929-31'.
49 'Meeting to determine future Coal Oil Policy', 3 March 1932, CR 59/8/124.
50 As (35), 1st refce.; J. H. Cotton, 'Note for Dr. W. H. Coates', 4 May 1933, CR 59/14/63; J. H. Wadsworth to General Purposes Committee, 'Oil from Coal', 5 July 1932, CR 59/14/116.
51 Chairman's Meeting, 23 May 1932, CR 59/8/124.
52 A. E. Hodgkin, 'Coal Hydrogenation', 12 Oct. 1932, p. 2, CR 59/8/124.
53 As (35), 1st refce., p. 9.
54 L. Patrick, 'Oil from Coal at Billingham as ICI's own Enterprise', 2 March 1933, CR 59/8/279.
55 Lord Melchett to Chairman, 21 Oct. 1932, 'Coal-Oil Process', WP 'Hydrogenation/Oil from Coal'.
56 As (52), p. 3.
57 As (54).
58 A. E. Hodgkin, 'General Notes on the Coal Oil Position', 14 Nov. 1930, CR 59/8/124; L. Patrick, 'Proposals for Discussion with H.M. Government', May 1932, CR 59/8/124; A. Hudson Davies, 'Effect on Employment of Building and Running a Coal Oil Factory', 2 Nov. 1932, CR 59/8/134.
59 A. J. P. Taylor, English History 1914-1945, Clarendon Press, Oxford, 1965, p. 405.
60 Sir H. McGowan to N. Chamberlain, 7 July 1933, CR 59/8/124.
61 Chairman to Board, 'Manufacture of Petrol by Hydrogenation', 6 July 1933, CR 59/8/124, vol. 2.
62 Hansard, Commons, 17 July 1933.
63 As (61), p. 7.

The Dyestuffs Cartel and the
Fine Chemical Industry

(i) A DUCK NOT SO LAME

As FERTILIZERS COLLAPSED and oil-from-coal was pushed forward to take their place, so, in another part of the field, efforts were being made to settle the future of another of ICI's major interests: dyestuffs. Here we move to a very different scene and—though the point was not at all obvious at the time—towards the course of ICI's post-war development.

The Billingham fertilizer plant represented a development of the kind of heavy chemical industry which Brunner, Mond had understood so well and from which, in Alkali Group, ICI continued, throughout the twenties and thirties, to draw more profit than from any of their other activities. Just as the alkali industry relied on one outstanding feat of technical development, the Solvay ammonia-soda process, so the fertilizer industry relied on another, the synthetic ammonia process, to provide a limited range of fairly simple products to sell in great bulk at a moderate profit, over a long period, without much call for innovation. This was the heavy chemical industry as the older generation of alkali men understood it, and oil-from-coal was intended to be another development in the same tradition.

Dyestuffs and the fine chemical industry in general had quite different ways. Products were far from simple, value in relation to bulk was high, and profits were expected less from an enormous volume of sales than from the vigorous exploitation of monopoly rights while patents lasted. Innovation was essential to provide a continuing flow of worthwhile patents, for profits would drop dramatically as rights expired. It was a business heavily dependent on research. Most of the research would be abortive, and the cost of it would have to be borne by the comparatively few successes.

The fine chemical industry was rooted in organic chemistry, and for developing it within ICI the most obvious starting point, though not the only one, was the Group based on the old British Dyestuffs Corporation of Manchester and Huddersfield. BDC, at the time of the merger, did not look like a promising start for anything at all. By contrast with Brunner, Mond's fifty years' unbroken success, BDC had a short history of Government interference, management squabbles, and financial disaster ending in the reduction of capital, in 1926, which prompted Mr. Justice Eve to refer to 'money of the country . . . invested in what I call rotten undertakings'.[1] BDC was in direct competition with IG, the success of which had begun with the crushing of the British dyestuffs industry in the eighteen-seventies. BDC had the disturbing knowledge not only that IG was very much stronger in world markets, but that so many British users of dyestuffs preferred IG's dyes that BDC could hardly hope to stay alive at home without the protection of the Dyestuffs (Import Regulations) Act 1921 which was due to run out in 1931. Altogether, there was far too much uncomfortable truth in a jingle sung at a Blackley smoking concert as late as the mid-thirties:

'It must be right,
It's bound to be,
The IG can't go wrong.'[2]

Neither McGowan nor Mond had the slightest confidence in BDC. McGowan, just before the merger, had brusquely refused McKenna's suggestion that Nobel Industries, in the national interest, should take the ailing Corporation over. Mond, in 1929, made his opinion clear with equal brutality:

I have always looked on our acquisition of BDC as largely a bargaining factor in our relations with IG and others. . . . Giving all possible credit to the staff and organisation of the ICI it is difficult to believe that with the experience we have had in this business, with a relatively small output, we are really equal competitors with the IG. . . .[3]

BDC was granted no direct representation among the original executive directors of ICI and the appointment of Ashfield, the former Chairman, as a 'lay Director' was no real compensation, formidable though Ashfield undoubtedly was.

The only executive Director of ICI at this time who has left on record anything approaching a true view of the potential importance of BDC is John Rogers. Writing in reply to Mond's memo just quoted, in May 1929, and no doubt reflecting opinion within BDC, he remarked: 'The Dyestuffs industry undoubtedly is the main organic section we have', and 'it appears to me absolutely necessary that we should be unfettered as to our technical advance, both because of the importance of the Dyestuffs industry, and of the possibility of all sorts of lines leading from the attack on the organic side of the business by concentrated research.'[4] It was no doubt on the strength of opinions like these that Rogers, in July 1928, had been appointed Chairman of a new Management Committee for BDC.[5]

BDC's main customers were in the depressed textile industries, especially cotton, and in 1927 it was working at about 50 per cent capacity.[6] The inevitable result—an assault on the standing charges of the business, especially at head office—was sufficiently drastic to make an impression still perceptible forty years later.[7] At the same time measures were put in hand to concentrate manufacture by closing some of BDC's works, which meant that the people who worked in them, as the Departmental Annual Report for 1928 put it, 'were liberated'.[8] In 1926, just before the merger, BDC had bought a majority interest in Scottish Dyes, founded by James Morton in 1915 to manufacture light-fast vat dyes previously imported from Germany, and in January 1928 the holding was increased to 100 per cent. The concentration of manufacture at Blackley (Manchester), Huddersfield, and Scottish Dyes' works at Grangemouth was by then nearly complete, and John Rogers became Chairman of a group with just under £3m. capital employed.[9] The capital employed at Billingham at the same time, by contrast, was approaching £20m.

On most important sides of its business, ICI had a monopoly or a close approach to one, but in dyestuffs, not so. As Alfred Mond regretfully put it, in 1929, 'What appears to be the matter with the Dye industry is that it has not really been merged. The BDC really represents only a part of the home trade, and not a sufficiently preponderating part to make it possible to ignore the other makers.'[10] In 1927, BDC and Scottish Dyes supplied about 20m. lb. of the 44m. lb. of dyestuffs produced in the United Kingdom, giving them about 45 per cent of the market.[11]

Among the other makers the most irritating was L. B. Holliday & Co. Ltd., flourishing impudently at BDC's gates in Huddersfield, having been founded by Major L. B. Holliday on money paid to him by British Dyes, BDC's predecessor, for his shares in Read Holliday & Co. Ltd.[12] Less wounding to BDC's *amour propre* was Clayton Aniline Ltd., wholly owned by Ciba, the Swiss dyestuffs firm which was a member of the 'Swiss IG'*—the only effective competitor, in world markets, of IG Farbenindustrie.

IG itself presented a constant threat. The Dyestuffs Act allowed the import of dyes if British makers could not supply an equivalent, either through ignorance of the technology or because of blocking patents, or if their prices were greatly above the foreigners' prices. Imported dyes were usually the newest, most desirable and most profitable, and as a consequence, by 1930, British-made dyes provided 93 per cent of the weight of dyes used in the United Kingdom but only 81 per cent of the sales value.[13] Moreover there was always the possibility that IG would one day decide to outflank the protection of the Dyestuffs Act by building works in the United Kingdom. Hoechst had done it before the war

---

* Not a subsidiary of IG Farbenindustrie.

to get round the Patent Law Amendment Act, 1907,[14] and Hoechst's successors might do it again.

The business of the dyestuffs maker consisted essentially in minute attention to detail. It was possible for a small firm to make quite a good living by concentrating on a narrow range, but for ICI to be taken seriously as a competitor of IG it was necessary to market several thousand different products, varying not only in chemical composition but also in fine shades of colour and physical state to suit different materials—a problem which grew as artificial fibres developed. In 1927 it was pointed out that cellulose acetate ('Celanese'), when first brought out, 'could not be dyed despite the existence of 2,000 to 3,000 dyestuffs'.[15]

It followed that quantities produced were tiny by the standards of other branches of ICI, and indeed of manufacturing industry generally. Blackley and Huddersfield worked on a continually interrupted succession of small batches, and anything more different from the continuous processes of Winnington or Billingham, with their six- or seven-figure tonnages, could scarcely be imagined. Customers bought from hand to mouth, carrying no stock and demanding almost instantaneous delivery (they were in the fashion business), so that 'bulk in the dyestuffs industry', as C. J. T. Cronshaw, of BDC, observed in 1929, 'was represented by a few tons'.[16]

For the dyestuffs side of ICI to become even the 'bargaining factor in our relations with IG', which seems to have been all that Mond expected of it, and still more for it to become the hopeful centre of 'concentrated research' which John Rogers foresaw, the indispensable requirement was technical competence and technical brilliance must be the aim. The Dyestuffs Act, by creating a protected market in the British textile industry—still, in spite of its troubles, one of the largest national textile industries in the world—provided the security needed for building up the kind of research effort which had been at the root of German success and was indispensable for developments just coming into view. ICI's dyestuffs research, small by IG's standards, was very considerable indeed by the standards of contemporary British industry. By 1929 seventy research chemists, nearly half of them less than five years from the university, were working at Blackley and Huddersfield and a dozen more were working for Scottish Dyes at Grangemouth.[17]

Dyestuffs research had been carried on more or less intensively since the end of the Great War, and in spite of 'a considerable reduction in staff' during the post-war depression, results were beginning to show:

Of the five outstanding developments in dyestuffs chemistry . . . during the last decade [ICI told the Dyestuffs Industry Development Committee in 1929], namely, Naphthol Ice Colours, Caledon Jade Green, Aminoanthraquinone derivatives for Acetate Silk (Duranol Colours), Indigosol products,

Soledon Colours, three are British discoveries and patents. The British Industry may, therefore, claim to be fast approaching a stage of development not inferior to that attained by the more advanced of its foreign competitors.[18]

That extract comes from a document of determined optimism, intended for a Government committee, but when due allowance has been made it can be set against Mond's lack of confidence in BDC. BDC was evidently beginning to emerge from the shadow of fifty years' neglect of the British dyestuffs industry and could already show solid technical achievements. 'The only way BDC could get into the dyestuffs business on anything approaching a substantial scale', as Cronshaw was prepared to testify some twenty years later, 'was to invent its way in. The IG never took much notice of BDC until they found out that BDC could invent.'[19]

(ii) THE DYESTUFFS CARTEL

IG's poor opinion of BDC had caused them to make demands which had wrecked the negotiations of 1927 (pp. 45–6 above) for a general agreement with ICI. Nevertheless dyestuffs, like nitrogen, remained one of the principal subjects on which both groups desired agreement, and in 1929 an ICI Chairman's Conference went so far as to resolve 'that an Agreement on Dyes was the corollary to an Agreement on Nitrogen'.[20] The trouble in both industries was the same: over-capacity.

In many countries after the Great War, and nowhere more than in Great Britain, dyestuffs industries were developed for reasons of economic nationalism without much regard to demand, so that by the mid-twenties the world's capacity for making dyestuffs was about twice as great as before the war but consumption was about the same.[21] International competition was hindered by the protection of national markets, as in the United Kingdom, and in this situation, so similar to the nitrogen situation, the instinct of ICI and IG, both dependent on foreign trade, was to seek agreement in whatever markets remained open, rather than tear each other's trade to pieces with uneconomic price-cutting.

Between 1927 and 1929 ICI and IG remained at arm's length, though they were never out of touch and co-operated in a limited way, particularly over prices in the nitrogen industry. Barred, for the time being, from an agreement with IG, ICI put pressure on the Swiss to sell their holding in Clayton Aniline, so as to give ICI a closer approach to monopoly in the British dyestuffs industry. Mond made much of the fact that ICI controlled Clayton's supply of raw materials, but the Swiss were not greatly impressed by his hint of blackmail, nor by Wadsworth's either when he wrote: 'I am sure you will not think that anything we do is in any way unfriendly.'[22] The Swiss were very confident of their technical ability and their close relationship with IG.

In 1929, while ICI were negotiating the Patents and Processes Agreement with du Pont (chapter 3 (iii)), they began to negotiate seriously with IG on the question of nitrogen. Nitrogen, plagued by overcapacity, was bound to bring with it thoughts of dyestuffs. 'We cannot', wrote Mond on 23 May 1929, 'watertight the activities of ICI and consider every branch a separate entity.' He recognized that the attempt to gain control of Clayton had failed ('due to the amour propre of some of the Swiss makers'), and went on:

The proposal which I put forward is the formation of a new British Dyes Company, which would embrace BDC, Clayton, and possibly British Alizarine: this Company to exchange its shares for those of the respective constituent members . . . and owing to our preponderance we should obviously control it. Representation of the various interests could be given on the Board of this new Company, and possibly the Swiss could be induced to feel that they were not asked to surrender part of their undertaking to ICI. This might make things easier. It might also be possible to give the IG some interest in this new undertaking and obtain their cooperation, technical and commercial, in this manner.[23]

This was what Mond had meant when he said, in the same paper, that he regarded the acquisition by ICI of BDC 'as largely a bargaining factor in our relations with IG and others'. In making his proposal he followed the same line of diplomacy as before the merger. The great end of policy, in Mond's view, was a settlement with IG, and in attaining it he was prepared, if necessary, to treat BDC as expendable. Devoted as he was to the heavy chemical industry and to the Billingham project, he took no account of the potentialities of the organic side of ICI. John Rogers drew Mond's attention to them, but not with sufficient force to make him change his mind.

Mond's colleagues accepted his proposal, hoping it would help towards the much desired 'ultimate agreement with the German-Swiss Group for the territorial division of markets on similar lines to those under review for the proposed Nitrogen Agreement'.[24] Mitchell investigated the implications of the scheme, and on 27 June 1929 a strong team from ICI, headed by Mond and McGowan, put it to Bosch and Schmitz of IG with a definite offer of IG participation in a jointly owned dyestuffs company which, as Mond frankly said, 'would entail removing BDC from its present position in ICI organisation, but it might get over the Swiss difficulty'.[25]

The plan made no appeal at all to Bosch. He was negotiating for an agreement with the Swiss and the French and he probably had no intention of irritating the Swiss by dealing at the same time with ICI separately. He turned the joint company down flat, and when Mond, evidently with a touch of desperation, asked 'whether the IG could not influence the Swiss, because we [ICI] definitely did not want competition with Swiss and IG all over the World', he simply said that IG had no shareholding in the Swiss Dye Industry and 'an agreement would be more possible when some of the older Swiss directors had passed away'.

Three times in the nineteen-twenties IG came close to taking a share in the British chemical industry. In 1924 they were prevented by the British Government. In the summer of 1926 Orlando Weber got in their way (and Alfred Mond's) and ICI came into existence instead. In 1929 they refused the offer. No similar offer was ever made again, and ICI remained in full ownership of their dyestuffs business. Quite quickly, this turned out immensely to ICI's advantage, but it was an advantage, as we have seen, which they worked quite hard to forgo.

Though rebuffed over the joint company proposal, ICI still wanted an agreement with IG. They had convinced themselves that a show of strength in the nitrogen industry, when they had gone ahead with construction at Billingham in spite of IG's objections, had brought IG to terms. Perhaps a hard line might pay in the dyestuffs industry, too.[26] They had already chosen their ground for a fight. It lay in the field of patent law.

IG's patent policy was aggressive. In Germany they opposed all ICI's applications on principle and were said 'to have so terrified the examiners in the German Patent Office that they do all they can to avoid trouble with the IG'. ICI, by contrast, began by letting IG applications for British patents go unopposed 'because', said Slade, 'we did not wish to do anything which would prejudice negotiations in the future'.[27] This passivity seems to have encouraged IG, who made more and more British applications—200 in 1925, 370 in 1926, 830 in 1927[28]—and many were 'selection patents', which represented a device, of dubious legality, for extending patents which would otherwise have lapsed. Some of IG's applications affected Billingham, but most concerned dyestuffs. Billingham had some success with opposition to German applications in 1928, and early in the same year, long before the negotiations for the dyestuffs merger, it was decided in ICI that as a matter of general policy IG's advance must be checked.

In the dyestuffs business the chief threat lay in 'user patents' which covered the processes used in dyeworks, so that BDC's customers might find themselves facing actions for infringement if they used BDC's products rather than IG's. The most important user patents were in Naphthol AS or Azoic Dyeing, which produced colours fast to light and washing, and in which the UK turnover was about £120,000 a year—too much for ICI to see denied to them without a fight. The original patents dated from 1912–13, and since 1923 Griesheim Elektron, the original owners, subsequently absorbed into IG, had been applying for selection patents in order to prolong the life of their rights.

ICI, advised by Stafford Cripps, KC, began proceedings against IG's Naphthol AS selection patents in the spring of 1928. Mond warned IG of what ICI were about to do, protesting that ICI were acting 'with no feeling of antagonism . . . and without any desire to prejudice the friendly relations which . . . exist between us, and . . . I am desirous that the matter should be treated . . . as a friendly case between us commenced solely with the object of testing the legal position.'[29] Schmitz, to whom Mond addressed himself, replied in an equally friendly tone and hinted at a settlement.[30]

On 1 November 1928 ICI filed a petition for revocation against three patents (193834, 193866, 199771) in the Naphthol AS field. The case was heard before Mr. Justice Maugham, sitting with an assessor, between 21 November 1929 and 21 January 1930, taking twenty-seven days in court, and Maugham gave judgment for ICI on 13 and 30 March 1930.

We succeeded [wrote William Morris, ICI's Solicitor] in getting a judgment in our favour on every point of law and fact, which . . . laid down for the first time the legal principles which it was necessary for . . . Selection patents to satisfy to be valid, which was one of the main objects with which we embarked upon the proceedings.

The Case will probably cost the IG about £30,000, and the effect of the Judgment is to make very nearly the whole of the Selection patents which were taken out by the IG subsequently to those the subject of the proceedings, clearly invalid, and consequently the whole Naphthol AS field of dyes is now open to us.[31]

This was a famous victory. It put an end to IG's main method of protecting their past technical supremacy just at the moment when ICI, in Cronshaw's later phrase, were beginning to invent their way in, especially in dyestuffs for acetate silk. ICI's bargaining strength improved immensely and the results were immediately noticeable. The judgment had no direct bearing on ICI's growing technical competence, but it seems to have jolted IG finally into recognizing it. ICI, hitherto so tame in dyestuffs matters, had shown that they could turn round and bite, and in presenting their case they gave a display of technical skill even more important, in its effect upon IG, than the purely legal aspects of the case itself. At the same time ICI's self-confidence grew. Before the Maugham judgment, broadly speaking, they had made the approaches and been coldly received. Afterwards, that was not the case at all.

By the time Maugham gave judgment the German, Swiss, and French dyestuffs makers had for some months been joined in the cartel agreement which was under negotiation when ICI put forward the joint company plan of 1929 (p. 188 above). The parties to the agreement were IG, Swiss IG (Ciba, Sandoz, and Geigy), and the French Compagnie des Matières Colorantes, generally referred to as Kuhlmann,* which with its subsidiary St. Clair du Rhône† and its 50 per cent associate St. Denis,‡ controlled upwards of 90 per cent of French dyestuffs production. It was with the three-party cartel as a whole rather than with any individual member that ICI would have to come to terms.

Before the judgment was two months old, in May 1930, R. T. Duchemin, Chairman of Kuhlmann, was complaining to Mond of an ICI invasion of the French dyestuffs market. ICI were getting round the French tariff by building plant near Rouen in which they could process dyes imported in high concentrations, and they were also cutting prices, both very unfriendly acts in the over-supplied dyestuffs trade of 1930. Duchemin threatened to get his own back in England. Mond's reply was aggressive:

On general principles I have to point out that we have no agreement whatsoever with the Continental Group to which you belong regarding the sale of dyes. . . . Your Firm has chosen to ally itself with the Germans in the matter of the Dye Industry, without in any way endeavouring to communicate with us to see whether or not it was possible for any mutually beneficial arrangements to be made between the leading industries of the two countries to whom the problem of national defence to which you refer was a matter of mutual interest. It is therefore a matter to be regarded more as a question of competition with a continental cartel than an Anglo-French problem. . . . I would . . . like to enter a protest against the paragraph in your letter in which you refer to your action in this country. May I candidly say that it is not a line of argument which is likely to prove useful in dealing with my Company in endeavouring to come to a mutually satisfactory arrangement. . . .[32]

This was all very sharp, but the letter as a whole was an invitation to seek a settlement. It is probably not a coincidence that on 1 July 1930 von Schnitzler, Molnar, and Weber-Andreæ, all of IG, were at Millbank talking to Mitchell and Wadsworth, and that their tone, for representatives of IG, was quite new:

The Germans [Mitchell reported] frankly admitted that—as distinct from the other dyes producers—we have not only achieved material individual progress, but have made substantive contributions to the development of the art of dye manufacture in general . . . and that this must receive due recognition in the conclusion of any arrangement with them. . . . There can be no

* Compagnie Nationale des Matières Colorantes et Manufactures des Produits Chimiques du Nord Réunies, Etablissements Kuhlmann.
† Compagnie Française des Produits Chimiques et Matières Colorantes de St. Clair du Rhône.
‡ Société des Matières Colorantes et Produits Chimiques de Saint Denis.

doubt that the Germans are seriously perturbed at our progress and apprehensive lest we may encroach still further on their remunerative fields. They particularly referred to the acetate silk problem. . . . They also realise clearly that it is useless for them to rely on patent protection or on the attainment of a favourable or monopolistic position in this country through any relaxation of the Dye Act restrictions.[33]

In this frame of mind the Germans were prepared to come much further to meet ICI than in the past. They agreed, in Mitchell's words, that 'it would be entirely proper for us [ICI] . . . to claim our freedom as regards further developments in the industry'—that is, that ICI might advance towards the newer and more rewarding branches of the dyestuffs industry instead of confining themselves, as the Germans had demanded in 1927 (p. 45 above), to 'the staple articles'. They conceded, too, 'that from a prestige point of view our aspirations with regard to the British Empire were not unreasonable'. This, again, was a long step beyond the Germans' previous position, which had been that the British dyestuffs industry was barely fit to operate in its home market and should certainly not presume to move outside it.

The Germans said they would be obliged to act with their cartel partners in coming to any formal agreement, but 'the view was expressed that in the general interest both sides should explore every possible avenue in order to reach an understanding, which would prepare the way for a general world agreement on the dye question.' Mitchell then, according to his own account, 'stated that it seemed to me it would be better if we did not actually join in their cartel, but that a separate agreement should rather be made with us by or on behalf of the cartel. I felt it would be wise if we could keep outside of the cartel obligations which, although I have no knowledge of them, I imagine must be detailed and troublesome.'

Mitchell, at this first unofficial and preliminary meeting, had stated the basis on which a dyestuffs agreement was eventually founded, after negotiations stretching over more than eighteen months. It is unnecessary to follow them in detail. ICI throughout played from their newly won position of strength. They kept up sufficient pressure on the French market to make sure that Duchemin's nerves were never entirely at rest.[34] At home they ostentatiously prepared to take advantage of the Maugham judgment by setting up plant in the field which it covered. At the same time, Rogers suggested to von Schnitzler that it might be best if ICI gave up opposition to IG's remaining British patents in the Naphthol AS field, because although the opposition was almost certain to be successful, 'such success might do no more than simply throw open in too clear a manner such fields to others [Holliday, perhaps?] besides ourselves'.[35] Later, when the two sides met (April 1931), the ICI representatives suggested 'that the best interests of both companies were to preserve the maximum appearance of validity of the patents and the maintenance of the already highly profitable selling prices; and this led us to propose that IG and ICI should divide the UK market on a quota basis.'[36]

The main thrust of ICI's policy was to persuade the cartel to agree to an ICI quota based on the weight of dyestuffs sold by all makers within the British Empire, but not territorially confined to it. This was an unusual demand for ICI to make, since their almost unvarying demand in international negotiations was for a territorial monopoly within the Empire, leaving the rest of the world to foreigners. In this case there were special circumstances, the most important, probably, being that within the Empire there was no demand for some of the most sophisticated and profitable dyestuffs. Besides that, the Swiss and Germans were well established within the Empire and there could be no question of dislodging them, because they supplied some dyestuffs outside ICI's range. Therefore ICI wanted a compensating right to sell in Swiss and German markets. Mitchell's estimate of the quota he was demanding, equivalent to British Empire trade exclusive of the trade of non-cartel makers, was 27,745 tons, which was 7,412 tons more than total British production in 1929.[37]

This demand caused a sensation. Köhler of IG, talking to ICI representatives at Frankfurt in April 1931, said he nearly fell off his chair when he saw it. 'Fancy,' he went on, 'you want 40m. marks' extra business from us.'[38] ICI remained difficult to shift and sufficiently aggressive in price competition to keep the other side very anxious for 'closer, friendly co-operation . . . which, while safeguarding the respective clientèles [of ICI and the cartel], would make it possible to stabilise, i.e. raise, prices'.[39]

The agreement so long in contemplation was signed on 26 February 1932. It was intended to last from 1 January 1932 until 31 December 1968—that is, to all intents and purposes it was intended to be permanent. It established a 'Cartel Management' responsible for 'all decisions coming within the scope of this Agreement'. The Management was to consist of a Board of Directors and an Advisory Council, and the Directors, who formed the executive body, were to be nominated by IG (three), Swiss IG (three), Kuhlmann (two), and ICI (two), making ten altogether. The Cartel was domiciled at Amsterdam.

The main business of the Cartel was to operate a quota system based on the value (not weight) of the joint turnover of the Continental Group and of ICI in the 'reference period' 1 January 1930 to 30 June 1931. ICI's quota, under these arrangements, was fixed at 8·43 per cent of the total. It had to be recognized that circumstances might change during the life of the agreement, and in that case the Cartel Management was required 'to take steps to re-establish the equilibrium' or to arrange compensation. Clearly, however, a major change would have required full-dress negotiations between the principals rather than a decision of their nominated representatives. The agreement did not, and the point was insisted upon with some vehemence from time to time, do away with competition among the parties. The quota was a maximum permitted figure. It was not guaranteed and it had to be striven for, though not by price-cutting, since the great object of the agreement was to prevent that.

Against outsiders the members of the Cartel presented a united front. They were required 'to refrain from all measures which might encourage outside competition', and after consultation amongst themselves, and acting through the Cartel Management, they might take measures towards gaining control over outsiders or reducing their competitive power. A special provision was made for ICI to take over Holliday, and the Continental Group had their named takeover objectives also.

The existence of the agreement probably could not have been kept secret and no attempt was made. It was announced to the world in a press notice issued by IG on 1 March 1932. But although the existence of the agreement was well known, its provisions were not. 'The general contents as well as the different stipulations of the contract', said a note circulated by Rogers to the General Purposes Committee of ICI, 'are to be treated as strictly secret towards unauthorised persons and especially towards customers.'[40]

Within ICI the agreement was often referred to as 'the 4-Party Agreement', emphasizing that ICI had associated with the Continental Cartel but had not joined it. The continental organization was much tighter and it provided for exchange of technical information between the parties. There was in general no exchange of information and ICI never sought it, which is probably a further indication of ICI's growing self-confidence, the more so since IG still had a long lead in scale and comprehensiveness of research.

From ICI's point of view the four-party agreement was a victory, for in arriving at the figure of 8·43 per cent for ICI's quota of the Cartel's trade, the Continental Group had agreed to forgo business worth £195,000—£90,000 in the United Kingdom, £75,000 in the Near and Far East, £30,000 in Australasia. ICI forced this concession by following a consistently hard line from the Naphthol AS litigation onward, but it would not have been credible without the backing of BDC's developing technical competence. Once again it was shown that in the world chemical industry, as in world politics generally, foreign policy needed a base of demonstrable strength at home.

By the mid-thirties, it appeared to the dyestuffs management in ICI that the effect of the four-party agreement had been to stabilize the trade and create sufficient security for the parties to develop their products in competition with each other. 'The agreement', says a document of 1936 intended for the General Purposes Committee of the ICI Board, 'has merely abolished price competition; it has left the parties free to compete in the search for novelty, in the attainment of higher standards of quality and in every direction in which service to the customer is involved. Against the suggestion that under this convention the consumer is exposed to the risk of mercenary exploitation, the above affords a convincing argument.'[41]

The 'convention', in fact, far from guaranteeing Dyestuffs Group a sheltered life, ensured that it would be exposed to international competition far fiercer than the competition any other Group had to face. One consequence was much greater consciousness than was usual in ICI of the importance of advertising and publicity. The Dyestuffs management themselves, in 1936, said they were 'in no doubt whatsoever that the tendency of trading agreements has been to place greater emphasis than ever on the technical aspects of competition, particularly on novel lines of development.'[42]

In two years of intensive diplomatic activity, driven by pressures generated by the state of world trade, ICI entered into three major international agreements: the Nitrogen Cartel of 1930, the IHP Agreement (sometimes called the Hydrogenation Cartel) in 1931, and the Dyestuffs Cartel in 1932. Along with a multitude of other agreements at home and abroad, including the Solvay and Alkasso alkali agreements inherited from Brunner, Mond and the du Pont Agreement developed from the foreign policy of Nobels, they provided a stable framework within which ICI's business could recover from the Billingham disaster. At the same time, in the late thirties, the first shadowy outlines began to form of ICI's development in mid-century, which was to turn it into a very different enterprise, but in 1939 the business was forced off its course again by the outbreak of war.

We shall carry these matters further in Parts III and IV of this volume. Meanwhile we must consider the general development of ICI's overseas trade and the course of events at home which led up to rebellion against McGowan in 1937–8.

REFERENCES

[1] *ICI* I, p. 447.
[2] J. D. Rose, Levinstein Memorial Lecture, *Chemistry and Industry*, 28 Nov. 1970.
[3] A. Mond, 'Dyes', 23 May 1929, CR 75/107/2.
[4] J. Rogers, 'Dyes', 29 May 1929, CR 75/107/3.
[5] ICIBM 575, 24 July 1928.
[6] Departmental Annual Reports 1927, p. 52, 'Dyestuffs and Intermediates (BDC)'.

[7] 'BDC Ltd., Economy Programme', Papers Relating to Board Minutes, vol. 2, p. 238A, LDHC.

[8] Departmental Annual Reports 1928, p. 269, 'Dyestuffs'.

[9] 'Report on Dyestuffs Business', 27 March 1929, p. 1, CP 'Dyestuffs Industry'.

[10] As (3), p. 2.

[11] As (9), p. 5.

[12] ICI I, p. 273.

[13] Report of the Dyestuffs Industry Development Committee, Cmd. 3658/1930, p. 7.

[14] ICI I, p. 266.

[15] BDC Ltd., 'The Dyestuff Industry in Terms of Numbers', Oct. 1927, CR 75/2–1C/–.

[16] Minutes of Meeting of Sales Control, 1 Jan. 1929, CR 75/22–3.

[17] BDC and SD, 'Report presented to the Dyestuffs Industry Development Committee', 1929, pp. 42, 46, CR 75/3–1B/R1.

[18] As (17), p. 48.

[19] USA v. ICI et al., 'Preliminary Outline and Draft of Dr. Cronshaw's Testimony', Box 173, File 173, LDHC.

[20] CCN 4 June 1929.

[21] Address by Sir Harry McGowan, Manchester, 25 Oct. 1927, ICI Secretary's Dept.

[22] A. Mond at meeting with Swiss IG, 13 April 1928, CR 75/55/1, also letter to Solvay, 7 Dec. 1928, CR 75/76/1; Wadsworth to Brodbeck (Ciba), 24 July 1928, CR 75/52/8.

[23] A. Mond, 'Dyes', 23 May 1929, p. 3, CR 75/107/2.

[24] As (20).

[25] CCN 27 June 1929.

[26] 'Notes on informal discussion 31 May 1929', CR 75/21/–.

[27] R. E. Slade to G. P. Pollitt, 22 Dec. 1926, CR 4/15/1.

[28] 75/94/–; M52, 27 March 1928, LDHC.

[29] A. Mond to H. Schmitz, 14 May 1928, CR 4/15/–.

[30] H. Schmitz to A. Mond, 18 May 1928, CR 4/15/–.

[31] W. Morris, 26 March 1930, CR 75/24/24.

[32] R. T. Duchemin to A. Mond, 19 May 1930, CR 75/16/65; Mond to Duchemin, 26 May 1930, CR 75/16/65.

[33] H. J. Mitchell, 1 July 1930, CR 75/24/40.

[34] Corres. of Nov.–Dec. 1930 in CR 75/16/65.

[35] J. Rogers to G. von Schnitzler, 19 Feb. 1931, CR 75/24/91.

[36] Notes of Frankfurt meeting, 15–16 April 1931, CR 75/24/91.

[37] H. J. Mitchell to G. von Schnitzler, 26 Feb. 1931, CR 75/24/40.

[38] As (36), pp. 10 ff.

[39] IG and Swiss IG to H. J. Mitchell, 7 May 1931, CR 75/109/109.

[40] Draft Agreement, Continental Cartel and ICI, 16 Feb. 1932, CR 79/109/134; see also 'Memo for meeting of the General Purposes Committee with . . . the [Dyestuffs] Group', 6 Feb. 1936, GPCSP; Dyestuffs Group 'Agenda for Meeting with the General Purposes Committee', 7 Feb. 1936, '4-Party Cartel Main Dyestuffs Agreement', GPCSP.

[41] As (40), 2nd refce., p. 16.

[42] Ibid. p. 8.

CHAPTER 12

Overseas Trade in the Thirties: ICI and Economic Nationalism

(i) THE EXPORT TRADE

IN OVERSEAS TRADE it can be taken for granted that one reward of building up a conspicuously successful export business will be a demand, at the receiving end, for local manufacture to compete with it. If local manufacture is not truly competitive—if, as is likely, the costs of a local manufacturer's small-scale plant are much higher than the costs of an exporter's large-scale plant, if the quality is lower, if local conditions are not conducive to a particular industry—then there will be a demand for protection which will probably be met by high import duties, import quotas, exchange control, or some other device. In the end, by one means or another, the successful exporter will be obliged to choose between losing his trade altogether or setting up local plant himself.

If he chooses to manufacture locally he then finds himself in an exposed position as a foreign investor. Success may well bring obstacles to the transfer of profits or threats of expropriation. If he is wise he will seek to naturalize his business by taking local interests into partnership.

ICI and their forerunners had always had to live with these problems. In the nineteenth century both Nobels and Brunner, Mond had been faced with protectionist policies in British colonies, the South African Republic and the USA, and in ICI they were aware from the start that policies of this sort would become more and more widespread as one country after another began to hanker after industrial self-sufficiency. The way the process worked was described by L. J. Barley, in general terms, in 1934:

... economic nationalism so far as the chemical industry is concerned starts with much simpler products than soda ash, nitrogenous fertilisers, and petrol from coal. The usual sequence is sulphuric acid, superphosphates, aluminium and other sulphates, soap, crude pigments, disinfectants and simple agricultural chemicals based on sulphur and arsenic. Tanning, low-grade textiles, steel rails and paper pulp may follow and the earliest ICI export products to be affected are generally caustic soda and chlorine.[1]

Against this background ICI came into existence with a fine

imperialist flourish—'Imperial in aspect and Imperial in name', as the founders put it[2]—to concentrate British strength in the markets of the world generally and in particular within the British Empire. Economic nationalism was already strong in the twenties, and it was reinforced in the thirties by trade depression, monetary chaos and, finally, preparations for war, so that from 1934 onward it was increasingly frequently discussed at high levels in ICI.

'One of the major problems which faces the Company at the present moment', wrote the second Lord Melchett in 1934, 'is the question of the extent to which it is desirable to increase the capacity of our factories in this country or alternatively erect factories in the markets to which we are today exporting.'[3] He pointed out that most overseas plants would be 'basically uneconomic', but that was immaterial because overseas governments were determined to have them and 'the cost factor, except in very extreme cases, is in no way decisive'. Since building plant overseas was inevitable, he suggested that so far as possible it should be timed to fit in with the expansion of demand at home so that plant in the United Kingdom would not have to be closed when plant overseas came into production.

'The character of our Overseas trade', wrote Pollitt in 1935, pursuing the same theme, 'is rapidly undergoing a complete change. In each country in which we trade, we are finding ourselves less and less able to sell goods manufactured in Britain and more and more compelled to look upon our . . . [foreign companies] . . . as manufacturing concerns from which we derive an investment revenue only.' The revenue was becoming more and more difficult to bring home and, as for capital

TABLE 10

EXPORT TRADE OF ICI 1934–1937

	Total Sales[a]					
	Great Britain and Ireland		Other Markets		Total[b]	
	£000	%	£000	%	£000	%
1934	26,955	64·34	14,942	35·66	41,897	100
1935	29,863	65·77	15,539	34·23	45,402	100
1936	32,762	68·43	15,111	31·57	47,873	100
1937	39,677	67·63	18,992	32·37	58,669	100

a Includes all ICI products except sales through SA Azamon in Spain. Presumably excludes goods sold for other manufacturers through the 'Foreign Merchant Companies' and 'venture trading'.
b Total ICI sales figures calculated in 1970 (Appendix II) show smaller figures but the same tendency.
Source: Deputy Treasurer to Finance Committee, 8 Aug. 1938, FCSP vol. xxxviii, p. 4151.

overseas, it 'must inevitably be finally lost, as there will exist no way of bringing it back'.[4]

ICI had a great deal at stake. The value of export sales in the mid-thirties (see Table 10 opposite) ran consistently a little below or a little above one-third of total sales. Exports were chiefly alkali, nitrogenous fertilizers, and dyestuffs; and the effort put into the negotiation of the Nitrogen Cartel and the Dyestuffs Cartel is a measure of the importance attached to the last two. Exports of alkali, running in the mid-thirties at about 340,000 tons a year, represented about one-third of the output of ICI's most prosperous Group which,[5] partly by reason of being the largest exporter of alkali in the world, made an indispensable contribution to the profits of ICI as a whole. The possibility that the economic policy of governments abroad, as for instance in Australia, India, and South America, might force ICI to put up alkali factories in their territories had to be taken very seriously indeed, for these factories would inevitably be uneconomic and they would take trade away from the very efficient plant at Winnington.

On the capital side there were considerable sums at risk if, as Pollitt gloomily supposed, they were all destined to be lost. Moreover, under pressure for local manufacture, capital invested in this way was likely to increase rather than diminish. In 1937 there appears to have been some £11·5m. capital employed overseas, representing perhaps 12 per cent of total capital employed in ICI (£94·5m.).* Of the £11·5m., some £3·1m. was employed in 'Foreign Merchant Companies', wholly-owned subsidiaries of ICI engaged in the export trade. The remainder, representing the kind of investment most likely to grow, was in manufacturing companies in Canada, South Africa, Australasia, and South America. The Canadian and South American interests were jointly owned with du Pont, the South African with De Beers, and in Australia a number of local companies were associated with ICI.[6]

Calculations of the capital employed in Great Britain making goods for export show a total of £18·5m., arrived at on the same basis as the £11·5m. quoted above. It seems, therefore, that in the late thirties ICI had some £30m. capital employed either overseas or in plant at home manufacturing for overseas markets. It was distributed as shown on the following page.[7] Each of the Foreign Merchant Companies was wholly owned by ICI and registered in the country where it operated. They were all selling agents, not manufacturers, established either where the export trade was too important to be entrusted to independent agents or where there seemed to be some prospect that it might become so.

---

* The figure of £11·5m. and the figure of £94·5m. are from different sources and the comparison between them is a risky one. The relationship, however, does not seem inherently improbable.

	ICI Capital employed in Overseas Trade 1937	
	Empire Markets £m.	Foreign Markets £m.
Home Groups	10·8	7·7
Foreign Merchant Companies	1·2	1·9
Canadian company (CIL)	3·3	
Australian company (ICIANZ)	2·4	
South African company (AE & I)	1·5	
South American companies		1·2
	19·2	10·8

The original companies had been established by Brunner, Mond in China, Japan, India, and Australia in the early twenties.[8] Their main purpose, of course, was to sell alkali and, later, fertilizers, but they had been allowed—indeed, encouraged—to take other manufacturers' goods as agents. Some, such as dyestuffs, were part of the stock-in-trade of the chemical industry. Others were not, and 'venture trading', as it was called in ICI, could lead the merchant companies along curious by-ways. 'In India', says a report of 1933, 'a large proportion of the Venture trade is in old newspapers. On these the margin is small but the business is of assistance to the sale of ICI products.'[9]

More 'Foreign Merchant Companies' were set up after ICI was formed. Late in the thirties there were about a dozen, some of them very small, operating in Asia, the Middle East, and South America as well as in the territories covered by Brunner, Mond's original companies. Some prospered, but the Group as a whole suffered both from the general difficulties of overseas trade and from peculiar disadvantages of its own.

To take the second point first, one of Brunner, Mond's leading motives for setting up selling companies abroad, rather than running export trade through branch offices, was to avoid United Kingdom tax on profits already taxed overseas, and that motive was no less strong after ICI took over. To avoid home taxation there must be no control, though there would certainly be influence as a major shareholder, from home.[10] But the local Directors were all used to control from home and their careers depended on the goodwill of the home authorities, represented by men of the stamp of J. G. Nicholson and Sir Harry McGowan. They could hardly avoid regarding themselves, inwardly, as branch managers, and indeed as late as 1936, in the opinion of P. C. Dickens, the Foreign Merchant Companies were 'merely sales offices abroad'.[11]

This ambiguous position was not a good basis for forceful management. Trouble arose from it in Brunner, Mond's day and again under

ICI. McGowan, prompted by criticism of the China Company made by Pollitt, descended on Shanghai in 1933 and spent three weeks there, rending the local management. 'The trouble', he wrote, 'lies primarily in the control exercised until quite recently from London. . . . China was run as if it were an Area Sales Office at say Bradford, instead of a country nearly 10,000 miles distant. . . . I found that the local Directors were literally afraid of their positions if they criticized adversely any proposals from London.' Elsewhere he remarked: ' . . . in all my experience I have not come across a Board of Directors with less idea of their responsibility.' He left behind him V. St. J. Killery, later a Director of ICI, with the ominous instruction: ' . . . the whole Board must be under your observation and unless they are inclined to put more hustle into their jobs they cannot remain as Directors.'[12]

Pollitt's criticisms of overseas staff in the thirties were not confined to the China Directors. Between February 1932 and May 1933 he travelled round the world, visiting ICI establishments in the Middle East, A'rica, Australasia, the Far East, and the countries between China and India. He was not impressed with the men on whom the Foreign Merchant Companies relied to carry out their essential function—selling. He found them not so good as their competitors and lacking in technical knowledge of ICI's products.[13] Pollitt blamed ICI's methods of selection and training, but it is probably also true that really able young men in ICI, scientifically educated, had no desire to go to these far distant selling jobs when they could find much more congenial employment, with much better prospects, at home. Whatever the cause, the effect was bound to be depressing.

The Foreign Merchant Companies Group was peculiarly exposed to the penalties of success. Whenever a market—Argentina, perhaps, or Brazil—showed real promise, the local ICI selling company would be turned into a manufacturing company and removed from the Group, with the paradoxical result that success in the export trade, which was what the Group was in business for, could be guaranteed to ruin the Group's profitability, leaving it with the struggling companies and depriving it of those which established themselves. As an administrative device, no doubt, it was sound enough, but it cannot have been good for morale.

The Foreign Merchant Companies' Group did not handle the whole of ICI's export trade. More than half went through other channels, particularly the jointly owned companies in the British Dominions and South America, which we consider below (p. 203). Nevertheless, in 1936, against total export sales of about £15m., the Foreign Merchant Companies' turnover was £6·7m. and in 1937, against £19m., £8·6m.[14] Their financial results, consequently, were bound to have a considerable bearing on the results of ICI as a whole.

The Foreign Merchant Companies' results, whichever way they were looked at (several ways were tried), were miserable. In 1934/5, taking account of manufacturing plant in Great Britain used for export production, capital employed in the Foreign Merchant Companies' business was reckoned at £11·5m. On that they earned £350,000 gross. By the time they had been charged with Obsolescence and Central Services the net profit was £8,557, or 0·074 per cent on the capital employed.[15] McGowan's consternation at what he found in China in 1933 is now, perhaps, all the more readily understandable.

When later figures, at Coates's insistence, were calculated on the basis of marginal costs they were not much more encouraging. In 1936 the Foreign Merchant Companies earned £69,340, representing 0·7 per cent return on capital employed, and in 1937, £332,976, representing a return of 2·6 per cent.[16] The proportionate increase in profit was impressive but the absolute figures remained deplorable, comparing very poorly with ICI's total figures—not over high—of 8 per cent return on capital employed for 1936 and 9 per cent for 1937.

The ICI Board, contemplating the Foreign Merchant Companies' sad showing, were not in doubt about the general underlying causes, as distinct from weak management in particular cases, such as McGowan had diagnosed in China. The Foreign Merchant Companies were suffering from the world-wide economic and political disorders of the thirties: disorders which persisted after the worst of the depression had gone by and grew worse, in some respects, as the forewarnings of world war became more insistent. Coates, in July 1939, summarized the situation: 'The small profits made by Merchanting Companies abroad on trading account have in the main been due to Japanese competition and numerous other obstacles to international trade.'[17] Among the 'numerous other obstacles' Coates could expect his readers to take for granted exchange control, import restrictions, high protection, and the rest of the apparatus of economic nationalism, and he did not go into detail, except to add 'war conditions in . . . three countries'—Spain, Palestine, and, from ICI's point of view much the most important, China.

In China war and Japanese competition went together. Indeed it might be said that Japanese competition took the form of war. Everywhere, Japanese producers were less amenable to control by cartel than any other major producers in the world chemical industry. Efforts were made to get them to make agreements, but without much success, and their uninhibited competition ravaged ICI's export profits not only in China and the Far East generally but also in India, which European and American competitors, in general, were content to leave to ICI. There, in 1937, ICI's export trade—there were no ICI manufactures as yet—was showing only a fraction over 5 per cent on the capital employed.

(ii) THE IMPERIAL INHERITANCE

If, about 1935, anyone had asked McGowan to justify the word 'Imperial' in ICI's title he would probably have pointed first not to any of the Foreign Merchant Companies, even the big one in India, but to ICI's operations in South Africa, Australia, and Canada. They showed how it was possible, in reasonably settled political conditions, to come to terms with protectionist policies, for in all these countries the governments were trying to move from heavy dependence on primary production towards industrial self-sufficiency, and they were relying on protection to do it, much after the example set by the United States in the nineteenth century.

Before 1939 the process had gone further in Australia and Canada than in South Africa, but in all three it was obliging ICI to establish local manufacture in co-operation with local interests. That, as Melchett, Pollitt and others observed, was often less profitable and always more troublesome than simple export, but in an increasingly protectionist world it was becoming the only way to do business overseas on a scale large enough for a company as big as ICI. As a consequence, it was indicative of the way in which ICI's trading policy overseas would develop, within and outside the Empire, from the thirties onward.

Just as ICI inherited a large export trade in alkali from Brunner, Mond, so from Nobels they inherited manufacturing enterprises in South Africa, Australia, and Canada. All three countries were important markets for mining explosives, and as such they had attracted Nobels' attention from the very start of the dynamite business. In each country, for reasons discussed in Volume I, Nobels had found it expedient to set up manufacturing companies, even though that meant depriving the factory at Ardeer of profitable export trade.[18] These explosives plants provided, in each country, the basis from which a diversified business in various branches of the chemical industry could be developed as opportunities presented themselves or local pressure became too strong to resist.

ICI's policy in each country was fundamentally the same. Put in its simplest terms it was a policy of making sure that as much as possible of the development of the chemical industry should be kept in the hands of ICI and their local associates, so that their position in their own territories would be as dominant as the position of ICI themselves in Great Britain. In establishing themselves in this way ICI were greatly helped by their general understanding with du Pont and IG that they would keep out of ICI's way in British territories if ICI kept out of du Pont's way in the USA and out of IG's way in Europe. While that understanding lasted there was no likelihood of competition from du Pont or IG in the Dominions, but there was always the possibility of a threat from local interests, other European and American firms (particularly in synthetic ammonia), or, in Australia, the Japanese.[19]

If, from time to time, this policy meant putting up uneconomic plant, then so be it. Better that than letting some outsider in, if there was any likelihood that he could persuade the Dominion Government to grant him protection. Therefore if the Australians, say, wanted alkali plant in a relatively unattractive location, ICI would build it rather than see anyone else do so, and the same argument would be extended to other activities because, as B. E. Todhunter put it in 1934, 'Australia will inevitably aim at becoming self-contained . . . and it may be necessary to sacrifice some immediate profit in order to consolidate our future position.'[20]

To partners overseas ICI offered capital resources, management, and technical help, including patent licences. The basis of the bargains which were made was that ICI agreed to make each overseas company the main agency for developing ICI's manufacturing interests in its territory, and in return the overseas company agreed not to compete with ICI, being specifically barred from the export trade.[21] These limitations, especially on exports, eventually became indefensible, but at the time they were negotiated, in the embryonic state of Dominion chemical industries and in the general conditions of world trade, they cannot have been much of a hardship, and the benefits offered in return were substantial.

In 1924, in South Africa, McGowan brought off his penultimate achievement in merger-building before the final triumph of ICI. He persuaded De Beers to overcome long-standing suspicions and merge their explosives interests with Nobels' in a 50 : 50 joint company, African Explosives and Industries.*[22] The deal, surrounded with the cigar-scented opulence in which he loved to move, gave McGowan enormous satisfaction. Speaking in 1935 to the Board of ICI, he said 'the first really constructive meeting' leading to the merger was at a dinner at the Savoy given by Solly Joel (Lieut-Col. Solomon Barnato Joel (1865–1931), South African 'financier and sportsman'[23]) to himself and H. J. Mitchell,[24] and he greatly enjoyed his entrée to the world of South African mining magnates, interwoven as it was with the governing circles of the Union. He travelled frequently, magnificently, and zestfully in South Africa and he was well content to let Sir Ernest Oppenheimer of De Beers have the Chair of AE & I, with himself as deputy, rather in the same way as he let Sir Alfred Mond have the first place in ICI.

In spite of personal civilities the equal partnership with De Beers was not easy. In the early days De Beers were less enthusiastic than ICI to

*The name was changed in 1944 to African Explosives and Chemical Industries Ltd.

increase their investment in the joint company.[25] During the war, through force of circumstances, the position was reversed. At all times there might be differences of opinion about manufacturing policy, as against imports, in South Africa. De Beers, being as strong in South Africa as ICI in Great Britain, could take a high line, and in the last resort, on matters of local production, they would usually be able to rely on support from the South African Government.

The basis of the partnership was an agreement between ICI, De Beers, and AE & I, signed on 2 April 1930, which took the place of an agreement between Nobels and the other parties. It was negotiated, for ICI, largely by H. J. Mitchell, and as an instrument of policy it is a perfect example of its kind: the kind, that is, by which the men in power in the world's chemical industry between the wars regulated their trading relationships. It was essentially a market-sharing agreement relating to the entire African continent between the Cape of Good Hope and 10°N latitude. Within that area, broadly speaking, ICI and AE & I were to exchange technical information and were not to compete. Outside it, AE & I were not to operate at all. To ICI's later regret, the agreement was perpetual.[26]

The core of AE & I's business was in explosives plant at Somerset West and Modderfontein, serving the largest market for mining explosives in the world. The mine owners, led by De Beers, had long ago shown how powerful they were, and trade with them was based on co-operative contracts which guaranteed them a share in the benefits of any savings in costs achieved by improved efficiency at the explosives factories.[27] In 1930 the importance of the mining explosives business was emphasized again when ICI began to urge on their reluctant partners the necessity of putting up synthetic ammonia and ammonia oxidation plant, for nitric acid, in South Africa. Nowhere else in the world did ICI propose such a thing of their own accord, but in South Africa they were determined to be independent of imported nitrate.[28] The plant began to operate in 1932.

Gold-mining accounted for about 70 per cent of South Africa's exports and paid for the import of most of the manufactured goods the country needed. Apart from mining, almost the only other industry of any consequence in South Africa between the wars was farming, and here, on the face of it, was another attractive market for AE & I. Superphosphate and nitrogenous fertilizers would run easily with explosives, and an agreement had been made in 1927 for the manufacture by AE & I of various other agricultural products belonging to Cooper, McDougall & Robertson.[29] The farmers' voting strength made agriculture important politically as well as economically, and the industry was subsidized. 'These subsidies', said McGowan in 1935, 'will be continued and that will result in an increasing demand for fertilizers

205

and other products of our factory at Umbogintwini, near Durban, where we make all kinds of fertilizers, locust poisons, etc . . . a very profitable part of the activities of AE & I Ltd.'[30]

AE & I supplied about 50–60 per cent of South Africa's entire requirements of fertilizers, but the political importance of the farmer, for AE & I, was double-edged. On the one hand the Government would make sure, by subsidies, that he remained able to buy fertilizers. On the other, by refusing to protect AE & I against foreign imports, the Government would see that the farmer got his fertilizer cheap. It was politically impossible for AE & I, with its well-known backing from ICI, to withdraw from the fertilizer trade altogether, although that desperate course was thought of, and so in this trade AE & I was obliged to operate in the unaccustomed chill breeze of foreign competition.[31]

Whatever the troubles of the fertilizer side may have been, AE & I, taken as a whole, was one of ICI's most successful enterprises. By 1936 the turnover in manufactures, at £4m.,* was 47 per cent greater than in 1927,[32] and as well as that there was a considerable import trade, notably in cyanide for the goldmines, which ICI had no wish to see manufactured in South Africa because of the effect on Billingham. The range of products made at Modderfontein near Johannesburg, Umbogintwini near Durban, and Somerset West at the Cape covered explosives and accessories (£3m.), fertilizers and other agricultural products, including some of Cooper, McDougall & Robertson's (£600,000), and various acids and other chemicals. The issued capital, £2·5m. in 1927, stood at £5m. in 1926, and the profits were:[33]

Year to	Issued Ord. Capital £m.	Profit £	Dividend %	Bonus %
31.12.27	2·5	364,449	10	
28		436,947	10	
(9 months) 30. 9.29	2·75	377,456	10	
30. 9.30	3·25	489,935	10	
31	3·5	469,193	12½	1
32		452,920	12½	2½
33		789,636	12½	9
34		878,028	12½	12½
35	5·0	900,010	12½	5
36		930,141	12½	6

Australia in the thirties was much further advanced, industrially,

* It is not clear whether this figure includes *all* manufactures.

Harvesting salt at the ICI solar evaporation salt fields at Dry Creek, South Australia, for use at Osborne Alkali Works (page 211).

An early photograph of AE & I's plant at Modderfontein (page 205).

Arthur B. Purvis, President (1925–41) of Canadian Industries Ltd.

Sir Lennon Raws, Chairman (1934–47) of ICIANZ.

Sir Ernest Oppenheimer, Chairman (1931–57) of AE & I, later AE & CI.

than South Africa. Mining and farming were both important, but there was a more varied industrial output and less dependence on imported manufactures. That this arose from a long tradition of protectionist policies (as well as enforced self-reliance during the Great War) no one doubted, and by the time ICI came into existence all political parties were committed to protection and manufacturers were well organized and spending freely on propaganda. Pressure was brought to bear on public authorities to place orders in Australia if they could, and it was not difficult for Australian manufacturers to get revenue duties increased to protectionist levels. 'The effect of all this', said B. E. Todhunter in 1927, 'is that . . . the importer is constantly harassed by the imposition of dumping duties and applications to the Tariff Board by manufacturers for increased duties.'[34]

Todhunter's general conclusion was that the Australians meant to be self-sufficient, that Government policy would always be directed to encouraging Australian industry, and that if ICI delayed too long in changing from export to local manufacture they would find 'that as a result of inducements offered by State Governments or the Development Commission or the Tariff Board, local capital, either alone or in conjunction with Continental or American interests, may start local manufacture and thus make it difficult for ICI to regain the market'.[35]

The obvious policy, and the one which Todhunter recommended, was for ICI to go into partnership with local interests after some version of the pattern set by Nobels in Canada and South Africa. Nobels had already made the first moves before the merger by transferring the ownership of their various Australasian interests to Nobel (Australasia) Ltd., a company formed for the purpose and registered in Victoria with issued capital of £1·1m. Except for £3,000, the capital was all held within the Nobel group, being issued in exchange for interests taken over by the new company, but there seems from the first to have been an intention to bring in Australian money, and the Australasian nature of the enterprise was made plain by appointing Directors who had independent positions of their own in Australia or New Zealand. The Managing Director was Sir W. Lennon Raws (1878–1958), an accountant who at the time of his appointment was Manager for Elder Smith in Melbourne and had been Chairman of the Metal Exchange, Chairman of Melbourne Chamber of Commerce, and Chairman of the Federated Chambers of Commerce of Australia.[36]

The assets of the new company came from the businesses in explosives, safety fuse, and ammunition carried on in Australia and New Zealand by various companies in the Nobel group. More than half its supplies were imported from the United Kingdom, but it owned three factories near Melbourne, at Deer Park, Footscray, and Spotswood. Deer Park

H

could make 2,500 tons of explosives a year, Footscray 17m. sporting cartridges, and Spotswood 1,600,000 coils of fuse. With the spent acid from the explosives cycle at Deer Park they could make up to 20,000 tons of superphosphate a year, and in Australia, where farming was so important to the economy, an interest in fertilizers was bound to become an important bargaining point. There were also plans afoot for a joint interest with du Pont in the manufacture of leathercloth and cellulose finishes at Deer Park. The turnover of Nobel (Australasia) Ltd. in 1926 was about £1·1m., providing about 10 per cent return on the paid-up capital after allowing a manufacturing profit to the factories in the United Kingdom.[37]

After ICI was formed Todhunter remained in charge of Australasian affairs, and for the rest of his career Australasia formed something of a private fief into which McGowan might be invited but over which no one else—certainly not H. J. Mitchell, that other great overseas potentate—would be allowed any degree of authority. Through Todhunter's agency ICI, in Todhunter's words, 'followed even more vigorously the Nobel policy'.[38] That, first of all, meant going into partnership with local interests, and by April 1928 Todhunter was reporting that in this way ICI had bought into the fertilizer business in South Australia, New South Wales, and Queensland, had combined with the largest firm of Australian paint manufacturers (British Australian Lead Manufacturers Ltd.) to put up factories in New South Wales and South Australia for nitrocellulose lacquers, and had bought from du Pont a controlling interest in Australian by-product ammonia companies. At the same time, the joint ICI–du Pont leathercloth company had been formed in Victoria and a factory was being built at Deer Park.

Along with these developments, the principle of the holding company, exemplified in Nobel (Australasia) Ltd., was extended to cover the interests brought into ICI by other parties to the merger, especially Brunner, Mond (Australia) Ltd., a company established chiefly to sell alkali. What Todhunter called 'an Australian ICI' was formed early in 1928, ready to start business as soon as might be convenient. It was called ICI (Australasia) Ltd.—altered in April 1929 to Imperial Chemical Industries of Australia and New Zealand Ltd. (ICIANZ)— and it was expected to be capitalized at about £5m.

'The intention', Todhunter explained, 'is that it [the new company] should take over all the existing ICI interests, combining therewith such of the more important local fertiliser and chemical manufacturers as may be disposed to enter such an association on reasonable terms, and that the combination should proceed to establish the manufacture of synthetic ammonia with the fertilisers and other products that flow therefrom.'[39] This was why ICI had taken shares in existing Australian fertilizer businesses, all of which dealt in superphosphate, and had bought the Americans out of ammonia companies. Todhunter thought the possibilities of developing the Australian fertilizer business were 'almost unlimited', but he said the future would lie with 'complete balanced fertilisers with a suitable nitrogen content' rather than with superphosphate, although a good deal of propaganda and development work would be necessary to get nitrogen established in Australia. Technical knowledge ought to be pooled and 'manufacture should be in the largest possible units', all of which required centralized control.

The parallel with contemporary ICI planning in the United Kingdom is striking. There is the same optimism about a rapidly growing fertilizer industry based on synthetic ammonia, the same readiness to envisage missionary work among farmers sceptical of the gospel of nitrogen, the same insistence on the economies of scale. There is a hopeful suggestion that oil-from-coal may follow ammonia, though in a later document Todhunter proposed leaving it outside the scope of ICIANZ. The development by ICIANZ of the heavy chemical industry —'caustic soda, chlorine, hydrochloric [acid] etc.'—is taken for granted. A replica of ICI is being planned for Australia, with one significant exception. 'I have not included British Dyes in the proposed arrangement', said Todhunter in May 1928, 'as their business does not appear sufficiently cognate to be profitably combined with fertilisers and heavy chemicals.'[40] In Australia, as in Great Britain, no place was prepared within ICI for the organic chemical industry.

In linking ICI with Australian interests, Todhunter's plan was to work with three main partners in fertilizers and chemicals: Cuming Smith & Co. Pty. Ltd., the Mount Lyell Mining and Railway Co. Ltd., and the Broken Hill Group. Broken Hill came into the picture rather later than the other two. Todhunter described it as 'undoubtedly the most influential financial Group in Australia', looking for a home for about £1½m. 'loose cash' and having ideas 'tending in the direction of the nitrogen and oil from coal industries'. 'If we associate ourselves with them', he said, 'we should . . . render the Australian chemical company [ICIANZ] immune from any influential Australian competition.'[41]

Todhunter left for Australia, to construct the amalgamation of interests, in September 1928. He thought that probably the four main groups—ICI, Cuming Smith, Mount Lyell, and Broken Hill—would take equal shares in ICIANZ and that other Australasian businesses would be brought in 'as and when it appeared desirable'.[42] That, however, was not to be. To Todhunter's exasperation Cuming Smith and Mount Lyell ('both concerns are extraordinarily timid and very parochial in their outlook')[43] turned out very difficult to bring to terms ('they have had an easy time making profits on a simple business'),[44] largely because they were nervous of ICI's nitrogen ambitions—in which, perhaps, they showed a sounder instinct than Todhunter was willing to admit.

Involved and acrimonious negotiations lasted several months. Todhunter fell off a horse and injured his right foot, improving neither his mobility nor his patience. The matter was at length resolved by combining the chemical businesses of Cuming Smith and Mount Lyell with ICI's superphosphate business (based at Deer Park) in Commonwealth Fertilisers and Chemicals Limited, in which ICIANZ, in 1930, held 4·22 per cent of the issued capital of £2·6m. That satisfied Cuming Smith and Mount Lyell and left ICI able to get on with things that really interested them.

ICIANZ started trading in March 1929. A year later it had two wholly-owned subsidiaries, representing the old Nobel and Brunner, Mond interests. It had substantial minority interests in three fertilizer companies, including one which took care of the interests of makers of by-product ammonium sulphate. It had majority holdings in two ammonia companies and a large minority holding in a third. It had no holding in companies formed to deal with Australian business in paints, leathercloth, metals, and oil-from-coal. They were linked directly with ICI in London, but there was Australian capital, in varying proportions, in all of them (see Figure 13).

The issued capital of ICIANZ itself, in March 1930, was about £2¼m. ICI London held 91 per cent, which was a long way from Todhunter's original conception of a company in which ICI would hold only 25 per cent. Rights of subscription to fresh capital issues, however, had been granted to Cuming Smith, Mount Lyell, Broken Hill, and the Baillieu interests, which would gradually reduce ICI's preponderance, and by the time war broke out in 1939 ICI held 78 per cent of ICIANZ's Ordinary capital in issue at that time.

Thus, by 1930, ICI had developed the comparatively simple structure of Nobel (Australasia) Ltd. into a network of interlocking interests designed to serve a diversified manufacturing business which was intended to dominate the growing Australian chemical industry. A considerable import trade continued, but there was no doubt in anyone's mind that the future lay with Australasian manufacture in association with Australasian interests, under Australasian management and very largely under Australasian control. Nothing of this would have come about if it had not been for the pressure generated by the Australians' determination to develop their own economy, for under conditions of free trade, or even with moderate revenue duties, it would have remained more economic to import commodities such as explosives and alkali, manufactured on a large scale in the United Kingdom, than to manufacture them on a small scale in Australia. The setting up of ICIANZ, like the setting up of Solvay Process Company in the USA many years before, was a victory for protection, and the parallel be-tween the development of the Australian and American economies, in this respect, is very close.

The point was forced home by the curious history of alkali manu-facture and ammonia synthesis in Australia. The developing industries of the country provided an important market for soda ash, taking 25,000 to 30,000 tons a year in the early thirties.[45] It was entirely an import trade and Brunner, Mond first, and then ICI, proved to them-selves expensively and conclusively that alkali manufactured in Australia would always be much more expensive than alkali brought from Winnington, because of the natural advantages of the Winnington site and the scale of manufacture there. They proved the same point to the Australian authorities, and yet within two years of having done so ICI came forward with proposals for an alkali plant. The reason was that if they did not put one up someone else would certainly do so, and he would get the protection necessary to make it viable. ICI therefore set up plant themselves (it was working by 1940), applied for and got protection like any other Australian manufacturer, and gave up the export trade from Winnington.[46]

The story of ammonia synthesis was similar. ICI's early enthusiasm was shattered by the Depression, and when the matter was discussed in London and Melbourne, between 1934 and 1937, London was always against the proposal because Australian plant could never be large enough to compete with imports. In 1937 McGowan visited Australia and attended a Board Meeting of ICIANZ. He sent a cable home which explains concisely how the matter ended:

. . . Ammonia synthesis fully discussed Board Meeting Monday Sir George Julius [a Director of ICIANZ] . . . made it clear that Government determined to be self-sufficient in this essential and that if we did not erect a plant others would STOP . . . We have informed Government of our decision to go ahead.[47]

That was that. The plant was put in hand, aiming at a capacity of 3,000 short tons of ammonia a year—a mini-plant indeed—and including provision for nitric acid and ammonium nitrate. It was completed in 1939, but ammonia synthesis did not begin until after war had broken out. By putting up ammonia plant in Australia ICI were aggravating a situation—world over-capacity—which elsewhere, through the opera-tions of the Nitrogen Cartel, they were trying to alleviate. However, in industry as in public affairs, politics is the art of the possible, and if contradictory policies were the price of survival, contradiction there would be.

'The ultimate objective of ICIANZ', said Todhunter in 1934, 'should be to become the dominating factor in the Australian Chemical

Industry, and to occupy the same position in the Commonwealth as AE & I and CIL hold in the African and Canadian chemical industries.'[48] By 1938 ICIANZ had advanced far enough towards that position to make a public issue of £A1m. in 5% Preference shares, all of which were taken up, and by the time war broke out ICI, through ICIANZ, had control or a large minority interest in a wide variety of manufacturing enterprises, covering explosives and safety fuse, ammunition, fertilizers and heavy chemicals, leathercloth, paints, alkali, and synthetic ammonia. There was even an interest in the Commonwealth Aircraft Corporation—taken, again, for political reasons. 'We had no desire to enter the aircraft industry,' McGowan told Lord Weir in 1937, 'but as one of the most important companies here we had to do so in order to show the Commonwealth Government that we were taking our share in the defence of Australia.'[49] When ICI was set up, the merging companies were making nothing in Australia but explosives, safety fuse, and sporting ammunition, plus a large and profitable import trade. The transition from import to manufacture was not regarded with enthusiasm in London, but it was accepted as the necessary price of a substantial ICI presence in Australasia.

ICI's business in Canada, like those in South Africa and Australia, was built on Nobel foundations. The original company, Canadian Explosives Ltd., was the result of McGowan's earliest exercise in merger-making. It took the form of a partnership with other interests, and this Canadian partnership was a very special one. The major shareholders were ICI and du Pont. In 1928, after they had bought an 8 per cent holding from Atlas Powder, they became the owners, in equal shares, of well over 90 per cent of the issued share capital, which amounted to some $15·9m. This made the Canadian business much the largest joint undertaking of ICI and du Pont. It lay at the heart of the ICI–du Pont alliance, which itself was the central fact of ICI's foreign policy between the wars.

ICI's policy in Canada was expansionist from the first: more so, perhaps, than du Pont's. The group of companies headed by Canadian Explosives Ltd. (CXL) had broadened their interests beyond explosives after the Great War, and in 1927 a new company was set up to act purely as a holding company. It was called Canadian Industries Ltd., thus giving notice that the broadening process was to continue. Melchett and McGowan, in North America in the autumn of 1928 for negotiations with du Pont which led up to the Patents and Processes Agreement of 1929, made their intentions aggressively clear. They undertook to back CIL in gaining 'its proper place in industry', and Melchett, according to G. W. White's recollections in 1945, said 'it was ICI's intention to develop Industry in Canada with the ultimate object of local manufacture. . . . If du Pont wished to come in . . . they would be welcome, but the scheme would go forward whether du Pont co-operated or not. . . .'[50]

In a policy of this kind, ICI could rely on support from the President of CIL, Arthur D. Purvis. In 1927, aged 37, he had already been President of CXL about two years, having between 1910 and 1924 been in the service of Nobel Industries in South America, South Africa, and the USA. His drive and ambition—he was thought to be aiming to become Prime Minister of Canada—did not make him popular, particularly at a slightly lower level than his own in management, but there was no doubt about his views on the expansion of CIL. Speaking to McGowan and Lammot du Pont, among others, in 1928, he said that 'the logical future of CIL lay in going into all the lines practicable in which its parent companies could give it technical assistance'.[51]

The Canadian chemical industry, on the doorstep of the USA, was more advanced than the chemical industry of South Africa or Australia and also more of a battleground for industrial power politics. Allied Chemical, in 1925, had bought Brunner, Mond's 40 per cent share in the alkali works at Amherstburg and, with it, the trade marks which had formerly belonged to Brunner, Mond (Canada). Mathieson Alkali and Union Carbide either had or showed signs of wishing to have a presence in Canada. The interests of ICI and du Pont themselves, although they were allied, were no more identical than the interests of allies usually are.

There was also the purely Canadian interest to be considered. In South Africa and Australasia the local interest was powerfully represented, but in Canada the minority stockholders could hardly hope to make their view prevail. They were simply, as Lammot du Pont once observed to McGowan, 'investors who wish to place their money . . . with the ICI-du Pont combination'.[52] That was well known in Canada, with the result that CIL was regarded with some suspicion and any policy which could be represented as contrary to Canadian national interests would be likely to attract unwelcome attention from the Canadian Government. Moreover Purvis was quite shrewd enough to use the threat of Government intervention in his own bargaining with the majority stockholders.

From 1928 onward Purvis pushed CIL's manufacturing interests rapidly outward from the Nobel nucleus—still very important—of explosives and cellulose finishes. Before the end of 1928 itself, partly as a blocking move against Mathieson Alkali and partly as a step towards a merger (which never came off) with the Canadian interests of Allied Chemical, CIL bought control of the Canadian Salt Co.[53] In the same year du Pont sold to CIL control of the Canadian Ammonia Company and the Grasselli Chemical Co. of Hamilton, Ont.*[54] Grasselli's chief

* Du Pont had taken over Grasselli in the USA to give themselves an interest in heavy chemicals.

product at Hamilton was sulphuric acid, which CIL intended to use in superphosphate. CIL's view of the fertilizer industry, supported by ICI, was cheerful in 1929 and they formed a Fertilizer Division with plans for plant at Hamilton and at Beloeil, PQ. Canadian consumption of fertilizers, said Worboys early in 1930, had gone up 300 per cent since 1925, to 220,000 tons a year, and 'the time appears ripe for a firm of the standing and organization of CIL to enter the field'.[55]

CIL's new fertilizer plants came into operation in the summer of 1930, just in time for the full force of the Depression. All through the early thirties the Fertilizer Division lost money, but by 1937 the outlook was brighter. In that year Purvis, evidently under attack from his shareholders, defended the fertilizer policy, which by then had led CIL to the purchase of three more companies besides Grasselli. He used an argument often heard from ICI's associates overseas:

Had the Company omitted to enter this field, others would have gone in and would probably have proceeded to make acids, to make ammonia . . . and thence to compete with CIL's profitable explosives industry; moreover it was no disadvantage to the Company to be interested in an industry which did not return a high yield at the same time as it was engaged in the production of other and more profitable lines.[56]

By the late thirties, then, the main structure of the manufacturing side of CIL was very much like that of AE & I and ICIANZ. The Nobel base had been used for expansion into fertilizers, heavy chemicals, and paints (CIL bought a paint company in 1935 and formed a joint company, Canadian Titanium Pigments, in 1936). At the same time there were other developments which had no parallel elsewhere. In 1931 CIL went into cellophane under licence from du Pont—something which ICI themselves never did—and towards the end of the thirties began moving towards association with the Shawinigan Chemical Co. which, with cheap power, was important in the electro-chemical field.

Besides manufactures, CIL, like AE & I and ICIANZ, did a large import trade, and it was the major stockholders' settled policy that CIL should manufacture nothing which either ICI or du Pont could more profitably put on the Canadian market.[57] Melchett, in 1928, made it clear that what ICI chiefly wanted to concentrate on was trade in dyestuffs, which he said he wanted to develop 'to a very much greater degree than . . . hitherto'.[58]

The British share of the Canadian dyestuffs market had fallen from 35·2 per cent by quantity in 1917, when there was no German competition, to 3·4 per cent ten years later. By that time the Germans were back with 27·1 per cent of the market. The Swiss had 12·9 per cent and the Americans 55·4. Melchett was therefore setting ICI a formidable task of recovery, not likely to be achieved without vigorous selling.

Melchett's forward policy was unwelcome to du Pont, who had no wish to see CIL selling imported ICI dyestuffs in competition with du Pont's own. After about a year's delay they reluctantly agreed to a scheme intended to divide dyestuffs business equally between themselves and ICI.[59] Equal shares were not achieved (Table 12) but CIL's sales of dyestuffs and related products increased rapidly. The sales figures, however, only represented about 11 per cent of the Canadian market, and CIL, wanting much greater things, suggested the ultimate heresy: price competition between ICI and du Pont products. ICI supported the policy, du Pont did not, which shows where the balance of advantage lay, but with some reservations du Pont agreed to it.[60]

TABLE 11

CIL: SALES OF PRODUCTS OF ICI DYESTUFFS GROUP

	Dyes	Other Dyestuffs Group Products	Total
	£	£	£
1927	12,865	513	13,378
1928	16,234	8,910	25,144
1929	17,056	10,386	27,442
1930	18,776	13,123	31,899
1931	21,064	15,580	36,644
1932*	44,498	11,647	56,145
1933	58,451	5,716	64,167
1934	76,971	3,627	80,598
1935	66,486	5,018	71,504
1936	76,665	5,915	82,580

Source: USA, CIL 2nd Draft, p. 42.

TABLE 12

CIL: DYESTUFFS SALES—DIVISION OF TRADE, BY VALUE

	Du Pont Products	ICI Products
Jan. 1930 to June 1931		
(annual rate)	%	%
	55·73	44·27
1932*	43·25	56·75
1933	36·09	63·91
1934	29·32	70·68
1935	37·75	62·25
1936	41 (approx.)	59 (approx.)
1937	46·00	54·00

Source: As above, p. 43.
* Imperial Preference was introduced in autumn 1932.

In the autumn of 1932, under the scheme of Imperial Preference which was the chief outcome of the Ottawa Conference, dyestuffs from ICI could be imported into Canada without paying the 10 per cent duty imposed on dyestuffs from outside the Empire. That swung the balance so far against du Pont's dyestuffs that du Pont contemplated withdrawing from the Canadian market altogether.

That prospect, earlier, might have been very pleasing to ICI, but in 1932 it was not at all what they wanted. Under the regulations of the Dyestuffs Cartel, which had come into force earlier in the year, ICI's quota in Canada could either be their own dyestuffs sales there or 50 per cent of CIL's dyestuffs sales, whichever was greater. The effect of du Pont's withdrawal (or their own) would simply be to give them a lower quota of sales in Canada under the Cartel agreement, with no guarantee of making good the loss anywhere else.

ICI therefore fought hard, even to the extent of selling at a loss, to keep their own dyestuffs trade in Canada, and they did their best to see that du Pont stayed in the Canadian market even after Imperial Preference made it uneconomic. This tortuous policy was the curious outcome of interaction between public and private protectionism—that is, between Imperial Preference and the Dyestuffs Cartel—and many years later ICI found it an embarrassment in preparing their defence to the anti-trust action *USA v. ICI et al.* The best the Legal Department could make of it was: 'There was one enemy—the IG—and one of the things to do was to fight for the business and fighting for the business was one part of the job. . . . It was ICI policy to establish CIL in as large a range of dyestuffs as possible.'[61]

The management of the dyestuffs business in Canada between the wars shows how the intricate international diplomacy of the chemical industry could impinge on the affairs of a national market. It was governed by the relations between ICI, du Pont, and IG, and the purely Canadian aspect of the matter seems hardly to have been considered, except in so far as Canadian interests were served by ICI's determination to invigorate the dyestuffs trade. Even then, they might have been equally well served by uninhibited competition from IG. It is difficult to believe that ICI and du Pont would have felt themselves free to act as they did if they had been accountable, as ICI were in South Africa and Australasia, to powerful and well-connected local shareholders.

The general theory underlying the relationship between the two major shareholders and CIL, as du Pont understood it, was set out by Lammot du Pont in a letter to McGowan:

The theory back of the CIL operation, as far as ICI and Du Pont are concerned, is expressed in the old saying: 'Canada for the Canadians', meaning

—the industrial operations of the partners in Canada are intended to be conducted through CIL. CIL was not set up to do anything else and has, we believe, never been considered so.

If the above is correct, it seems to us to follow . . . that CIL shall stay in Canada and not spread out into other countries, either by laying down plants, exporting their products or licensing under their processes, unless both ICI and Du Pont believe it is advantageous to so spread out and then only to the extent and for the time and under the conditions that ICI and Du Pont agree upon. [62]

This was a point of view which ICI broadly agreed with, since it placed CIL, in Canada, in much the same position as AE & I in Africa and ICIANZ in Australasia. It was certainly restrictive, but the major shareholders justified the restrictions by reference to the advantages CIL derived from association with du Pont and ICI.

To Purvis, there is no doubt that the restriction on CIL's freedom of action was irksome. He looked on it, evidently, rather as the Founding Fathers of the USA looked on the old colonial system, and he did his best to find loopholes. That was not very difficult, because the contractual basis of the relationship (two agreements made before the ICI merger) was out-of-date and vague.

Purvis's struggle for independence alarmed du Pont, especially when CIL went into the ammunition business in South America. After long discussion, a new agreement between CIL and the major shareholders was drawn up in 1936. Purvis made his opinion plain:

whatever reservations and anxieties CIL has expressed [clearly they had expressed plenty] in regard to the terms or workability of the new document agreed upon by ICI and du Pont . . . the CIL organization will . . . do its best to make it a practical success. [63]

With Purvis thus disclaiming, in advance, responsibility for the contents of the new agreement, it was signed late in 1936. The basis of it was 'that both ICI and du Pont will regard CIL as the vehicle for all Canadian undertakings, whether manufacture or sale, unless adequate reasons can be shown for following some other procedure in specific cases';[64] and by its terms, which covered all products made by CIL, ICI and du Pont granted CIL exclusive rights in Canada and Newfoundland in exchange for a similar grant of rights, by CIL, in the territories reserved to du Pont and ICI in the Patents and Processes Agreement of 1929. There was a provision that nothing in the agreement should prevent the grantor of a licence from selling the licensed product in the territory of the grantee, but the proviso never seems to have been put into effect and in the Complaint in *USA v. ICI et al.* it was described as 'camouflage'.

How nervous CIL's principals were of CIL escaping from Canada was demonstrated in 1937 when there was a proposal for CIL to enter a joint company with Shawinigan for making methanol. There was a possibility that methanol would be exported, particularly to Great Britain, where it would have Imperial Preference, and both the major shareholders took fright.

Ethylene glycol is probably next on the list [White wrote to Coates], and a whole range of products could subsequently emerge . . . from the grouping together of cheap power, ample capital resources, carbon monoxide from the carbide furnaces . . . hydrogen . . . acetylene, etc., etc. . . . ICI would find it very difficult indeed to resist export unless it is clearly understood and accepted at the outset. . . . You are probably aware it is regarded both by ICI and du Pont as important not to set up precedents for CIL to break out of Canada.[65]

Confined though CIL might be to the Canadian market, its results during the worst years of depression, apart from a sharp drop in 1932, were steady and even buoyant.[66]

	Gross Sales $ million	Profit from Sales $ million	Earnings per Share (common) $
1930	18·3	3·7	5·09
1931	18·3	3·4	4·65
1932	16·3	2·4	3·65
1933	18·3	3·5	4·63
1934	22·8	5·1	6·43
1935	24·7	4·5	5·85

Those results, in those years, would no doubt have been widely considered an adequate answer to criticism of the major shareholders' policy.

The same, broadly, may be said of each of the companies we have considered in this section—the companies at the imperial heart of ICI's overseas business in the first dozen years of its existence. Each was built up, from similar foundations, to expansion and diversification. Each had to face the depression years, and each was stronger at the end of them than at the beginning. Their strength was founded on local investment and, whatever criticism may be made in detail of the use of it, it contributed not only to the profitability of ICI but also to the economic recovery of the host countries. ICI's imperial purpose was not ill-served, especially when it is looked at against the background of ICI's difficulties, in the thirties, at home.

(iii) SOUTH AMERICA

Outside the Empire there was another region where traditions of British trade and investment were almost as strong: South America. ICI's interests, derived from Nobels, Brunner, Mond and UAC, lay chiefly in Chile, Bolivia, Argentina, and Brazil.

These were all thinly populated countries heavily dependent on international trade in primary produce hard hit by the Depression. Chile (population 5m.)[67] and Bolivia (3m.) were both important centres of mining—nitrate and copper in Chile, tin in Bolivia—which made them attractive markets for Nobels and other makers of explosives. Industrialization had hardly touched them, and it had not gone far in Brazil or Argentina, though in each the Government was trying to encourage industry by protectionist policies.

Brazil, the largest country in South America and larger in area than the United States, had a population in the thirties of about 46m. It was sometimes held to be 'the country of the future', but its European connections lay rather towards Italy and Germany than towards Great Britain. The great South American stronghold of British interests was in the Argentine Republic (population 13m.), where there had been heavy British investment in public utilities and which had a large trade in beef with the United Kingdom. Both Brazil and Argentina were important markets for soda ash.

In South America, as in the British Empire, ICI in the thirties faced the problems of changeover from export to local manufacture. South America, however, unlike the British Empire, was not a market which du Pont or IG were willing to regard as ICI's 'exclusive territory'. They both meant to share it, along with other competitors local and foreign. Moreover IG, with interests world-wide, could threaten aggression elsewhere, even within the British Empire, to get what they wanted from ICI in South America.

The politics of the South American trade, consequently, could never be divorced from the politics of ICI's other interests. Moreover, although the South American market was important to ICI, the markets of the Empire at all times had priority, and in the last resort any South American associate of ICI was likely to find itself, as was once rather bitterly observed, 'a pawn in the larger game that the shareholders play'.[68]

That remark came indirectly from the management of Compania Sud-Americana de Explosivos, a Chilean company in which ICI and du Pont ('the shareholders') each had a 42½ per cent holding, the remainder being held by Atlas Powder Company.[69] It was in Chile that the power politics of the world chemical industry were conducted at their crudest.

CSAE owned an explosives factory at Rio Loa, put up, as McGowan, before ICI's day, publicly said, to placate the Chilean Government.[70] In the depressed conditions of the mid-thirties, when it was working at 15 per cent capacity with a much reduced staff, it could have supplied the entire Chilean demand, and probably the Bolivian demand as well. The Chilean Government sometimes inquired why it did not—'a question', in the words of G. W. White, 'that has had to be answered with diplomacy, on more than one occasion'. No doubt, because at the time when Rio Loa was so under-employed a market-sharing agreement was in force between CSAE and IG, as well as other European explosives makers, and 48 per cent of the explosives used in Chile were being imported through a tariff wall put up by the Chilean Government for the express purpose of protecting Rio Loa.[71]

TABLE 13

CHILE: MANUFACTURE AND IMPORT OF EXPLOSIVES, 1929–1933

	1929	1930	1931	1932	1933
Manufacture:		tons of 2,000 lb.			
by CSAE	4,085	3,154	2,236	644	604
Imports:					
by CSAE	243	161	272	287	159
from Norsk Hydro	2,626	1,001	650	11	315
Dynamit AG	912	1,028	219	111	145
Trojan Powder (USA)	272		100		
Belgians					35
Total	8,138	5,344	3,477	1,053	1,258
CSAE Manufacture as % of total	50	60	70	60	50

Source: G. W. White to H. J. Mitchell, Report on CSAE, 24 May 1934, USA, CSAE III (ii).

This contradictory state of affairs was brought about by IG's latent power to harass ICI and du Pont in markets far away from Chile. IG had two associates in the explosives trade—Dynamit AG and Norsk Hydro—and their consistent policy was to keep these companies' home factories busy in the export trade, not to go in for local manufacture. As far back as the early twenties IG had refused a share in CSAE which ICI and du Pont would have been glad to see them take up, and in the thirties, under Nazi pressure, their attitude became still harder. The Germans, said Dr. Muller of IG, had to export 'at any price', and if DAG's quota of exports to Chile were reduced that might cause the closing of their factory in Germany, which would be 'an untenable position in relation to [Hitler's] Government'.[72]

Muller, in negotiating a new quota agreement in 1935, was polite to ICI ('extremely friendly, helpful and understanding', said their chief representative) but he left no doubt at all about his intentions if he did not get his way. 'Threats were not used but such markets as Mexico and India were quietly mentioned.'[73]

The interests of the Chile company, to the continuing distress of the local management, were therefore quite deliberately sacrificed to the wider interests of ICI. 'The Chile Company', it was observed in 1940, 'of course have never wanted any Agreements but it is for the Principals to dictate their Policy.'[74] The Policy was dead by that time, killed in the war, and it was not revived, but in the thirties the quota system was regarded by the older generation on the explosives side of ICI as being the traditional way—almost, the sober, godly and righteous way—of doing business, and companies associated with ICI were expected to play their part in a properly altruistic spirit: 'All the time', wrote the Chairman of the Explosives Group, an elderly Scot, in 1939, 'I have been against any company . . . being allowed in the long run to think only of themselves. My upbringing has been the general interest and even [the Chile Company] should have a little of that spirit.'[75]

The most remarkable aspect of the whole affair seems to have been the helplessness of the Chilean authorities. The President of CSAE, himself a Chilean, disapproved of arrangements with competitors and evidently did his best to know as little as possible about them, but they were bound, sooner or later, to attract the unfavourable attention of the Chilean Government. In the end it was the prospect of anti-trust legislation in Chile, rather than the outbreak of war itself, which caused CSAE to break off relations with the local representative of DAG. The rupture was conducted ceremoniously and with courteous expressions of regret. ICI (New York) wished CSAE to 'assuage to marked degree any feeling on the part of DAG that you might be taking advantage of their present position of being unable to supply'—the Royal Navy was stopping them—and the final letter, written to DAG in Chile in December 1939, said 'We regret exceedingly that present circumstances make it necessary to terminate our long and harmonious association. . . .'[76]

CSAE was a single-product company in what was very nearly a single-product economy, so far as international trade was concerned. There was never any suggestion of extending the major shareholders' interests beyond explosives towards fertilizers and the chemical industry generally. Nor were there powerful local interests to consider.

In Argentina and Brazil the situation was more complex. There was some trade in explosives, and Nobels brought into ICI a 60 per cent holding in an Argentine ammunition firm, *Cartucheria Orbea*, but the main foundation of ICI's activities at the time of the merger was a flourishing import trade in alkali and chlorine products built up by

Brunner, Mond and UAC. This trade was in itself evidence of industrial growth, and in Argentina and Brazil, as in the British Dominions, there was the possibility of developing a diversified business in various branches of the chemical industry.

The resemblance to the Dominions went further. In Argentina and Brazil there were powerful local interests whom it was wise to conciliate. 'Foreigners trading in a country like the Argentine', wrote H. J. Mitchell in 1934, 'are well advised to have on their side Argentinians standing well in social and political circles, if only to remove the doubts which generally exist as to the *bona fides* of the foreign enterprise. Many managers of foreign-owned concerns expressed this view . . . that while it was a matter of little moment in past years the developing economic nationalism makes it advisable to have well-known and trusty natives associated with our concern.'[77]

'Natives', whether 'well-known and trusty' or otherwise, were not the only people to be considered. Du Pont and IG, as well as other foreigners, were also on the South American scene, active in various branches of the export trade.

Probably the most important and widespread export business was in explosives. In South America outside Chile it was handled for ICI, du Pont, and the associates of IG, and for as many other makers as could be induced to join a quota agreement, by Explosives Industries Ltd., a selling company in London which until 1938 was jointly owned by ICI (37½ per cent), du Pont (37½ per cent) and IG (25 per cent). In 1938 du Pont, fearing an anti-trust suit, withdrew, leaving ICI and IG sole shareholders in the proportion 60 : 40, but an understanding remained. 'The various trade agreements', says a letter of July 1938, ' . . . will be carried on and . . . du Pont, although no longer partners in EIL, will do nothing to interfere with their due fulfilment.'[78]

In the situation they found in Argentina and Brazil the policies of ICI and du Pont were similar, and similar also to the policy of ICI in the British Dominions. Each took local partners—ICI more readily than du Pont—and began to build up a group of manufacturing enterprises in the chemical industry, alongside their import trade, which could expand and replace imports in the industrial economy which the Government of each country was striving to create.

This course of action might have brought ICI and du Pont into competition. The US Department of Justice was later to say that it should have done. In fact, in the circumstances of the thirties, it brought them into alliance. In Argentina in 1935 and in Brazil in 1937 joint ICI–du Pont companies were set up—Duperial of Argentina and Duperial of Brazil*—which were modelled closely and consciously on

* *Industrias Químicas Argentinas 'Duperial' SA Industrial y Commercial* and *Industrias Químicas Brasileiras 'Duperial' SA.*

the company accepted as the pattern of what ICI–du Pont co-operation ought to be: Canadian Industries Limited.

TABLE 14

ICI SALES IN ARGENTINA, 1928–1931

	1928	1929	1930	1931
Major Products	£	£	£	£
Alkali and Chlorine products	287,513			243,774
Explosives and accessories		32,887	14,791	
Sporting Powders		18,597	16,459	
Ammunition (Orbea)		180,787	107,766	
Sulphuric Acid (Rivadavia)		75,211	102,637	
Minor Products				
Metals		4,317	4,029	
Electrodes		453	202	
Leathercloth		690	2,111	
Nitrogen Products and				
Fertilizers		1,491	5,151	
Cartridges, &c.		893	1,097	
Lighting Trades' products		600	600	

Source: FD 'Argentine II', 22 Nov. 1932. There is nothing to explain why the alkali and chlorine figures for 1929 and 1930 are not given, or why sales of other products are not listed for 1928 and 1931.

TABLE 15

ICI INVESTMENT IN ARGENTINA, 1932

	ICI Holding Sterling at par	% of issued Capital	Cost to ICI
	£		£
Cartucheria Orbea Argentina SA	165,390	89·4	217,121
ICI SA Com. e Ind.	152,174	100	147,915
SA Ind. e Com. 'Rivadavia'	52,174	40	52,594
Barraca Amberense SA	4,348	100	4,348
	374,086		421,978

There was also £24,377 on loan at 5 per cent to Rivadavia.

Source: FD 'Argentine II', 15 Nov. 1932.

ICI's group in Argentina, when the soundings for the merger with du Pont's began, included a wholly-owned selling company set up to handle alkali and other imports from the United Kingdom, an interest in *Cartucheria Orbea*, the largest single investment, which had been increased from 60 per cent to just short of 90, and a 40 per cent holding in *SA Industrial e Commercial Rivadavia*. This was the centre of the embryo manufacturing group. The company made or traded in acids, sulphur, sulphur products and fertilizers: in other words very much the sort of chemical products that might be manufactured in an agricultural country passing through the early phases of industrial development.

ICI's partners in Rivadavia were Solvay et Cie and a company called Bunge & Born controlled by Alfredo Hirsch. Hirsch was an elderly cripple in poor health whom Henry Mond in 1937 described as 'the richest man . . . and broadly speaking the most influential individual' in Argentina. He had made his fortune in grain but in the early twenties, foreseeing the collapse of the grain trade and having funds to spare, he had put Bunge & Born into chemicals. As a competitor he would be troublesome. He could easily buy technical assistance from Europe and his firm had a reputation for being 'extremely harsh in their business dealings . . . good and useful friends, but extremely dangerous enemies'.[79]

In 1932 du Pont bought two Argentine businesses which brought them directly into ICI's way, since one supplied Rivadavia with carbon bisulphide and the other was a direct competitor in sulphur. Even with these acquisitions, du Pont's Argentine investment, about £155,000, was much smaller than ICI's and their 1932 turnover was £148,000 against ICI's £633,000. The two provocative purchases may have been intended to jolt ICI towards a joint company, and by October 1932 the idea was being discussed between Lammot du Pont and McGowan in Wilmington. Over the next six months or so the full implications were worked out and early in 1934 H. J. Mitchell visited Argentina.[80] Mitchell concluded 'that the country is bent on industrial development and therefore it offers a promising field of manufacturing enterprise in which ICI will need to keep in the forefront.' Moreover he thought it would be a good moment to announce an ICI–du Pont joint venture, which would 'induce others not precipitately to enter their field' and— as between ICI and du Pont themselves—would 'prevent conflict of interests outside the Argentine field'.[81]

Mitchell saw the President of Argentina, General Don Agustín P. Justo, and explained that the joint company would be an Argentine company which the Argentine Government could rely on for help. The President took no exception to the idea of foreign ownership, and in March 1934 ICI and du Pont announced their intentions publicly, though the joint company did not come legally into existence until

15 May 1935. By that time Solvay, willingly, and Hirsch, not so willingly, had been bought out of Rivadavia, largely at du Pont's insistence, so that in Argentina, as in Canada, there was no strong body of local investors, although there were Argentinian directors and the shareholders were careful to appoint an Argentinian President. The capital of Duperial—15m. paper pesos—was held equally by ICI and du Pont.[82]

The constitution of Duperial was modelled on the constitution of Canadian Industries Limited, but without the vagueness of definition which allowed Purvis to claim an embarrassingly wide field of action. Duperial's Board were told to 'investigate and exploit all opportunities for . . . local manufacture' to replace imports. The shareholders claimed a right of veto on schemes put up by Duperial but they promised not to be influenced 'by any effect which a decision to introduce such local production would have on their profit and loss account'. In other words, they agreed to take their profits from Argentina by way of dividends from Duperial, not from money made in the export/import trade.[83]

'The whole theory', it was laid down, 'is that the Argentine company shall be regarded as the vehicle of industrial effort for ICI and du Pont in Argentine', and accordingly Duperial was to have exclusive rights, within certain territories, to patents, processes, and information belonging to ICI and du Pont and to the sale of the major products made by the shareholders which were still imported. Duperial's exclusive territory was Argentina, Uruguay and Paraguay. 'As a natural corollary to the grant of these exclusive rights', the document goes on, 'it follows directly and as a matter of course that the Argentine company must stay within its own territory and not spread out into other countries. . . . [It] would have no rights whatsoever to trade outside its own territory, either in its products or in processes, patents or other developments it may evolve, such becoming automatically the property of ICI and du Pont for disposal as they may jointly agree, and against such compensation (if any) as they . . . may elect to be payable to the Argentine company.'[84] That was plain enough. Argentina for the Argentine company, even at the shareholders' expense, but abroad Duperial was not to go, even with its own inventions, if there were any.

Moreover, the rights conferred on Duperial were restricted to 'certain definite lines of industry at present being merged'. Beyond that, the shareholders were careful to make no more than a conditional promise. They 'expressed to each other the intent—without conferring any right on the Argentine company—that other lines of industry would be conducted in Argentine through the medium of the Argentine company so far as external or other commitments may permit and provided it is not shown to be disadvantageous for some reason or other to the partner possessing the rights.' Altogether the supremacy of the shareholders and the primacy of their interests was unambiguously asserted. Duperial was left with no excuse not to know its place. As between the shareholders and the subsidiary, 'God'—or some other power—'made them high or lowly And ordered their estate.'

In Brazil in the early thirties, as in Argentina, ICI and du Pont moved towards a fusion of interests. The import trade of the two companies was no more directly competitive than in Argentina and ICI's business, once again, was considerably the bigger of the two. For the purpose of the merger, when it came, du Pont's Brazilian business was valued at 1,760 contos and ICI's at 6,807. The calculation was recondite, but the result presumably expresses the ratio between the sizes of the two undertakings.[85]

ICI's main import trade into Brazil was in alkali and chlorine products and in 'acids and derivatives, including sodium sulphide'. They came in through ICI (Brazil),* set up in 1928, which also imported a large range of 'minor ICI products' and engaged in considerable 'venture trading', the value of which, in 1935, amounted to about two-thirds of the value of the 'major ICI products' imported. Du Pont's trade was chiefly in dyes and related products, 'clar-apel' (transparent cellulose film), and finishes, with a very small trade indeed in leather-cloth and rayon. Explosives were imported into Brazil by Explosives Industries Limited.[86]

In Brazil the onset of the Depression, combined with political instability, produced grave weakness in the currency, and the Government imposed severe exchange control, though they negotiated an agreement with the German Government which gave Germans considerably more favourable terms than other exporters in Brazil. Among traders of various nationalities—British, American, German, Japanese—there was sharp competition, and as a group they were no more popular in Brazil than elsewhere. C. S. Robinson, sent to investigate the Brazilian market for ICI in May 1934, concluded that if ICI wished to go on doing business in Brazil they had better go into local manufacture. It would not do for them to go on being seen by the Brazilian Government merely as importers.[87] He was backed by Ralph Olsburgh, the head of ICI (Brazil): 'Importations are looked upon as a drain on the national economy. . . . I consider it, therefore, expedient that we establish a small vested interest in Brazil, so as to create the impression in official quarters that we are interested in the country's industrial development.'[88]

These recommendations were swiftly followed up. In January 1935, for 4,100 contos (rather under £45,000 at the free rate of exchange:

* *Companhia Imperial de Industrias Químicas do Brasil.*

considerably more at the official rate),* ICI bought a 50 per cent interest in an ammunition business owned by Count Matarrazzo (an Italian title) and his family.[89] Remington Small Arms bought the remaining 50 per cent. In March 1935 ICI took a 95 per cent interest, for £74,000, in a leathercloth business which had formerly been offered—at a lower price—to du Pont. It became *Cia Productos Nitro-cellulose Azevedo Soares* with Soares himself, the founder, as President, who had only brought himself to sell because he was convinced that if he did not ICI would put up plant themselves in competition.[90]

The Matarrazzo ammunition company, reconstructed, became *Cia Brasileira de Cartuchos SA*. In 1933 du Pont had taken a 60 per cent interest in Remington, and fear of an ICI–du Pont combination had played a large part in persuading Count Matarrazzo to sell. For their part ICI and du Pont much preferred a purchase to a fight because it gave them a foothold inside the Brazilian tariff wall, and it allowed them, they hoped, to soothe Brazilian xenophobia by keeping Costa-bile Matarrazzo on the Board alongside representatives of ICI and Remington.[91]

The partnership with Remington in the arms company brought ICI and du Pont very close in Brazil. In June 1935 a strong du Pont delegation, headed by Lammot du Pont, discussed co-operation, in London, and by November 1935 du Pont were reported 'most anxious that the merger [of du Pont and ICI interests in Brazil] should take place'.[92] It was arranged in the USA the following spring by W. H. Coates and L. H. Mitchell, H. J. Mitchell's son. On 1 October 1936, ICI (Brazil) and du Pont do Brasil began co-operating and they were formally amalgamated when *Industrias Químicas Brasileiras 'Duperial' SA* was established on 4 May 1937. Duperial (Brazil)'s capital was 14,000 contos. It took over the entire capital of the Soares business, including Soares's remaining 5 per cent, but the ammunition company was left as a separate undertaking.

Duperial of Brazil's first President was Ralph Olsburgh of ICI (Brazil). The Vice-President was a du Pont nominee, and of the seven members of the first Board, two were Brazilian. For matters of high policy there was a Shareholders' Committee, as there was for the Argentine Duperial. The members were the Director of du Pont's Foreign Relations Department, W. R. Swint, and the Manager of ICI (New York), G. W. White. The company's rights and the limitations upon them do not seem to have been set out in such careful detail as for the Argentine company, but clearly much the same was intended, for among the matters reserved to the shareholders for their decision were

---

* The free rate of exchange was £10·9 = 1 conto (the Brazilian unit of exchange), but the official rate was £17·7 = 1 conto.

'entrance into any major activity' and 'entrance into any trade policy arrangement with competitive concerns affecting territory outside Brazil, or affecting commodities not at present manufactured or sold by the merger company'.[93]

Along with Canadian Industries, the two Duperial companies represented the most ambitious attempts made by ICI and du Pont to manage joint enterprises. The industrial politics of South America were complex, the interests of the two partners were not identical, and their styles of management differed. Coates once told du Pont, not perhaps altogether tactfully, 'that ICI's wider experience in overseas management . . . had led us to give a wider latitude to the man on the spot than is yet perhaps the practice of du Pont, and for us to care more for essentials than for details . . .'.[94]

It is probably fair to say that the fundamental difference between du Pont's outlook and ICI's sprang from the fact, which Coates hinted at, that du Pont were engaged in far fewer overseas markets than ICI. As a consequence they regarded the South American undertakings far more single-mindedly than ICI, who were at all times concerned with the interrelationship between South America and their other markets. As McGowan observed to Lammot du Pont in 1937:

We must avoid a perhaps not unnatural tendency to regard Duperial Argentine (and Brazil) as another CIL—monopolistic to ICI and du Ponts. Canada is an Empire market, and ICI's right of pre-emption has been generally recognized and respected by its friendly European competitors— the same cannot be expected in ex-Empire markets, and du Ponts should recognize the vulnerability of the Argentine to attack from Europe, particularly Germany and Italy.[95]

The outbreak of war in 1939 radically altered the situation in South America, partly by interrupting trade with Europe and partly by making it impossible, or at least very difficult, for ICI to finance new developments. Consequently the kind of development which ICI and du Pont had had in mind for the two Duperials was brought to a halt. One new manufacturing venture—rayon—had been started in Argentina, to be carried on by Ducilo, a subsidiary of Duperial in which the principals had been obliged to allow Hirsch to take a 15 per cent interest. In the alkali industry, in 1939, ICI and Solvay, to their later regret, had committed themselves to a joint company (Electroclor) with IG. In Brazil there had been a great deal of talk—also, some of it, with IG—but no action in the alkali field. In general, however, the two Duperial companies still depended heavily—in Brazil, almost entirely— on the import trade. A beginning in local manufacture had been made, but no more, and progress, interrupted by the war, was later to be turned in a different direction altogether by the launching of anti-trust proceedings against ICI and du Pont in the USA.

(iv) WAR AND RUMOURS OF WAR

Colonel Pollitt in 1937 was no longer a full-time Director of ICI, but he could still ask awkward questions and make sure he got an answer. He was worried about ICI's foreign trade under the growing pressure of economic nationalism, political instability, preparation for war, and war itself. He suggested, with some support from B. E. Todhunter, a complete withdrawal from foreign trade outside the Empire and the re-investment of repatriated capital at home. When Coates implied that he was a dangerous extremist, he asked for figures 'showing how much better off ICI would have been had it had no trade whatsoever in overseas countries, other than Empire countries, for the years 1931 to 1937'.[96]

Coates plainly thought this was ridiculous—'apart from the time and labour . . . the information desired was not available'—but he could hardly ignore Pollitt altogether. Amongst the figures which he eventually produced, well over a year after Pollitt had asked for them, were the following figures of return on ICI capital employed:[97]

	Return on capital employed:	
	in Empire trade	in foreign trade
	%	%
1935/6	9·66	1·32
1936/7	10·34	3·12

These figures are a commentary on ICI's foreign policy during the first dozen years of the company's existence. It was the policy of establishing the British Empire as ICI's 'natural market', which foreigners did not invade, and of leaving foreigners alone in theirs. By 1937 it operated through what Pollitt called 'the network of agreements . . . so complicated that it is no longer possible for any individual to be fully cognisant of all the considerations necessarily pertinent to a decision on any proposal'.[98]

These agreements were not proof against all the tribulations of the thirties, particularly the political ones. During 1937 and later the pressure of Nazi policy caused ICI to write down or sell all their interests in continental Europe. Over the world as a whole the 1937 figures for return on capital employed, analysed in rather greater detail, show how widespread was the pattern which so disturbed Pollitt:[99]

1936-7	Return on capital employed:	
	in Empire trade %	in Foreign trade %
Home Groups	7·42	2·43
Foreign Merchant Companies[a]	4·35	3·77
ICIANZ	7·20	
AE & I	34·16	
CIL	13·48	
Joint companies:		
with du Pont in S. America		6·29
with Remington in Brazil		9·66
Return over all	10·34	3·12

a These figures, based on the book value of ICI's investment overseas and taking no account of capital employed on export production at home, give an optimistic view of the Foreign Merchant Companies' results.

Pollitt no doubt based his opinion of the superior merits of Empire Trade on the outstanding success of the companies in the Dominions, especially South Africa and Canada. But the returns from South America were also encouraging, considering that the Duperials and the joint venture with Remington had only just been set up and were in a very early stage of development. The real trouble, as we saw in Section (i), lay in the export trade, and especially, though the point does not emerge from the table above, in the Middle East, India, and the Far East.

The reason for ICI's varying performance in different parts of the world now begins to suggest itself, and it is not entirely a matter of the British Empire v. the Rest. The markets where business was paying well or reasonably well were those where the cartel-makers either worked fairly amicably together, as in Canada and South America, or, by agreement, kept out of each other's way, as in South Africa. In other words, in these areas the traditional policy of the leaders of the world chemical industry, inherited from the nineteenth century and adapted to the conditions of the nineteen-thirties, was paying off, although the shadow of anti-trust activity was already beginning to cloud du Pont's horizon.

In India and the Far East, by contrast, there was the almost un-inhibited competition of the Japanese to face. The whole scene was complicated, as it had not been in days gone by, by economic national-ism and political instability which in some places—Spain, Palestine, China—had already come to the point of war. Economic nationalism, in spite of Pollitt's forebodings, could probably be dealt with, and indeed had been in the British Dominions. The threat of war was another matter.

And thus matters stood in 1939. The whole elaborate edifice of ICI's cartel-regulated foreign trade, centred on the British Empire, stood like the British Empire itself, firm, imposing, on the brink of collapse. But no one expected the collapse of either, except perhaps the Japanese.

REFERENCES

[1] L. J. Barley, 'ICI Policy with regard to Economic Nationalism', 14 March 1934, CR 436A/13/13.
[2] Sir A. Mond and Sir H. McGowan to the President of the Board of Trade, 5 Nov. 1926, CR 93/1/–.
[3] Lord Melchett, 'Overseas Factories', 26 March 1934, CR 436A/13/13.
[4] G. P. Pollitt, 'Overseas and Home Trade Development', 19 March 1935, CR 93/94/304.
[5] 'Discussions à Winnington', 14 June 1935, p. 2, Archive de la Section A., Solvay et Cie.
[6] Treasurer, 'Export Trade of ICI', 9 June 1939, CR 179/7/260.
[7] Ibid.
[8] ICI I, p. 335.
[9] S. P. Leigh to London Committee, FMC Group, 'Venture and Agency Business', 12 May 1933, p. 3, CP 'Foreign Merchant Companies'.
[10] ICI I, p. 337.
[11] P. C. Dickens to W. H. Coates, 'Foreign Merchanting Companies', 20 Oct. 1936, CP 'Foreign Merchant Companies'.
[12] Sir H. McGowan, 'Memorandum on Visit to the Far East—China', 13 Dec. 1933, CR 93/6/254R; Sir H. McGowan to V. St. J. Killery, 21 Nov. 1933, CR 17A/27/27.
[13] G. P. Pollitt, 'Recruitment and Training of Staff', 27 June 1933, CP 'Staff Training and Recruitment'.
[14] Treasurer to Overseas Committee, 'Foreign Merchanting Companies', 22 April 1938, CP 'Foreign Merchant Companies'.
[15] D. M. Stephens to Treasurer, 12 March 1936, CR 93/94/295R; Stephens to President, 13 March 1936, CR 93/94/295R.
[16] As (14).
[17] Treasurer to Finance Committee, 'Export Trade', 5 July 1939, covering Coates, 'Foreign Trade', July 1939, p. 4, CP 'Export Trade'.
[18] ICI I, p. 171.
[19] B. E. Todhunter, 'ICI Export Trade', 2 May 1939, p. 4, CR 93/348/484.
[20] B. E. Todhunter, 'Development in Australia', 4 June 1934, TP 'Minutes and Memos re Meetings in London 1934–37'.
[21] African Explosives and Industries—'Synopsis of Draft Agreement', 12 Dec. 1929, CR 413/5/14; L. du Pont to Sir H. McGowan, 11 Dec. 1933, CR 537A/10/54.
[22] ICI I, p. 403.
[23] Concise Dictionary of National Biography 1901–1950, Clarendon Press, Oxford, 1958, p. 237.
[24] 'Report by the Chairman on his Visit to South Africa', 25 March 1935, CR 743/1/18.
[25] G. Ormsby Pearce, 'African Explosives and Industries Ltd.', 12 Nov. 1928, FD (African) 36/3 D and E.
[26] As (21), 1st refce.
[27] ICI I, p. 403.
[28] H. J. Mitchell, 'Report on Visit to South Africa', 31 March 1930, CR 743/1/1.
[29] L. J. Barley, 'Cooper, McDougall & Robertson', 24 Sept. 1930, CR 53/4/4.
[30] As (24), p. 5.
[31] 'Notes in regard to the Activities of AE & I', 6 Oct. 1932, p. 10, FD (African) 36/25.
[32] AE & I—History of the Ten Years 1927–36, 2 Feb. 1940, p. 11, FD (African) 36/25.
[33] Ibid. p. 1.
[34] B. E. Todhunter, 'Note on Australian Policy and Development', April 1927, p. 2, TP A 12/4.

[35] Ibid. p. 5.
[36] B. E. Todhunter, 'Report on Visit to Australia and New Zealand 1925-26', Feb. 1926, p. 13, TP.
[37] As (34), p. 6.
[38] B. E. Todhunter, 'The Position and Policy of ICI in Australia', 3 April 1928, p. 2, TP A 11/5.
[39] Ibid. p. 3.
[40] B. E. Todhunter, 'ICI (Australasia) Ltd.', 22 May 1928, TP A 11/5.
[41] B. E. Todhunter, 'Australian Company—Capitalisation', 19 July 1928, p. 2, TP A 11/5.
[42] Ibid. p. 2.
[43] B. E. Todhunter to H. J. Mitchell, 7 Feb. 1929, TP A 6/11.
[44] B. E. Todhunter to H. J. Mitchell, 28 Feb. 1929, TP A 6/11.
[45] B. E. Todhunter, 'Soda Ash Australia', 10 March 1932, TP A 2/3.
[46] B. E. Todhunter, 'Manufacture of Soda Ash in Australia', 20 Jan. 1932, TP A 2/3; Sir H. McGowan to Sir L. Raws, 'Soda Ash', 17 March 1932, TP A 2/3; ICIANZ, 'Alkali Manufacture in Australia', 1 Oct. 1932, TP A 2/3; J. L. S. Steel, 'Memo on Alkali Manufacture in Australia', 17 June 1933, TP A 1/2; B. E. Todhunter to Sir L. Raws, 12 June 1933, TP A 1/2; ICIANZLM, 7 June 1934; Todhunter–Taylor, cable, 16 July 1934, TP A 1/2; C. Prichard, 'Addendum to Report entitled "Alkali Manufacture in India and Australia"', 18 April 1934, TP A 1/2; ICIANZLM, 25 April 1935; I. G. Gilbert, 'Summary', 20 May 1937, TP A 1/4; Bennett and Prichard, 'Revised Estimate of the Total Expenditure', 22 Nov. 1938, TP A 1/4; Sir L. Raws, 'ICI Alkali (Australia) Pty. Ltd.', 21 April 1937, TP A 1/4; Minister to Sir L. Raws, 12 Dec. 1935, TP A 1/4.
[47] C. S. Robinson, 'Synthetic Ammonia Manufacture', 9 Feb. 1937, TP A 3/3; Lord McGowan to B. E. Todhunter, 10 Feb. 1937, TP A 3/3.
[48] As (20).
[49] Lord McGowan to Lord Weir, 9 Feb. 1937, Weir papers.
[50] 2 Oct. 1928, CR 537/2-4/1; USA, CIL 2nd Draft, p. 33.
[51] 6 March 1928, CR 537/3/3.
[52] L. du Pont to Sir H. McGowan, 11 Dec. 1933, CR 537A/10/54.
[53] On Canadian Salt, see Purvis to H. J. Mitchell, 5 Aug. 1928, CR 28/12/73 (4).
[54] As (50), 2nd refce., Appx. A.
[55] W. J. Worboys, 30 Jan. 1930, CR 537A/5/5.
[56] Meeting at Wilmington 19 Oct. 1937, CR 537A/22/22.
[57] ICI/DuP/CIL Meeting 4 Dec. 1930, CR 537A/1/26.
[58] As (50), 2nd refce.
[59] Ibid. pp. 34b ff.
[60] Ibid. p. 37; as (57).
[61] As (50), 2nd refce., p. 47.
[62] As (52).
[63] 'Minutes of Meeting . . . ICI and DuPont', GPCSP, 28 July 1936.
[64] L. W. B. Smith to Central Administration Committee, 11 Jan. 1937, CR 595/29/29.
[65] G. White to W. H. Coates, 5 March 1937, CP 'CIL/Shawinigan'.
[66] ICI Foreign Department Reports, CR 537A/16/38.

Sect. iii

Much of the material on which this section is based comes from documents and commentaries prepared for the defence to the action *USA v. ICI et al.* See Source Material, p. 522 and Abbreviations, p. xiii.

[67] Population figures are from the *Statesman's Year Book* 1940. Since South American censuses were rare events the population figures are estimates, the figure for Bolivia being an estimate of 1935 based on a census of 1900.
[68] G. W. White to H. J. Mitchell, 28 March 1934, USA, CSAE III (ii), p. 21.
[69] *ICI* I, p. 397.
[70] Sir H. McGowan, Address to Nobel Industries' shareholders, 22 Sept. 1922, quoted *ICI* I, p. 397.
[71] USA, CSAE III (ii), pp. 23, 19.
[72] Ibid. pp. 32-3.

[73] Ibid.
[74] USA, CSAE VI 1940, pp. 6, 8.
[75] J. Laing to G. W. White, 20 Oct. 1939, USA, CSAE V, pp. 22–3.
[76] USA, CSAE V, pp. 30–1.
[77] H. J. Mitchell to Sir H. McGowan, USA, DA I, p. 76.
[78] USA, Legal Dept. 'Write-Up' EIL II, pp. 268–9.
[79] Lord Melchett, 'Report on Visit to South America—Argentina Part I', 7 June 1937, MP 'South America'.
[80] S. P. Stotter, 22 Nov. 1932, 13 April 1933, FD 'Argentine 11'.
[81] As (77), p. 75; H. J. Mitchell to J. Crane, n.d., CR 520/10/18R.
[82] Foreign Dept. Report, 17 April 1931, Administrative Committee Papers; Administrative Committee minute 156, 27 April 1931; H. J. Mitchell to A. Hirsch, 18 July 1932, FD 'Argentine 11'; USA, DA II, pp. 36–7; H. J. Mitchell to Sir H. McGowan, 17 Jan. 1934, USA DA II, pp. 20–1.
[83] 'Merging of Argentine Interests', USA, DA II, p. 30.
[84] Ibid.
[85] 'Memorandum of Agreement reached in Wilmington, May 26th, 1936', CP 'South America, Brazil, Dr. Coates' Report on Visit'.
[86] Ibid.
[87] GPCM 825, 21 Nov. 1934.
[88] R. Olsburgh to H. J. Mitchell, USA, DB I, p. 25.
[89] See: USA, DB; GPCM and GPCSP; FD 'Matarrazzo Cartridge Factory'; FD 'CBC 1—Administration'.
[90] GPCM 910, 27 March 1935; R. Olsburgh to H. J. Mitchell, 26 Feb. 1935, FD 'Azevedo Soares 2—Finance and Board'.
[91] As (89).
[92] Trevor Bell to H. J. Mitchell, 19 Nov. 1935, USA, DB I, p. 69.
[93] As (85), p. 73.
[94] W. H. Coates, 'The Duperial Companies', 5 July 1938, USA, DA II.
[95] Lord McGowan to L. du Pont, 4 Oct. 1937, CP 'South America, Argentine. Bunge & Born. Relations with Duperial'.
[96] G. P. Pollitt, 6 July 1937 and 19 May 1938, CP 'Export Trade'; B. E. Todhunter, 'ICI Development Policy', 21 May 1937, GPCSP; B. E. Todhunter, 'ICI Export Trade', 2 May 1939, CR 93/348/484.
[97] Treasurer to Finance Committee, 'Export Trade', 5 July 1939, covering Coates, 'Foreign Trade', July 1939, and Treasurer's Memo, 'Export Trade of ICI', 9 June 1939, CP 'Export Trade'.
[98] As (96), first refce., p. 2.
[99] As (97).

Dictatorship in the Thirties:
the Barons' Revolt

(i) THE DICTATOR SUPREME
McGOWAN'S MOST IMPORTANT SERVICES to ICI were probably rendered in the first seven years of his Chairmanship, between the beginning of 1931 and the beginning of 1938. Great decisions of policy were taken, on the elaboration of the cartel system, on the founding of joint companies with du Pont, on the rescue and transformation of Billingham, all at a time when the industrial and financial structure of ICI had been gravely weakened by the Depression.

To get the measure of McGowan's achievement it is not necessary to pretend that the entire series of decisions was his alone, or that all the decisions were flawless. Some, particularly the policy of oil-from-coal at Billingham, lie open to serious criticism. But they do represent willingness to face reality, to accept the fact that the early investment policy of ICI had gone irretrievably wrong, and to take measures, some of them harsh, to repair damage which otherwise might have broken the newly founded merger. The responsibility lay upon the Chairman, and he neither ran away from it nor attempted to push it on to his colleagues. Having demanded supreme power he was prepared to exercise it, and his imperiousness was equalled by his strength of purpose.

We have surveyed ICI's response to the Depression in preceding chapters of this Part. At the same time the first signs were appearing of the technological revolution which we shall consider in Part IV, but that lay outside McGowan's sphere of competence and he played a less decisive part in it. What we have now to consider is how opposition developed towards McGowan and eventually broke out into open rebellion.

As an autocrat, McGowan was selective. He did not attempt detailed control of technical policy. As a consequence the technical development of ICI, during the thirties, was very much a matter for the management of the Groups, which suited them, since most of the individuals concerned, particularly in the powerful Alkali Group, were technical men first and commercial men second.

McGowan's personal influence, therefore, bore most heavily on commercial and financial policy, including ICI's selling prices, and the control of all these matters was kept firmly at Millbank. The Groups, under this system, could not be held fully responsible for their profitability, with he result explained by J. G. Nicholson in 1936:

So long as the Chairman dictates the selling prices, so long as ICI controls their finance, so long as the Groups are dependent on an ICI purchasing organisation, the Groups cannot regard themselves as solely responsible for their profit and loss account.[1]

There could hardly be a more succinct account of McGowan's dictatorship at its height. Every key decision—even, if only formally, every technical decision of any importance—was made at the centre. At the centre sat McGowan, advised by other Directors but issuing instructions on the widest matters of Company policy as personal decrees and relying on senior managers, below the level of the Board, to carry them out. ICI was as near to being run by one man as any business of its size could be, and the experience sank deep into the corporate consciousness.

McGowan's autocratic style, with enthusiastic support from another autocrat, Nicholson, was institutionalized on the selling side of ICI, mainly because sales policy, at home and abroad, was governed in minute detail by the great array of trading agreements built up over the years, and especially in the early thirties. These agreements regulated how much might be sold, in what parts of the world, and at what price. With their elaborate case law and rules of procedure, some of it running back for many years, they were the furniture of ICI's world in the thirties. It was the kind of furniture which McGowan and his contemporaries had grown up with, and they were accustomed to moving with great care to avoid knocking any of it over. The elaborate diplomacy of managed markets was one of McGowan's main preoccupations, and it dictated the nature of ICI's selling organization. It was centralized. Very little liberty was allowed to the Groups, or to the sales offices overseas in case they broke agreements or caused them to come into conflict.

Prices lay at the heart of the matter, and McGowan himself laid down the rules for fixing them. Where they were not 'fixed by agreement with competitors'—itself a matter for the central management— the Groups, acting with the Home Controller of Sales Organization, were required to recommend standard selling prices, standard average net naked realization prices, and limits of permissible variation of the latter. These recommendations were to go up by way of the Sales Committee 'for approval by the General Purposes Committee [of the Board] or by me'. McGowan, that is to say, reserved the right to have the last word, and even when fast action was required to meet competition, he ordered 'that whenever time permits, my approval for the variation proposed is obtained through Mr. Nicholson'.[2]

McGowan's ascendancy was at its peak by 1936, when he was treating his colleagues on the Board with something approaching open contempt. Certainly he does not seem to have regarded them as in any way his equals, with joint responsibility for the Company's policy and a right to be consulted on it. On the contrary, he claimed an unfettered right to make decisions, and he expected the Board's support—'obedience' would perhaps be the better word—whether or not they agreed with what he had decided.

These points emerge from the report of a meeting of full-time Directors held on 9 March 1936 to consider a report by Nicholson on the sales organization. The Chairman was on holiday, so Mitchell presided, having become President since the death of Reading at the end of 1935. Mitchell told the meeting that he regretted they had had such short notice, but 'before leaving for his holiday, the Chairman left instructions . . . that he would expect on his return (which is next Monday) to have placed before him proposals . . . considered by all those who would, or might, be affected by their adoption'.

That might be thought a sufficiently short way with Directors, but the Chairman, speaking through Mitchell, went further. He was not bound to accept any recommendations and there was no mention of any further consultation. 'It will still', Mitchell said, 'be for the Chairman in his wisdom to adopt or reject or amend our proposals but in so far as he does adopt or amend them he will be entitled to expect—and I am sure will demand—the loyal carrying out on the part of everyone of the plan which is finally evolved in all its detail.'[3] Could an absolute monarch demand more, short of invoking Divine Right?

McGowan's dominance of ICI affairs was general, but one particular aspect of major policy was peculiarly his personal creation: the heavy investment of ICI funds in the shares of other large enterprises, especially General Motors. There was no intention, and indeed no possibility, of gaining control, and the result was that the value of a large proportion of ICI's cash reserves depended on Stock Exchange prices, quite beyond ICI's influence. Even while McGowan's power was at its height, in 1936 and the early part of 1937, criticism began to fasten on this unorthodox policy both within and outside the Company.

The origins of this policy, like so much else in the early days of ICI, ran back to McGowan's time in Nobel Industries.[4] Put shortly, the chief motive force had been McGowan's faith in the future of the motor industry, which accounted for the large investment in General Motors and for other holdings as well, particularly in Joseph Lucas. To this had been added, before the ICI merger, the chance effects of merger operations in the USA and Germany, which had brought a holding in

Allied Chemical to Brunners and a holding in IG to Nobels. Various other investments, including a large one in International Nickel and a small one in du Pont (which du Pont themselves knew nothing of), made up the rest of ICI's portfolio.

Up to 1929, while the Stock Exchange stood high, especially in the USA, and ICI were hungry for capital, there were large sales. The profit on these sales was not, as the law then stood, taxable unless ICI were considered to be in business as financiers. The Inland Revenue contended that they were, and the contention was only fought off, with great difficulty, after four years' negotiations. Even then, the Revenue might try again.

As part of ICI's defensive campaign against the Revenue it was considered wise to liquidate Nobel Industries, because after the merger the company had no function but to hold and manage investments, so that it might with good reason be held taxable as a financial institution. Accordingly Nobel Industries went into voluntary liquidation on 28 September 1928. The immediate tactical point, against the Revenue, was made, but the consequences turned out very embarrassing indeed for ICI and therefore, indirectly, for McGowan, the author of the investment policy.

The root of the trouble lay in the valuation of Nobels' marketable investments, chiefly GM shares, for the purposes of the liquidation. They were valued at the middle market price on 28 September 1928, and that showed GM at $84·2, against $30·75 in Nobels' books. On this basis, and on the basis of high valuation of other assets, a 'surplus capital profit' of £9,264,000 was calculated. This was a mythical figure, because it was based mainly on one day's market prices and none of the investments from which it was supposed to have arisen were in fact realized. Nevertheless it was taken into ICI's books and used to create various reserves, including a Central Obsolescence and Depreciation Fund of £2·5m. which caused trouble with ICI's auditors in 1931.[5]

GM shares rose to $91¾ in 1929—their highest inter-war figure—and then collapsed. At their lowest point, in 1932, they were at $7⅝. They recovered to $77 in 1936, but they fell again in the bad year of 1938, and all the time they were swinging very widely, but nothing was done, in these years, to write down their value in ICI's books. 'The various reserves which were created', wrote Dickens in 1937, 'depended not on any realized capital or revenue profit but on a purely fortuitous market valuation of an American investment not subsequently maintained.'[6]

When McGowan became Chairman there was nothing to be done about ICI's marketable investments except hold on and hope for better times. There was no need, fortunately, to sell at the disaster prices of

1932-3 because during the worst three years of depression, up to the end of 1933, ICI, far from being short of cash, was accumulating it through reducing stocks and stores and cancelling or postponing capital expenditure.[7] Consequently the first tasks of reconstruction, after the worst was over, could be directed at paying off Debentures of subsidiary companies, writing off unproductive assets (at the Annual General Meeting in 1937 McGowan said £6·9m. had been applied to that purpose), and to getting rid of the Deferred shares which had been created at the time of the merger in anticipation of vast profits from Billingham (p. 19 above).

By 1936 £4m. had been applied to redeeming subsidiaries' Debentures. Times were getting better, and ICI was beginning to need cash again. At once the question of the investments—above all, the GM shares—arose. Was it sensible to continue to hold them, when the market was back to a reasonable level and ICI needed cash? More, could the Directors justify a policy of holding reserve funds in Ordinary shares, which might be at the bottom of the market when cash was badly needed? In the bold days before 1929, the possibility of a really bad fall in GM had seemed remote and McGowan's vision as an investor had been much praised. The experience of 1929–32 had made him look perilously like a speculator, and there was no doubt that the confidence of the Directors had been shaken. But the policy represented by the GM holding was McGowan's own, and it would take a bold man to criticize it to his face. H. J. Mitchell suggested that Henry Melchett should undertake the task, and Melchett agreed.

Melchett presented his views to McGowan in a paper dated 19 March 1936.[8] In support, he enclosed a note from Dickens which said: 'The policy of a company should be to invest its money in its own concerns which it manages itself. It should also have a certain proportion of liquid cash . . . invested in such a way that every pound sterling invested will be returned when required as a pound sterling, and this can be achieved by investment in fixed-dated Government and other guaranteed securities.' This was a statement so uncompromising as to be foolhardy. Melchett himself was much milder. He suggested that keeping *all* the cash in gilt-edged securities would produce a low return ('a drag on the trading profits of the Company') and also make it impossible to expect 'the natural capital appreciation which would follow the natural progress of the world'. He concluded that it might be 'a reasonable provision to keep a quarter or one third of our £6 million in gilt-edged securities'. After this conciliatory opening, no doubt carefully drafted, he came to what he said was his main point:

It does not seem sound policy to continue to hold large quantities of equities and to keep the Company short of cash. Such a policy was doubtless justifiable in the past, in years when our investments stood at figures far below their book value or the value at which we appraised them, but when as today they stand above their book value, I suggest that it is better to realize some of our securities so that the normal business of the Company can be carried on without the handicap of shortage of ready funds.

He provided a list of suggested sales designed to raise about £2·5m., leaving about £6m. still invested in Ordinary shares.

At the end of July 1936 the Finance Committee—McGowan, Mitchell, and Coates (Melchett was away)—considered a programme of selling much more drastic than the one put forward earlier by Melchett. It included the sale of 290,000 GM shares, leaving only 100,100 in the portfolio, and the sale of 100,000 out of 120,000 shares in International Nickel. The Committee 'agreed to make decisions later in the year'[9] and the Board confirmed their Minute. The end of the year came and the end of the year passed, but no decisions of consequence were taken because McGowan would not take them. Although pressed by the other members of the Finance Committee throughout 1937, he would allow nothing but relatively minor sales of Lucas and Murex shares. The holdings in GM and International Nickel remained intact. No matter what the Board might say, McGowan was determined not to sell.

In this frame of mind McGowan sailed for Australia. At Fremantle on 19 January 1937 he found a letter from the United Kingdom High Commissioner telling him that he had been offered a peerage. Next day he wrote by hand to his Personal Assistant, L. H. F. Sanderson:

Well, Sandy, the long expected Honour has come at last. . . . In all modesty this Honour will give much pleasure to Industry in general and our own people in particular, and to those who know me. They will appreciate that I have neither intrigued for, or bought it. I have written Mother about it but poor old dear I'm afraid it will be all over her head. . . .

I've no doubt there will be a great number of congratulations . . . I would . . . like to reply by cable to any congratulations from Cabinet Ministers or Government officials. . . . Never mind the expense—the Coy will pay as I am away on their business.

Of course I am proud of the Honour. I wouldn't be human if I weren't. It is certainly an encouragement to anyone with brains and ambition.[10]

Like other despots, before and since, McGowan at the height of his career was to find that his absence bred revolt. Even as he wrote his exultant letter, in the Orient liner *Oronsay*, 'rolling like Hell in the Bight', the Directors at home were nerving themselves to confront him, when he returned, with a challenge to his autocratic behaviour.

(ii) THE DICTATOR UNSEATED

In January 1937, while McGowan was at sea, E. J. Solvay, in London

I

for a Board meeting, found Mitchell anxious to talk about the Directors' discontents. They wanted to sell ICI's GM shares; McGowan didn't. McGowan, said Mitchell, wanted to get his elder son on to the Board and was seeking to make matters easier for himself by getting rid of Melchett on a tour of South America. Would Solvay help to frustrate McGowan's machinations? Would he join the Finance Committee?[11]

Solvay was in London again a month later. Mitchell was in poor health and John Rogers, perhaps prudently, was absent 'en congé de repos'. Three lay Directors—Lord Weir, Lord Colwyn, and Sir Christopher Clayton—went on from where Mitchell had left off last time. They said there were too few independent Directors and complained of McGowan's dictatorial behaviour. He made appointments to the Board without consulting the lay Directors. He didn't abide by committee decisions (it was presumably his refusal to sell GM shares they were thinking of). The question of making his son a Director came up again and Solvay said he would oppose the appointment on principle. Mitchell, a little late, perhaps, said he was against the notion of one man being both Chairman and Managing Director and he said that the other Directors had no real power except on the day of a Board meeting, so that their independence, as Solvay put it, was 'toute rélative'. That may have been true enough, but the Directors had not had the courage to oppose the handing over of power to McGowan which Mitchell had led them into.

McGowan was due back in England in April 1937, in time for his son's wedding. During the remainder of his absence the underground movement against him gathered force and coherence. In March Solvay met Weir, Colwyn, and Clayton in Weir's London flat. They had not yet taken the whole Board into their confidence, which is no doubt why they met away from Millbank, protesting to each other, as they did so, that they were neither criticizing McGowan nor conspiring against him. The next day Solvay reported the proceedings to Mitchell. They had settled on two main demands, both aimed at greater independence for the Board. They wanted the lay Directorate ('représentants du public') strengthened and the appointment to the Finance Committee of a member unconnected with the management of ICI, but Solvay had told them that he, personally, would not serve because he thought they needed an Englishman of the highest public standing—another Reading, in fact.

As soon as McGowan got back, Mitchell faced him with the demands of the three lay Directors. What McGowan said, unfortunately, is not recorded, but in substance he was compliant. The lay Directors should certainly be reinforced, and one of them should come on to the Finance Committee. More, there should be a pause for reflection before the appointment of any new Director. Whether anything was said about young McGowan is not clear, but Mitchell, in conversation with Solvay, hinted that he had similar ambitions for his son, and he went on to say that neither his son nor McGowan's had any right to a Director-ship—their fathers, after all, were 'les serviteurs de l'ICI et non les propriétaires'. ˙

Solvay, presumably encouraged by McGowan's amiability, opened the whole matter to Nicholson and Coates. Nicholson was sceptical. McGowan's attitude might change from one day to another; they must wait and see. Coates was hurt. Why hadn't Solvay spoken to him before? Because, said Solvay, he had wanted to avoid giving the impression that his intervention in the matter was the result of intrigues.

The Directors, having made their demonstration, seem to have been content to await results. Nothing changed. McGowan's agreement as Managing Director came up for renewal in July 1937. Here was an opportunity, if the Board had been looking for one, to force the issue, but the agreement was renewed for another three years on the old terms. In the autumn McGowan left for North America.

Early in November, Melchett asked E. J. Solvay to spend a week-end with him. He wanted a long discussion of 'matters of the first importance to ICI'. It is fairly clear that among the matters for discussion there was a new element in the situation: namely, rumours that McGowan had been speculating heavily and unsuccessfully in shares and commodities, and that he might be bankrupt. The question of replacing him had been raised, and the candidate for the succession was Melchett himself.

As the rumours spread, the old grievances flared up—the composition of the Finance Committee; the appointment of new Directors; McGowan's nepotism; McGowan's despotic behaviour generally. Mitchell came out plainly against McGowan, and Colwyn supported him. So did Melchett. Weir and Ashfield, Solvay thought, were friendly to McGowan.

McGowan came back from the USA on 15 November 1937. John Rogers, deputed by the rest of the Board, met him that day and told him what was being said about his speculations. McGowan responded with dignity, speed, and frankness. He told John Rogers at once that 'he was prepared for the fullest possible investigation into his private financial affairs with a view to satisfying his colleagues that nothing he had done had in any degree prejudiced the position of the Company'.[12] Four days later he met the entire Board in Lord Weir's flat and offered to let them nominate an accountant (they chose Sir William McLintock) who should examine McGowan's affairs and 'advise the Board whether, if a full disclosure of all these matters was in possession of the Board, the Board would in his opinion be satisfied that they could . . . continue to give their full support to Lord McGowan'. If not, then Sir William would tell 'a small committee of the Board' precisely what his reasons were.

McGowan was obliged to reveal speculation, on his own behalf and his family's, on a scale which would pass belief if it were not recorded in Sir William McLintock's reports. McGowan's activities were at their height in April 1937. He had credit, then, from brokers of £1·9m., supported by securities and commodities on open account and by other securities pledged. His commitments for stocks, shares and commodities (chiefly rubber) amounted to £1·7m., so that they were comfortably covered by his credit so long as markets moved his way. Instead, they fell. McGowan had to sell quickly and heavily. By 30 November 1937, when McLintock investigated, he had reduced his liabilities to £155,000, but his credit with his brokers was exhausted and he owed them £63,000. He was close to bankruptcy, but he was avoiding it by borrowing from his family, by the desperate expedient of surrendering a life policy in Lady McGowan's favour, and—most of all—by accepting a bank guarantee of £60,000 from Lord Camrose.[13]

Even graver, potentially, than the threat of bankruptcy were the implications of some of McGowan's share transactions. Throughout the time when he was resisting the Finance Committee's desire to sell part of ICI's holdings in General Motors and International Nickel, he himself was dealing in the shares of both companies. At the peak of his activities he and his relations held 6,500 shares in General Motors and 16,800 in International Nickel, with a combined market value of £300,000. ICI was a much larger shareholder in each company (390,000 shares in GM; 120,000 in International Nickel), and it was in McGowan's interest as a private investor that ICI should not drive share prices down by selling. Was this the reason why he had stopped the sales of ICI's holdings?

On 9 December 1937 the Directors met again in Lord Weir's flat. McLintock told them he could not recommend them to continue to give their full support to McGowan. Mitchell, at great length, said McGowan had lost most of the confidence his colleagues had in him and they would never again put up with his 'despotic attitude'. Ashfield, Melchett, Pollitt, Rogers, and Weir were nominated to hear McLintock's reasons in detail. A few days later, their conclusions having been conveyed by all of them personally to McGowan, two of them met the full-time Directors.[14]

The Committee's main conclusion was simple. McGowan must go— on very good terms, if possible, but go he must. The only question was whether he would resign or whether they would have to dismiss him. If they did that he might bring an action for breach of his new agreement but its validity, the Directors thought, was doubtful, because if they had known about McGowan's speculations when it was being negotiated, in the summer of 1937, they would probably not have entered into it. The full-time Directors promised their unanimous support to the Committee.

The Committee needed it. A lesser man than McGowan, his enormous speculations a total loss and his career apparently hopelessly in ruin, might have gone away quietly with the very generous terms which the Committee were anxious he should accept. Not McGowan. He meant to fight. Hence, presumably, the unmistakable air of apprehension which runs through Mitchell's notes of the Directors' deliberations.[15]

Not that there seemed to be much real cause for alarm. The Directors' case against McGowan was overwhelming and they had their candidate for the succession: Melchett. He was willing, there is no record of opposition to him, and plenty of evidence of support. No one, on the other hand, seems to have contemplated any possibility of McGowan staying in office. Mitchell wrote confidently 'of the irrevocable conclusion'.

On Christmas Day 1937 Lord Melchett, still under forty, had a heart attack. The second Lord Reading, his brother-in-law, told Solvay on 3 January 1938 that he was 'far from well', and it soon became known that he would be away from business for a year. He appointed Reading his alternate in ICI.

From this moment the Board began to go into reverse. No more was heard of the 'irrevocable conclusion', and although the Board more than once reassured itself, by taking legal advice, that there were excellent grounds for dismissing McGowan, it showed no eagerness to do so, and it gradually came to be taken for granted that he would remain Chairman. The turning point seems to have been a report from the committee deputed to negotiate with McGowan, dated 7 January 1938. The committee reported Sir William McLintock as saying that there was no reason to suggest that McGowan's speculations had 'adversely affected the Company'. The committee themselves then said they found 'no ground for suggesting . . . that the advice which the Chairman gave to the Company during the period of these speculations would have been different . . . if he had not then been engaging in these or any speculations'. They went on to say that he 'ought not to have engaged in these speculations', but although they discussed the possibility of dismissing him and the Board's power to do so, they nowhere said that he ought to be dismissed.[16]

Quite the contrary. They suggested that 'the present plan of [ICI] organisation should be altered in important particulars' and went on to say: 'The determination of the Chairman's services before this reorganisation had been carried out would prejudice the carrying out of this plan of reorganisation and should if possible be avoided.' 'The presence and active co-operation of the Chairman', they said, 'would be of value in carrying out this reorganisation', but it was not possible to say in advance quite what duties would be required of him. 'Hence', they concluded, 'we do not think that it would be in the interests of the Company to negotiate with the Chairman either with a view to his immediate retirement or with a view to assuring him a position for any lengthy period under what must be described as the untried conditions of the reorganisation plan.'

What an extraordinary turn-round! Why should anyone who knew McGowan suppose that he would give his 'active co-operation' to a scheme of reorganisation which could only be aimed at curbing his powers? Surely it would be much better to get him out of the way. What made the Committee change its mind so suddenly?

The answer seems to be that although McGowan might have been persuaded to hand over to Melchett, he would willingly hand over to no one else, and that although his legal position was weak, he never-theless had considerable scope for obstruction, if he chose. With Melchett unexpectedly knocked out, he did choose, remarking to his dentist, it is said, 'They can't do without me!' Whether they could or not, active hostility from the Chairman was not something the Board could lightly contemplate, for the crisis in ICI's affairs could not be entirely hidden, and damaging rumours were abroad, which would get worse the longer the problem was unresolved.

That seems to be the most likely interpretation of the obscure passage in the Committee's report—doubtless intentionally obscure—which deals with the point:

If the Committee were faced with a situation which simply involved the selection of an individual . . . our task would assume a totally different aspect from that which confronts us to-day. But we are faced with the fact that Lord McGowan now occupies the position of Chairman and Managing Director; and before the Board can feel able to deal with his position with a free hand and with the least possible injury to the Company we strongly feel that some preliminary reorganisation . . . is necessary.

McGowan, then, would remain. The decision was not universally popular. 'We feel', said the Committee, 'that the views of individual Directors on the suitability of the Chairman to remain the leader of this Company must necessarily vary in some degree.' Mitchell, cer-tainly, must have felt the ground giving way under him, and Coates decided to seek 'the protection of Counsel's advice'.The Committee, in finding no ground for supposing that McGowan had let his private interest interfere with his duty to ICI, might have been right in their mind-reading, but it was 'at least reasonably . . . open to a hostile critic' to suggest otherwise. Were the arrangements to ensure

McGowan's solvency adequate? Could he be relied on to keep any promise he might make not to speculate any more? It was said he had broken such a promise given in the past. 'One would expect', Coates concluded, 'some logical link between the first orally conveyed conclusions of the Committee and the final conclusions embodied in the written report, but it is difficult to find it.'[17] It is, indeed.

The Committee's central proposal was to abolish the office of Managing Director. Instead, the 'central executive authority' would be vested in a Management Committee. The full-time Directors would become executives again, each being chairman of a committee responsible for one of 'the major branches of the Company's activities'. The members of the Management Committee would be the executive Directors, with at least two lay Directors, and the Chairman of the Company, ex officio, in the Chair. The Chairman of the Company would still be McGowan, but on a three-monthly agreement, at a lower salary, and having given 'the most definite possible assurances against engaging in any further speculation'.

When Mitchell saw these proposals he was deeply disturbed. 'The scheme . . .', he said, 'does not give the full-time Directors any more authority than they have had in the past and therefore does not meet one of our very real and recurring problems, that of the present GPC being faced with decisions which may not have had any "collective" consideration. It still (on paper) leaves the staff with a right—not infrequently exercised—of by-passing the whole administrative machinery . . . and seeking for decisions of the Chair by direct approach.'[18] He was convinced that it was a mistake to make the Chairman of the main Board the Chairman of the Management Committee also. By this arrangement, said Mitchell, 'he advises the body which passes judgment on a decision of another body led by himself. This position has never been favoured by me. It can so easily hamper free expression and cause embarrassment.'

This was no doubt a delicate way of saying that nobody on the Board was prepared to stand up to McGowan. Certainly if the object of the scheme was to assert the authority of the Board against the authority of one individual, and if that individual were to be McGowan, there was a grave risk that it would be self-defeating for the reason that Mitchell pointed out—that McGowan was to be both Chairman of the Board and of the Management Committee. No one who knew him could doubt that he would attempt to impose his authority in the future as he had imposed it in the past, whatever the constitutional position might be, and no one could be sure that he would not succeed. Mitchell thought he would, and no one was better qualified to judge.

Mitchell was overridden. The scheme suggested by the Committee was elaborated but not substantially altered. We shall return to it in a future chapter and all that need be said here is that when the new organization was brought into effect, on 22 March 1938, it provided for a Management Board, six Executive Committees for managerial functions, a Salaries Committee, and a Groups Central Committee for 'the co-ordination of the activities of the Operating Groups'.[19] McGowan ceased to be Managing Director, remained Chairman, and became Chairman of the Management Board. The only concession to Mitchell was that it was made clear that the two positions were separate, although for the time being held in plurality—a sufficiently meaningless distinction, it may be thought, given the personality involved.[20]

McGowan signed a new agreement, terminable at six months' notice. His salary remained, but his commission was reduced from one-third of 1 per cent of ICI's profits to 0·2392 of 1 per cent. His total emoluments in 1938, a bad year, were £10,000 lower than in 1937, but they soon picked up again: he was not seriously impoverished. As a condition of the agreement, he was required to sign 'an assurance that so long as I remain Chairman of the Company I will engage in no speculation of any kind'.[21]

The Board did its best, in drafting the document on the new organization, to make it clear that McGowan had given up the supreme executive power. What was more, they obliged him to sign a letter to senior staff which he cannot have liked at all:

I send you herewith a Memorandum setting out the changes in the higher organisation of the company which have been approved by the Board.

You will observe that they are designed to relieve me from the heavy burden of daily duties which have fallen on my shoulders during the past seven years. To make that object fully effective, I shall be glad if you will particularly note that I do not wish in the future to be looked to for any executive instructions. Such instructions must be obtained . . . from the Chairmen of the appropriate Executive Committees, who will be in constant consultation with me and be the channel for the expression of my views.[22]

The King had been forced to grant a constitution. The magnates—officially—were in control.

Within twelve months, in 1937–8, McGowan was raised to the peerage and a few months later brought to the edge of spectacular ruin. His Board turned unanimously against him. He faced what seemed to be inevitable dismissal and was saved by chance—Melchett's illness. Then, single-handed, he negotiated himself back into a position not, in substance, very much less powerful than the one the rebels had tried to turn him out of. He was 63, but he had no mind for retirement. He saved himself by sheer coarse vitality, and although his behaviour in prosperity had been deplorable, his fortitude in adversity compels admiration.

The principal casualty of the barons' revolt seems to have been H. J. Mitchell. He withdrew from executive duties, and in 1941 he died, aged 64. In the year of his death his name appeared for the first time in *Who's Who*. The brief entry says:

MITCHELL, Harold John, CMG 1939; Member Overseas Settlement Board since 1936; *b.* 1877; *s.* of John Edmond Mitchell; *m.* 1900, Clara Maud, *d.* of Thomas Mortlock Powell; one *s.* two *d.*

Not a word about ICI. He must have been very bitter.

REFERENCES

[1] J. G. Nicholson, 'Sales Report', 30 Jan. 1936, pp. B8–9, CP 'Organisation Commercial and Sales 1936'.

[2] Sir H. McGowan, 'Organisation and Administration of Sales', 22 March 1933, p. 9, CP 'Memos 1933'.

[3] 'Organisation and Administration of Sales in the Home Trade', Meeting 9 March 1936, CP 'Organisation Commercial and Sales 1936'.

[4] *ICI* I, p. 383.

[5] W. H. Coates, 'Nobel Industries Limited Liquidation', 4 July 1928, FCSP, p. 355, 'Liquidation of Nobel Industries', 15 Nov. 1928, FCSP, p. 571; ICIBM 2086, 5 March 1931.

[6] P. C. Dickens, 'Marketable Investments', 15 Oct. 1937, MP 'Finance Committee 1935–39'; the same repeated, 7 March 1938, FCSP XXXVI, p. 3829.

[7] Lord Melchett, 'Financial Policy', 5 Dec. 1933, CP 'Capital (4) 1933–34'.

[8] Lord Melchett, 'Investments', 19 March 1936, MP 'Finance Committee 1935–39'.

[9] FCM 1723, 29 July 1936.

[10] Letter lent by Mr. L. H. F. Sanderson.

[11] Much of the narrative is based on material in the archives of Solvay et Cie, Brussels.

[12] 'Minutes of a Special Meeting . . . 19 Nov. 1937', SCP.

[13] Draft 'Summary of Facts', 31 Jan. 1938, SCP; Sir W. McLintock to Special Committee of ICI Board, n.d. (after 12 Jan. 1938), SCP.

[14] 'Minutes of a Special Meeting . . . 9 Dec. 1937', SCP; see also Solvay archives.

[15] Notes, n.d., by H. J. Mitchell, SCP (File H. J. Mitchell).

[16] Report, 7 Jan. 1938, SCP.

[17] W. H. Coates, 'Case for Counsel', 18 Jan. 1938, SCP.

[18] H. J. Mitchell, draft letter, 13 Jan. 1938, SCP.

[19] J. E. James, 'Organisation', 22 March 1938, Bunbury folder 15.

[20] 'Minutes of a Special Meeting . . . 1 March 1938', SCP.

[21] Draft Agreement, 11 March 1938, SCP; Lord McGowan to Directors, 22 March 1938, SCP.

[22] From the Chairman, 1 March 1938, CP 'Organisation General 1934/38'.

PART III

THE GREAT DISTRACTION:
REARMAMENT AND WAR 1935–1945

"NOW WOULD BE THE TIME TO CALL THE POLICE, KID—HUH, HUH!—IF THERE WERE ANY POLICE."

CHAPTER 14

Warlike Supply

In september 1933 the *Evening Standard* published a cartoon by David Low showing a group of villainous figures labelled with names prominent in the armaments industry: 'Krupps', 'Vickers', 'Schneider', 'Skoda'. In the background, plump and self-satisfied, appears 'Imperial Poison Gas Industries Limited'. The drawing expresses the rising public hysteria of the day. Terror of 'the next war', as early as 1933, was making people unwilling to face the risks of preventing the outbreak. In this situation it was convenient, safe, and flattering to the self-righteousness of the liberal conscience to blame 'the arms racket' for the rising tide of violence in the world. One result was the appointment, in 1935, of a Royal Commission on the Private Manufacture of and Trading in Arms. From evidence given to it on behalf of ICI, by McGowan and others, it appears that at the time when Low drew his cartoon trade in war material amounted to less than 2 per cent of ICI's turnover and they were making no war gases at all.[1]

Low's innuendo—that ICI grew fat on the profits of a flourishing trade in poison gas—was false. It was true, nevertheless, that ICI's importance to the country's capacity for making war, over a much wider field than Low indicated, was very great. The small output of war material in 1933 simply represented a lack of demand which it was no part of ICI's policy to stimulate. Rearmament had not begun, and it had scarcely begun when the Royal Commission heard ICI's evidence in February 1936. It is ironical that when so much heat was being generated about the iniquities of 'the armaments industry', that industry was torpid. The real threat to peace lay far outside the Royal Commission's terms of reference.

It was unrealistic, as the Royal Commission found out, to consider 'the private manufacture of arms' in isolation, because the industrial base of modern war runs much wider than the manufacture of lethal ironmongery and the products directly associated with it. It runs throughout the economy, and the strategic importance of ICI arose from ICI's dominance of the chemical industry.

The chemical industry was pre-eminently one in which it was impossible to disentangle warlike from peaceful purposes. Merely by continuing its normal activities of supplying materials to other industries it made an indispensable contribution to war production, not least as the plastics industry developed. Among the more important direct applications, ammonia synthesis supplied the basis of nitric acid for explosives; dyestuffs technology could be applied to explosives, war gases, pharmaceuticals; petrol could be produced by hydrogenation; many products required in peacetime industry, such as chlorine and cyanide, could become chemical weapons or essential ingredients of them. In general, it might be said that almost any branch of the chemical industry could be converted from peaceful to warlike applications; some certainly would be; others—such as the production of specialized war gases—had no peaceful application at all.

Consciousness of the wartime importance of the chemical industry, as we have seen in Volume I and in earlier chapters of this volume, had influenced Government policy from the time of the Great War onwards, particularly in relation to the dyestuffs industry and the formation of ICI itself. ICI from the start had a recognized place in the British defence Establishment, and Lord Weir, who besides being a Director of ICI was one of the Government's chief advisers on warlike supply, bracketed ICI with Vickers, Woolwich Arsenal, the Royal Dockyards, and the Royal Ordnance Factories. They were all, he considered, parts of the 'professional armaments industry', making arms and munitions as part of their regular peacetime business and being thereby qualified to train other firms in war production if the need for expansion arose, as it would do as soon as war became likely.[2] It would be entirely false, nevertheless, to suggest that anyone in ICI wanted war for the sake of profit.

Beyond the conventional boundaries of the chemical industry there was a much wider field in which ICI became important to the war effort of Great Britain. The war required the application of many branches of pure science to practical problems involving large-scale technology, but in British industry of the late thirties it was unusual to find pure science and large-scale technology under the same roof. There were scientists in the universities and technologists in industry, but no other organization in the country, academic or industrial, privately owned or owned by the State, could offer anything like ICI's broad command of scientific talent and the technological resources of heavy industry. Any problem which required that combination was therefore likely to find its way to ICI and many did, especially in the development of weapons.

As well as working for the Government in their own establishments, ICI were called upon to lend experts to the Government. During the period of rearmament and war, 3,345 trained staff and workpeople were sent to run Government factories, and in 1942 a list compiled by J. E. James showed two ICI Directors, five Chairmen and Directors of Groups, and 33 other staff, including 15 chemists, 'seconded to

Ministries'.[3] A very hungry eater of men was the atomic bomb project, which ICI were concerned with from the first.

Although ICI had been founded with warlike supply in mind, that had never been one of the principal purposes of the business or an important source of profits, and when in the early thirties the threat of war began to loom ICI was inadequately prepared for it. Successive Governments, ever since 1918, had run down the Army and RAF and kept them small and sparsely equipped, and building for the Royal Navy had almost ceased under the terms of the Washington Naval Treaty of 1921. In these circumstances the arms and munitions industries had been allowed to run down too, including the once-flourishing business in cordite at Ardeer and the ammunition side of Metals Group at Birmingham.

All that was active in ICI before rearmament began was a cordite plant at Ardeer producing perhaps thirty tons a week for foreign buyers and a small ammunition plant at Witton similarly engaged. By way of reserve capacity there was Government-owned plant standing by at Ardeer, unused since the Great War, and plant for small arms ammunition similarly preserved at Witton. Moreover at Witton they had kept up with new developments, and as new weapons had appeared they had manufactured ammunition for them.[4] Apart from small-scale experimental plant here and there in other Groups, including experimental plant for gas, this was the entire extent of ICI's operations in the 'professional armaments industry' in the early thirties.

By this time the international situation had moved from the apparently secure tranquillity of the late twenties, in which disarmament looked sensible, to an atmosphere of growing menace. In 1931 the Japanese attacked Manchuria. In 1933 Hitler came to power in Germany. In 1935 the Italians invaded Abyssinia. The consequence of these three events and of others linked with them was that by 1935 Great Britain was faced with a strategic nightmare: the possibility of having to fight two wars at once—against Japan in the Far East and against Germany in Europe—with Italy glowering on the long, long lines of communication to India, Malaya, and Australasia.

The British Government's first decision to rearm on anything like a serious scale was taken in the spring of 1935. Even then Stanley Baldwin, who replaced MacDonald as Prime Minister in June, was unwilling to push matters very hard until he had tested public opinion in a General Election. The Election, in November, was fought less on rearmament than on matters of home policy, but it gave Baldwin the majority he sought. A year later he interpreted it as 'a mandate for doing a thing'—rearmament—'that no one, twelve months before, would have believed possible'.[5]

It was thus not until towards the end of 1935 that ICI, in common with the rest of British industry, began to be seriously concerned with preparations for war. From then on, under the force of successive shocks from Hitler's policy, the pressure rose rapidly, until the demands of the rearmament programme, in the last twelve months or so before war broke out, began to shoulder ICI's own affairs to one side.

In 1935, to carry out the rearmament programme, the Government had at its disposal a severely rundown armaments industry and a chemical industry which, though in much better shape for war than in 1914, would need considerable modification and expansion in warlike directions. How was expansion to be achieved to meet, first, the demands of rearmament and, secondly, the very much greater demands of war itself?

The Government had put the problem, as far back as 1933, to three industrial advisers: Sir James Lithgow, Sir Arthur Balfour, and Lord Weir. Their solution, for which Weir was largely responsible, was to create, as he put it, 'what has been called a Shadow Munition Industry'.[6] When the Committee of Imperial Defence first accepted the idea of a shadow industry, in May 1934, it was aimed chiefly at engineering, but the principle could be applied to any kind of manufacturing activity. Firms outside the professional armaments industry were to be asked to prepare themselves to make war material which the 'professionals' would design. Weir's own firm, for instance, undertook to make 25-pdr. gun carriages, and motor firms undertook the making of aero-engines. Plant and buildings were provided by the Government, to remain Government property and to stand by until required, being used until then only for 'educational' orders. These were the 'shadow factories' which became a feature of the British landscape in the later thirties. The professional armaments makers would supply expert advice, but in general it was expected that the 'shadow' firms, having been 'educated', would run the factories themselves as agents for the Government.

In the chemical industry, dominated by ICI, the functions of supplying expert advice and of running war factories tended to coalesce to such a degree that ICI, though part of the professional armaments industry, became also, through the various Groups, the Government's largest agent. For that reason the 'agency factory' requires more than a passing glance. Like many of the British administrative devices of the Second World War, it had been invented in the war before, essentially to solve the problem of financing the production of war material on a gigantic scale.

In the Great War it soon became apparent that if private firms were to put up capital for huge, specialized war factories they would run a grave risk of heavy losses after the war, when most of the plant would be useless. They would have to make sure they earned enough to pay for war plant while the war was still going on, by charging very high prices to earn very high profits. This was precisely what du Pont, in the USA, did as long as they were supplying powder to foreign governments, chiefly the British, French, and Russian. They made each contract pay for itself, so that their prices were enormous; so were their profits, and their plant was very quickly paid for. Then when the USA came into the war, in 1917, du Pont were able, without any loss to themselves, to offer powder to their own Government at prices lower than in peace-time, because by then they were working from very large plants, fully written-down at no expense to the American taxpayer.[7]

This solution did not offer itself in Great Britain in either war. There were no foreign governments, desperate for supplies, who could be induced to pay for plants put up by private firms. Moreover in the thirties the Government was determined to prevent the kind of 'profit-eering' which had caused such bitterness in the Great War. The 'agency factory' solved the problem. The Government found the money and private firms did the work, but without either the opportunity of blood-stained profits or the risk of post-war loss. They were invited to under-take a public duty, for a sufficient but not exorbitant fee, and to drop it as soon as it was no longer required. All Government planning, before and during the war, was conducted on the principle that suppliers' profits should be eliminated by agency agreements or else very strictly controlled.

This policy was entirely acceptable to ICI. The instinct of those in charge of the business, in time of emergency, was to sacrifice private interest to public duty, and they never saw war contracts as an oppor-tunity for normal commercial profit-making. Many of them—Melchett, Pollitt, Slade, Gordon, Sampson, and others—had fought in the Great War. They no doubt shared the loathing of the troops for profiteers, and they would have been horrified to see ICI laid open to any kind of justified suspicion.

ICI's negotiators went out of their way, in making agency agreements with the Government, to avoid payment based on output or on cost of production 'because', in the words of one of the official historians of the war, '[ICI] did not wish to be open to the charge of being financially interested in the prolongation of hostilities or in enhancing the price of products under its control.'[8] Fees for construction and management of agency factories were settled, early in the war, on a basis (see Table 16 below) which represented, in the view of the official historian, a good bargain for the Government. For managing the factories to which the scale of management fees applied, ICI's fees averaged about 1·2 per cent of the capital cost of the factories to the Government.[9] The Ministries chiefly concerned (Supply and Aircraft Production) never-theless continually pressed for lower charges. In August 1941 the sliding scale of construction fees was replaced by a flat fee of 1 per cent, and in 1943 ICI agreed to forgo even that on contracts under £500,000. Meanwhile, with effect from 1 January 1941, management fees were put on a flat rate of ½ per cent, reducing income from that source from £350,000 to £175,000.[10]

TABLE 16

SCALES OF WARTIME FEES

(a) Construction Fees
     On a sliding scale from 4 per cent of the estimated capital cost of contracts under £1m. to ½ per cent on contracts over £7m.
(b) Management Fees
     Annual Payments as follows:
     1·5 per cent of total fixed capital expenditure under £1m.
     plus
     1·0 per cent of total fixed capital expenditure between £1m.–£2m.
     plus
     0·5 per cent of total fixed capital expenditure over £2m.

Source: Stevens, ch. 4, pp. 7–8.

The small return accepted on agency work during the war was no mere formal token of patriotic goodwill. The work was on such a scale that if it had been paid for at anything approaching commercial rates the income from it would have made an important, instead of negligible, contribution to ICI's revenue. Between 1937 and 1939, as the Government's largest agent, ICI undertook to build and manage eighteen factories, and before the end of the war the total had risen to twenty-five.[11] They included factories, some very large, for explosives, propellants, and ammunition; for ammonia and ammonium sulphate; for chlorine; for light alloys; for aviation petrol (the largest of all); for poison gas (the heaviest investment); and for numerous chemical materials needed in other manufactures. The total cost of ICI agency factories, between 1936 and 1939, was almost as great as the cost of the Royal Ordnance Factories. Throughout the war, alongside the agency work, sometimes on the same site, a great deal of Government work was going on in ICI's own plant.

During the period of rearmament and war ICI invested some £58m. on behalf of the Government against about £20m. on their own account. At the same time direct sales of ICI products to Government rose from £1·2m. in 1937 to £13·7m. in 1941; and in 1941 it was estimated that indirect sales to Government amounted to about £34·8m., giving a total of Government business of £48·5m. against sales to other

TABLE 17

CAPITAL EXPENDITURE 1938–1944

	ICI	ICI on behalf of Govt.	Total
	£000	£000	£000
1938	3,017	2,793	5,810
1939	2,484	4,581	7,065
1940	3,501	14,353	17,854
1941	3,829	15,931	19,760
1942	3,275	12,243	15,518
1943	2,692	5,928	8,620
1944	1,486	2,647	4,133
	20,284	58,476	78,760

Source: ICI BR (Technical Director), 31 May 1945.

customers of £43·8m.[12] This business, unlike agency business, was done on commercial terms, but the civil servants who negotiated them insisted—quite properly, from their point of view—on heavy discounts, which combined with wartime taxation to prevent any rise in ICI's net income, which was smaller in 1945 (£4·7m.) than in 1939 (£5·2m.).[13]

The Second World War was the greatest source of business that ICI had ever had, but it never looked like an attractive business proposition. On the other hand, as we shall see in Part IV, it got in the way of things that did. From ICI's point of view the years from 1935 to 1945 were the years of the Great Distraction.

REFERENCES

[1] Minutes of Evidence taken before the Royal Commission on the Private Manufacture of and Trading in Arms, Fifteenth Day, Wednesday 5th February 1936, HMSO, London, 1936, pp. 439, 443. 'Profits', p. 443, is presumably a slip for 'turnover', p. 439. Committee of Imperial Defence, Sub-Committee on Defence Policy and Requirements, Industrial Production—Memorandum by Lord Weir, 27 Jan. 1936, p. 2, Weir Papers.
[3] Lord McGowan, 'ICI's Technical Achievements during the War', speech at AGM, 24 May 1945, p. 3; J. E. James, 'ICI Staff seconded to Ministries', 21 July 1942, CR 93/34/439.
[4] WR, Bradley, 'History of the War Effort of the Metal Group', Draft Preliminary Report, Aug. 1942.
[5] Taylor, p. 387.
[6] As (2).
[7] Chandler and Salsbury, ch. 13.

8 W. Ashworth, *Contracts and Finance*, HMSO and Longmans, 1953, p. 156.

9 Stevens, ch. 4, p. 7 and generally; FCM 169/39, 31 Oct. 1939; as (8).

10 ICIBM 7696; MBM 1282, April 1943; Agreement with Ministry of Supply 25 Feb. 1942, with Ministry of Aircraft Production 1 Oct. 1942, CR 93/767/796.

11 W. Hornby, *Factories and Plant*, HMSO and Longmans, 1958, p. 449.

12 A. J. Quig and H. D. Butchart to Government Contracts Committee, 'Prices and Profit Margins', 18 Jan. 1944, CR 179/24/24; W. H. Coates to Government Contracts Committee, 8 Jan. 1944, CR 179/24/24; H. D. Butchart to Government Contracts Committee, 12 Jan. 1944, CR 179/24/24.

13 ICI published accounts.

CHAPTER 15

War Production

(1) PREPARATIONS

THE KIND OF WAR MATERIAL which ICI were asked to produce, from 1936 onward, depended on the kind of war which the Government thought the nation was likely to have to fight. Production on a large scale could not start until some years after the decisions to put up factories had been taken, and so it followed that the general pattern of ICI's wartime activities was settled before 1939, according to the notions of warfare held by the Government and their professional advisers at that time.

In the Government's planning it was taken for granted that as soon as war broke out there would be devastating air raids, probably including gas attacks. It was assumed that the best defence—perhaps the only defence—would be what the Air Staff called the 'powerful offensive': that is, the bombing of Germany by an RAF 'striking force' based in the Low Countries and France. On land, the Government desperately hoped to avoid committing a large British force to the kind of fighting the Army had endured barely twenty years before, and the more optimistic advocates of the 'powerful offensive' suggested that the war might be won by air power alone.[1]

These assumptions led to concentrating the rearmament programme on bombers for the 'powerful offensive', of which none existed in 1935; on ground defences, including early forms of radar, against air attack; on poison gas, for reprisals; and on very large quantities of military explosives and ammunition, since in so far as land fighting was provided for it was conceived of in terms of the heavy bombardments and infantry tactics of 1914–18. The Royal Navy, though ageing, was large and would in any case require much time and money for modernization. It did not figure so prominently in the rearmament programme as the RAF, and the Army came a poor third behind the other two.

This pattern of thought can be seen in the way ICI's activity in rearmament took shape. The earliest Groups to be involved were those with peacetime interests in the armaments business—Explosives Group and Metals Group—and the first agency factories that came into production, as well as expanded output by the two Groups themselves, showed the planners' concern for the 'powerful offensive' and for anti-aircraft defence. Incendiary bombs were being made under an agency agreement in an Explosives Group factory at Linlithgow by the spring of 1938, and various contracts with Metals Group for ammunition and cartridge cases, about the same time, included large provisions for anti-aircraft calibres. Cordite was fundamental to the whole ammunition programme, and plans for expanding Ardeer's capacity were being discussed as early as 1935. By the time war broke out Ardeer could produce about ten times as much as its peacetime output—that is, about 300 tons a week.[2]

These plans were coming forward at a time of great technical advance in the design of aircraft. Wood-framed, fabric-covered biplanes were giving place to metal monoplanes built chiefly of wrought aluminium alloys. Here was a field which Metals Group, at a meeting held at Witton in May 1936, decided it ought to be in, not merely as the Government's agent for the temporary purposes of rearmament and war, but for the permanent good of the business.[3]

ICI's policy for light alloys, accordingly, was in sharp contrast to policy in aspects of rearmament which had no peacetime application. Metals Group acted independently and aggressively, deliberately avoiding reliance on agency agreements or on Government finance, which in any case was not very readily on offer in the field of light alloys until after the Munich crisis of September 1938. In November of that year, when the Government's aircraft programme was building up towards wartime intensity, A. J. G. Smout wrote an account of the development of Metals Group's policy and indicated future intentions. He addressed himself to the members of Metals Group's Board, but the letter was probably mainly intended for the central Management Board of ICI who were being asked to sanction capital expenditure.[4]

Smout did not even then write as if he were contemplating war, but with the Air Ministry's increased programme he expected demand for light alloys to go on rising, at least until 1941, and after that there would be 'a permanent tonnage for replacement purposes . . . quite apart from commercial outlets in other forms of transport (road, rail and sea)'. To meet the promise of this situation Metals Group, up to the time when Smout was writing, had relied on a twofold policy.

They had spent a comparatively small sum (£78,000) on 'temporary measures based on existing plant which will give rapid production . . . with a relatively high profit'. They could thus keep the Air Ministry happy 'by supplying every ounce of Light Alloy material at the earliest possible date'. In the process the temporary plants ought to earn enough to pay for themselves. At the same time the Group was building permanent plant at Holford, 'which will place the Group in a strong competitive position when the present rearmament programme eases down'. They had so far spent £195,000 on the first section of this plant.

What Smout was asking for was about £64,000 more for the temporary plant, because Vickers were short of formed strip for Wellington bombers, and between £180,000 and £200,000 to finish the permanent plant. The total expenditure, he said, would by the turn of the year 1940/41 give a total capacity for strip, sheet, rods and tubes of about 6,000 tons a year which, given an average selling price of £250 a ton, would raise turnover from its existing figure (after only two years' operations) of £450,000 to £1½m. Smout assumed that 'after the rearmament boom is over orders will fall by two thirds'. The Group would then close its temporary plant—fully written down—and 'we shall then . . . be left . . . with a modern plant at Holford with a capacity of 2,250/2,500 tons and an order book of approximately 1,700 tons, whereas our competitors will be left with order books only sufficient to keep their big plants concentrated on one centre at about 25 to 30 per cent capacity.'

By the time Smout wrote, the Air Ministry had submitted RAF Expansion Scheme 'M', the last and most ambitious expansion scheme devised before war broke out. It was intended to produce a 'striking force' of 1,360 bombers by March 1942, and also 800 fighters, far more than in any previous programme. With reserves, that meant producing over 12,000 aircraft—an impossible target for the aircraft industry under ordinary peacetime conditions, and ordinary peacetime conditions were no longer contemplated. And as the new expansion scheme went on to paper, the planning of previous years was beginning to show results in production, reflected in the rate of spending on the RAF, which overtook Army spending in 1938 and naval spending the following year.

Metals Group's activities, as suppliers to the aircraft industry, were swept along in the gale. They were asked to design and put up a shadow factory for light alloys, at Landore, Glamorganshire, which would have a capacity—8,000 tons a year—much greater than the Group's own. Orders to John Marston and Excelsior, both makers of radiators and other components for aircraft, rose rapidly until in 1938 an order from Vickers for 500 sets of fuel-tanks for Wellington bombers, worth £400,000, required £100,000 capital expenditure on a new factory.[5] At about the same time, in another request for capital, the Group Board observed: 'Excelsior and Marston have between them orders valued at £750,000, almost 18 months' single shift output. £250,000 of this have been booked in the past four months, and there appears to be plenty of business still available.'[6] The engineering trades were beginning to recover from depression before rearmament seriously began. War completed the cure.

Air raids caught the public imagination as one of the worst of the prospective horrors of war. So did poison gas, and most people expected both as soon as war broke out. The Government certainly did, and the whole civil population, including babies, were issued with gas masks in neat cardboard boxes. The Government's response to the possibility of gas was not purely passive. If the enemy used it, so should we, and no one had very much faith in the various declarations against it that were made from time to time. Moreover it was thought, perhaps rightly, that if the Germans knew the British would retaliate, they would never use gas themselves. Gas therefore had to be made, and the best people to undertake the research and development needed, and perhaps some of the production, were in Dyestuffs Group of ICI, as had all along been recognized when the wartime importance of the dyestuffs industry was under discussion. Five tons of a tear gas, code-named KSK, had been made for the War Office in 1926 and the plant had been kept in existence.

The war gases contemplated in the thirties were not only the chlorine and phosgene of the Great War. Indeed, only one 'gas' used in that war —'mustard gas', a liquid—was prepared for use in the war of 1939. The others—KSK, BBC (brombenzyl cyanide), DA (diphenylchloroarsine), and DC (diphenylcyanoarsine), organic compounds related to dyestuffs, had never been used in action and as matters turned out never were.

Dyestuffs Group was concerned with all of them. As with other branches of war preparation, the work included research and development, engineering service in the building of agency factories, the manufacture of intermediates, and some quantity production. In the early years of the war 30 tons of KSK were made, 145 tons of BBC, 650 tons of B.001 (phenylarsinic acid), an intermediate for DA and DC, and 500 tons of thiodiglycol, for mustard gas.[7] Plans for much larger production, especially of mustard gas, to be carried out in factories run by General Chemicals Group, were put in hand, but although plant was got ready it was never put to use. No doubt the Government had scruples about the use of gas. No doubt also, as A. J. P. Taylor has pointed out, they were well aware that high explosive is a much more effective killer, weight for weight, than any of the gases available to them.[8]

Some constituents of military explosives, like those of war gases, had affinities with dyestuffs and were peacetime products of the Group, required in its own business. Dimethylaniline, for instance, a dyestuffs intermediate, was also required for tetryl, used in detonators and the primer charges of shells. Carbamite (diethyldiphenylurea), a stabilizer for cordite, was also produced in peacetime by the Group, though in small quantities.

The demand for all these things increased very much as war approached, being inflated by Government planning based on experience of the heavy bombardments of the Great War. Early in 1939 plans were made for raising the capacity of the dimethylaniline plant, in peacetime 800–900 tons a year, to several times that figure, and the carbamite plant, from 1938 on, was several times enlarged on an even more generous scale. We shall examine the wartime production of explosives in the next chapter.

In the official mind, and in German minds as well, no part of ICI was more closely connected with British war potential than the great complex at Billingham, even though in peacetime military demand for its products was negligible. It had ammonia plant, petrol plant, plant for producing hydrogen, methanol, chlorine, bleaching powder, and a whole range of chemical products indispensable in war. At Billingham, as in the renascent dyestuffs industry, the chemical preparedness of Great Britain showed up in the thirties in striking contrast to the chemical vacuum of 1914.

Billingham, tactically, was not a good site. Even in 1919, with the limited range of aircraft in those days, some people had pointed out that perhaps it was unwise to place so essential a part of the country's war machine so close to German bomber bases, for even then the possibility of another German war was not inconceivable. By the thirties, the position was much worse, and Kenneth Gordon, in 1936, wrote: 'Billingham is so vulnerable to air attack that it can be neglected as a source of supply in the event of war.'[9] By that time, in fact, the range of aircraft was increasing so fast that before long it would not matter much where in the British Isles works were built, because the bombers would be able to get there.

In the spring of 1936 the Government asked Lord Weir, in great secrecy, 'to take up with Imperial Chemical Industries the problems raised by the exposed situation of Billingham and to make proposals . . .'. Having discussed the matter with John Rogers, Weir told the Government that it would be sensible to consider building duplicate ammonia plant, smaller than at Billingham, in a safer place. In April 1937 a decision was taken to put up ammonia plant at Bishopton in Scotland; a year later, at Dowlais in South Wales. Preparations went forward for nitric acid plant, some of it to German designs; for hydrogen plant for filling barrage balloons; and for two large ammonium nitrate factories, one to be run by Dyestuffs Group at Huddersfield and the other at Pembrey in South Wales. All this planning began to show results in production from the latter part of 1938 onward into the early years of war.

Weir took the opportunity of reporting on Billingham to force the issue of the strategic value of oil-from-coal. 'What value', he asked, 'do the Committee of Imperial Defence put on a domestic supply of petrol in war in reducing our dependence on imported spirit? If the answer be that the possible domestic supply . . . can only be a trifling percentage of our war needs and that we are content to rely on keeping the ocean free, then we should dismiss it from this problem.'[10]

This was precisely what the Government decided to do, acting on the recommendation, published in February 1938, of a sub-committee of the Committee of Imperial Defence under Lord Falmouth:

> ... that in general a policy of depending on imported supplies with adequate storage is the most reliable and economical means of providing for an emergency; and they cannot recommend the reliance of the country in war time on supplies of oil from indigenous sources especially established for this purpose.[11]

The trouble was what it had been all along—that the oil-from-coal process was not a true commercial proposition. Even with the 8*d.* preference granted in 1934 and renewed in 1938, it could hardly be made to pay unless coal prices fell. Instead, they rose, and ICI told the Falmouth Committee 'that even if the guarantee of protection were unlimited they would not extend their hydrogenation commitments for at least another two years, when they would have derived full benefit from their Billingham experience'.[12]

Nothing that happened in the next two years gave them any reason to change their minds. Coal, the original basic material, turned out to be unpractical both economically and technically, and was soon given up in favour of creosote, which gave a much more manageable flow through the plant; but even then, by the spring of 1939, Kenneth Gordon—by nature an optimist—had to admit that the results of four years' experience, both with coal and with creosote, had been disappointing. The net ICI profit he expected for 1939, after all charges

TABLE 18

OIL WORKS BILLINGHAM—CAPITAL AND PROFITS, 1936-1938

	Capital Employed[a]	Gross Trading Profit	Net ICI Profit or Loss[b]	Net Return[c]
	£	£	£	£	%
1936	4,905,818	292,798		17,314	
1937	4,707,331	369,009	94,669		2·01
1938	4,320,731	433,598	119,933		2·78

Notes: a 'Rolled-up' capital of petrol plant and services including phenol plant and butane filling station.

b Gross trading profit *less* Profit & Loss items, Central Services, Obsolescence.

c After providing obsolescence and central services.

Source: K. Gordon, 'Billingham Petrol Plant—Report on Future Policy', 31 March 1939, CR 59/2/435.

including Central Services and Obsolescence, was £247,059, showing a return of 5·65 per cent on the written-down capital value of the petrol plant—£4,373,731.[13]

With figures like that, and no real hope of improvement, it is impossible to say how long, in peacetime, the Billingham petrol plant might have gone on. As matters were, it was put to use during the war producing petrol—'pool' petrol—of the standard wartime grade, and it survived for some years after, but in 1958 it was finally closed. Thus faded, to its long-drawn-out and melancholy end, the last Imperial dream of Alfred Mond.

In the late thirties, however, despite the discouraging cost figures, the outlook for hydrogenation petrol seemed far from hopeless. It was too expensive for ordinary motor spirit but it was particularly suitable for treatment with tetra-ethyl-lead and, later, for blending with iso-octane to produce high-grade aviation spirit. Moreover, equally suitable petrol before catalytic cracking was fully developed could not readily be obtained from other sources, so that the high production cost of hydrogenation petrol was less of a disadvantage.

A convenient source of high-grade spirit was needed because designers of aero-engines, working up higher and higher compression to improve their power–weight ratios, needed fuel of higher and higher 'octane number' to avoid 'knocking' caused by pre-ignition under pressure. Aero-engines in the early thirties worked on fuel of about 77 octane (No. 1 Motor Spirit, for comparison, was 66–70), but by 1935 all new engines coming to the RAF were designed for 87 octane fuel, and by the beginning of 1939 all high-performance aircraft, meaning especially the new eight-gun fighters, were designed with engines to run on fuel of 100 octane.

The Falmouth Committee, therefore, with strong support from the Air Ministry, put in a confidential recommendation alongside their published conclusions. They suggested setting up plant for making aviation spirit by hydrogenation. They were probably influenced by the large German hydrogenation programme going on at the time and by the difficulty of getting high-grade spirit from Great Britain's normal sources of supply. They were certainly influenced by evidence from ICI showing that their petrol, uneconomic for ordinary motor spirit, was particularly suitable as base for fuel of 87 and higher octane numbers.

ICI's position, even with the support of the Falmouth Committee, was difficult. Relations with the oil companies, perennially suspicious of ICI's activities in their trade, had to be handled with the greatest care, and in particular any suggestion of a separate deal between ICI and the Air Ministry had to be avoided.[14] The Air Ministry, for their part, greatly disliked exchange of information between ICI and IG under the IHP agreement, especially when, late in 1938, ICI discovered a new field of catalysts.[15] ICI were gravely embarrassed by the Ministry's pressure. The whole working of the international chemical industry depended on scrupulously honouring agreements, and moreover the national interest in the matter, whatever they might say at the Air Ministry, was not clear. A great deal of information had come from IG and much more might be lost than gained by refusing to go on with the exchange of information. The problem was eventually solved only by the outbreak of war.

With these obstacles in the way and with officials at the Air Ministry seriously overworked in harassing and unfamiliar conditions, matters moved forward too slowly for the more ardent spirits in ICI. 'With financial and technical fairies playing battledore and shuttlecock,' one of them complained in February 1939, 'it seems as though some time must yet elapse before any decision is reached.'[16] By that time proposals were beginning to emerge, over lunch with Frederick Godber, one of the Managing Directors of Shell, for co-operation between Shell and ICI in running a hydrogenation plant for aviation spirit, to be built in the United Kingdom but not at Billingham, because of its vulnerability to bombing.[17]

From the ideas discussed at this lunch there eventually emerged the largest agency factory that ICI were concerned with. It was a hydrogenation plant to be run by ICI and Shell jointly, using materials supplied by Trinidad Leaseholds. In September 1939 the three partners set up Trimpell Limited to manage the undertaking.[18] The plant was put up at Heysham, Lancashire. There was deep water for tankers, water supplies and electricity were readily available, and it was fairly safe from bombing. It was intended to have a capacity of 200,000 tons of fuel a year, working on imported gas oils which, besides being cheaper and more easily available than creosote, had a higher hydrogen content. The spirit produced was to be blended with tetra-ethyl-lead and iso-octane to give aviation petrol. The plant opened in 1941.

Heysham was expensive, as hydrogenation always was. The Government put £6·2m. into Heysham, more than into any other single plant that ICI were agents for. Without ICI's persistence it seems doubtful whether it would ever have been built, for Shell were lukewarm and Godber indicated that they went into it mainly to please the Government.[19] It was as much a part of the drive for rearmament as the cordite factory at Ardeer or the war-gas experiments at Dyestuffs Group, and it had no peacetime attraction either for ICI or Shell. It was the last expression of Alfred Mond's original vision of hydrogenation as a contribution to British military strength, and as such we shall see what became of it in the later years of the war (p. 275 below).

The outbreak of war in 1939, unlike the outbreak in 1914, took no one by surprise. ICI had been under increasing pressure for four or five years, and in some departments of the business, since the Czech crisis of September 1938, there had been very little time for anything but war preparations.

In London heavy bombing was expected as soon as war started, if not slightly before. Elaborate plans had been made in ICI, and in the last days of peace the Board and Head Office Departments dispersed, chiefly to Welwyn and Mill Hill, but with outliers in Slough, Surrey, and the North of England.[20] Two of the Directors, Nicholson and Todhunter, set up offices in their own houses, one in Surrey and the other in Essex, and Holbrook Gaskell departed for Frodsham, Cheshire, to act as a kind of Viceroy in the North. With their Head Office staff thus deployed, and with the staff as a whole much reduced by the demands of the armed services and Government departments, ICI faced the strains of war.

(ii) ORGANIZATION

In the spring of 1938 (p. 246 above) Lord McGowan wrote a letter disclaiming his former personal authority. 'Executive instructions', he said, were no longer to be sought from him. The dictatorship of the thirties was at an end. McGowan, against all probability, had ridden out the rebellion of the Board and survived as Chairman, but he was a severely shaken man. Even his brass-bound self-confidence had been dented. The Directors, for their part, though still in awe of him, were determined to assert their authority as individuals and as a Board against his as sole Managing Director. That office was abolished, and from that day to this ICI has never again had a single Chief Executive. The organization with which ICI went to war, which we must now examine, took the form of an elaborate system of interlocking committees. It came into effect on 22 March 1938.[21]

The Board, says the document describing the new organization, 'should consist of (a) Executive Directors, and (b) other Directors, eventually in approximately equal numbers'. At least a month's notice was required before any new Director was elected, a hit at McGowan's former practice of springing appointments on his colleagues without warning. McGowan's own position, as Chairman, was carefully defined. 'The Executive Directors', wrote J. E. James, the Secretary, 'consider it will be desirable that the Chairman . . . issue instructions . . . with a view to freeing himself from those daily duties which he has hitherto undertaken as the Managing Director of the Company.' McGowan did as he was told, which is no doubt a measure of his state of shock.

Shock, no doubt, would pass, and two other appointments provided McGowan with ready stepping-stones back to power as his old instincts began to reassert themselves. Very unwisely, in H. J. Mitchell's opinion, the Board made McGowan Chairman of a new body, the Management Board, authorized to exercise 'the functions . . . usually assigned to the office of a Managing Director'. They also nominated him to the Salaries Committee, a small body to deal with all salaries over £3,000 a year. The Management Board consisted of all the executive Directors and three non-executives, Ashfield, Pollitt, and Weir. Perhaps they thought they could control McGowan by weight of numbers, but in case that turned out to be difficult the Board as a whole required the Management Board to submit for the decision of the full Board proposals for capital expenditure over £100,000 and any matters on which the Management Board was not unanimous.

The executive Directors, under the new constitution, put themselves back into day-to-day administration which they had given up in 1930 (p. 140 above) because it was then said to leave them with too little time to help the Chairman and President with 'major questions of policy'. In 1938, implicitly reversing that judgement, they not only took on, as a body, major questions of policy in the Management Board but, as individuals, each Director took a function of the business and was specifically charged with the management of it. 'In this task', says James's document, rather quaintly, 'he will be fortified by a Committee consisting of himself as Chairman and three other Executive Directors.'

No one, it soon turned out, knew quite what was meant by 'fortified', but that was no obstacle to the setting up of seven 'Executive Committees' with Chairmen as follows:

Commercial—J. G. Nicholson   Overseas—Lord Melchett
Finance—W. H. Coates   Research—John Rogers
Personnel—H. O. Smith   Technical—H. Gaskell

Development (set up Dec. 1938)—J. H. Wadsworth

Each of the executive Directors except B. E. Todhunter served on at least two other committees besides his own, but the Chairman and President served on none. Todhunter, who was 73 in 1938 and evidently semi-retired, though still classed as 'executive', served on the Finance, Overseas, and Technical Committees but was not Chairman of any.

The duties of each Executive Committee were set out in a 'Charter' —ICI relished the dignified terminology of constitutional law—and each Charter had an identical preamble. 'The Chairman of the . . . Executive Committee', it said, ' . . . will carry on the management of the business of the Company falling within the scope of this Charter and will be assisted therein by the other Executive Directors who are members of that Committee.'[22] The Committee Chairman, however, was not obliged to call meetings of the Committee, and indeed, in the interests of prompt decision, he was rather discouraged from doing so, although he was enjoined to 'cause his colleagues on the Committee to be informed of his decisions of particular interest' and to 'seek their advice in all matters on which he is not prepared to assume sole responsibility'.[23] How, in these limited circumstances, were the members of an Executive Committee to 'fortify' the Committee Chairman? As the 1938 organization began to take effect, that was a question more readily asked than answered.

With the Directors thus reasserting themselves, what of the operating Groups? The general functions and powers of each Group remained unchanged but the methods of central control were considerably altered to make way for the re-entry of the Directors into individual managerial functions. The Groups remained substantially as they had been set up in 1931, except that Paint and Lacquer Group was added in 1936 and Plastics early in 1938. Just after the reorganization Salt Group was set up, being based chiefly on the Salt Union Ltd., that forlorn late Victorian merger,[24] which became a 100 per cent subsidiary of ICI in 1937. The Central Administration Committee was replaced by a Groups Central Committee, 'which should devote its attention to the co-ordination of the activities of the Operating Groups and to tendering advice to the Management Board on the major policies of the Company which may affect the Groups'.[25] Wadsworth was at first put in as Chairman but at the start of 1939 H. D. Butchart took his place, and after that no Directors served on the Groups Central Committee, which consisted of the Group Chairmen and the Secretary and Treasurer of ICI.

The Group Chairmen, in spite of being required to concern themselves with co-ordination and major policies, were not told ICI's gross profit figures.[26] The GCC, nevertheless, was on paper a powerful body, for it had independent powers over capital expenditure up to £5,000, at that time a high figure (the corresponding figure for the Groups was £2,000)*. Its position, however, was flawed by the chartered autocracy of the Executive Committee Chairmen. 'Each Group', the instructions ran, 'should be responsible to the Management Board, through the Chairmen of the respective Executive Committees, for matters within their several sections and through the Chairman of the Groups Central Committee for general purposes, including Group co-ordination.'[27] How were Group Chairmen to find their way through that administrative maze?

They do not seem to have tried very hard. Instead, under the stress of war, they conducted their affairs with growing independence, seeking retrospective sanction for decisions already taken rather than submitting

---

* In February 1940, by Management Board Minute 482, the powers of the Groups were increased to £5,000 and those of the GCC to £10,000.

them for approval beforehand. Communications with London were slow—the journey from Billingham, in 1943, took 6 hours, against $3\frac{1}{2}$ in peacetime—the questions to be settled were urgent, the Groups knew their own affairs better than anyone at Millbank, and in the north-west there was in any case a decentralized administration under Holbrook Gaskell. In their dealings with Government departments— and other dealings did not matter much during the war—the Groups operated as independent businesses rather than as constituents of ICI, and the experience made a deep impression in the Groups themselves and centrally. 'This tendency', wrote Nicholson in 1943, 'may gradually lead the Group Boards to consider their own sectional interests as paramount instead of those of ICI as the major unit.'[28]

The executive Directors, forgetting or ignoring the experience of 1926-30, grievously overburdened themselves. The new network of Executive Committees was imposed upon an already luxuriant growth of committees. As early as April 1938, before the new system was fairly going, Coates calculated that he was on six committees (Chairman of two), Nicholson was on twelve (Chairman of seven), and the other six executive Directors (leaving out the Chairman and President*) had between three (Rogers) and seven (Melchett) committees each.[29] At the General Meeting in 1939 McGowan felt called upon to make a public defence. 'There has possibly been some misapprehension', he said, 'of this system of Executive Committees. It has been said that nothing can be done without a Committee Meeting, so that Directors, fully occupied in this machinery, have no time for individual duties or responsibilities. That entirely misconceives the position.' Did it?

At the beginning of 1939 the Board tried to lighten its own load by appointing five 'Executive Managers'[30]—W. A. Akers, F. W. Bain, W. F. Lutyens, A. J. Quig, and F. Walker—with the intention that before very long they should become Directors, as within two years all but one (Walker, who died) did. They were accordingly assigned to various Executive Committees and they attended meetings of the Managment Board, although until early in 1940 they were required, at the insistence of Dickens, the Treasurer, to withdraw when profits and writings-off were being considered.[31] The office of Executive Manager was exalted but insubstantial. They were management trainees at a very high level and referred to themselves as 'the Board School'. The experiment was never repeated, and the five first appointed were also the last.

Early in 1940 the problem of the overloading of the executive Directors was tackled from another direction, and by a rather round-about route. It had been foreseen that Deputy Chairmen might be

---

* H. J. Mitchell resigned as President (though not from the Board) in 1938 and the office lapsed until McGowan was appointed Honorary President after retiring.

Above: Nitrocellulose plant at Dumfries, 1940 (chapter 15 (iii)).

Loading cotton linters into one of the nitrators at Dumfries.

The Blacker Bombard (page 277).

The Projector Infantry Anti-Tank (PIAT gun) (page 277).

appointed, and about the turn of the year 1939–40 McGowan began to push the idea on the ground that if he were out of action 'the Deputy Chairmen would form a rallying point for the maintenance of the Company's affairs'.[32] He at first suggested two Deputies, Nicholson and Melchett, and later added a third, John Rogers. The other Directors were not at all pleased. Why should two or three of their number be made more equal than the rest? Moreover, although they did not put the point on paper, they no doubt foresaw, quite correctly, that the Deputy Chairmen would become an Inner Cabinet round McGowan, whose personal power would grow. No doubt McGowan saw that, too.

They wrote to each other; they talked; they had lunch at the Reform Club; they hoped, without much confidence, that McGowan might put the whole thing off.[33] He did nothing of the sort. He persisted, and he won. A Committee of the Board, appointed at McGowan's instance on 8 February 1940, satisfied itself 'that under present conditions the Executive Directorate is overloaded and requires strengthening'. Accordingly the Committee recommended 'that a sufficient number of the Executive Directors should be relieved of all responsibility for the administration of the Executive Committees and other routine work, and be appointed as Deputy Chairmen'. The three were to be Nicholson, Melchett, and Rogers. The Committee also recommended F. W. Bain and A. J. Quig for the Board. McGowan, probably not without a degree of malicious satisfaction, suggested that perhaps the normal requirement of a month's notice for the appointment of new Directors might, just this once, be waived, 'as the Directors proposed are known to the Board'.[34]

In this way an office new to the ICI Constitution—Deputy Chairman —was created, and it became permanent. It is easy to see why. The Management Board created in 1938 was intended to be an executive body, but it was too large. Sooner or later, if an individual Chief Executive were not acceptable, an executive group would be sure to emerge, small enough to sit round a table and take decisions. The Committee of 1940 had no intention of creating one—indeed, they were careful to enjoin the Deputy Chairmen 'to act in a consultative capacity', not as an executive group—but by 1943 Nicholson was quite casually referring to 'your "Inner Cabinet"' in a draft letter to McGowan.[35]

The organization of 1938 was intended to restrain McGowan and reinstate the executive Directors. The Directors, however, had bungled their opportunity. If they really wished to rid themselves of McGowan's domination they should have sacked him when they had the chance. As long as he was in office he would draw lines of power towards himself as a magnet does with iron filings. The new organization gave him plenty of scope and there was no one strong enough to stop him. Within

271

K

a year or two of his disgrace, in 1937–8, the King had come into his own again.

Not that his position was quite what it had been in the thirties. The Groups were even further removed from his direct control, but then he never sought to control them in detail, being no doubt well aware that the task would be impossible. What he did seek was the last word on ICI's high policy, especially overseas. In that sphere, during the war, there was still no one who would challenge his authority.

Complaints about the new organization began almost at once, chiefly, as we have seen, about the proliferation of committees. As soon as war started matters were made worse by the dispersal of head office departments and the departure of numerous individuals, including some of the most important, into government departments and the armed services. In October 1939, no bombs having fallen, the Board, the Secretary, the Treasurer and the Solicitor came back to London, though they soon had to leave Imperial Chemical House, requisitioned by the Government. Other departments remained scattered, however, until ICI finally regained possession of its own offices in 1952.

There were provisions for reviewing the 1938 organization, and if war had not broken out it would almost certainly have been drastically altered within two or three years. As matters were, revision had to stand over until 1944. The cumbersome machinery of 1938 had to carry ICI through the greatest activity and fiercest strain of the war, which would have distorted and hampered even a well-designed and compact scheme.

(iii) PRODUCTION

    (a) *Agency Factories*

'. . . at first in a trickle, then in a stream, and finally in a flood, deliveries will take place.'[36] Thus Churchill, describing how war production was likely to build up after planning started. Table 19 opposite shows the progress of ICI's construction contracts, as Government agents, from trickle to flood. New construction, in spate by 1941, dropped rapidly away afterwards, being virtually complete by the end of 1942. By that time the stream of deliveries was flowing strongly, and agency work was mainly concentrated on the massive task of managing the factories built in previous years.

Over £58m. of public money was invested in agency factories run by ICI operating Groups during the war. It is a striking sum, but misleading if it gives the impression of a unified, centrally directed war effort. The Groups worked independently on very dissimilar projects. To get a picture of the scale and diversity of their activities as Government agents we must glance at each major Group separately.

The Government's heaviest investments in ICI agency factories went

TABLE 19

GOVERNMENT CONSTRUCTION CONTRACTS
(EXCLUDING RESEARCH), WITH EXPENDITURE
OFFICIALLY AUTHORIZED AND ACTUAL OR ESTIMATED
DATES OF STARTING PRODUCTION, AS AT
30 SEPT. 1943

Production to start	Expenditure authorized
	£
1937	5,000
1938	2,502,000
1939	6,721,000
1940	5,802,000
1941	21,538,000
1942	14,471,000
1943	4,198,000
1944	226,000
No date given, or cancelled	2,723,000
Total authorized	58,186,000

NB: These figures, produced in 1943, do not agree exactly with figures produced in 1945 and quoted in Table 17 (p. 257 above).

Source: Papers submitted to Government Contracts Committee 1943.

TABLE 20

GOVERNMENT CONSTRUCTION CONTRACTS—
EXPENDITURE BY GROUPS

	Official Government Authorization as at 30 Sept. 1943
	£
General Chemicals Group	19,234,457
Fertilizer Group	16,807,842
Explosives Group	11,769,188
Metals Group	5,605,345
Alkali Group	2,821,682
Dyestuffs Group	276,074
Plastics Group	200,278
Leathercloth Group	14,475
Special Weapons Group	1,456,894

Source: Papers submitted to Government Contracts Committee 1943.

through four Groups (Table 20)—Metals, Explosives, Fertilizers, General Chemicals—largely according to plans laid before the war. Thus the two biggest agency projects in Metals Group, small-arms ammunition at Standish (£1·8m. authorized) and light alloys at Gowerton (£1·8m.), as well as two factories for small-arms ammunition managed but not built by the Group, had been planned before the war although production did not start until 1940 at Standish and 1942 at Gowerton.[37]

In ammunition and explosives ICI's agency activities were planned to fit in with the Government's own manufacturing operations in the Royal Ordnance Factories. About one-third of the total output of small-arms ammunition, at the peak, was coming from agency factories, all but one run by ICI.[38] Explosives Group, still making mining and blasting explosives in its own plant, became responsible as a Government agent for nearly all the military high explosives and propellants that were made in the United Kingdom outside the ROFs.[39] A great deal, from 1940 on, was produced abroad, chiefly in North America, but from January 1942 to June 1945 production in the United Kingdom of cordite, TNT, and tetryl was divided between ICI (including agency factories) and the Royal Ordnance Factories as follows:[40]

	ICI + Agency Factories	Royal Ordnance Factories	Total	ICI share
	Short tons			%
Cordite	63,396	117,292	180,688	35
TNT	49,551	283,761	333,212	15
Tetryl	2,298	12,151	14,449	16

The Fertilizer Group, with its production of hydrogen, ammonia, methanol (the only plant in the country), and petrol, and with its command of the heavy engineering associated with high-pressure technology, was among the most versatile of the ICI Groups. Its resources were applied in many ways unforeseen before the war, to which we shall return. Most of the expenditure on agency plant, nevertheless, followed pre-war plans, particularly for putting ammonia plant and plant associated with it, such as nitric acid plant, in places less likely than Billingham to be bombed. Ammonia plant, planned before the war, opened in September 1939 at Mossend, Lanarkshire, in April 1942 at Dowlais, S. Wales, and in December 1942, combined with ammonium sulphate plant, at Prudhoe, Northumberland. An ammonium nitrate plant for producing one of the components of amatol (the other being TNT) opened at Huddersfield, in Dyestuffs Group's works, in September 1939 and at Heysham in May 1943. Nitric acid concentration plants were put into various Royal Ordnance Factories, and hydrogen plant, for barrage balloons, was built at Billingham, Cardington, and Weston-super-Mare.[41]

Fertilizer Group was responsible for the largest single agency project in any ICI Group: the £6·2m. hydrogenation plant at Heysham, built and run in uneasy partnership with Shell (p. 266 above). Its outstanding achievement was the development of Victane, a fuel for piston engines which made fighters fast enough to catch flying bombs in 1944. The importance of hydrogenation as a source of supply of aviation spirit, however, was never more than marginal, and scarcely even that after the Americans came into the war. Shell never really liked Heysham and before it had reached the capacity it was designed for it was partly turned over to ammonia, to be produced by ICI alone.

In all, the Ministry of Supply financed 62 agency factories for munitions production, chiefly explosives and ammunition, at a total cost of £72m. £19·2m., the largest sum spent on agency factories for any group of products (see Table 21 below), was for 'chemical defence', meaning poison gas and, presumably, stores and equipment for defence against it. The country's entire stock of gas was made in agency factories, and the largest were in a group of six built by the General Chemicals Group of ICI.[42]

TABLE 21

MINISTRY OF SUPPLY AGENCY FACTORIES FOR MUNITIONS PRODUCTION

	Number of Factories	Capital Cost
		£ thousands
Explosives, propellants	8	10,444
Explosives materials	11	7,691
Small arms ammunition	5	9,106
Cartridge cases	10	5,350
Shell and fuses	7	2,188
Filling	6	15,867
Small arms	3	1,395
Signals and transport	5	999
Penicillin	3	1,946
Equipment and stores	2	76
Chemical defence	12	19,200

Source: Hornby, 'Factories and Plant', p. 159 (Reproduced by permission of the Controller of Her Majesty's Stationery Office).

TABLE 22

GENERAL CHEMICALS GROUP—
FACTORIES FOR TOXIC PRODUCTS AND INTERMEDIATES

	Date of Starting	Official Government Authorization	ICI's Construction Fee
Toxic Products:		£	£
Randle	April 1938	2,502,544	64,850
Valley	January 1941	3,161,671	80,000
Springfields	February 1942	2,624,061	26,143
Intermediates:			
Rocksavage	September 1939	2,378,645	52,381
Wade	April 1940	1,318,103	13,055
Hillhouse	June 1941	4,777,365	45,947

Source: Government Construction Contracts 30 Sept. 1943, in Papers submitted to Government Contracts Committee 1943.

This demand for gas, none of which was used, meant that more money for agency factories was spent through General Chemicals Group than through any other group in ICI—altogether, up to September 1943, about £18m., of which £16·8m. was for 'toxic products' and their intermediates. The six factories came into production between 1938 and 1941, and a report made in 1946 says that a large number of weapons were filled. Most of the basic research had been done by the Government, but ICI developed a new, very stable form of mustard gas and made improvements to the process, including a new type of reactor which avoided the construction of an entire factory at a cost of £1m.[43]

The war itself, its shocks, its disasters, the deficiencies which it revealed, and the unexpected tactical demands which it made, particularly for the invasion of Europe in 1944, produced projects uncontemplated in peacetime and hence not provided for in the agency arrangements discussed above. Leaving on one side for a moment the greatest of them, concealed under the code name 'Tube Alloys', there were developments in land warfare which were brought together under the general heading 'special weapons'.

They were directed particularly at the destruction of tanks, a subject grievously neglected before the war, and at the rather similar problem of breaching concrete fortifications. For these purposes Explosives Group produced a range of plastic explosives which would spread out over the surface of a target and stick to it. These explosives, intended for use at close quarters in battle, had to be proof against accidental detonation by shock, especially from small-arms fire. Nobels' Explosive 808, blasting gelatine desensitized to the point where a hit from a rifle bullet would not make it go off, was flung in 'Flying Dustbins'—40 lb. projectiles—against fortifications on the invasion coast at ranges between 75 and 100 yards.

Projectiles required projectors. The development of these, chiefly an engineering job, might be thought to belong to the makers of guns, but in fact most of the work was done by ICI. A report written in 1943 makes the reason clear:

In 1940 when invasion seemed imminent, provision of a weapon which could deal with tanks in confined surroundings became of the utmost urgency. Rapid production was essential but the arrangements made by the Ministry of Supply with a group of sub-contractors resulted in delay and to save the situation ICI was asked to take over manufacture both of weapons and of projectiles.[44]

The weapon here referred to was the Blacker Bombard, a 29 mm spigot mortar invented by Lieut.-Col. L. V. S. Blacker, a distinguished and eccentric Irish soldier with a long record of warfare and travel in Asia including, by his own account, 'two visits to summit of Mt. Everest 1933'. He had visions of arming infantry with his bombard and sending them far to the rear of the enemy in helicopters. He said the device would 'win a war as surely as other novel weapons have always won wars in the past'.[45] It did no such thing, but 19,000 were made and came to hand when anti-tank weapons of any kind were very scarce. 'When first made available', it was remarked in 1943, '. . . it was the only effective close quarter weapon against tanks and its production in quantity . . . was due entirely to the efforts of ICI.'[46]

Between April and November 1942, working from prototypes supplied by the War Office, ICI developed another weapon which Col. Blacker claimed to have invented. This was the PIAT (Projector Infantry Anti-Tank).[47] Fired from the shoulder at close range, it pierced tank armour with a cone-shaped charge of Nobels' Explosive 808.[48] 115,000 Piats were ordered, all from ICI, and the contract was completed seven weeks early. ICI also undertook the production of 9m. rounds of Piat ammunition.[49] McGowan in 1945 spoke of 'this little gun' and the 'affectionate regard in which it is held by the individual soldier'.[50] His choice of phrase is perhaps curious, especially since his next remark, that six VCs had been won with the Piat, suggests the desperate circumstances in which it was likely to be used. Its effectiveness, however, was not in doubt, and the way in which it was produced illustrates the general competence of ICI in warlike supply. 'This was something', said McGowan, 'which could only have been achieved . . . by an organization combining first-class knowledge in the use and manufacture of metals, with a staff which included chemists, physicists, metallurgists and engineers accustomed to working together as a team.'

The Groups in ICI chiefly concerned with special weapons were Explosives, Metals, and Fertilizers. The functions of the first two are self-evident, but why the third? The answer appears to lie in the wide experience gathered at Billingham, since the early twenties, in engineering, and especially in the practical application of theoretical principles to large-scale production. About 1942 the Government's arrangements for the design and development of weapons were drastically reorganized and 'a distinguished production engineer, Mr. F. E. Smith of ICI, was brought into the position of Chief Engineer Armaments Design'.[51] Smith had been Chief Engineer at Billingham and became Sir Ewart Smith, FRS, a post-war Director of ICI. He brought with him to the Ministry of Supply 'a number of highly qualified engineers and designers' and it was natural that the Group should become closely associated with the engineering side of armament design and production. Smith's appointment came about through a crisis in one department: the Ministry of Supply. It also reflects, no doubt, general concern with the quality of management, as well as specialized technical competence, which spread through the supply departments as the war went on. 'The better-managed firms', says an official historian, 'were singled out and loaded with contracts to the point of overloading.'[52]

Most of the 'special weapons' developed in ICI during the war were new in design but not in principle, being unorthodox applications of well-known methods and materials. The Piat was a specialized gun and Explosive 808 would have been perfectly comprehensible to Alfred Nobel. Moreover the importance of all these devices was tactical, not strategic, and the same might be said of other wartime developments, such as vegetable dehydration plant built by General Chemicals Group and soup-heating cartridges invented by Explosives Group. None of these devices was ever thought of as a war-winner on its own, unless by Col. Blacker, though without the 'special weapons', taken as a group, the task of the British forces would have been even harder.

Besides taking part in these developments, ICI was an essential partner in another project which made them all look insignificant. This was 'Tube Alloys', the code name for the British organization for the development of the atomic bomb. Before we examine that, however, it will be convenient to survey the war work done in ICI Groups, not as Government agents, but in their own accustomed fields of activity.

(b) ICI Groups

Agency factories represented the extra work which ICI Groups were required to undertake as part of the professional armaments industry.

At the same time, in war as in peace, ICI remained the largest supplier in the country of a wide range of industrial materials, and the only supplier of some of the most important. Contracts were made directly with Government departments for most of the normal products of the Groups, and as the war effort mounted indirect sales for war purposes— sales, that is, to firms themselves under contract to the Government— became even larger than the direct sales.

We do not know [wrote Coates in 1941] the ultimate consumers of alkalis, dyes, general chemicals, fertilizers, lime, paints, etc. even in the form we sell them. . . . Fertilizers may go into grass and that into a cow, the milk of which may be consumed by the Army, and any resultant cheese by the Navy, the final meat by the Air Force, and the more expensive offal by a mixture of civilians and members of the Forces. . . . The warp and woof of the modern chain of production is so intricate that it is impossible to unravel it without such a volume of effort and expense as to be impracticable.[53]

An attempt, nevertheless, was made to unravel the 1941 sales figures. They showed direct sales to the value of £13·7m. and indirect sales of £34·8m., giving total sales of £48·5m. to the Government against sales to all other customers of £43·8m.[54]

ICI's sales figures were much greater during the war than before. The total figure from 1934 to 1938, both years inclusive, covering the climb out of depression and the relapse of 1938, comes to £233m., against sales of £499m. for the war years 1940–1944, showing an increase of 114 per cent. Prices, over the same interval, rose considerably. The index of prices for intermediate products, taking 1938 as 100, stood at 167·5 in 1944, having been 83·9 in 1935, and that perhaps gives some measure of the rise in prices received by ICI, though it is impossible to be precise.

Even allowing for the rise in prices, it appears from the sales figures that ICI's output—agency factories entirely apart—was rising during the war, and such tonnage figures as we have, chiefly from Billingham, point the same way. This was at a time when management was stretched very thin, with the simultaneous demands from the armed forces, Government service, the running of agency factories, and the company's ordinary business. Labour was hard to get and plant was difficult to maintain. That output, under these circumstances, should have risen is a comment both on the intensity of ICI's war effort and on the under-employment of British industrial resources in the last years of peace, even when, to contemporary eyes, the depression seemed to have passed away and activity bred of rearmament was beginning.

Profits in wartime were greater than in peace. The total figures, after depreciation but before taxation and fixed loan interest, came to

TABLE 23

SALES, PROFITS, CAPITAL EMPLOYED, PRE-WAR AND WARTIME,
AND NET INCOME AFTER TAXATION, 1937–1939

ICI Sales

	£m.			£m.	
1934	39	1940	78		
1935	43	1941	94		
1936	44	1942	103		
1937	54	1943	111		
1938	53	1944	113		
Total	£233m.		£499m.		Increase 114%

ICI Profits

	£m.			£m.	
1934	7	1940	14		
1935	7	1941	15		
1936	7	1942	17		
1937	8	1943	13		
1938	8	1944	12		
Total	£37m.		£71m.		Increase 92%

Average Annual Capital Employed

1934–8 £95m.	1940–4 £99m.	Increase 4%

Source: Figures supplied by Merger Accountancy Section printed in Appendix II.

Net Income after Taxation 1937–1939
£m.

1937	5·9
1938	5·4
1939	5·2

Source: Financial Statistical Record published with Twentieth Annual Report, 1946, pp. 8–9.

£37m. for the years 1934–8 and to £71m. for 1940–4, giving an increase of 92 per cent, against an increase in sales of 114 per cent. The bare figures, however, give a misleading impression. Wartime taxation, through Excess Profits Tax, was designed to deflate wartime profits, which it did to great effect. The company's net income after taxation, published in post-war accounts, was:

1940 £3·7m.	1942 £3·8m.	1944 £4·3m.
1941 3·2	1943 4·0	

These figures are considerably lower than net income figures, as published in the Annual Report for 1946, for the years 1937 to 1939 (Table 23). In 1946 just over £2m. appears in the accounts as EPT post-war refund, but even when that is allowed for it is plain that ICI, along with most other British companies, did not grow rich on the profits of war. The Government saw to that.

The Second World War, like the first, gave a strong push to industrial development outside Europe, which was reflected in the war activities of Canadian Industries, ICIANZ, and African Explosives and Chemical Industries.* Each had a long history of local manufacture and each, during the years between the wars, had diversified outwards from its original base in the explosives industry. During the war all these companies acted, like ICI at home, as Government agents, as well as extending their own operations.

Canadian Industries set up a subsidiary, Defence Industries Limited, to make explosives, ammunition, and other supplies. At its peak, in 1943, it employed 32,000 people. In Australia the drive towards self-sufficiency, already strong, was reinforced by the Japanese destruction of the British Empire in the East, and ICIANZ, as well as making ammunition and other supplies in its own plant, built as a Government agent four ammonia and nitric acid plants and works for making ammonium nitrate, cordite, and military explosives, besides operating Government-owned factories for rubber chemicals and sulphonamide drugs. The South African company, reversing the usual order of things, lent fifty of their technical staff to the home company to train staff and workers in agency factories.[55]

In ICI's foreign relations generally, the outbreak of war shattered the network of cartel arrangements with IG (chapter 22 below). The dyestuffs trade, in particular, was thrown wide open. Not only was ICI no longer bound by quota arrangements, but the British blockade stopped German dyestuffs from getting to their markets overseas, particularly in India. As in 1914, great visions arose of seizing these markets for the British dyestuffs industry, meaning principally Dyestuffs Group of ICI, and this time there was no great deficiency in technical knowledge to be made up, though much more manufacturing capacity would be needed, at home and overseas, if the Germans were to be permanently kept out.

Expansion in wartime needed Government consent. The case for expanding dyestuffs capacity was put as a measure of economic warfare, and an ambitious three-year programme, requiring £8·5m. capital expenditure, later inflated to £9·9m., was worked out as a matter of urgency, not without a sidelong glance at du Pont, who might equally

---

* Formerly African Explosives and Industries. The name was changed in May 1944.

be expected to take advantage of the absence of the Germans from their usual markets. Carrying out the proposals, said the document outlining them, 'involves large commercial risks which the Board of ICI are prepared to face provided they are assured of the full support and close collaboration of H.M. Government'.[56] The scheme had at first the general support of the Board of Trade. It was based on doubling the output of dyestuffs and providing for 80 per cent increase in related products, and it required £6·5m. for fixed assets and £3·4m. for working capital. In November 1941, after £3·1m. had been spent, the authorities withdrew support and Dyestuffs Group was left to do the best it could with a severely mutilated version of the expansion scheme. The increased output actually achieved was 19 per cent in dyestuffs and 49 per cent in related compounds.[57]

Whether this foray into economic warfare added much to the harassment of the Germans may be doubted, but it was launched, as similar measures in the Great War had been launched, with an eye to post-war business also, and it certainly helped ICI to increase their export trade in dyestuffs, which rose from £1m. in 1938 to £3·7m. in 1941 and £5·4m. in 1944.[58] It led ICI also to an undertaking, later to prove embarrassing, to put up a dyestuffs plant in India, the market which they chiefly wished to capture from the Germans.

If Dyestuffs Group was regarded partly as an instrument of economic warfare, so also, in a different way, was the agricultural side of ICI's activities based on the fertilizer plant at Billingham. On that side, from Alfred Mond's day onward, commercial motives had been intertwined with motives of public service, and Sir Frederick Keeble (p. 104 above) had foreseen a grand, not to say grandiose, imperial future for the research and advisory service, centred on Jealott's Hill, which Mond brought him into ICI to build up. Depression put an end to those ambitions, but the notion of ICI as the farmer's friend, if only the farmer could be persuaded to see it that way, made a powerful emotional appeal, and even after Mond's death agricultural research never entirely lacked defenders when the commercial case against it looked strong. Then, in the later thirties, the argument for coming to the rescue of British agriculture was reinforced by the growing menace of war.

ICI's Agricultural Adviser just before the war was (Sir) William Gavin, who on the outbreak of war joined the Ministry of Agriculture as Chief Agricultural Adviser. He and others in ICI, including Slade, were led by the results of ICI research to form strong views on methods which British farmers, in the national interest, ought to be persuaded to take up. First, not unnaturally, ICI's experts believed in intensive use of nitrogen fertilizers. More specifically, their experiments in the late thirties convinced them that fertilizers could be used on grassland to prolong grazing in spring and autumn and that grass could be preserved by a method of drying which ICI put a great deal of effort into developing. Finally, Slade, in searching for a source of starch to make up the balance of foods on a farm using dried grass, took up straw, and when the threat of war became serious in September 1938 he started work on straw pulping at Jealott's Hill.[59]

ICI's agricultural experts, with the enthusiastic backing of Melchett and, ultimately, of McGowan, pressed their ideas on the Ministry of Agriculture through the Agricultural Research Council, a body of eminent scientists responsible to the Cabinet rather than to the Ministry. Their main emphasis was on grassland management for 'early bite' (spring grazing) and for grass for preservation. ICI put money and effort, not without a certain glow of public-spirited virtue, into the development of grass-drying machinery, but that was never very successful (ICI were inclined to blame official apathy) and by the time war broke out they were more inclined to stress the importance of growing grass for silage. In the spring of 1940 they wanted the Government to mount a campaign for 1m. acres of 'early bite' and 2m. acres of silage.[60] About ½m. acres for each, they thought, might be achieved in 1940–1, saving about 400,000 tons of imported balance cake.

In this plan, clearly, there was an element of self-interest. Grassland needed nitrogenous fertilizer and the complete programme would take 150,000 tons of sulphate of ammonia. The opinion of the ICI Committee which studied the matter, nevertheless, was that 'the immediate encouragement of silage making is more a National problem than a pressing ICI problem'. ICI could sell, in 1940, 'every ton of fertilizer which it can produce', and in any case 'the magnitude of the problem is such that unless it is treated by the Government as a National problem, ICI effort would have comparatively little effect'. ICI's experts, that is to say, considered themselves distinterested advisers of high standing ('their opinions', said McGowan a little later, 'are not lightly disregarded in any quarter of the world'[61]) and they were affronted when their views were coolly received in the Ministry of Agriculture.

In 1940 McGowan opened an offensive publicly with a letter to The Times. A flurry of correspondence with the Minister (Sir Reginald Dorman-Smith) followed, conducted on McGowan's side in a tone of magisterial reproof: 'Whatever plans', he told Dorman-Smith, 'might be in existence or in contemplation . . . had not . . . been put forward in such a manner as to be common knowledge to the farming community, or indeed even to the members of my own staff who were in close touch with your Ministry.'[62] At lunch on the day after that letter McGowan evidently told Dorman-Smith that his Ministry had not shown enough drive in putting forward ICI's ideas on fertilizer usage, silage, grass drying, or straw pulping. The Ministry had instead been conducting a campaign for the ploughing-up of grass, but ICI thought it should have been accompanied by a drive to increase the yield of the arable land created and to improve the management of the grassland which remained.

Sir Reginald seems to have been disappointingly unreceptive, but when Churchill formed his Government on 11 May 1940 R. S. Hudson, who was much more inclined to ICI's way of thinking, took over the Ministry of Agriculture and held it for the rest of the war. Moreover, at about the same time a committee reported on the utilization of grassland. It had been appointed by the Minister—presumably Dorman-Smith—to work in co-operation with Sir George Stapledon, and two of its members, Col. R. W. Peel and S. J. Watson, came from ICI. A memorandum from them formed the main basis of the committee's report.[63]

Under Hudson the Ministry launched a campaign for a million tons of silage. It was run by County War Agricultural Executive Committees, relying heavily on help from ICI which had been offered to the Ministry by McGowan. The campaign got under way in June 1940 and by November Watson could write: 'At Jealott's Hill and at outside centres, instructors have been trained and have in turn trained others. . . . Our wide experience of all aspects of grassland management and the utilization and feeding of the various products has been placed at the disposal of the authorities and is being fully utilized.'[64]

Towards the end of 1941 Hudson approached McGowan ('My dear Harry') for help with 'a general campaign for the improvement and fuller use of grass', to be run alongside the silage campaign. He proposed a committee including Sir George Stapledon, Gavin, and, from ICI, S. W. Cheveley, and he asked for about 40-50 men 'hitherto employed on the silage campaign'. 'I need hardly tell you', McGowan replied, 'that we are in complete agreement, not only with the need for such a grassland campaign at this time, but that we must come in if you need us to help in every way possible.'[65]

Alongside their campaign for grassland management, ICI tried hard to persuade the Ministry of Agriculture to advertise the merits of straw pulping, so as to get farmers to follow ICI's method of treating straw with caustic soda and feeding the result (washed) to cattle as a substitute for maize or barley. ICI were sure the animals would like it, but the Ministry was not, so ICI's full lobbying-power was switched on. The scheme was put to R. S. Hudson as soon as he took office, and as well as that the Board laid siege over five or six months in 1940-1 to other centres of power. Melchett had Lord Woolton, the Minister of Food, to dinner, described the 'trumpery opposition' in the Ministry of Agriculture, and asked for his support. He also approached Lindemann who 'was surprised that Lord Woolton had not shown greater interest'.[66] Evidently Woolton was not prepared to embroil himself in an inter-departmental quarrel with Hudson, and both Ministers politely advised ICI to go ahead with their excellent and public-spirited ideas—alone.

ICI's biggest gun, McGowan, was therefore trained on the War Cabinet. In December 1940 McGowan wrote to Sir John Anderson to whom, as Lord President, the Agricultural Research Council was responsible. In his loftiest tone, saying that he had 'been several times in touch with Hudson', McGowan appealed for Sir John's 'help in a matter of considerable National importance which is becoming very urgent'. He made it plain that he thought 'a division of opinion on the Agricultural Research Council' was causing the Ministry of Agriculture to obstruct a process which ICI had already shown to be valuable. 'If', he said, 'the Agricultural Research Council proceed to repeat all the work and all the experiments which we have done in order to arrive at the same results they will not be able to give an answer . . . until it is too late.'[67]

'I find', Anderson replied, 'that there is no question of any division of opinion on the Agricultural Research Council. . . . The hitch . . . arises from a divergence of experimental results . . . of a fundamental character.'[68] This was a frosty answer, but nevertheless by June 1941 the Agricultural Research Council had decided 'that the process is one which the Government would be fully justified in encouraging',[69] which suggests that the Lord President had, after all, listened to McGowan. Before the end of the war several thousand farmers were pulping straw by ICI's process, even though, as a friendly official remarked in 1944, 'some amount of technical opposition has been encountered, and . . . the Ministry's sponsorship was inevitably limited in scope'.[70]

ICI's campaign to get their views on agricultural policy adopted shows every sign of careful planning, and might be taken as a textbook example of how to get things done within the British political system. Although ICI held a strong position of widely acknowledged authority, their influence was disliked both within the Ministry of Agriculture and outside. Speyer, in 1944, remarked on 'prejudice against ICI both in official circles and among farmers' which he attributed to 'jealousy and suspicion' aroused by 'the outstanding work of . . . Jealott's Hill'.[71] A little later he reported that Lord Teviot, in a letter to the Director of Rothamsted Experimental Station, had hinted that ICI's contribution of £10,000 to Rothamsted's Centenary Fund 'possibly . . . accounted for the attitude of Rothamsted towards certain problems'.[72]

Strong opposition brought strong tactics. ICI, as we have seen, used many channels of influence, public and private, and in the process they showed no unwillingness to play Ministers off against each other (but the Ministers refused to be played) or to go over any heads that might get in the way. The sincerity of their motives is not in doubt, except in so far as a cynic might murmur 'What's good for ICI is good for the country', and their success was remarkable. In certain matters of wartime farming ICI, without any constitutional standing in the matter, came close to directing the policy of H.M. Government.

War broke out when ICI was rather less than thirteen years old. The business had already come to dominate a wide range of British production, both of finished goods and industrial supplies. ICI's influence, direct and indirect, spread throughout the economy. Every business of any size in the country must have been to a greater or less degree a customer or a supplier of ICI, and many were both. Wartime demand threw this looming presence into sharp relief. Without the resources which ICI commanded—the resources, particularly, of General Chemicals, Explosives, Alkali, and Metals Groups—the British industrial effort in the Second World War would have had to take a very different form. Indeed, whether one looks at the system of agency factories or at the output from ICI's own plant, the risky adjective 'indispensable' seems for once not only tempting but justifiable.

In 1939 the industrial pattern of ICI, allowing for the distortions introduced by the nitrogen disaster, was still much as it had been conceived by the founders in 1926, and the profits relied heavily on old-established products, especially alkali. But there were already signs of change, mostly in an organic direction.

Polythene and 'Perspex' were both commercial possibilities on a small scale and numerous other plastics were in prospect. 'Ardil', a textile based on nut protein, was being developed, an agreement on nylon had just been signed with du Pont, and in 1943 the Ministry of Supply asked ICI to examine a fibre-forming polymer, developed by the Calico Printers' Association, which was the origin of 'Terylene'. In these developments—plastics and fibres—lay the origins of a new style of heavy chemical industry, though no one foresaw it yet.

At the other end of the scale of size, in fine chemicals, Dyestuffs Group in 1939 was making, though not yet marketing, pharmaceuticals, and on the outbreak of war moved at once into the production of drugs formerly imported from Germany, particularly the anti-malarials 'Mepacrine' and 'Pamaquin', equivalent to the German 'Atebrine' and 'Plasmochin'. Later on, the Group developed the first process for manufacturing penicillin by surface culture. A Pharmaceuticals Group, in other words, was taking shape, and in a technically related field, research was in progress in 1939 which before the end of the war produced 'Gammexane' and 'Methoxone', foreshadowing later developments in insecticides and agricultural chemicals.

In developments such as those outlined in the last two paragraphs the best hopes for the future lay, but the immense demands of war made anything like large-scale production impossible unless, as in the case of 'Perspex' and polythene, there was an immediate and urgent war application. There was, nevertheless, continuity of development from the late thirties right through to the early fifties. It represented the natural progress of the business towards a technical revolution as great as the revolution of the eighteen-seventies in which the forerunners of ICI had their origins. Progress of this kind was hindered, not helped, by the war. It is best studied as a continuous episode, and it is the subject of Part IV of this volume.

(iv) 'TUBE ALLOYS'

The atomic bomb was born in the USA, but it was conceived in Great Britain of Anglo-German parentage. Early in 1940 two émigré German scientists, O. R. Frisch and R. Peierls, set out, in three foolscap pages of typescript, the theoretical possibility of building an atom bomb.[73] In April 1940 a committee of eminent scientists—the MAUD Committee—under G. P. Thomson was set up to see whether there was any likelihood of making a bomb in time to use it during the war. The members of the committee entered the project, as they later put it, 'with more scepticism than belief', but in July 1941 they reported that the thing could be done,[74] and early in September Churchill, advised by the Chiefs of Staff, took the basic decision to go ahead.

The industrial implications of this decision were not clear when it was made, but over the following eighteen months or so it became evident that something had been started which it was beyond Great Britain's wartime strength to finish. Great Britain was shown up, in glaring and painful light, as a junior partner to the USA, easily dispensable, and the facts were not easily accepted by those on the British side who knew them. They persisted for a long time in the fiction that an all-British bomb could be made during the war if necessary. The British side of the bomb project was increasingly coloured by jealousy, frustration, and a bitter sense of belittlement, and it inflicted deep psychological scars on many who took part in it.[75]

For transforming scientific theory into industrial fact, in the conditions of wartime Britain, ICI's resources were unmatched. Other firms were called in—Metropolitan-Vickers for heavy engineering, British Thomson-Houston and General Electric for electro-magnetic work, Lund Humphries and Sun Engraving for specialized applications of printing technology—but ICI alone had the scientific and industrial range needed to comprehend and co-ordinate the central processes, especially the separation of U_{235} from uranium. Right from the start, no one who knew anything of the project was in any doubt about that, and at its second meeting the MAUD Committee called ICI into consultation.

Lord Melchett, whose connections among scientists were widespread, was ICI's main link with the project at the level of the Board, which dealt with the matter through its Secret War Committee—McGowan, Melchett, Nicholson, Pollitt. W. A. Akers, one of the Executive Managers, R. E. Slade, head of ICI Research, and his assistant, M. W. Perrin, worked closely with the MAUD Committee, and in ICI generally many others, as time went on, were employed on various aspects of atomic work, though few, if any, apart from the small group at the centre, knew the full significance of what they were asked to do.

The explosive material in the bomb the MAUD Committee was considering was to be uranium 235, an isotope present in natural uranium to the extent of 0·7 per cent. Nearly the whole of the remainder is the heavier isotope uranium 238. The industrial processes required to produce an atomic explosive, as conceived in 1941, started with the preparation of a gas, the mixed hexafluoride of U235 and U238. The hexafluoride of U235 would be separated out and the final step would be the recovery of U235, as metal, from combination with fluorine.[76]

ICI's work for the MAUD Committee was done on a very informal basis, especially when it came to payment. J. P. Baxter, Research Manager of General Chemicals Group, is reported to have paid for making a sample of uranium hexafluoride out of 'spare' research money, recovered from the Government much later, and in the autumn of 1940 Melchett arranged for Canadian Industries to grant $5,000 to the National Research Council Laboratories of Canada when it looked as if work there was going to be held up for lack of money. The first formal research contract with ICI was placed in June 1941, when work for the MAUD Committee had already been going on for over a year.

The preparation of uranium hexafluoride required some research and reconditioning of plant, but it presented no really formidable difficulty. ICI supplied three kilograms to the MAUD Committee and estimated that for full-scale operations a plant to make 450 kg. a day could be built in about eighteen months for £100,000 or so. The worst problem lay in the separation of the hexafluoride of uranium 235 from the mixed hexafluorides of the two isotopes. The MAUD Committee and ICI backed a separation process which depended on the very slight difference in weight between U235 and U238. If uranium hexafluoride were pumped through minutely porous membranes U235 would pass slightly faster than the heavier U238, and if the flow could be kept up for long enough and suitably controlled there would be a gradual build-up of U235, still in a gaseous form from which the solid metal would have to be extracted.

There were two formidable obstacles to designing plant for gaseous diffusion. The first was to produce millions of square feet of metal membrane with 160,000 holes to the square inch, allowing a tolerance of 10 per cent on the mean hole size. The second was the engineering difficulty of pumping highly corrosive gas through the thousands of stages which would be needed to produce the degree of separation required. Not only would the gas attack most of the materials which plant designers would normally expect to use, but a very great deal of power would be required. Moreover, and this was a weighty consideration in wartime Britain, the full-scale gaseous diffusion plant would present a large target to German bombers.

No confident prediction about building a plant could be made without some way of making the membrane. It came from the unlikely direction of printing technology. S. S. Smith, Research Manager of Metals Group, having observed that photographs are reproduced in print from plates bearing regular patterns of very small dots, asked Michael Clapham,* Manager of Metals Group's Kynoch Press, whether by the same process a pattern of very small holes could be etched. Etching was tried and failed: some weeks later Clapham had another idea. On plant belonging to his old employers in Bradford, Lund Humphries, he produced a prototype 'diffusion barrier' by an electro-deposition process originally used for offset lithographic plates and later adapted it to mass production.

In July 1941, when the members of the MAUD Committee reported their opinion that a bomb could be made, they appended to their findings, among other papers, a 'Note by Messrs ICI' which recommended separation of U235 by gaseous diffusion and suggested the outlines of a programme aimed at producing a bomb by the end of 1943. The capital cost of the diffusion plant was estimated at £5m.—considerably less than the cost of the petrol plant at Heysham—and the note ended: 'ICI is prepared to take executive charge of this work on behalf of the M[inistry of] A[ircraft] P[roduction].The arrangement would be similar to that already made in other cases with this Ministry.'[78] Just another agency factory, in other words.

ICI were prepared and indeed eager to help the Government with research, without worrying very much about payment, and to put up gaseous diffusion plant, but they did not see the bomb project as having any more commercial attractiveness than any other agency work. It was a matter of public duty. Where they did see a commercial future was in the work of two Frenchmen, H. von Halban and L. Kowarski, who had escaped to England in 1940 with the entire European stock, at that time, of heavy water—185 kg. of it. Halban and Kowarski

* Later Sir Michael Clapham, KBE, a Deputy Chairman of ICI.

needed heavy water in a nuclear reactor they were developing. If it succeeded it would not produce an explosion, like the bomb, but a controllable flow of energy for industrial use, particularly in generating electricity. It was considered, wrongly as matters turned out, to have no direct military application, but its commercial possibilities, as the work for the MAUD Committee went forward, began to attract ICI very strongly indeed. 'Halban's work', says a note apparently written late in July 1941, 'can lead to the development of a new source of power of great importance in peace and war and would make possible the re-orientation of world industry. It is essential that these ideas should be developed for the British Empire by a UK firm and ICI is prepared to do this.'[79]

Between June and October 1941 the small group in ICI who knew of the atomic project in its entirety worked hard to get the development of nuclear energy for industrial purposes in the United Kingdom handed over to ICI. They very nearly succeeded. Halban would have been very happy to accept the excellent terms, including a salary and a share of the profits from his process, which ICI offered him for his services. The plan then was that in agreement with the Ministry of Aircraft Production—a suitable agreement was drafted—he should move, as he wished to do, to the USA, where it was hoped that ICI's connection with du Pont would help him. The negotiations with Halban were carried out chiefly by Akers, Slade, and Perrin, who reported them to Melchett and he, in turn, to McGowan. Melchett took care to carry with him G. P. Thomson, Chairman of the MAUD Committee, and Lord Cherwell, formerly Professor Lindemann, Churchill's scientific adviser. On 23 July 1941 the Secret War Committee of the ICI Board agreed to start negotiations with the Ministry of Aircraft Production.[80] Somewhere about this time McGowan discussed the MAUD Report, characteristically over lunch, with Lord Hankey, Chairman of the Government's Scientific Advisory Committee. It is impossible to believe that McGowan would have missed that opportunity to take soundings.

McGowan, therefore, had ample reason to feel sure of his ground when, in October 1941, after negotiations with the Government had gone much further, he wrote what was evidently intended to be a clinching letter to the Minister of Aircraft Production, J. T. C. Moore-Brabazon, whom he addressed as 'Dear Brab'.

In view of your discussions with the Lord President [Sir John Anderson], I think that I ought to put in writing the substance of our conversation yesterday.

My Company has made a careful study of this new potential development of the use of Uranium for the production of energy, and we feel that it is one of extreme importance. . . .

In Dr. Halban this country is fortunate in having one of the leaders in this new development, but he can achieve nothing without adequate scientific and, in the near future, substantial technical assistance.

We do not think that the development of this process can be undertaken at a sufficiently rapid rate by the Department of Scientific and Industrial Research or any other Government organisation because . . . the whole of their scientific resources are being used, obviously rightly, in the investigations of more direct military interest. Likewise our Research Laboratories are almost entirely occupied in the investigation of problems connected with National Defence but I am confident that, with the help of our organisation in the United States, and our very close relations with Du Ponts and other large American companies, we can arrange for Dr. Halban to carry on his work there, where facilities for research of this kind are still available.

Apart from this it is our belief and experience that the normal procedure of Government research organisations is not best suited for a development of this kind because decisions, involving the expenditure of large sums of money, have to be taken at very short notice. To a considerable extent such expenditure is, in the early stages at any rate, in the nature of a gamble and it is not easy for an official body to spend money in this way whereas it is part of the normal research routine for an industrial company like mine.[81]

This proposal, conveyed with so fine a flourish of McGowan braggadocio —quite in the old manner—was turned down flat. The Scientific Advisory Committee's Defence Services Panel, reflecting the opinion of Professor Oliphant, Sir Edward Appleton, and other eminent scientists (G. P. Thomson was out of the country), refused to contemplate the development of nuclear power by private interests, and some rather nervous phrases about the Lord President in McGowan's letter to Moore-Brabazon give good reason for thinking that Sir John Anderson, though in the past and again in the future a Director of ICI, was himself against the idea.[82]

So ended ICI's bid to become the United Kingdom Atomic Energy Authority. It was made under a total misapprehension. McGowan had no idea—nor had anyone else—how big a thing he was offering to take on. It is hardly conceivable that ICI could have contained both nuclear energy and the post-war chemical industry, and a sustained attempt to do so might have been disastrous. Those who turned down ICI's offer deserve, though they probably did not seek, the thanks of ICI.

ICI's takeover bid for nuclear energy was made and rejected while the full-scale bomb project was taking shape. The MAUD Committee's report, which necessarily left great areas of uncertainty, was received by Tizard and others sceptically. In particular they thought ICI's estimate of the time that would be needed to produce a bomb was wildly optimistic, as indeed it turned out to be. Nevertheless Churchill took the basic decision to go ahead, on the advice of the Chiefs of Staff, early in September 1941, although still, inevitably, with a very imperfect conception of the immense industrial effort the bomb would require.

Sir John Anderson became the Minister responsible for the bomb project. His education as a scientist, as a young man, had included research on uranium, and he knew ICI well. As a civil servant in 1930 he had demolished their early proposals for Government help with petrol production (p. 174 above), but his performance must have impressed them, because in 1938 he was on the Board for a few months before he took office as Lord Privy Seal. Under the Minister, in what Anderson thought would be a part-time appointment, Akers was put in as the working head of the project, with a technical committee of scientists on which Slade served. The project was removed from the Ministry of Aircraft Production to the Department of Scientific and Industrial Research, where it was stowed away in the Directorate of Tube Alloys, a name invented, like the word 'Tank' in the Great War, to cast a screen of vague and not very interesting credibility about the whole undertaking.

Akers became a temporary civil servant and ICI were not, in the end, required to carry any general responsibility for nuclear developments, military or otherwise.[83] ICI's influence, nevertheless, remained strong, as it was bound to do, given ICI's industrial omnicompetence, and it was too strong for some people's liking. Among officials and scientists there was at first a good deal of opposition to Akers's appointment, because of his ICI associations, and although in the United Kingdom hostility died away, it flared up in the USA when Akers and Perrin were sent to negotiate co-operation. Americans, influenced perhaps by anti-trust proceedings against ICI, which were just beginning, found it impossible to believe that Akers, as a Government official, was not seeking post-war commercial advantage for his peace-time employers. The suspicion was unjust. Akers deeply resented it, and so did his British colleagues. Nevertheless General Groves, who came into atomic affairs midway through 1942 and took over supreme direction of the American effort (the Manhattan Project) the following year, was most unwilling to deal with Akers and, according to one ICI observer, seemed to pursue a vendetta against Akers and ICI. There seems no doubt that his hostility to Akers, because of Akers's ICI background, contributed to the breakdown of joint Anglo-American development of the bomb. In August 1943 the Americans made it clear that they would not accept Akers as Technical Adviser to the British members of the Combined Policy Commission set up under the Quebec Agreement governing Anglo-American collaboration in atomic affairs. Akers was sacrificed, Chadwick was appointed, and collaboration, on a very unequal basis, was resumed. Akers, with what good grace he could muster, which was considerable, for he was magnanimous by nature, remained at the head of Tube Alloys in the United Kingdom.[84]

Altogether, by 30 June 1944, the Government had authorized ICI to spend £950,000 on Tube Alloys research and about £870,000 had been spent. Most of the money had gone through Billingham and General Chemicals, with some through Metals and through Alkali Group as well. The total was greater than the money spent for the Government by ICI on all other wartime research projects put together, since the amount spent on Government research as a whole, up to 30 June 1944, is given as a little over £1½m.[85]

The written records of the work that was done were either destroyed at the end of the war or swallowed into official secrecy. It is plain, nevertheless, that for those principally engaged—Baxter at General Chemicals, S. S. Smith at Metals Group, Kenneth Gordon and C. F. Kearton* at Billingham, and others as well—the reward was bittersweet. The work they did, continued after the war, made fundamental contributions to British nuclear technology which have not gone entirely unacknowledged, even though some of those concerned feel that the acknowledgement might have been more generous.[86] On the other hand work done in Great Britain, once the Manhattan Project was fairly going, was of small consequence to the object immediately in view: that is, to build a bomb to win the war. At the personal level, work on 'Tube Alloys', however great its national value, was of no great advantage, within ICI, to the careers of those engaged in it, once it had been settled that ICI were not going to be allowed to take charge of the development of nuclear energy. The point was made brutally clear to Kearton, at least, and it is not accidental that he and others, including Perrin, parted from ICI after the war.

'Tube Alloys' left ICI, at the end of the war, with a difficult choice of policies. More was known about the problems of nuclear energy within ICI than within any other British organization, and ICI, in Akers's words, was 'in the position of being the only large competent chemical plant-designing organization in the country at present'.[87] Akers went on to conclude—he was writing in October 1945—that ICI would have to take a major share in building plant for Britain's peacetime nuclear energy programme. Taking the traditional ICI line, he saw it as a matter of public duty.

What Akers could also see, and so could others, was that this would be an enormous task, much bigger than it had looked when McGowan so blithely offered to take it on in 1941. It would severely strain ICI's engineering resources at a time when ICI's natural field for profitable development, the chemical industry, also badly needed attention. Here

---

* Later Lord Kearton, Chairman of Courtaulds Ltd.

was a conflict which could be represented as a conflict between public duty and commercial interest. The same conflict was to emerge in some of ICI's other activities, as we shall see when we consider the post-war development of the business.

REFERENCES

1 Corelli Barnett, *The Collapse of British Power*, Eyre Methuen, 1972, pp. ‡33–8; Reader, pp. 226–42; W. N. Medlicott, *Contemporary England 1914–1964*, Longmans, 1967, pp. 361–4; Taylor, pp. 411–13.
2 WR, 'Report on the History of ICI's War Effort (Explosives Division) by ICI Limited Nobel Division', 15 Oct. 1952; WR, Bradley, 'History of the War Effort of the Metals Group', Draft Preliminary Report, Aug. 1942; papers of 22 Jan. and 22 Feb. 1937 in CR 139/9/69.
3 As (2), 2nd refce.
4 A. J. G. Smout, 'Wrought Light Aluminium Alloys. Prospective Capital Expenditure in the near Future', 19 Nov. 1938, MBSP.
5 J. B. Nevitt to A. J. G. Smout, 27 Aug. 1936, CR 139/5/113; MBM 184(b), Annex I, 14 Dec. 1938.
6 As (5), 1st refce.; MBM 163(c), Annex I, 16 Nov. 1938.
7 WR, Bradley, 'History of the Dyestuffs Division War Effort', May 1947, p. 84.
8 Taylor, pp. 427–8n.
9 K. Gordon, 'Aviation Fuels at Billingham', 21 July 1936, AD 113,638(o).
10 Committee of Imperial Defence, Sub-Committee on Defence Policy and Requirements— Billingham, Report by Lord Weir, 28 May 1936 (DPR 90), Weir Papers.
11 Committee of Imperial Defence, Report of Sub-Committee on Oil from Coal, Cmd. 5665/1938, para. 301.
12 Ibid. para. 174.
13 K. Gordon, 'Billingham Petrol Plant—Report on Future Policy', 31 March 1939, I, CR 59/2/435.
14 L. Patrick, 'Air Ministry—Visit to Sir Hugh Dowding, 16 Aug. 1935', 19 Aug. 1935, WP; and other papers WP.
15 K. Gordon, 'Notes for Mr. Wadsworth', 3 Nov. 1938, CR 59/25/25; J. H. Wadsworth, 'Note of Interview with Sir Harold Hartley and Air Marshal Sir William Freeman 16 Nov. 1938', WP.
16 L. Patrick to K. Gordon, 2 Feb. 1939, CR 59/25/400.
17 J. H. Wadsworth to A. Fleck, 3 Feb. 1939, CR 59/25/400.
18 J. H. Wadsworth to Treasurer, 6 Sept. 1939, CR 59/25/400/1.
19 As (17).
20 Stevens, ch. 5, p. 4.
21 J. E. James, 'Organisation', 22 March 1938, Bunbury folder 15.
22 See, e.g., Overseas Executive Committee Charter, original signed by McGowan 28 June 1938, Bunbury folder 15.
23 As (21), p. 4.
24 *ICI* I, pp. 103–4.
25 As (21), p. 5.
26 A. J. Quig, 'Groups Central Committee', 21 March 1941, CR 93/395/400.
27 As (21), p. 6.
28 J. G. Nicholson to Post-War Committee, 'ICI Organisation', 30 March 1943, CP 'Organisation 1943–44'.
29 W. H. Coates to J. G. Nicholson, 'Organisation', 6 April 1938, CP 'Organisation General 1934–38'.

[30] MBM 183, 14 Dec. 1938.
[31] W. H. Coates to Lord Mcgowan, 16 Feb. 1940, CP 'Organisation 1939–42'.
[32] B. E. Todhunter to Holbrook Gaskell, 'Appointment of Deputy Chairman' with attached note, 5 Jan. 1940, CR 93/2/39.
[33] Letters and associated papers between Todhunter and Gaskell, 5, 6, 12 Jan. 1940, CR 93/2/39.
[34] Mcgowan to B. E. Todhunter, 5 April 1940, with attachment, CR 93/34/439.
[35] Draft of 22 April 1943, MP 'Organisation'.
[36] Winston S. Churchill, *The Second World War* I, Cassell, 1948, Appendix D, p. 539.
[37] 'Government Construction Contracts (excluding Research and Supply Contracts)— Official Authorisations, Estimates and Actual Expenditure at 30th September 1943', Papers submitted to Government Contracts Committee, 1943; WR/2638 (4/46).
[38] Hornby, p. 162.
[39] Ibid., esp. ch. IV (iv).
[40] WR, 'History of ICI War Effort, Explosives Division, 26 Aug. 1952'.
[41] As (37).
[42] Hornby, pp. 159, 161.
[43] WR, C. W. James, 'ICI War Effort', Dec. 1943; WR Miscellaneous Papers 1943–46.
[44] As (43), 1st. refce., 'Arms and Ammunition'.
[45] Reader, p. 304. Col. Blacker's initials wrongly quoted.
[46] As (44).
[47] *Who's Who*.
[48] M. M. Postan, D. Hay, and J. D. Scott, *Design and Development of Weapons*, HMSO and Longmans, 1964, p. 269.
[49] As (44); see also *Statistical Digest of the War*, HMSO and Longmans, 1951, Table 121.
[50] Lord McGowan, 'ICI's Technical Achievements during the War', speech at AGM, 24 May 1945, p. 3.
[51] Postan, Hay, and Scott, pp. 477–8.
[52] M. M. Postan, *British War Production*, HMSO and Longmans, 1952, p. 394.
[53] W. H. Coates, 'Supplies to H.M. Government', 19 Jan. 1944, CR 179/24/24.
[54] H. D. Butchart and A. J. Quig, 'Prices and Profit Margins', 18 Jan. 1944, CR 179/24/24.
[55] Lord McGowan, 'ICI's Technical Achievements during the War', speech at AGM, 24 May 1945.
[56] 'Proposed Programme of Dyestuffs Development', 15 April 1940, CR 75C/12/156.
[57] J. Payman, 'Post War Expansion of the Dyestuffs Group', 8 Oct. 1943, CR 75C/12/156; W. H. Coates, 'Curtailment of Dyestuffs Group Expansion Scheme', 22 July 1942, CR 75C/12/156.
[58] Stevens, 'Sales Statistics'.
[59] Melchett (?) to Lord Denham, 28 Nov. 1939, CR 153/37/229.
[60] W. J. Worboys to A. J. Quig, 'Grassland Development', 21 May 1940, covering Report, CR 153/7/165.
[61] Lord McGowan to Sir John Anderson, 19 Dec. 1940, CR 153/37/229.
[62] Lord McGowan to Sir Reginald Dorman-Smith, 9 April 1940, CR L153/110/25.
[63] S. J. Watson to R. M. Winter, 12 Nov. 1940, CR J153/14/262.
[64] Ibid.; and see Sidney Rogerson, 'National Silage Campaign, 1,000,000 tons the Aim', 12 June 1940, and 'National Silage Campaign', 9/10 July 1940, CR 153/4/262.
[65] Correspondence between McGowan and R. S. Hudson, 13, 14 Nov. 1941, CR 153/7/513.
[66] Lord Melchett to Lord Woolton, 31 May 1940, CR L153/37/229; 'Discussion with Professor Lindemann on 18th February 1941', MP 'Agriculture 1940–42'.
[67] As (61).
[68] Sir John Anderson to Lord McGowan, 27 Dec. 1940, CR 153/37/229.
[69] Ministry of Agriculture and Fisheries Circular to County WAECs, 23 June 1941, CR 153/37/229.
[70] V. E. Wilkins to S. W. Cheveley, 3 Jan. 1944, covered by S. P. Stotter to R. B. Brown, 13 Jan. 1944, CR 153A/59/69.
[71] F. C. O. Speyer, 'Consumption of Nitrogen Fertilisers in Great Britain', 10 Jan. 1944, MP 'Agriculture'.
[72] F. C. O. Speyer, 'Note of Conversation . . . 27th January 1944', CR 5/19/19.

[73] Margaret Gowing, *Britain and Atomic Energy 1939–1945*, Macmillan, 1964, Appendix 1, 'The Frisch-Peierls Memorandum'.

[74] Ibid. Appendix 2, p. 394.

[75] Ibid. generally; and private information.

[76] Gowing, Appendix 2, p. 413; Draft in papers belonging to Sir Michael Perrin; see also Gowing, pp. 48, 52, 62, 69.

[77] R. W. Clark, *The Birth of the Bomb*, Phoenix House, 1961, pp. 127–9.

[78] Gowing, p. 414; Draft in Perrin Papers.

[79] 'Nuclear Energy for Power Production (Halban Scheme)', Perrin Papers.

[80] Ibid., also other papers.

[81] Lord McGowan to Lieut.-Col. J. T. C. Moore-Brabazon, 10 Oct. 1941, CR 103/1/1.

[82] Gowing, pp. 98, 105; R. W. Clark, *Sir Edward Appleton*, Pergamon, Oxford, 1972, pp. 125–6.

[83] Secret War Committee Minute 18/41, 21 Oct. 1941, CR 93/795/795.

[84] Gowing, pp. 172–3, 176, 241; R. W. Clark, *Sir Edward Appleton*, pp. 127–8; private information.

[85] 'Research Expenditure on behalf of H.M. Government, Statement at 30th June 1944', CR 93/5/820.

[86] But see Margaret Gowing, *Independence and Deterrence, Britain and Atomic Energy 1945–52*, vol. II, ch. 17, 'The Use of Private Industry'. The author is indebted to Professor Gowing for access to her work in typescript.

[87] W. A. Akers to Chairman, 'Atomic Energy', 22 Oct. 1945, SCP.

CHAPTER 16

After the War . . .

Post-war planning started early in ICI: as soon, indeed, as the great effort to get wartime construction going was fairly under way (p. 272 above), leaving some leisure, however scanty, to think about the future. In January 1942 McGowan proposed and the Board agreed that a Post-War Committee should be set up. It met for the first time on 23 April 1942 with McGowan as Chairman and Coates, Lutyens, Melchett, and Nicholson as members.[1]

The Committee ranged widely over ICI's needs and prospects, generating reports on which a great deal of action came to be based. The Committee had a sharp eye for social, economic, and political pressures, in the country and the world at large, upon ICI policy, and in this chapter we shall examine the direction which it took over certain broad fields, particularly labour relations, research, company organization and the reconstruction of the Board, before turning, in Part IV, to the revolution in technology and the destruction of the cartel system with which the period covered by this History is brought to a close.

At the national level, too, in the mid-years of the war, a great deal of post-war planning was going on, marked by the publication of Sir William Beveridge's Report (*Social Insurance and Allied Services*) in 1942, the Government's public commitment to a policy of full employment in its White Paper *Employment Policy* (Cmd. 6527 of 1944), and the passing, also in 1944, of the Education Act of which R. A. Butler was the main architect. The general mood in the country combined somewhat overdone disgust at the recent past with romantic optimism about the near future, so that it came to be fairly widely assumed that the war was being fought almost as much for social justice as against Hitler. Churchill himself was sceptical of post-dated promises—he had seen them before—and economists were beginning to take a view of the country's difficulties which was far from optimistic, but of these things most people knew nothing. The nation's sense of common purpose was still strong, military victory was only one element in it, and on foundations laid much earlier the framework of the Welfare State was being built.[2]

Underlying the enthusiasm for ambitious social policy there was a movement towards the Left in politics, given force by the dramatic change in the ordinary working man's position brought about directly by the war. He was no longer a dispensable statistic in the dreary tables of mass unemployment but a person of consequence, whose services were sought by employers and whose interests were forwarded, at the Ministry of Labour and in the War Cabinet itself, by that incarnation of the power of organized labour, Ernest Bevin, 'the embodiment of all natural and unlettered men drawing upon wells of experience unknown to the more literate'.[3] At the Ministry of Labour one of his senior officials was R. Lloyd Roberts, seconded from ICI.

In the Post-War Committee they realized from the start that ICI's labour policy—centrally directed, as always—would have to be modified to take account of the new strength and strength of feeling of the payroll workers. Among the members of the Committee, Melchett concerned himself with labour matters, but the direct executive responsibility, from November 1940 to November 1944, lay with B. E. Todhunter. He was much the oldest Director—he was born in 1865—but he was neither inactive nor, in spite of a tendency to regard Alexander Fleck as a dangerous revolutionary, unreceptive to the idea of change.

In ICI, during the war, only about 80,000 man hours were lost in strikes,* and in Alkali, Dyestuffs, Leathercloth, Paints, and Salt, none at all.[4] In view of this remarkable record, no doubt owing a great deal to a long history of enlightened labour policy, the degree of unrest among ICI's wartime labour force must not be exaggerated. Nevertheless there was sufficient of a groundswell to be noticeable. 'Our men do not trust the Company. . . . They do not trust industry in general', said Lime Group's Labour Manager in 1942,[5] and it is no doubt a tribute to those in charge of ICI's labour policy that they were closely enough in touch with factory opinion to deal with discontents before trouble rather than after.

The points of friction centred, as they always had done, round two pillars of ICI's pre-war labour policy: Staff Grade and Works Councils. Both were valued in ICI for the very reason that made them offensive to organized labour: that they fostered loyalty to ICI rather than to the unions (chapter 4 (ii) above). Before the war, with the unions very weak and unemployment high, ICI could carry matters with a high hand, but the war changed that. Union membership was still far from universal in ICI, varying a great deal from factory to factory, but the power of the unions was growing and no one on the management side of ICI was looking for a fight.[6]

The chief objection to Staff Grade, from the labour side, was that selection was not based on fixed, objective qualifications but lay in the discretion of the management, leading to distinctions between man

---

* Of which 12,470 were lost in Explosives and 62,404 in Metals.

and man which most of those on the payroll considered quite unwarranted. 'It is obviously difficult', as Lloyd Roberts observed in 1945, 'for a worker with 30 years' service to appreciate why he is not good enough for Staff Grade while another worker of 5 years' service is promoted.'[7] With this view there was a good deal of sympathy among the management, which emerges in the records of the numerous conferences and discussions, between the central authorities and managers in the Groups, held during the mid-years of the war, and in other papers.[8] Melchett was sympathetic and Fleck, as Chairman at Billingham, seems to have scandalized Todhunter, in 1942, with a suggestion which had wide support—'that the principle of the staff grade scheme be extended in a wholehearted wholesale manner to cover . . . 100 per cent of ICI established personnel'. The number of Staff Grade men varied widely from factory to factory, but as late as the end of 1944, in ICI as a whole, there were only 12,879 of Staff Grade, 30·21 per cent of those eligible under the existing rules which stipulated an age of twenty-six and service of five years.[9]

What Fleck and many others felt, irrespective of pressure from outside opinion, was that the differences in working conditions between 'staff' and 'payroll' were indefensible. Fleck, putting the matter as provocatively as he could, pointed out that 'staff' were expected to work 'something under 40 hours a week' and were trusted to do so. Payroll workers, on the other hand, were required to work a 48-hour week and were so little trusted that their pay was calculated 'strictly in ratio to the hours spent at their place of work', at which, of course, they had to 'clock in' and 'clock out'.

Fleck's memorandum, coming from the Chairman of one of the most influential Groups, stirred up discussion, and after a meeting early in 1943 the main differences between 'staff' and 'payroll' conditions were summarized:

(a) Salaries of staff climb gradually to the maximum at which a post is assessed, whereas a worker receives the full rate for his job at once. . . . The majority of employees on the works payroll can only increase their income by winning promotion or by working overtime. . . .

(b) Labour is paid by the hour for hours worked, whereas staff is paid by the year. Only long absences, amounting to months, would involve a reduction of income to an established member of the staff with some years' service.

(c) A clerk of mediocre ability in a routine job eventually overtakes and earns more than a skilled tradesman.

(d) Staff employees contribute 5% and the company $7\frac{1}{2}$% to the Staff Pension Fund, whereas workers contribute $2\frac{1}{2}$% and the company 3% to the Workers' Pension Fund.

—with the result, inevitably, that workers' pensions were less attractive than those paid to staff.[10] On top of everything else, the staff got two or three weeks' paid holiday a year and the payroll only one: a distinction which was coming to be regarded as intolerable.

As the Staff Grade scheme, and especially the method of promotion, offended many payroll workers' sense of justice, so amongst the members of trade unions there was a feeling that Works Councils unfairly by-passed the unions as a channel of communication with ICI's factory management. The Councils were forbidden to discuss pay and hours of work, but that did not prevent suspicion and dislike, at any rate so long as membership of a union was not laid down as an indispensable qualification for election to a Works Council. ICI's official position—that the unions could not claim to represent all ICI workers, and that it was not ICI's policy to inquire whether employees belonged to a union or not—was not a line of argument the unions were prepared to tolerate with any pretence of goodwill.

The unions made their attitude to Works Councils plain enough,[11] as they always had done, but they concentrated their frontal assault on two points: the recognition of shop stewards and the establishment of the closed shop. To concede either would mean, for ICI, a turnabout in policy, but ICI's approach to labour policy was undogmatic and no one on the management side showed any disposition to take a rigid stand on principle. Two Groups—Explosives and Fertilizers—pressed as early as the summer of 1943 for the enforcement of a closed shop, with the recognition of shop stewards that would automatically follow.[12] For ICI as a whole, however, it was a difficult policy to enforce universally. In some Groups closed shops existed already, for all workers or for some trades, but in others barely half the men belonged to unions at all, raising difficult questions of representation if shop stewards were to be recognized. Moreover there were thirty-five unions to satisfy, raising boundless possibilities of strife if the alteration of policy were badly handled. On the whole Todhunter thought the whole explosive package would be best left unopened until after the war.[13]

The war caused ICI to look critically at conditions of work in the factories. In factories built for the Government, as Todhunter pointed out in 1943, 'a much higher standard of comfort and of amenities generally . . . has been set than had been normally current in the country before the war'.[14] The newer ICI plants—he mentioned polythene plant at Winnington, 'Mepacrine' plant at Grangemouth, and an intermediates building at Huddersfield—were up to the new standard, but older ones, at Blackley and elsewhere, were below it. It became accepted policy that factories built after the war would have to be designed with much greater regard for the comfort and convenience of those who worked in them. There is no evidence that ICI, mindful of their reputation as good employers, were unwilling to build to new and higher standards, but equally it is evident, from the report of a committee set up by Holbrook Gaskell in 1945, that they had been 'spurred on . . . by the arrangements existing in Government war factories' and that 'the workers also have become more critical'.[15]

In August 1945 Lloyd Roberts, back from Government service, gave Sir Frederick Bain, Personnel Director, his views on ICI's post-war labour policy.[16] He wrote just after the Labour Party had come to power, deeply suspicious of private enterprise and full of enthusiasm for nationalization. He suspected, however, that the unions might not be so convinced of the virtues of nationalization as the Party. 'They are far more concerned', he wrote, 'with the attainment of their own strictly practical ends than with the particular means of doing so. If private enterprise can do so, the natural conservatism of the Unions and of their members will be used to discourage any change from that system.' 'I suggest', Lloyd Roberts went on, 'that the right course in the country's interests . . . is to drive in still further the wedge between the political and the industrial wings, by a more active encouragement of the Trades Unions', and the first recommendation he made was 'to give substantially more "backing" to the Unions, in recognition not only of their coming political influence but of the fact that a strong Union movement wisely led is unquestionably a stabilising force in industry'.

In support of his recommendation he made an unflattering analysis of the unions as they then stood. They had been weak, he said, and unable to discipline their members, and had thus acquired—'justly'—a 'reputation for feebleness'. With very few exceptions, he said, they were 'badly led and inefficient. . . . They cannot agree among themselves and each is fighting for its own vested interest . . . without regard to the well-being of the workers in general.' He gave some faint praise to Ernest Bevin as a leader, and concluded '. . . basically they [the unions] are sound in their outlook and when taken adequately and honestly into the confidence of the employer, they can generally be relied upon to take a sane view. . . .' It would not be dangerous, therefore, for ICI to indicate whole-hearted support of the unions. 'On the contrary, the Company might expect to reap a substantial political advantage from the influence of a friendly Trade Union movement within the Socialist Party.'

Lloyd Roberts suggested that after a certain date ICI should undertake to engage only union members in the factories; that shop stewards should be recognized; that union dues should be collected through the pay packet; and that membership of Works Councils should be open only to union members. In return he hoped to get the unions to give up restriction of output, to arrive at reasonable demarcation agreements among themselves, to refrain from coercing non-unionists already in

ICI's employment, and to agree not to 'attempt to interfere in the day-to-day responsibilities of management'.

These suggestions, as Lloyd Roberts himself said, amounted to 'somewhat revolutionary changes in policy', and he admitted that he had become so convinced that the full goodwill of the unions would become increasingly necessary to ICI that he had had to bury what he called 'some of my own cherished beliefs'. He still hoped, though, as he had hoped for many years, to induce ICI's workers 'to make the Company, rather than the Unions, their first loyalty', and he suggested that post-war labour policy should be developed with that aim in view. 'I would remind the Board', he told Bain, 'that the worker of to-day, even on the docks, is not the illiterate clodhopper of the last generation. He has that little knowledge which is a dangerous thing, but it is enough to convince him that there was something wrong with his old conditions of employment, and in particular, with the system under which his own livelihood and that of his family were subject to the sudden threat of one hour or one week's notice.'[17]

It is unlikely that Bain was greatly shocked by Lloyd Roberts's suggestions. 'Revolutionary' though they were, most of them had been heard before in the very frank and thorough discussions of labour policy that had gone on in ICI since 1942. By the time the war ended, there was a great deal of agreement about what needed to be done, taking account of the new strength of labour and the unions, of the effect of wide-ranging social legislation, and of the nation's wartime hopes of better times.

Political pressure from outside had a strong and acknowledged effect in ICI, being most clearly discernible in the writings of Lloyd Roberts. There is no reason to think, however, that it was pushing those responsible for policy down a road they did not wish to travel. Rogers, Melchett, Fleck, Todhunter and others, particularly at the factories, were men broadly of a liberal turn of mind who were perfectly prepared to believe that the labour policy of ICI's early days, advanced though it might have been then, would have to be altered to meet the very different conditions of the post-war world. Benevolent despotism would have to give way to a far less authoritarian scheme of things, in which the distinction between 'staff' and 'payroll' was blurred, especially in matters of job security, holidays, and pensions; in which management would have to pay serious attention to joint consultation; and in which the power and position of the unions would have to be acknowledged and allowed. Full employment, and governments of both parties committed to maintaining it, had never been known before. Its effects would be far-reaching, and it was as well that the Post-War Committee had looked hard at problems of labour policy while the war was still going on.

Plastics factory at Hillhouse, 1947. An 'agency factory' (chapter 15 (iii)) built for the Ministry of Supply and taken over after the war by ICI.

Women washing casks at the Dyestuffs Works, Blackley, 1943.

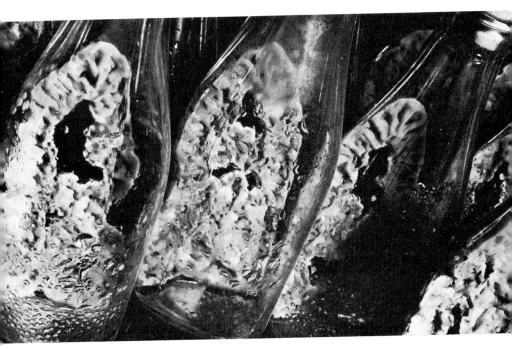

Penicillin production, 1944 (page 286).

Research policy needed looking at, too. Immediate pre-war developments right across the chemical industry—in plastics, in fibres, in pharmaceuticals and elsewhere—were showing the importance of well-directed research, but ICI's activities had nearly been throttled in the early thirties and had barely recovered by the time war broke out. ICI's inventive capacity, even so, was respectable, but there was an uneasy feeling in the business that it did not compare well with the performance of du Pont and IG.

In 1940 the proceedings of a conference of ICI research managers, a trifle complacent in tone, encouraged McGowan to think of drawing du Pont's attention to the achievements of ICI's research. J. Baddiley, Research Manager, later Director, of Dyestuffs Group, at once advised against doing so, and he was supported by Francis Walker. Both agreed that important discoveries, including 'Perspex' and polythene, had recently been made in ICI, but Baddiley doubted very much 'whether it is realised in some quarters how little real research work ICI does in comparison with the size of the company and in view of the intense competition of the IG and the Americans'. 'It would be dangerous', Walker added, '. . . to provoke comparison between ICI and du Pont's developments.'[18]

R. E. Slade agreed, in general, and blamed commercial agreements, particularly with Distillers Company Ltd. and with Unilever, for blocking promising lines of research. Later, he said that lack of semi-technical and pioneer plant hindered development to full-scale production, which in du Pont was very rapid.[19] Engineering services, much weaker in ICI than in du Pont, turned out to be a major handicap in ICI's post-war development. Sir Ewart Smith, who greatly admired du Pont's engineering, became Technical Director of ICI after Sir Holbrook Gaskell retired in 1946, and he directed his formidable energy to repairing the deficiency.[20]

Uneasiness about ICI's research was linked with misgivings about ICI's relations with the universities. Some resentment no doubt lingered about the abrupt dismissals of research staff in the early thirties, and there was certainly a feeling among academic scientists that the terms on which ICI made grants for research were illiberal.[21] Some people in ICI thought the reports of academic hostility exaggerated, but the terms on which grants were made were considerably eased early in 1943:[22] an indication, perhaps, that a general repair job on university relations would be put in hand when wartime pressures eased.

In 1944 Sir Henry Tizard launched an attack on ICI research, saying that ICI had shown 'a lack of originality and a lack of power to open new fields of research as compared with other countries generally and in particular with Germany and the United States'. The charges were not publicly admitted and a spirited defence was prepared, but the remarks

303

L

of so eminent a critic, chiming in with internal disquiet, could hardly be shrugged off. Melchett, ICI's chief liaison officer with the academic world, went to lunch at Oxford with Tizard and sought his advice.[23]

In July 1944 the papers announced that ICI were setting up eighty science fellowships, worth £600 a year each, in nine British universities. The holders were not to be tied to ICI in any way. In the early post-war years the number and value of fellowships were increased, until by 1951 there were ninety-two, worth £800 a year each on average, at eleven universities. The scheme, said McGowan in 1944, 'was the conception of Lord Melchett', and it was entirely in the Mond tradition of close relationships between industry and academic science. F. A. Freeth, recalled from exile, was given the congenial task of overseeing the disbursement of ICI's munificence to the universities.[24]

The Tizard episode no doubt stimulated the project for founding the ICI fellowships. It was certainly taken to support the case, put forward by Slade, Akers and others before the war, for a Central Research Laboratory on the lines of the du Pont establishment where Carothers discovered nylon.[25] They wanted to give specially chosen scientists the opportunity to work on fundamental problems undisturbed by the emergencies and purely commercial considerations which, they said, were continually distracting the attention of the staff of divisional laboratories.

In the Divisions the proposal was unpopular. It flew in the face of their established tradition of autonomy. 'Staff of such an institution', wrote Lutyens and Lawson, voicing Alkali opinion, 'would be cut off . . . from ICI's real business—the manufacture of chemicals.'[26] Nevertheless the laboratory, or rather five laboratories on the same site, came into existence in 1946 in the grounds of a Hertfordshire country house, The Frythe, at Welwyn. 'The outlook and atmosphere', wrote Akers, '. . . should be rather akin to that of our best university laboratories but with the additional material facilities which ICI is able to supply. It is believed that this policy will attract . . . some of the best of the younger academic research workers who would not consider joining an ordinary industrial laboratory.'[27]

Whatever suspicions may have been harboured in the Divisions about the ulterior purposes underlying the setting-up of a Central Research Laboratory, there seems to have been no thought at Millbank of diminishing the importance of the Divisions' own research. Indeed, quite the contrary. As early as December 1942 the Management Board approved a recommendation of the Post-War Committee that the Groups, as they then were, should plan to increase their laboratory capacity by 33 to 50 per cent, and that purpose held firm, being reinforced by the growing national consciousness of the importance of industrial research.[28] Scientific discovery, like social welfare, was to have high priority—a favourite word of the day—in the new, improved, post-war Britain, and early in 1944 the Research Director's Conference, discussing a passage on research and development intended for the Chairman's Statement to the Annual General Meeting, put on record that, 'in view of the present intense public interest in research, it was essential that the subject be dealt with in a thorough and impressive manner'.[29]

If labour policy and research after the war required the attention of the Post-War Committee, so, no less, did the problem of setting up a workable organization for ICI, which was evidently on the edge of a period of rapid growth after the enforced lethargy of the thirties. The plan worked out in 1938, after the overthrow of McGowan's dictatorship, was aimed at removing power from the hands of one man—McGowan—and giving it to the Directors, acting jointly and severally through an elaborate system of committees (p. 268 above).

There were complaints from the first about the scheme's complexity, and if it had been strictly followed it would no doubt have turned out unmanageably cumbersome. In fact, under force of circumstances, and because those required to work the system acted with good sense and goodwill, the 1938 organization was made to carry the main burden of the war, but the trick was done as much by neglect as by observance. The Groups, in particular, acted with an independence which the framers of the 1938 constitution had never meant to confer upon them, and which arose partly from the enforced wartime dispersion of Head Office departments. That was not very important during the war, when the policy the Groups were required to carry out was directed not by ICI but by the Government. It would become very important, however, as the direction of ICI's affairs passed back to ICI. It was abundantly evident, as we shall see in Part IV, that the technology of the chemical industry was developing rapidly and fundamentally, and investment decisions of capital importance would have to be taken as the strain of war began to ease. No one would wish to hinder the Groups unnecessarily in the management of their own affairs, but over the disposal of ICI's resources as a whole the centre must be able to exercise control.

The problem of organization, like other problems, was discussed at length over a period of years.[30] There was no master-mind at work, and there were sharp differences of opinion, but in the spring of 1943 J. G. Nicholson produced a criticism of the existing organization and a coherent plan for a scheme to replace it.[31] His paper bears the imprint of his own forceful personality, but it is a product of collective thought.

'The system in its present state', said Nicholson, tended 'to loosen control without . . . achieving the advantages of decentralised responsibility.' The Groups were tempted 'to consider their own sectional interests as paramount, instead of those of ICI as the major unit', and they might be influenced 'by their own immediate Profit and Loss Account, without sufficient broad-minded consideration for the future'. If that happened they could hardly be blamed, 'as their principal officials have been afforded too little opportunity of appreciating the problems which confront the Company as a whole, both in National and International affairs'.

TABLE 24

ICI GROUPS AND DIVISIONS AS CONSTITUTED WITH EFFECT
FROM 1 JANUARY 1944

Group A	Heavy Chemicals
	Alkali Division
	General Chemicals Division
	Lime Division
	Salt Division
Group B	Dyestuffs and Pharmaceuticals
	Dyestuffs Division
	Pharmaceuticals Division
Group C	Ammonia and Agriculture
	Billingham Division (formerly Fertilizers & Synthetic Products Group)
	Agricultural Division (later Central Agricultural Control)
Group D	Metals
	Metals Division
Group E	Explosives
	Explosives Division
Group F	Paints and Plastics
	Plastics Division
	Paints Division
	Leathercloth Division

Source: 'Organisation', 30 October 1943, approved by the Board 11 November 1943, operative on and from 1 January 1944, CP 'Organisation 1943–44'.

The main object of the new organization was to reforge a direct link, assumed to have been broken some time in ICI's early history, between the main Board and the factories. For this purpose the operating Groups, renamed Divisions, were brought together into six much larger Groups (see Table 24) based on 'more or less allied' interests in raw materials or finished products. The Divisions thus brought together kept their separate identities, with their own Chairmen and Boards, but they were represented on the main Board by Group Directors, each of whom was 'charged with the responsibility . . . for the direction, management and efficiency' of Divisions in his Group, being armed for that purpose with considerable executive powers, including authority for capital expenditure up to £20,000.[32]

Alongside the six Group Directors seven 'Functional Directors' were to sit: Commercial, Development, Finance, Overseas, Personnel, Research, and Technical. Each presided over a central department run by 'controllers' who had considerable powers, not closely defined, in the Divisions. The title 'controller' was not introduced without misgivings. 'It is the correct designation', J. E. James explained defensively, 'for the duties allotted and is in general use to-day, particularly in Government Departments'. That may not have gone very far to set Divisional minds at rest. The title was perhaps symptomatic of a determination which seems to be observable in the new organization to assert the power of the centre as paramount over the circumference.

This pattern of parallel, co-equal chains of authority from the Functional Directors and the Group Directors provided abundant opportunity for friction, as it does in any company with a similar organization. Whether or not friction would develop in ICI would depend on the personalities concerned, though there was one important organizational safeguard against it. This was the provision for a Chairman and Deputy Chairmen who, although full-time Directors, had no departmental duties. Their formal powers were no greater than those of their colleagues, but their position set them apart, and in 1954 S. P. Chambers said: '[The] work of guiding Executive Directors in the matters they bring to the Board and in the way in which they exercise their functions is undoubtedly the most important part of the work of Deputy Chairmen.'[33]

The administrative paraphernalia of the 1938 organization—the Management Board, the Groups Central Committee, the Executive Committees—were comprehensively tidied away. The Board as a whole—including the lay Directors—assumed the functions formerly vested in the Management Board which were those usually associated, as we saw above (p. 268), with the office of a Managing Director. The Groups Central Committee was abolished, being replaced by a quarterly meeting of Directors and Division Chairmen, and similarly the Functional Directors' former Executive Committees were replaced by less formal 'conferences' held once a month. The Chairman, too, had his conference, of all the full-time Directors, which in 1949 met on the day before each fortnightly Board meeting.[34] This conference had no place in the written constitution of the Company, its proceedings were informal, and very few records have survived. No doubt it was a very important decision-making machine indeed.

When the new organization was being designed, ICI had a depleted and ageing Board. At the beginning of 1943 there were thirteen full-time and five lay Directors. The thirteen full-time Directors included the Chairman and three Deputy Chairmen, leaving nine nominally available for departmental duties, but in fact one Deputy Chairman (John Rogers) and two Directors (Akers and Bain) were away on Government service. Of the full-time Directors five, including the Chairman and two of the Deputy Chairmen, had been born before 1880 and one of the five—Todhunter—dated from 1865. One Director—Melchett— was only 45 but he was in bad health. There were not enough Directors to fill all the positions proposed in the new scheme, and of those there were, fully half were beyond or very near retiring age. Of the lay Directors, Pollitt was 65 in 1943, Weir and Ashfield were both older, and although opinions differed on how many lay Directors were required, it was generally agreed that there ought to be more than five.

In the spring of 1943 the three Deputy Chairmen, with the new organization in mind, met at the Chairman's instance 'to consider the list of potential candidates for the Board, and the members of the Board already over age, or who would soon be over age in the immediate post-war period'.[35] The discussion must have been a delicate one, in view of the age of two of the Deputy Chairmen (Nicholson was 64 and Rogers 65) and of the Chairman (nearly 69) himself. Leaving these matters on one side, they decided to ask Todhunter (78), Holbrook Gaskell (65), H. O. Smith (60), and Wadsworth (56) to retire immediately after the war. 'The same problem'—age—they observed would soon arise with W. H. Coates (61), and 'the training of a successor . . . in the immediate post-war period should be regarded as urgent'. The question of new lay Directors they put to one side, on the ground that it would be bad for ICI's public reputation and for 'the general continuance of Board policy' to make too many changes at once.

As for McGowan, rising seventy, he had no intention of retiring and no one seems to have dared to suggest that he should. His Service Agreement was renewed for two years at the end of 1943 and then again for five years from the beginning of 1946. He had a salary of £25,000, fixed commission of £25,000, £2,000 as a Director, £5,000 general expenses and travelling expenses as required (he required a great deal). His pension was fixed at £10,000, with £5,000 for his widow, raised to £15,000 and £12,500 respectively in 1946.[36] McGowan frequently said that he hoped to die in harness, and he presumably regarded his pension arrangements simply as an insurance against bad health. He also expected to nominate his successor.[37]

My dear Henry [he wrote to Melchett in November 1944], Ah me, Time marches on, and I have now passed the allotted span of three score years and ten. Frankly I do not feel that great age, and I may not look it, but that does not get over the inescapable fact.

With the bewildering and complex problems ahead of ICI when the war is over, I hope I may be able to continue in sufficiently good health for some time to come, as Chairman, as while the expression is often used that this is the day of the young man, one cannot disregard the experience, extending over many years, of some one like myself.

It is not an easy matter, when the Chairman is in good health, and is willing to carry on, for the Board to choose his successor. At the same time, while the formation of ICI was my original conception, carried into effect with the collaboration of your revered father, my great friend, I am anxious indeed to ensure that the position of Chairman should be in the hands of one who will carry on the Company as successfully as your father did, and as I hope I have done.

You and I have been together now for many years, and as I think in the past I have not been unsuccessful in 'sizing' up people, I unhesitatingly say that at the appropriate time I would like to see you occupy my chair. May I make one reservation. In my opinion the Chairmanship of the Company must be an all-time job—not some part of a week. If your health does not permit of this, then the Board must decide what should be done.

I leave this letter with you to be submitted to the Board after I have gone, the effect of which is to recommend to them that you be elected my successor. . . . God be with you.

The Deputy Chairmen, during their 1943 discussion, drew up a list of eighteen potential Directors, of whom eight eventually reached the Board. Between 1943 and 1947 eight new full-time Directors were appointed (five from the 'possibles' of 1943) and four pre-war Directors resigned. Of these resignations three—B. E. Todhunter's (1944), Sir John Nicholson's (1945), Sir Holbrook Gaskell's (1945)—were planned. Lord Melchett, on the other hand, was forced into retirement by illness in 1947, at the age of 49, and two years later he died. The succession to McGowan's Chair lay wide open, and the time was coming when even McGowan must retire.

These changes altered the balance of the Board in age, in the representation of various sides of the business, and in technical strength. As the younger men came on the Board, they confirmed the swing away from the dominance of Nobels, so marked in ICI's early years, towards the ascendancy of Winnington. W. A. Akers and W. F. Lutyens, who became Directors in 1940, both came from Winnington. So did D. R. Lawson (1943), J. L. S. Steel (1945), and V. St. J. Killery (1947). Others were to follow. From other sides of the business, A. J. Quig had come in from Paints (with a Nobel background) in 1940 and F. W. Bain, in the same year, from General Chemicals. Alexander Fleck (1944) and Sir Ewart Smith (1945) were both Billingham men and A. J. G. Smout was from Metals. Dyestuffs—at last—were directly represented on the Board when C. J. T. Cronshaw joined in 1943.

It could be said with some justice that ICI's Board before the war was strong in commercial skill and experience but weak in technology. After 1945 that was no longer so. Most of the new Directors were by education scientists and had made their careers as technical men. One, Sir Ewart Smith, was an eminent engineer. If anything, in the revolution of the generations, the weakness may have been transferred to the commercial and financial side: certainly it was no longer technical.

The weakness revealed itself in 1947 when the question of a successor to Sir William Coates arose. Like his predecessor, Lord Stamp, he had come to Nobels, and thence to ICI, from the Inland Revenue, and there was no one else on the Board, or elsewhere, with the same grasp of economic theory and practical finance including, no doubt, the intricacies of taxation. That, at any rate, seems to have been McGowan's view. As soon as Coates said he wanted to retire at the end of 1948 (he did not, in fact, go until 1950) the Chairman had his candidate ready, and he was outside ICI altogether. He was 'a really brilliant Civil Servant', S. P. Chambers.[38] Chambers, like Stamp and Coates, had begun his career in the Inland Revenue. He had been Income Tax Adviser to the Government of India in the late thirties and then, returning home in the early days of the war, had become in 1942 Secretary and a Commissioner of the Board of Inland Revenue. He went to the British Element of the Control Commission for Germany in 1945 and when McGowan began to think seriously of him for ICI he was Chief of the Finance Division.

The first hint of Chambers as a possibility for ICI seems to have come from Sir James Grigg, Secretary of State for War, who mentioned him to McGowan some time in 1944 or 1945. McGowan saw Chambers early in 1947, 'was much impressed by him, and . . . asked, in a general way . . . whether he would be disposed to "burn his boats" in the Civil Service and join an Industrial Corporation'. Chambers was invited to lunch with the Board in May 1947 with all the force of McGowan's recommendation behind him, a firm offer was made, and Chambers saw his official superiors, Sir Edward Bridges and Hugh Dalton, Chancellor of the Exchequer. They both pressed him to stay in the public service, Dalton saying that he would be back in it soon, in any case, because the Government intended to nationalize ICI; but ICI's offer was an attractive one, and Chambers decided to accept.

What Paul Chambers was offered was an immediate seat, as an executive Director, on the Board of ICI.[39] No such appointment had ever been made before nor, up to the moment of writing, has any such ever been made since. The formal letter was written on 30 May 1947 and Chambers was elected a Director of ICI, after twenty years in the Civil Service, on 10 July. It was perhaps the most spectacular leap from the public service into private enterprise, in Great Britain, that had ever been made, and its results were to be no less spectacular than its beginning.

By the time the war ended the Boards of Divisions, like the main Board itself, needed drastic reconstruction, and for broadly similar reasons. The average age of Division Directors was high—52 in November 1944, with only three under 40—and numerous vacancies were to be expected through retirement on grounds of age or illness. Nicholson, reporting after consultation with Cronshaw, Fleck, Lawson, and Smout, thought 'the opportunity might now be taken to lay down *general* principles as to the future composition of Division Boards'.[40] Nicholson suggested, broadly, that for the larger Divisions a Board of nine members, including a Chairman and two Managing Directors, one technical and the other commercial, would be appropriate. Responsibility for the various 'functions' of the seven Functional Directors on the main Board should be covered, but there would be no need to have a separate Director for each function. The findings of Nicholson and his colleagues were accepted by the Board and between November 1944 and October 1945, acting as a Committee of the Board, they recommended the appointments and retirements necessary to carry them out.[41]

The investigation of ICI's top management was supported, at a lower level, by inquiries leading to recommendations for the post-war selection and training of managerial staff coming in. Non-technical graduates, it was suggested in 1943, should be engaged—as, after the war, they were—and the thought was put forward that 'no woman should be debarred from occupying a higher executive post merely because of her sex'.[42] In 1943 an Education Committee reported, having surveyed the field very widely within ICI, and from its deliberations emerged a recommendation, strongly backed by Quig, for an 'ICI Staff College'. It came into being after the war at Warren House, Kingston-upon-Thames.[43]

ICI faced the bleak dawn of peace, then, with a new organization, a thoroughly reconstructed management structure, the makings of a new labour policy, and a mass of reports, which we have not space to examine in detail, on a great range of problems which would fall to be dealt with once the war was over. The atmosphere was harsh. Private enterprise was out of favour, and large companies even more out of favour than usual, with nationalization a live threat. Supplies of all kinds, especially steel, were hard to come by at a time when new construction and reconstruction urgently needed to be done, and Government controls stretched into every nook and cranny of ICI's operations. Even if constructional materials had been more freely available, the use that could have been made of them would have been limited, for ICI was afflicted, for some years after the war, by a crippling shortage of engineers and draftsmen. And as the Board contemplated the tasks that lay before them, it was borne in upon them that the growth and complexity of ICI were beginning to set large problems of planning and control.

Nevertheless, the self-confidence of ICI's new management, generally speaking, was high. During the war ICI's normal development had been seriously hindered, but the versatility and indispensability of the business had been demonstrated unmistakably and bracingly: without ICI the country could not have carried on. Ahead, the peacetime years looked bright, once the immediate difficulties were out of the way. The technological revolution in the chemical industry, which had been gathering way since the early thirties, was beginning to promise very big things indeed. In ICI, freed at last from the great distraction of the war, they were eager and ready to take up the promise.

REFERENCES

1 Post War Committee Minutes, 23 April 1942.
2 Medlicott, pp. 465, 469; Maurice Bruce, *The Coming of the Welfare State*, Batsford, 1961, ch. vii.
3 Lord Francis-Williams, in the *Concise Dictionary of National Biography*.
4 R. Lloyd Roberts, 'Wages and Post-War Labour Policy', Table III, 7 Jan. 1946, CR 2/8/1.
5 'Notes of Meeting of Group Labour Managers . . . 12th July 1942', p. 29, CR A44/4A/5A (14).
6 R. Lloyd Roberts, 'Post War Labour Policy', 24 Aug. 1945, CR A/44/4A/5G.
7 Ibid. p. 10.
8 As (5), p. 7; Minutes of Meeting, 29 March 1943, p. 1, CR 93/7/800; 'Notes of a Discussion . . . with Group Labour Managers . . . 12 July 1943', p. 1, CR 209/7/800.
9 A. Fleck, 'A Goal for ICI Labour Policy', 14 Oct. 1942, CR A44/A4/5; as (6), p. 9.
10 R. A. C. Watson to J. Hay, RACW/EGB, 16 Jan. 1943, CR A44/A4/5A (5).
11 H. B. Trumper, 'The Outline of a post war Labour Policy for ICI', 5 March 1943, p. 7, CR A44/A4/5A (5); B. E. Todhunter, 'Post War Staff and Labour Policy', Dec. 1943, Post War Committee Papers.
12 As (8), 3rd refce., p. 3.
13 As (11), 2nd refce., p. 5.
14 Ibid. p. 6.
15 'ICI Amenities', Report of a Committee, 12 June 1946, p. 1, CR A26/4A/2AC.
16 As (6), pp. 2–4.
17 Ibid. pp. 7, 9.
18 'Notes of a Meeting of Research Managers . . . 5 April 1940', CR 50/110/110; J. Baddiley to F. Walker, 8 May 1940, CR 50/110/110; F. Walker, 'A Review of our Research and Development Effort in Relation to Du Pont', 17 May 1940, CR 50/110/110.
19 R. E. Slade, 'Comments on Mr. Walker's Review', 15 June 1940, CR 50/110/110; 'Research—Minutes of a Meeting . . . 2 July 1940', CR 50/110/110; R. E. Slade, 'Research Policy after the War', 21 Jan. 1943, CR 50/110/110.
20 F. E. Smith, 'Report on a Visit to the United States', ICIBR 31 July 1947.
21 H. O. Smith, 'ICI Policy in Regard to Universities and Professors', 13 Nov. 1942, CR 50/12/54; 'Notes of a Discussion . . . on 8 Dec. 1942 on the Relations between ICI and the Universities', CR 50/12/54.
22 'Relations between ICI and the Universities . . . Letters of Appointment', 22 Jan. 1943, CR 50/12/54.
23 Lord Melchett, 'Chemical Research', 10 May 1944, MP 'Research 1930–46'; Lord Melchett, 'Fundamental Discoveries in Chemistry', 24 Aug. 1944, with attached 'Notes on the Research Policy of the Chemical Industry' by Sir Henry Tizard, CR 50/138/138.

[21] 'List of Universities to which letters . . . have been sent', July 1944, CR 50/346/346; Lord McGowan to the ICI Board, 13 June 1944, MP 'Research Fellowships'; Press Notice, 25 July 1944, CR 50/346/346; Professor J. W. Cook, 'Scientific Research', *The Times*, 21 April 1951; ICIBM 13,383, 13 July 1950.

[25] 'ICI Central Research Laboratory', 2 Aug. 1944, CR 50/270/311.

[26] W. F. Lutyens and D. R. Lawson, 'Why the Institution of an ICI Central Laboratory would be a Mistake', 15 Dec. 1944, CP 'Research'.

[27] W. A. Akers, 'ICI Butterwick Laboratories', 17 Jan. 1946, CP 'Research'.

[28] MBM 1212, 30 Dec. 1942; Research Director's Conference Paper 73, 15 Feb. 1945, CR 93/5/67.

[29] ICIBR (Research Director), R 44/27, 22 Feb. 1944.

[30] W. F. Lutyens, 'Scheme for co-ordinating ICI's Present Interests and Future Developments in the Organic Chemical Field', 2 April 1941, CR 93/34/439; C. J. T. Cronshaw, 'Memorandum on Organisation', 30 May 1941, CR 93/34/439; Lord Melchett, 'Memorandum on the Higher Control of ICI', 27 Nov. 1942, MP 'ICI Organisation'; John Rogers, 'ICI Reorganisation', 5 May 1943, MP 'Post War Committee'.

[31] J. G. Nicholson, 'ICI Organisation', 30 March 1943, CP 'Organisation 1943–44'.

[32] 'ICI Organisation', 19 Nov. 1943, p. 7, CP 'Organisation 1943–44'.

[33] S. P. Chambers to Chairman, 5 May 1954, SCP.

[34] A. J. Quig, 'The Organisation of ICI', paper given at Nuffield College, July 1949, p. 6, CR 93/53/53.

[35] 'Notes of Conversation . . . 21st April 1943', MP 'Organisation'.

[36] SCP.

[37] Letter, 6 Nov. 1944, kindly made available by the late Lord Melchett.

[38] Lord McGowan to various Directors, 23 April 1947, SCP.

[39] Letter, 30 May 1947, SCP.

[40] Sir John Nicholson, 'Division Boards', 8 Nov. 1944, CR 93/3/900.

[41] ICIBM 10,189, 9 Nov. 1944; ICIBM 10,674, 11 Oct. 1945 and supporting document.

[42] As (11), 2nd refce., p. 16.

[43] Education Committee—Report to ICI Post War Committee, Nov. 1943, generally and Appendix E, issued as Appendix (3) to (11), 2nd refce.

PART IV

ORGANIC SYNTHESIS: THE EMERGENCE OF THE NEW CHEMICAL INDUSTRY

1927-1952

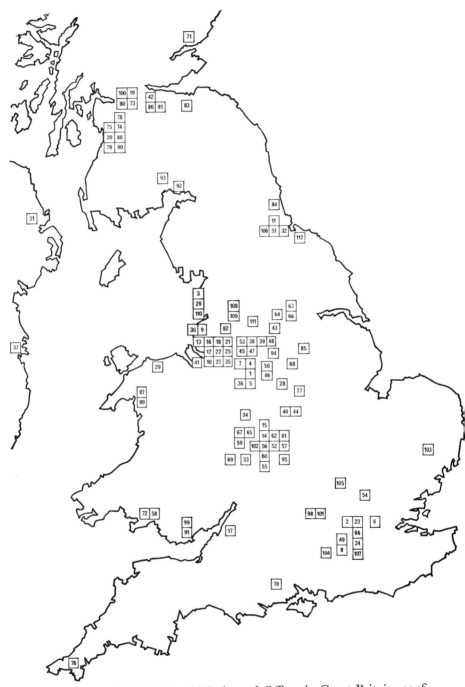

Figure 14. ICI Divisional Works and Offices in Great Britain, 1946.

Source: CR 5/35/35

GROUP A (Heavy Chemicals)

ALKALI DIVISION

1 Div. HQ & Winnington
2 Alfloc Water Treatment
3 Fleetwood
4 Lostock
5 Middlewich
6 Silvertown
7 Wallerscote
8 Sutton Works

GENERAL CHEMICALS DIVISION

9 Div. HQ Liverpool
10 Certain HQ Depts. Runcorn
11 Cassel
12 Castner-Kellner
13 Central Traffic
14 Chance & Hunt (Oldbury)
15 Chance & Hunt (Wednesbury)
16 Gaskell Marsh
17 Netham
18 Pilkington-Sullivan
19 St. Rollox
20 Walker
21 West Bank Power Stn.
22 Widnes Lab.
23 Chance & Hunt Office
24 Heat Treatment, Earlsfield
25 Pest Control, Randle
26 Hillhouse
27 Rocksavage

LIME DIVISION

28 Div. HQ Buxton
29 Raynes

SALT DIVISION

30 Div. HQ Liverpool
31 Carrickfergus
32 Clarence
33 Stoke
34 Tillington
35 Weston Point
36 Winsford
37 Dublin

GROUP B (Dyestuffs & Pharmaceuticals)

DYESTUFFS DIVISION

38 Div. HQ Blackley
39 Blackley
40 Derby
41 Ellesmere Port
42 Grangemouth
43 Huddersfield
44 Spondon
45 Trafford Park

PHARMACEUTICALS DIVISION

46 Div. HQ Alderley Edge
47 Distribution Centre, Manchester
48 Blackley Office
49 Earlsfield Depot
50 Veterinary Research Lab.

GROUP C (Ammonia Only)

BILLINGHAM DIVISION

51 Div. HQ & Billingham

GROUP D (Metals)

METALS DIVISION

52 Div. HQ & Kynoch
53 Broughton Copper
54 Eley Estate
55 Kings Norton
56 Kingston
57 Kynoch Press
58 Landore (BCM) Works
59 Plume Street
60 Selly Oak
61 Holford
62 Lightning Fasteners
63 Armley Road
64 Bradford
65 Fordhouses
66 Oldfield Lane
67 Paul Street
68 Sheffield
69 Steatite & Porcelain
70 Damerham Game Research
71 Riverside Brass
72 Gowerton

GROUP E (Explosives)

EXPLOSIVES DIVISION

73 Div. HQ Glasgow
74 Div. HQ Stevenston
75 Ardeer
76 Bickford Smith
77 Chesterfield
78 Crosslee Mills
79 Kyle
80 Maryhill
81 Regent
82 Roburite
83 Roslin
84 Sabulite
85 Westfalite
86 Westquarter
87 Cookes Explosives
88 Harbour Office, Irvine
89 Miners Safety
90 Portland Glass
91 Curtis's & Harvey
92 Powfoot
93 Dumfries

GROUP F (Paints & Plastics)

LEATHERCLOTH DIVISION

94 Div. HQ & Newton Works
95 Coventry
96 Wandsworth

PAINTS DIVISION

98 Div. HQ & Slough
99 Cardiff
100 Glasgow
101 Slough
102 Smethwick
103 Stowmarket
104 Leatherhead

PLASTICS DIVISION

105 Div. HQ & Welwyn
106 Billingham
107 Croydon
108 Darwen
109 Britannia Works
110 Hillhouse
111 Rawtenstall

112 Wilton (not a Division)

CHAPTER 17

Coal, Oil, or Molasses?
The Problem of Raw Materials

IN THE LATE NINETEEN-THIRTIES it was widely recognized within ICI, centrally and in the Groups, that new possibilities were emerging in the chemical industry. Quite how important these new possibilities would turn out to be, or even their precise nature, nobody could be sure of. What seemed certain was that very attractive opportunities lay in the wider industrial exploitation of organic chemistry—that is, the chemistry of carbon compounds—and its associated engineering, with emphasis on polymerization, condensation, catalysis, and high pressure.

The most spectacular advances seemed likely to be made in fibres and in plastics with the synthesis of materials which could surpass their natural equivalents, if any. News of nylon came from du Pont in 1937 and the first full-scale polythene plant started in 1939. Developments of this sort, however, barely reached large-scale production until the nineteen-forties. What had already developed, during the twenties and thirties, was a lively demand for solvents, synthetic resins, plasticizers, rubber chemicals, wetting agents, and ethylene glycol as anti-freeze. Industrial alcohol, during the thirties, was increasingly incorporated in motor spirit. These were all products which found their main markets in the industries catering for a rising standard of living which had flourished in the twenties and led the way out of depression in the thirties, particularly building, electrical goods, and the motor trade. They entered into the composition of paints, both for house decoration and for car finishes; of rubber for tyres and electrical insulation; and of thermo-setting plastics which found wide application in household goods, electrical gear, and motor-car parts. Ethylene glycol and the constituents of motor spirit were part of the same hopeful scene.

Plain on the technical horizon in 1939, though still some distance from the market place, there was a growing range of other developments. Sulpha drugs had already appeared. Pesticides and selective weedkillers were coming forward from research. The outlines of greatly extended control of the environment were discernible, though no one foresaw the scale on which new products would eventually be manufactured or the remoter consequences of their widespread distribution. Some were becoming available before the end of the war. In the drab

(ICI employees of the forties.) William Tyler, M.M., Acting Safety Officer in the ICI works at Runcorn during the Second World War, by F. E. Jackson, A.R.A.

(ICI employees of the forties.) W. H. Pickford, quarryman of Derbyshire, by Eric H. Kennington.

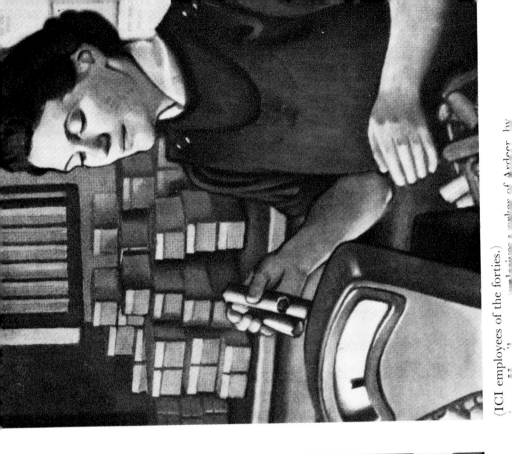

(ICI employees of the forties.)

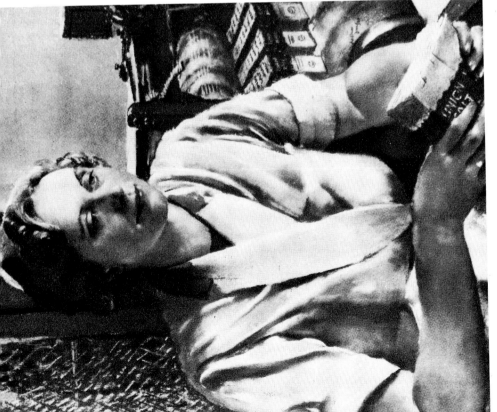

(ICI employees of the forties.)

(ICI employees of the forties.) George Robinson, a 'spreader' in the leathercloth factory at Hyde, Cheshire, by Eric H. Kennington.

(ICI employees of the forties.) James Cunningham, a varnish maker of Paints Division, by Miles de Montmorency.

(ICI employees of the forties.)
F. A. Kendrick, a scientific worker on penicillin by

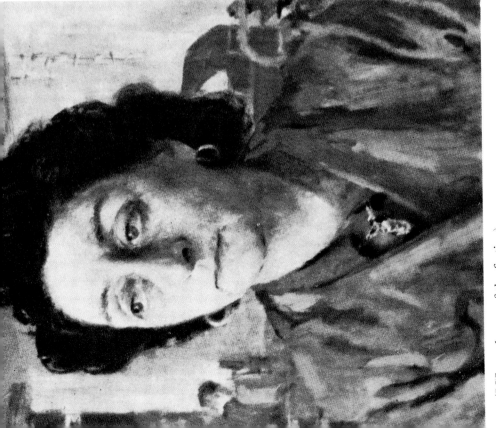

(ICI employees of the forties.)

and battered Europe of 1945 the new materials looked bright and promising, not least on girls' legs as nylon stockings, and in the world at large new drugs, insecticides, and aids to farmers looked like life-savers, as they were. For the time being that was enough.

What was needed as a starting-point for developments of this nature was a range of simple hydrocarbons. They would provide raw materials for the manufacture of substances with varying properties of strength, elasticity, solubility, insulation and so on. This idea in itself was nothing new. The dyestuffs industry and its offshoots had been based since the mid-nineteenth century on hydrocarbon chemistry, using chiefly 'aromatic' chemicals such as benzene, naphthalene, and anthracene derived from coal tar. This was the fine chemical industry, producing small batches of a large number of specialities made by elaborate multi-stage processes.

What was now proposed, and what by 1939 had already been done for a good many years in the field of solvents and related products, was to apply the same chemical principles to the treatment of a different class of hydrocarbon compounds. These were 'aliphatic' compounds not based on coal tar. Three basic materials, all gases, could be used to build up far more complex substances. They were acetylene, ethylene, and propylene. These materials were required cheaply and in large quantities as the basis of a new organic chemical industry which would work them up either into finished products or, more commonly, into the intermediate forms in which the manufacturers of finished products required them. They and their numerous derivatives became known, as early as the nineteen-twenties, as 'heavy organic chemicals'. The word 'heavy' indicated the tonnages which would be required. It was not related to chemical properties.

By the beginning of 1939 the branch of the chemical industry based on heavy organics had become, in the words of L. J. Barley of ICI's Development Department, 'of importance potentially as great as the remainder of the organic chemical industry', although not yet 'organised or even comprehended as a whole'. Du Pont, he thought, were putting about 75 per cent of their research expenditure into the heavy organic field, but there were not many companies concentrating on it. 'Great financial resources for research and plant development during the lean years which new developments almost inevitably pass through', said Barley, were 'necessary for success.'[1]

Barley's message is clear, and it is expressed also in papers written in 1938 and the early part of 1939 by Slade, Akers, and others.[2] A whole new field of the chemical industry was opening, and ICI would be obliged to make an investment decision potentially as important as the decision taken in the mid-twenties to invest in the fertilizer complex at Billingham. Nor could the decision be indefinitely delayed, for powerful rivals were poised to enter the field or had already entered it. Yet the need to make a decision could not have come at a worse moment, for the management of ICI, at all levels, was heavily and increasingly pre-occupied with rearmament and war, in which the development of heavy organic chemicals did not largely figure. As a consequence, for a period of five or six years ICI was obliged to give inadequate attention to the most natural and profitable lines of technical development. American competitors were under no such constraint.

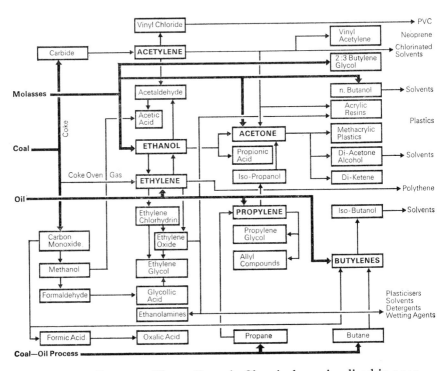

Figure 15. Routes to Heavy Organic Chemicals as visualized in 1944.

Source: Adapted from a diagram in 'The Importance of Ethylene & Propylene to ICI', 17 April 1944, CR 534/3/59.

Before any decision could be taken, the first problem was to pick the right route to the basic heavy organic chemicals: acetylene, ethylene, propylene. The choice was not simple. Three possible source materials could be used: coal, oil, molasses. Coal was the starting point of an established route *via* calcium carbide to acetylene. Oil, by the fractiona-tion and cracking of refinery gases, would yield ethylene, propylene, butylene, and much else. Molasses, fermented, yielded industrial alcohol (ethanol), valuable in its own right and also as a starting point for other products, including ethylene and acetone, an important solvent, and acetates.

Processes based on all three source materials were being worked before the war, and there was a great deal of overlapping between their products (see Fig. 15). Du Pont placed great faith in coal. They wished to avoid dependence on the oil companies, and in the late thirties they were working hard on carbon monoxide, telling J. W. Armit of ICI, in 1938, that 'in the field of carbon monoxide chemistry' they had 'an answer to anything the oil companies can do from hydrocarbon gases'.[3] IG, for their part, set up a factory during the war, at Gendorf, based on the semi-hydrogenation of acetylene (derived, *via* carbide, from coal) to ethylene. Choice between the three main source materials depended on a complex balance of technology, economics, politics—national, international, industrial—strategic planning, and the incidence of taxation. Choices that might be right for du Pont in America or IG in Germany might be quite wrong for ICI in Great Britain.

Coal, for a British firm, had obvious attractions. Ethylene might be derived from it by way of coke-oven gas: other products by du Pont's route through carbon monoxide. Acetylene could be derived by way of calcium carbide, although that required large supplies of cheap electricity. For ICI, with large interests in hydrogenation, coal was especially attractive. The by-products of the petrol process could yield ethylene and propylene, and it had always been foreseen that some parts of the petrol plant itself could be adapted to produce methanol. Strong pressure came from du Pont to rely on coal. 'A long-distance policy based on coal', du Pont told Armit, 'is sounder than one based on oil as there are periodic doubts whether the oil resources have anything but a limited life.' Barley, writing in 1939, took the matter as a foregone conclusion. 'Fundamentally', he said, 'coal is our ultimate raw material and failing oil refinery gases we should aim to base all our organic manufactures on it.'[4]

When Barley wrote, coal was still fairly cheap, there was still some hope that the coal-based petrol process would turn the economic corner, and very little oil was refined in Great Britain—hence his remark about oil-refinery gases. But even with all these influences in its favour, coal was still a doubtful starting material. Acetylene from carbide could never be produced in Great Britain at a price which would compete with prices in Canada and Norway, where they had plenty of cheap hydro-electric power. The shaky economics of hydrogenation depended on the preference over imported petrol, and the preference was a matter of Government policy, which might change. The coke industry in Great Britain was scattered, and du Pont's process *via* carbon monoxide was limited in scope.

What, then, of oil? By 'cracking' a mixture of volatile products which were a by-product of refining it was possible to separate out ethylene, propylene, and a large range of other hydrocarbons. In the USA, where petroleum and natural gas, its near relation, were readily available, there was no doubt about the attractions of 'cracking', no matter what du Pont might say. In the early twenties Union Carbide and Carbon Corporation had decided, as Barley put it, 'to take natural gas in West Virginia'—they had later gone on to buy refinery gases from the oil companies—'and to make [themselves] for the chemistry of aliphatic compounds what the IG Farbenindustrie had been in relation to coal tar derivatives'.[5] The success of ethylene glycol as an anti-freeze for cars and for cooling aero-engines had given them the money they needed for large-scale research, and by 1939 their subsidiary, Carbide and Carbon Chemicals, had a list of over two hundred products, some of which they were offering in 'tank-car lots'. In Great Britain they would have liked to work with ICI. 'Frequent friendly discussions', says a Development Department paper of 1941, 'have taken place.' Nothing happened, however, 'largely on account of our commitments with du Pont'.[6]

By 1939 the oil companies themselves were knocking hard on the door of the chemical industry and looking for partners inside it. In Great Britain there was one obvious partner: ICI. In 1932 the Anglo-Iranian Oil Company suggested joint activities, but they could not offer enough gas to satisfy ICI and their price was high.[7] They came back again, from time to time, from the winter of 1937 onwards. Standard Oil made approaches, too. They too were turned away. The reason, as with Union Carbide & Carbon, was the du Pont Agreement. Du Pont always reacted violently to any possibility that technical information passed to ICI might leak to other firms through agreements for co-operation, and they had an especial dislike for Standard Oil. If ICI must co-operate with an oil company—and du Pont would much rather they did not—then let it be Shell, whose main centre of chemical operations in the USA was in California, about as far from Wilmington as it could be.[8]

However great the prospective attractiveness of coal and oil as source materials for heavy organic chemicals, there was in 1939 only one source material being used in Great Britain on any scale. That was the least likely of them all—molasses—and it had been chosen for a curious reason. It was used for preparing industrial alcohol which, unlike drinkable alcohol, was not subject to excise duty. Nevertheless the makers of industrial alcohol, Distillers Company Limited, had to submit to the inconvenience of close supervision by the excise authorities which prevented continuous working. As a consequence DCL had been allowed, since 1921, to claim an 'inconvenience allowance' of 5*d.* (about 2p) per proof gallon (8·3*d.* per bulk gallon of 66 over proof spirit). No government in the world, apart from the British, paid any allowance of that kind. It made industrial alcohol, in Great Britain but not elsewhere, much the cheapest source material for heavy organics, with the odd result that ICI became dependent for certain important chemicals on what McGowan referred to, rather contemptuously, as 'the whiskey trade'.[9]

In ICI the 'inconvenience allowance' was regarded as an unsound basis for manufacturing operations, because it could be withdrawn at any time if Government policy altered. This consideration must have been equally obvious in DCL, but it did not stop them, in the years between the wars, from setting up a considerable business in industrial alcohol and its derivatives. Alcohol was particularly rewarding in the mid-thirties, because in 1933 the preference in favour of home-produced motor spirit gave DCL yet another reason to bless the beneficence of Government. The preference was intended mainly to help ICI with the oil-from-coal process, but it was applied also to industrial alcohol blended with petrol, even though the alcohol was made from imported molasses which cost more than the corresponding quantity of imported petrol. DCL, therefore, enjoyed (the right word) exemption from excise, the 'inconvenience allowance', *and* tariff protection. Their business in power alcohol went ahead fast on the basis of an agreement with the Cleveland Petroleum Company under which alcohol was blended with benzene in Cleveland 'Discol' motor spirit.

This happy state of affairs lasted until 1938, when the preference on alcohol made from imported molasses was withdrawn. Even then the inconvenience allowance continued and alcohol remained cheap as a raw material. DCL were the main suppliers in Great Britain of solvents, including acetone, and other sugar derivatives, and they prospered on the rising demand. Much their largest customer was ICI, who accounted for about 50 per cent of the country's total solvent consumption.[10] ICI also relied on DCL alcohol for all the ethylene they needed. A plant was put up by ICI at Huddersfield in 1934 to use it as base material for ethylene, ethylene glycol, and ethylene oxide, and this plant supplied ethylene to the first polythene plant at Wallerscote, started in 1939.[11]

At Billingham ICI made methanol and they could perfectly well make solvents if they chose. more especially after the petrol process began to supply them with by-product hydrocarbons. The economics of this branch of the chemical industry, however, were distorted by the 'inconvenience allowance', and as long as that lasted it suited ICI, on the whole, to buy rather than to make. There was no danger of excessive prices because at DCL they knew that unless ICI got favourable terms they would go into solvents themselves. Moreover, DCL was a large customer of ICI, so much so that in the late thirties the balance of trade between the two companies was heavily in ICI's favour:

| ICI purchases from DCL, 1937 | £258,462 |
| DCL purchases from ICI, 1937 | £347,682[12] |

From this point of view also, therefore, ICI had good reason to keep clear of making solvents.

DCL, for their part, were wary of ICI and in general avoided competition, though they had interests in carbon dioxide, which brought them into the same field as ICI's 'Drikold' (solid carbon dioxide), and in plasterboard, based on calcium sulphate, which was a direct competitor with ICI products in building materials. Moreover they had an interest in a salt firm, Charles Moore, evidently in order to make it plain to ICI that they could start making alkali instead of buying it from ICI if they saw any reason to do so. On the whole, however, during the twenties and thirties, relations between the two groups were friendly and were governed by agreements on sales and prices, including an important agreement, signed in 1930, which regulated the manufacture of acetic acid.

The policy no doubt served ICI's commercial interests well enough, especially during the years of depression when there was nothing to spare for a policy of expansion and the last thing anyone wanted was to offend a good customer. But it got in the way of technical progress. Speaking in 1940 of the higher alcohols,* which could readily have been made at Billingham and which would have provided materials for a business in solvents, R. E. Slade said that in the late twenties they had been made in ICI laboratories 'and to a limited extent on our methanol plant . . . but the subject was not pursued because what we had already done enabled us to conclude a satisfactory agreement with the Distillers and so not only work on Higher Alcohols themselves but also on their chemical derivatives was allowed to drop. In the meantime du Ponts had started work on the subject and had pressed on with it so that when we became interested again some eighteen months ago they were by then some distance ahead of us.'[13]

Manœuvres like this, in which promising scientific work was used as a negotiating weapon, and dropped when it had served its purpose, were exasperating and discouraging to Slade and others concerned with research, especially when they were accused of not doing enough original work to justify their keep. No doubt such tactics confirmed scientists in their poor opinion of the use made of science in industry, particularly if their political views were towards the Left. Nevertheless,

* Alcohols in the higher boiling range, between 130 and 230 degrees C.

the use thus made of research was commercially eminently justifiable, and it certainly served to keep down the price of ICI's supplies.

As the Depression passed and the demand for heavy organics became more insistent, the matter of higher alcohols came up again. In October 1937 Kenneth Gordon put to Akers a proposal for plant at Billingham to supply ICI's demand for higher alcohols for paints and lacquers— rather less than one-third of the entire UK consumption. 'It would be difficult', said Gordon, 'to imagine . . . a more natural and logical extension of the present manufactures at Billingham', but 'the ICI does not wish to embark on any capital expenditure not absolutely necessary' and 'the enterprise involves competition with the distillers.' Before they spent any more money at Billingham or went to du Pont for 'the most up to date information' (which ICI would have had for themselves if their research had not been stopped), Gordon wanted to know 'whether either or both of these reasons is regarded at present as a valid objection to continuing further development'.[14]

Starting from this point, the whole subject of heavy organic chemicals was surveyed in ICI during 1938 and 1939, under the general direction of Holbrook Gaskell. The principles which should guide the formation of policy were discussed at a meeting, at which John Rogers presided, in August 1938. 'It is desirable', said the minutes, 'that this field should be developed by ICI in collaboration with all those firms reasonably interested, rather than that competition should be allowed to start.'[15]

That would mean discussions with the oil companies and with Union Carbide, which was likely to be a long and uncertain business. 'The Oil Companies', Akers remarked, 'are not noted for much co-operative spirit amongst themselves'—that is, they did not share ICI's distaste for competition—and, as far as ICI themselves were concerned, 'the du Pont position would have to be cleared'.[16]

DCL also had to be consulted, but with them the way was clearer, largely because ICI and DCL were customers of each other and because it was very obvious that if no agreement were reached, ICI could enter DCL's field without one. By the end of 1938 terms had been settled. A final draft agreement was laid before ICI's Management Board by J. G. Nicholson on 12 December.[17] The two groups had made elaborate arrangements, in a manner then thoroughly characteristic of the chemical industry, for keeping out of each other's way. The agreement defined a 'DCL Characteristic Field', an 'ICI Characteristic Field' and a 'Common Field' (see Table 25, pp. 326-7), and laid down that 'in the case of well established products the term "characteristic" shall in practice imply exclusivity'. If, therefore, either party should wish to enter the other's field, there would have to be consultation. As the agreement put it, 'the party contemplating the new manufacture shall not commit itself . . . without exhaustive prior discussion with the

TABLE 25
DISTILLERS COMPANY LTD./ICI AGREEMENT, 2 FEB. 1939
PRODUCT FIELDS

D.C.L. CHARACTERISTIC FIELD

1. POTABLE SPIRITS
2. YEAST
3. ADHESIVES BASED ON STARCH
4. GLASS BOTTLES—The ICI interest in this field arising from their control of the Portland Glass Co. is recognised.
5. PHARMACEUTICALS AND OTHER PREPARATIONS BASED ON YEAST
6. ALCOHOL DENATURANTS—The ICI interest in this field is recognised.
7. CHEMICALS

    (a) *Aliphatic Alcohols*

Ethyl alcohol—including	Iso. Butyl alcohol
Methylated Spirits	Amyl alcohol
N. Propyl alcohol	Diacetone alcohol
N. Butyl alcohol	

    (b) *Aliphatic Aldehydes*

Butylaldehyde	Aldolaldehyde
Crotonaldehyde	

    (c) *Esters*

Various Esters of:	Amyl alcohol
Ethyl alcohol	also:
Propyl alcohol	Aceto Acetic ester
Butyl alcohol	Oxalacetic ester

    (d) *Aliphatic Ketones*

Acetone	Mesityl oxide
Methyl ethyl ketone	

    (e) *Miscellaneous*

Cellulose triacetate	Acetaldol

ICI CHARACTERISTIC FIELD

1. EXPLOSIVES AND BLASTING ACCESSORIES FOR ALL INDUSTRIAL AND MILITARY PURPOSES
2. NON-FERROUS METALS AND ALLOYS AND ARTICLES MADE THEREFROM
3. LIGHTNING FASTENERS
4. AMMUNITION
5. PAINTS AND LACQUERS
6. LEATHER AND RUBBER CLOTH
7. DYESTUFFS, PIGMENTS AND COLOURING MATTERS, INCLUDING PRIMARIES AND ESSENTIAL INTERMEDIATES
8. PHARMACEUTICALS—EXCEPT THOSE BASED ON YEAST
9. ADHESIVES—EXCEPT THOSE BASED ON STARCH
10. CHEMICALS

    (a) All Inorganic Chemicals including Cyanides—
       Distillers' interest in Sodium Chloride and brine purification products arising from their ownership of Chas. Moore is recognised.

    (b) Primary Hydrogenation Products including Methanol

    (c) Organic Acids and their Salts

Formic Acid	Picric Acid
Oxalic Acid	Chloracetic Acids
Benzoic Acid	Maleic Acid
Phthalic Acid	

TABLE 25

DISTILLERS COMPANY LTD./ICI AGREEMENT—*continued*

(d) Miscellaneous Organic Chemicals

Ether	Chloracetophenone
Ethyl Nitrate	Dimethyl Acetal
Glycol Cellulose	Tetra ethyl lead
Methyl Cellulose	Chlorinated Benzene and derivatives
Nitro Cellulose	Chlorinated Toluene and derivatives
Benzyl Cellulose	Chlorinated Naphthalene and
Gelatine	derivatives
Osseine	Chlorinated Ethylene and derivatives
Formaldehyde	Chlorinated Acetylene and derivatives
Metaldehyde	Chlorinated Ethane and derivatives
Benzaldehyde	Halogenated methane and derivatives
Urea	'Alloprene'
Phthalic Anhydride	'Cereclor'
Nitrophenol	Phosgene
Benzoyl Chloride	Leucotrope
Benzoyl Peroxide	

(e) Synthetic Organic Chemicals used in the Textile Finishing Industries, and in the Rubber, Oil and Photographic Industries.

(f) Synthetic Organic Chemicals used as Detergents, Wetting and Emulsifying Agents, and as Corrosion Inhibitors for metals.

(g) Sulphonated Oils used in the Leather Trade and Synthetic Organic Tanning Agents.

(h) Synthetic Organic Chemicals used as Flotation Agents in the Mining and Allied Industries.

11. SYNTHETIC RUBBER AND RUBBER-LIKE POLYMERS

12. PEST CONTROL PRODUCTS AND PLANT GROWTH REGULATING AGENTS

13. SYNTHETIC RESINS

Substituted Acrylic Acid Series resins and moulding powders

Glyptal resins and compositions

Polymerised Ethylene and Polymerised Isobutylene, resins and allied and/or derived products

COMMON FIELD

1. CARBON DIOXIDE

2. BUILDING MATERIALS

3. ACETIC ACID

Also: Acetic Anhydride
Acetaldehyde
Paraldehyde

4. ETHYLENE OXIDE

5. ETHYLENE GLYCOL

6. MOTOR FUELS

7. SYNTHETIC RESINS

(a) Phenolformaldehyde resins and moulding powders

(b) Urea formaldehyde resins and moulding powders

(c) Styrene resins and moulding powders

(d) Vinyl resins and moulding powders

Source: ICI Legal Dept.

other party [in order to] reach an agreement which, while recognising the natural aspirations of both parties, will react to the maximum natural benefit of the parties and which will not stand in the way of technical advance and/or logical economic development.' The Agreement was signed in the first place for two years but 'it is the intention of the parties that the Agreement shall be perpetual'. What could be more gentlemanly? This agreement was the last of a long line of major market-sharing agreements. On technical grounds it was out of date as soon as it was made, being based on far too narrow a view of the rapidly expanding prospects opening in the heavy organic industry. It was no sooner in force than ICI gave notice under its terms, not much to DCL's liking, that they proposed to put up a plant for 600 tons a year of higher alcohols, thus forcing a broad review of 'the mutual spheres of interest of ICI and Distillers, with a view to making proposals for rational and, in suitable cases, joint production of products within the range'.[18] That review was interrupted by the outbreak of war. In 1944, when ICI's view of the whole subject of heavy organic chemicals had altered and broadened, they came to the conclusion that if the DCL agreement were to break down ICI could regard the position 'with equanimity'.[19]

REFERENCES

[1] L. J. Barley, 'The Development of Heavy Organic Chemicals including Plastics by ICI', 16 Feb. 1939, DECP.

[2] Papers on heavy organic chemicals:
  (i) W. A. Akers, 'Heavy Organic Chemical Field', 2 Aug. 1938, CR 261/11/11.
  (ii) 'UK Heavy Organic Field—Meeting to discuss Policy', 3 Aug. 1938, CR 261/11/11.
  (iii) R. E. Slade, 'Proposals for the Development in ICI of certain heavy organic chemicals (including Plastics)', 6 Feb. 1939, DECP.
  (iv) As (1).
  (v) L. J. Barley, 'Development Policy in the Aliphatic Organic Chemical Field with special reference to Raw Materials', 26 April 1939, DECP.

[3] J. W. Armit, 'The Du Pont Company's Views on Chemical Products from Hydrogenation Gases', Nov. 1938, p. 2, CR 261/2/13.

[4] As (1), p. 13.

[5] Ibid. p. 4.

[6] E. D. Catton, 'ICI Relations with other Companies interested in the UK Market', 8 Aug. 1941, p. 9, CR 261/11/11.

[7] As (2 i), pp. 2–3.

[8] Correspondence between F. Walker and E. J. Barnsley and associated papers, 26 June 1940, CR 261/11/11.

[9] Lord McGowan to Lord Melchett, 4 April 1941, MP 'Post War Problems'.

[10] As (2 i), p. 1.

[11] As (6), p. 3.

[12] J. W. Armit, 'ICI and DCL Relationship', Feb. 1938, CP 'Distillers'.

[13] Research—Minutes of a Meeting held 2 July 1940, CR 50/110/110.

[14] K. Gordon to W. A. Akers, 14 Oct. 1937, CR 45/14/14.

[15] As (2 ii), p. 2.

[16] As (2 i), pp. 5–6.

[17] J. G. Nicholson to Management Board, 'Distillers Company Ltd and ICI: Relationship', 12 Dec. 1938, MBSP.

[18] A. Fleck, 'Proposal for a Plant to manufacture 600 Tons/Year of higher Alcohols', 20 March 1939, MBSP; J. H. Wadsworth to DEC, 25 Jan. 1940, CR 45/14/39.

[19] 'The Future of ICI/DCL Relations', 20 July 1944, CR 161/10/10.

CHAPTER 18

Dyestuffs Group in the Thirties

ORGANIC SYNTHESIS, in ICI before the Second World War, was chiefly a matter for Dyestuffs Group. The history of the Group, going back before the foundation of ICI, was unhappy; ICI's main effort, early on, went into the nitrogen industry; and Dyestuffs Group, until the early thirties, was regarded as being of no great value, except perhaps as a bargaining counter.

Then the nitrogen industry collapsed and new prospects of organic synthesis, almost simultaneously, began to open up. Dyestuffs Group already made its living from the fine chemical industry, so that rubber chemicals, drugs, pesticides, and synthetic resins looked like natural developments. On the heavy organic side the chemical technology—though not the engineering—was related to the Group's traditional activities. Therefore as soon as the importance of the new developments in the chemical industry began to be appreciated Dyestuffs Group moved from the circumference to the centre of ICI's activities. We must examine the early stages of the transition.

Dyestuffs Group would have been in no position to develop at all, in the early thirties, if it had not been for two measures in restraint of international trade, one devised by the British Government and the other by the world's main dyestuffs makers. The British Government, in 1920, passed the Dyestuffs Act to close the British market against German competition, and when the Act was in danger of being allowed to expire, in 1930, its life was saved, against the wishes of the textile trades, by a campaign expertly directed, in Parliament and elsewhere, by Henry Mond and the public relations staff of ICI.[1] The dyestuffs makers protected themselves with the Cartel Agreement of 1932 (p. 194 above), which acknowledged the newly established technical competence of British dyestuffs makers and allowed them to compete with the Germans and the Swiss on the quality of their products without being harassed by unregulated price cutting.

With the home market protected by statute while foreign trade was regulated by the cartel agreement, ICI's dyestuffs business prospered. Figures prepared in 1937 (Table 26) show sales value of all products not far short of twice as great in 1936 as in 1930, and a rise of about 45 per cent, over the same period, in the average price realized for dyestuffs. Costs were well under control—at Grangemouth, in three years, the number of workers fell from 842 to 545 while production doubled[2]—and the recipe for success was complete. 'Within the last six years', the Group management told the ICI Board in 1936, 'the financial results of the Group have crossed over from a loss to a continually improving profit.'[3] The improvement was from a loss of £205,897 in 1930 to a profit of £938,444 in 1936, and that figure was more than ten times the profits of the late twenties (Table 27).

TABLE 26

ICI DYESTUFFS GROUP

SALES AND AVERAGE REALIZED PRICES, 1930–1936

	Total Sales	Sales of Products other than Dyestuffs	Average realized Prices, Dyestuffs only
	£	£	d. per lb.
1930	2,202,535	541,312	16·028
1931	2,516,228	581,197	16·742
1932	2,972,817	748,523	19·305
1933	3,584,507	926,445	22·418
1934	3,785,873	980,596	23·268
1935	4,069,513	1,111,343	23·491
1936	4,262,156	1,219,528	23·000

Source: Dyestuffs Group Memo prepared for a meeting between the General Purposes Committee and the Group, 20 July 1937, GPCSP.

TABLE 27

ICI DYESTUFFS GROUP

CAPITAL EMPLOYED AND PROFITS EARNED, 1928–1936

	Capital Employed	Profit[a]	Loss
	£m.	£	£
1928	4·0	96,393	
1929	3·8	90,046	
1930	4·1		205,897
1931	4·4		51,567
1932	4·5	452,521	
1933	4·4	636,728	
1934	4·3	818,779	
1935	4·3	900,895	
1936	4·7	938,444	

a Profits are before Income Tax, Central Services and Obsolescence.

Source: 'Dyestuffs Group: Statistical Information', 12 Dec. 1938, CP 'Dyestuffs General'.

Since the figures in both Tables are from contemporary sources they do not agree exactly with figures in Appendix II, calculated in 1970. In particular, profit figures in Appendix II are shown before tax but after other charges.

ICI's dyestuffs business prospered in this way in spite of deep depression among its largest group of customers, the British textile industries, especially cotton. Between 1928 and 1935 there was a great increase, behind rising tariffs, in the manufacture of textiles in countries which had been large importers from Great Britain. Moreover, the Japanese competed fiercely and successfully for whatever trade remained. In India, the biggest British market of all for cotton textiles, the output from cotton mills in 1935–6—3,571,000,000 yds.—was a record, and imports fell to 946,000,000 yds., against 3,000,000,000 in 1913–14. In the foreign trade of colour-using industries as a whole, taking account of woollens, silks, linoleum, leather, and paint as well as cotton, the export value of British goods fell from £164,184,000 in 1928 to £74,595,000 in 1933, and then rose slightly to £79,334,000 in 1935.[4]

British textile industries took 72 per cent of Dyestuffs Group's home sales in 1935.[5] The textile-makers' grief, nevertheless, was to some extent ICI's gain. Foreign makers who were ruining firms in Britain had to get their goods dyed, and there were not many suppliers of dyestuffs in the world. ICI was one. In 1935 ICI's trade in exported dyestuffs, at about £1m., was half as great as their trade at home, and it included turnover of £250,000 in India, £170,000 in China, £110,000 in Australia, and £70,000 in Canada. These were the largest markets.[6]

Foreign trade was less profitable than trade at home. The cartel agreement protected ICI against price-cutting by the continental makers and there were arrangements with the Americans in China, but there was nothing to restrain the Japanese, and in China and India prices had to be reduced to match theirs. Compensation, however, was provided by economic nationalism in smaller markets: '. . . we have interesting though small developments [in 1935] taking place in Egypt, Turkey and other minor countries, where national ambition to become self-contained is leading to an increased consumption of dyestuffs for producing their own requirements in textiles and coloured papers etc.'[7]

However irritating the Japanese might be, then, the prospects of the Dyestuffs Group abroad, in the mid-thirties, were hopeful, largely because the Germans and Swiss, like Dyestuffs Group itself, were restrained by the cartel agreement from price competition. That had the effect, as we have already had occasion to notice briefly, of pushing competition firmly towards technical development. 'There can be no doubt whatsoever', says a document of 1936, 'that the tendency of trading agreements has been to place greater emphasis than ever on the technical aspects of competition, particularly on novel lines of development.'[8]

This development, by the mid-thirties, was running a good deal wider than dyestuffs themselves and their intermediates. It is central to our theme—the changing nature of ICI's business—and we must now examine it in some detail.

Dyestuffs Group in 1936 was very much stronger technically than any British dyestuffs maker had ever been. Nevertheless, the management were well aware of the challenge which the Group still faced:

Our main problem is the highly competitive character of research work in organic chemicals. The IG believe so whole-heartedly in this field that their research effort is preponderating in this field. This is indicated by the patents taken out by the IG, almost three-quarters of which are in fields in which the Dyestuffs Group is interested, i.e., dyestuffs, intermediates, textile chemicals, rubber chemicals, and synthetic materials for paints and varnishes, including three important fields closely related to the above, in which ICI up to the present, has taken no active part, i.e., pharmaceuticals, photographic chemicals, and synthetic rubber. [9]

The object of this 'competitive research' was to exploit what Dyestuffs Group called 'the liability of organic chemical invention to create profitable monopoly'—or to prevent other firms from doing so. 'Painful evidence' of IG's strength was 'furnished by the frequent experience of finding that when we [Dyestuffs Group] do succeed in opening up a new line of work, the IG are already there, setting up the inevitable patent barrier.' [10]

IG showed what they could do with aggressive research in 1934, when Dyestuffs Group launched 'Monastral' Fast Blue, the Group's invention and the first of the phthalocyanine pigments. The patent position was unassailable by direct assault, so IG outflanked it and in the same year launched 'Heliogen' Blue B, a competing phthalocyanine pigment made by methods which ICI's patent lawyers could not touch—'a revelation', as Dyestuffs Group remarked, 'of the resources they are able to throw into any opening for research, whether provided by their own initiative or by that of their competitors'. [11] ICI's hopes of 'profitable monopoly' were blasted, and the best they could do—a much less lucrative option—was to come to an agreement with IG, in 1936, for cross-licensing of rights and exchange of technical information on phthalocyanines. [12] This was technical competition, as fostered by the cartel agreement of 1932. By comparison with price competition it lacked nothing in ferocity and it was certainly much better calculated to stimulate inventive powers.

Dyestuffs Group's own research was cut back in 1930 along with other research in ICI. Even in its diminished state it held such promise of 'new discovery and invention' that in 1933 it was decided to spend an extra £20,000 spread over the years 1933 to 1935, and over the same three years the research staff was increased by about twelve chemists a year, which must have represented a total increase of nearly 50 per cent. At the same time the Group was making grants for fundamental research work in the universities. In 1936 it was subsidizing 41 research students, supervised by ten professors, and 32 other 'academic gentlemen'. The development of 'Monastral' Fast Blue, among other things, the Group said, would have been impossible without this kind of work, which was gratifyingly cheap in comparison with the Group's own research activities.[13]

In the three years 1933 to 1935 87 new products were added to Dyestuffs Group's manufacturing range and 912 more products, worked out in the laboratory, were approved for manufacture. These products included about twice as many 'novelties'—new inventions—as imitations of competitors' goods, whereas between 1930 and 1932 the ratio had been the other way round. The greater originality, the Group considered in 1936, was a direct result of the 1933 decision to spend more money.

TABLE 28

DYESTUFFS GROUP

EXPENDITURE ON RESEARCH AND PATENTS, 1932–1935

	Research	Patents
	£	£
1932	97,000	12,204
1933	99,000	11,024
1934	107,000	12,701
1935	119,000 (approx.)	13,028 (approx.)

Source: Memo prepared to serve as a background for the Meeting between the General Purposes Committee and the Group, 11 Feb. 1936, Annex C, GPCSP.

The management of Dyestuffs Group must have been well aware that within ICI there were powerful critics of research expenditure, ready to pounce on any activity that could not be shown to earn its keep. 'Novel developments', they said in 1936, 'are generally slowish in getting under way, but when successful mount up to considerable figures.'[14] The financial returns from research expenditure are notoriously difficult to calculate, but the kind of business which Dyestuffs Group were advocating in the thirties was a small-scale model of the chemical industry that was to develop on a very large scale between ten and twenty years later, in so far as it relied on the vigorous exploitation of patent rights while they lasted.

In the development of dyestuffs themselves, the cartel agreement limited ICI's possibilities of expansion within the bounds of the 8·4 per cent sales quota (p. 194 above). In the mid-thirties Dyestuffs Group

333

had not, in fact, managed to claw its way up, past continental competition, to the permitted figure, but even if it managed to do so the extra turnover would only be about £250,000 a year, a prospect which the Group management did not find very exciting. Moreover, as they pointed out to the ICI Board, 'our selling activities . . . are so hedged round with regulations and restrictions, limiting our sales of this product to certain customers, and preventing our active competition in others, that it becomes a very complicated problem'.[15]

For these reasons, by 1936, other products of organic chemistry were beginning to look more attractive than any but the most profitable dyestuffs. 'When we consider products other than dyestuffs,' runs the document just quoted, 'our scope is almost unlimited, and it is not an exaggerated anticipation to suggest that the date is not far distant when the sales of these materials overshadow dyestuffs themselves.'

'These materials' divided broadly into two classes: those that were dyestuffs' intermediates and those that had no functional connection with dyestuffs at all, and in 1935 Dyestuffs Group's sales of all products other than dyestuffs came to about £1·1m. against dyestuffs sales of nearly £3m.[16] Sales of intermediates, at about £330,000, were important and, being free of cartel restrictions, energetically promoted, but it was from the other class of products that the most attractive developments were expected.

Sales of the most promising of these products were rising fast all through the early thirties. None, with the exception of rubber chemicals, were really large sellers, even by Dyestuffs Group's standards, but some of the rates of increase shown in figures prepared in 1936 were dramatic:[17]

Sales of:	1928 £000	1931 £000	1933 £000	1935 £000
Rubber Chemicals	91	170	300	301
Resins and Lacquers	—	3	22	53
Detergents and Wetting Agents	1	4	20	49
Phthalic Anhydride	9	11	26	45
Agricultural Products	—	3	16	28
Pharmaceuticals	4	12	8	13

Rubber chemicals went chiefly to tyre makers, especially Dunlop, as vulcanization accelerators, anti-oxidants, plasticizers, and for other purposes in the manufacture of tyres. Synthetic rubber itself the Group did not make, though the management watched IG and du Pont in this field with considerable nervousness, and in 1936 'the view was advanced that it would be thoroughly unsound for ICI to defer any longer an attempt to contribute to technical development in the synthetic rubber

(ICI employees of the forties.)
G. H. Garner, a canal worker at Oldbury, by
B. Fleetwood-Walker, A.R.A.

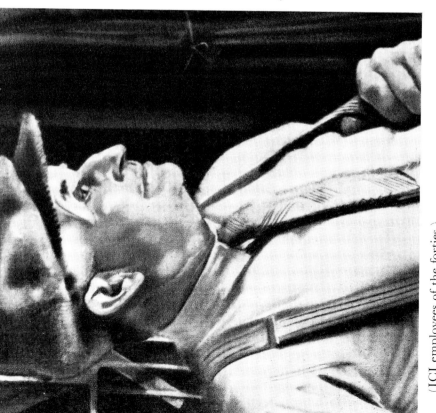

(ICI employees of the forties.)
John Lugg, F.R.H.S., head gardener at Jealott's Hill,
by Eric H. Kennington.

In the Intermediates Department, Dyestuffs Works, Huddersfield, 1935.

A pair of buckles made from the first 'Perspex' ever produced (page 346).

field'.[18] A year later they were still hesitating, and indeed the attractions of synthetic rubber, with the natural produce of Malaya readily available, were by no means self-evident. For the Germans, under the threat of being cut off from the East in wartime, the matter was different.

Synthetic resins and lacquers, organic and inorganic pigments, went to paint manufacturers, including ICI, who supplied both house decorators' materials and finishes for cars. Paints and lacquers brought ICI, somewhat gingerly, fairly close to the retail customer, but a direct advertising approach, it was felt, would annoy decorators and their suppliers, who were the most important customers in the household trade—hence, as late as the early fifties, the slogan, in consumer advertising, 'Say Dulux to your Decorator'. The day of 'do-it-yourself' was not yet. The textile industry bought detergents and wetting agents, chiefly for scouring wool. There was no thought of selling them to housewives.

'Agricultural Products' were chiefly pesticides, which provided an early example of the organizational difficulties caused by the complexity of new branches of the chemical industry. Two other Groups—General Chemicals and Fertilizers—were interested in pesticides besides Dyestuffs. Their welfare was watched over by an intricate network of committees, an arrangement which pleased none of the Groups and hindered rather than helped development, since no Group felt inclined to put its whole weight behind a project which it did not wholly control.

Outside ICI altogether, pesticides fell in territory disputed between ICI and Cooper McDougall and Robertson, with whom on other matters ICI somewhat uneasily co-operated. In October 1937 a perpetual agreement between the two companies came into force. ICI took over Coopers' manufacturing activity in pesticides and their factory at Yalding in Kent, and a joint (50–50) company, Plant Protection Limited, was set up to sell the output.[19] 'The result', the General Purposes Committee hopefully minuted, '. . . would be that competition between ICI and Coopers would be eliminated, there would be a common selling effort . . . in a direction in which ICI is inexperienced and there would be one manufacturer instead of two.'[20] In fact, the arrangement was most unpopular with the ICI Groups concerned, particularly Dyestuffs Group, and the joint ownership of PPL, which lasted until 1953, was never harmonious. It was an early illustration, of which British Nylon Spinners was a later and larger one, of the dangers inherent in partnership agreements, and it showed up the absence, in ICI, of any unified organization capable of taking charge of the projects coming forward, in the organic field, in the late nineteen-thirties.

335

M

By 1936, then, Dyestuffs Group had built up a list of widely diverse products, serving widely diverse interests, which gave an indication both of the kind of opportunities and the kind of problems with which ICI would increasingly be faced as the chemical industry developed. This list of products was not the outcome of long-range planning on a grand scale, like the activities at Billingham. Indeed it is difficult to maintain that there was any planning at all, when in 1936 Dyestuffs Group itself could say to the ICI Directors:

We are faced . . . with the problem of deciding whether to pursue an opportunist policy in exploiting whatever situations present themselves as apparently favourable, with a quick change of ground in the absence of early promising results, or definitely to restrict the ground we cover, leaving some subjects deliberately untouched. This till recently has been our policy, for example, with regard to synthetic rubbers.[21]

Products of this kind came to be made, in the first place, because they fitted the technology of the dyestuffs industry at a time when the dyestuffs industry in Britain was struggling, by any means that offered, to survive in a world that could get more dyestuffs than it needed. The worst was over in the mid-thirties, though scarcely long enough over to be comfortable, and with slightly more money available, the promise held out by the versatility of dyestuffs technology was even greater.

Then, after the making of the cartel agreement in 1932, the technical argument was reinforced by the tactical one. If the Group wished to expand its turnover outside quota limits it could do so most conveniently with products that were not dyestuffs, and this was the immediate reason why they were taken up with enthusiasm in the mid-thirties. It was an opportunist policy, and there is no evidence that anyone at the time was looking far ahead. They were pleased enough with what they had. In July 1937 McGowan told the assembled management of Dyestuffs Group, under their Chairman, Francis Walker, 'that they had come together remembering the excellent results which had been achieved by the Dyestuffs Group in the past year, and at a time when the future seemed full of promise. He stressed the importance of new developments and the rewards which would come from fresh enterprise.'[22]

REFERENCES

[1] Lord Melchett, 'History of getting the renewal of the Dyestuffs Act, 1930', 29 Nov. 1933, MP 'Dyestuffs Act 1930-33'.
[2] 'Memorandum prepared to serve as a background for the meeting between the General Purposes Committee and the Group', 11 Feb. 1936, p. 11, GPCSP.
[3] Ibid. p. 1.

[4] Figures from Dyestuffs Group, 'Memorandum prepared as a background for the meeting between the General Purposes Committee and the Group', 20 July 1937, GPCSP; see also A. R. Smith, 'The economic position of the British cotton and wool industries', 19 Oct. 1938, CP 'Dyestuffs General'.

[5] As (2), p. 18.

[6] 'Agenda for meeting with the General Purposes Committee on 11 Feb. 1936'—Dyestuffs Group Brief for Delegate Directors of the Group—'Sales Development', p. 3, GPCSP.

[7] Ibid. p. 3.

[8] As (2), p. 8.

[9] Ibid. p. 7.

[10] Ibid. pp. 5–7.

[11] Ibid. pp. 7–8; see also D. W. F. Hardie and J. D. Pratt, *A History of the Modern British Chemical Industry*, Pergamon, Oxford, 1966, p. 164; J. D. Rose, Levinstein Memorial Lecture, *Chemistry and Industry*, 28 Nov. 1970.

[12] F. Walker to Central Administration Committee, 'Phthalocyanines', 15 July 1936, CR 75/33/33, gives the substance of the agreement.

[13] As (2), pp. 4–6.

[14] Ibid. pp. 4–5.

[15] As (6), p. 3.

[16] As (2), p. 3.

[17] Ibid. Annex C.

[18] 'Notes on Meeting of the General Purposes Committee held 11th February 1936, attended by the Delegate Directors of the Dyestuffs Group and Others', Agenda No. III (iii), GPCSP.

[19] H. D. Butchart to Central Administration Committee, 'Plant Protection Ltd', 14 July 1937, CR 53/101/101.

[20] GPCM 1134.

[21] As (2), p. 8.

[22] 'Notes on Meeting of the General Purposes Committee held 20th July 1937 . . . attended by the Delegate Directors of the Dyestuffs Group and Others', p. 1, GPCSP.

CHAPTER 19

Mid-Century Materials 1: Plastics

(i) THERMO-SETTING HESITATIONS: THE ORIGINS OF PLASTICS GROUP
'PLASTIC MATERIALS'—the phrase comes from an ICI document written in 1927[1]—were another manifestation, like the products of the Dye-stuffs Group, of the transformation which was beginning to come over the chemical industry in the thirties. They were (and are) materials produced artificially to compete with natural materials, especially wood, metal, rubber, leather, and various textile fibres. They found a market either because the natural material was dearer or because the synthetic product had properties which the natural material lacked. They belonged to the organic side of the chemical industry, they were at their most profitable while they were covered by patent rights, and as a consequence they needed substantial and continuous research backing.

These implications were not all immediately obvious. In October 1927 a committee under L. J. Barley was appointed 'to report to the Executive Committee upon the future of ICI in relation to the plastic materials industry'. What their report describes is not a field for research and innovation but a promising market for materials which ICI was already producing or could produce: 'The Committee', says the report, 'considers the Plastic industry as a whole to be one of great potentialities, using raw materials and processes which belong essentially to the chemical industry of this country, the chief of the raw materials being products of fundamental importance to ICI.'[2] What Barley's committee was talking about was a fairly narrow range of products which relied either on nitrocellulose or on materials such as formaldehyde and urea which the Billingham complex was well adapted to supply.

Of these products the oldest was celluloid, which had been on the market in one form or another since the eighteen-fifties. It was a horny material which could be put to all kinds of uses, from knife handles and shabby-genteel stiff collars to photographic film, but it suffered from being regarded, with some justice, as a cheap and nasty substitute for other materials and it was highly inflammable. This was not surprising since it was almost an explosive, being based on nitrocellulose and closely related to cordite. For this reason it might have been attractive to ICI as a support for Nobels' flagging trade in explosives, but by 1927 the market was over-supplied with celluloid and there were better plastics available.

These were the plastics which chiefly interested ICI. They were synthetic resins, which had applications both as solids and as liquids. Glyptal resins, based on glycerine and phthalic anhydride, were chiefly of consequence in paints and lacquers. The resins which could be adapted to produce solid constructional materials were compounds of formaldehyde with urea or phenol. It was these on which the earliest foundations of ICI's plastics business were laid.

From either phenol formaldehyde or urea formaldehyde, with a woodmeal 'filling', powder could be produced which would set hard in a heated mould and keep its shape when it cooled. Phenol formaldehyde, which for many years had been marketed as 'Bakelite', was always dark in colour, but urea formaldehyde was colourless and translucent and the mouldings produced from it, from 1928 onwards, under the British Cyanide Company's brand 'Beetle', could be coloured according to taste or left translucent. From ICI's point of view, the importance of both materials was that formaldehyde (from methanol) and urea (from ammonia and carbon dioxide) were both attractive products for Billingham, especially after the nitrogen market collapsed and the Billingham management was looking for every conceivable way of putting plant to use. The technology of these plastics would present no difficulty. It was familiar at Blackley and at Billingham.

What may have been even more important than these considerations, although in 1927 it was not stressed so heavily within ICI, is that in the late twenties and early thirties the trend of the market was strongly in favour of bakelite and other thermo-setting plastics. In some parts of the country there was deep depression, but elsewhere standards of living were rising, because earnings in general held up while prices fell, and consumer-goods industries did well while basic industries did badly. These were the years when car ownership was becoming commonplace in the middle classes; when the point was being reached at which nearly every household in the country would have a wireless set; when household electrical gadgets were selling well. Moreover, in the thirties there was a building boom, with the accompanying demand for electric light switches, door knobs, and other household equipment, to say nothing of the 'fancy goods', as ICI called them, which lay about in the new houses when they were built.

In all the industries which supplied these demands there was scope for plastics, partly to replace wood and metal, partly to make things which had never been needed before, or never in such quantities, such as motor-car parts and electrical fittings. It was essentially a consumer-goods market—precisely the sort of market in which ICI felt least at home and of which it was inclined to be slightly contemptuous. In the late twenties most of ICI's energies—and capital—were being directed to quite different and more highly regarded purposes, especially the improvement of agriculture in the British Empire. Plastics, to minds brought up in the traditions of the heavy chemical industry, were hardly to be regarded as a worthwhile branch of the chemical industry at all, and that no doubt partly explains why ICI's approach to plastics was so slow that even as late as 1935 Hodgkin complained to the General Purposes Committee that ICI, having 'come late into the industry', was still 'tailing along in the rear'.[3]

The appointment of Barley's committee, in 1927, was prompted by a Franco-German approach. Rheinisch-Westfälische Sprengstoff, an old associate of Nobels, now controlled by IG, and Société des Matières Plastiques, an offshoot of the French Nobel company, both had interests in celluloid and in the newer plastics, which they regarded with more hope for the future. Dr. Müller of the German company and M. Le Play of the French one, meeting A. G. Major of ICI in Paris in July 1927, wanted to know whether ICI would co-operate with them if they extended their activities to British markets. This they intended to do in any case, but, as Müller put it, 'he was . . . following out the general agreed principle that all friends, before taking any steps in each others' territory, should consult with one another'.[4]

In Great Britain celluloid was made by the British Xylonite Co. Ltd., old-established (1877), strong in assets, but not, in 1927, flourishing, because continental competition had driven prices down from 4s. (20p) a lb. to 2s. 6d. (12½p) in eighteen months and profits had dropped from £45,000 in 1924 to £20,000 in 1926.[5] The main makers of bakelite were the Damard Lacquer Co. and Mouldensite Limited, both controlled by the Bakelite Corporation, an American company in which ICI had an indirect minority interest. There were several smaller firms making bakelite, including one—Synthite Limited—controlled by the Distillers Company Limited, but none of them was highly regarded in ICI. Erinoid Limited had a substantial business in plastics made from casein formaldehyde, and Beetle Products Limited, controlled by British Cyanide (British Industrial Plastics after 1936), were on the point of putting thio-urea formaldehyde on the market.[6]

Barley's committee suggested that in two or three years—that is, about 1930—the output of the entire British plastics industry, in 'saleable semi-finished products', might amount to 6,500 tons, giving a turnover of £1·4m., of which bakelite would account for £600,000. Figures of this order were unlikely to create much of a stir in ICI when sales from Billingham in 1927 were already £1·2m. and were expected to grow rapidly. The committee, nevertheless, pressed their point about raw materials: 'the probable demand for the more important, in particular nitrocellulose (2,000 tons) and formaldehyde (1,500), is so large that steps should be taken by ICI to safeguard its position as potential suppliers to the industry.'[7] On this basis the committee made

340

a recommendation in the expansive style characteristic of the Mond age in ICI. What they suggested was nothing less than 'the control of the [plastics] industry within the Empire'. The capital cost they put at £1m. in ICI shares and cash. The shares would be required because an essential feature of the plan was international regulation of the trade in celluloid—'an agreement between the chief manufacturing Companies in the world would be of great value in limiting production and obtaining higher prices and profits'[8]—and for that purpose ICI must gain control of British Xylonite. Moreover, control of British Xylonite would give ICI an entry into the more rewarding market for thermo-setting plastics.

British Xylonite refused to be controlled. Their Managing Director was willing to issue to ICI, in exchange for ICI shares, enough new shares to give ICI a 20 per cent interest, with a seat on British Xylonite's Board and an agreement by British Xylonite to buy raw materials exclusively from ICI, but further than that he would not go, and he wanted from ICI an undertaking not to 'enter the celluloid field'. Mitchell recommended the bargain to the Executive Committee for a bundle of reasons thoroughly typical of the industrial politics of the day. He was in favour of it partly because of the raw material agreement ('unless the industry were controlled important foreign Companies and/or UK interests would be induced to enter the manufacture of these raw materials'), partly because ICI 'should have an interest in an industry which may only be in its infancy', and partly because British Xylonite's demand for industrial alcohol, added to ICI's own, would strengthen ICI's position in relation to Distillers, who supplied the alcohol.[9] Throughout these years, as ICI moved towards the organic side of chemical industry, they found themselves increasingly aware of the Distillers' power and influence. As the negotiations developed, British Xylonite valued their shares at more than ICI was prepared to pay.[10] 'It has been found impossible', the Executive Committee recorded in February 1929, 'to carry out the ideas originally suggested for obtaining control of the industry within the Empire',[11] and there the scheme broke down, never to be renewed.

Later in 1929 the Executive Committee took another negative decision. The Safetex Safety Glass Co. came to ICI with a proposal that ICI should put £50,000 behind Safetex's splinterless glass, based on sheet glass and cellulose acetate film, which was much cheaper than the better-known Triplex product based on plate glass and celluloid. ICI turned Safetex down, saying that their glass was not good enough for the motor industry. Then, 'with the objects of extending nitrocellulose sales, applying benzyl cellulose [which was being developed by Nobels] to celluloid and safety-glass manufacture and ensuring that so far as possible processes for making celluloid etc. from new products are brought before us', a meeting was arranged between ICI, Pilkingtons, Triplex, and British Xylonite. The Executive Committee made their views on policy quite clear beforehand: 'In view of the substantial purchase of soda ash by Pilkington Bros and the desirability of increasing our sales of nitrocellulose to British Xylonite Co., the Committee recommended that ICI refrain for the present from participating in *any* safety glass proposition except possibly Triplex (which, however, is not yet before us). . . .'[12]

By the autumn of 1929, then, ICI had taken two steps (in different directions) towards the plastics industry, and two steps away again, the first by deciding not to go on with the British Xylonite purchase and the second by deciding that the goodwill of existing customers was worth more than the problematical profits of new enterprise. The Executive Committee at this time seems to have looked at the plastics industry almost entirely in relation to the existing structure of the business, particularly on the heavy chemicals side. If plastics could serve that by providing a market for its products, well and good, but there must be no remotest possibility that plastics might put any existing interest at hazard. Plastics as a promising field for development in their own right, despite Mitchell's faint noises, seem scarcely to have registered on the Executive Committee's mind, however highly they may have been regarded lower down.

These decisions left ICI with an unambitious little programme, sanctioned by the Executive Committee in February 1929 at the same meeting at which they killed off Barley's grand imperial design. It provided £10,000 for a 200-ton phenol formaldehyde (bakelite) plant at Huddersfield, where research was already going on, and not more than £1,000 to buy technical advice from British Thomson-Houston, who were already making moulding powders for their own use in the electrical industry. An exclusive licence for glyptal resins was to be sought, also from BTH, and Dyestuffs Group were to investigate various alternative materials for moulding powders.[13]

Three years later, in the spring of 1932, Francis Walker, Chairman of Dyestuffs Group, surveyed the plastics scene. He remarked on a small pilot plant (about 150 tons) just coming into operation at Ardeer, where they were investigating benzyl cellulose, and went on to Dyestuffs' major interest, phenol formaldehyde. Demand, depression notwithstanding, was healthy, being variously estimated to lie between 4,000 and 8,000 tons a year (Barley, in 1927, had forecast about 3,000 tons in 1930). The small plant at Huddersfield could not give 'the range of quality and colour now demanded by moulders'. 'While ICI has considerable potentialities in this industry,' Walker concluded, 'at the present moment they do not amount to anything. . . . What I am, therefore, aiming at in the course of the next nine to twelve months is to really establish ourselves in the industry, but on a comparatively modest scale, and as the result of the experience so gained, to size up the position and decide whether the moulding powder industry is worthy of a considerable investment on the part of ICI.'[14] Walker might have added that the decision he was seeking had been taken and then reversed in the negotiations over the purchase of an interest in British Xylonite, and that as a result ICI, in 1932, were not much further forward than they had been nearly five years before.

In the summer of 1932 Kenneth Gordon of Billingham, never one for caution, moved boldly in to outbid Francis Walker and Dyestuffs. Not for Gordon the 'comparatively modest scale'. In a 'Suggested ICI Programme for Plastics' he said that the UK market for phenol formaldehyde moulding powder was going ahead so fast—he put the current demand at 8,000 tons a year, worth £750,000—that ICI ought to put up plant capable of 2,000 tons a year 'at the most advantageous site'. As to what the most advantageous site was, Gordon left his audience in no doubt at all. He produced figures which, he said, showed an advantage for Billingham over Huddersfield of 0·45d. (0·2p) a lb. of finished product if synthetic phenol were made at Billingham and 0·225d. (0·1p) if phenol were bought, for both sites, at 6d. (2½p) a lb.[15]

In support of his proposal, Gordon said that as matters stood ICI's plastics interests were spread over too many Groups. Dyestuffs were making phenol formaldehyde resin and glyptal resins. Explosives Group were investigating benzyl cellulose and a plasticizer for nitrocellulose, tricresyl phosphate. At Billingham they were thinking about urea formaldehyde. General Chemicals were looking into vinyl resins. On the raw material side, Gordon pointed to Billingham's capacity for methanol, phenol, urea, and other things; to General Chemicals' interest in chlorine; to Ardeer's interest in nitrocellulose for celluloid, made outside ICI altogether. 'New plastics', he said, 'and new uses for plastics are continually being discovered, and if ICI is to play a leading part . . . research is called for.' Let it be centralized (at Billingham, naturally) under a General Manager, and let us have 'an organisation . . . to hold together these various interests, and to plan a forward policy for ICI plastics'. Billingham, he assumed, would be 'the future headquarters of the ICI plastics industry'—an assumption which two other Groups at least (Explosives and Dyestuffs) could be relied upon to dispute vigorously if it were made in their hearing.

Francis Walker, somewhat hesitantly, and Kenneth Gordon, most emphatically, seem to have been the first people in ICI to look at plastics mainly as potential profit-makers in an expanding market, rather than simply as an outlet for the products of other Groups. Gordon certainly seems to have been the first to call attention to the

343

fundamental weakness of ICI's plastics effort: that it was dispersed among the Groups with no central authority. The weakness arose, no doubt, from the fact that plastics had not been of great importance to any of the founder members of ICI, so that no natural home or centre of development had been provided for plastics in the original organization of the company, nor was there anyone at the centre who had strong reason to feel any particular concern for them. Once the original ICI organization hardened into a network of vested interests, as it did very quickly, this situation became very difficult to alter.

At this time, ICI were looking for some way of buying into the plastics industry. During the summer of 1932 J. H. Wadsworth talked to the Managing Director of Bakelite Limited, but he turned ICI down.[16] By the beginning of September John Rogers was in touch with Croydon Mouldrite Limited, described as 'the next most important firm' to Bakelite and able to face Bakelite's competition. It was controlled by the British Goodrich Rubber Co. Ltd., who were at first unwilling to part with a majority interest. This ICI were determined to have, partly to give them a free hand in developing Croydon Mouldrite, partly to prevent the fruits of development from going too freely to outside shareholders. ICI made a firm offer for 51 per cent of Mouldrite's capital in January 1933, accompanied by the threat of powerful competition if it were turned down. 'The Mouldrite Company today is obviously not of paramount importance to either the Goodrich Co. or ICI', said Francis Walker in the letter making the offer, and Sir Walrond Sinclair,* to whom it was addressed, evidently decided that Walker was right.[17] On 22 June 1933 ICI became the owners of a 51 per cent interest in Croydon Mouldrite, at a cost of about £33,500.[18]

What did ICI get for this modest outlay? Not, primarily, premises, plant, or technical knowledge. Croydon Mouldrite owned a cramped site in Croydon, regarded with distaste by the local authority, on which they could manufacture, by 1935, about 2,000 tons of phenol formaldehyde powders a year. On another site in Welwyn, acquired while negotiations with ICI were going on, they attempted to manufacture urea formaldehyde powders by the Kelacoma process, which turned out useless soon after it was bought. By 1935, at Croydon, in an 'ill-designed and badly located factory', they could turn out 250 tons of urea formaldehyde powders a year by a process developed with the backing of ICI. ICI's general view of these assets was summed up, before the interest in Croydon Mouldrite was bought, by A. E. Hodgkin, writing of the British plastics industry as a whole: 'Plant is primitive, and such tricks of the trade as there may be are due to incomplete understanding of the processes involved; both of these disabilities

_____

* Lieut.-Col. Sir Walrond Sinclair, KBE (1880-1952), Chairman (later President) of the British Tyre and Rubber Co. Ltd., with other Directorships and Government appointments.

ought to be overcome if we make the effort.'[19] What ICI hoped they had bought with Croydon Mouldrite, again in Hodgkin's words, was 'goodwill and an immediate footing in the market', which they would otherwise have had to build up for themselves. In other words, they had bought time. 'We are already late', said Hodgkin, 'and have a long way to catch up if we now start afresh by ourselves; and at this stage time is important.'[20]

As well as buying a place in the plastics industry, ICI had to provide themselves with some means of co-ordinating the activities which different Groups were carrying on. A. E. Hodgkin was asked to set up the new organization. According to his own account, the proposal was made by Henry Mond, the 2nd Lord Melchett. As Hodgkin turned to go, having accepted, he said to Melchett: 'By the way, what are plastics?' and Melchett replied: 'God knows, but you might show a little enthusiasm.' A curious exchange, the more so since Hodgkin and Melchett knew very well what plastics were, but it set the tone for a good deal of what was to follow.

In setting up his organization, Hodgkin acted on the principle that 'it is not feasible, or desirable, to concentrate the manufacture of plastics in any one group'.[21] That might have been debatable, but no doubt it took account of the realities of power within ICI. None of the Chairmen of established Groups would willingly give up any promising activity, and there was no one at the centre prepared either to persuade or to compel them to do so. In that situation, the best that Hodgkin could suggest was a system designed to 'create as little disturbance within the present order of things as is compatible with the effective tackling of a very complex situation'.

What emerged from this reasoning was the quaintly named 'Plastics Division'. It was not a Division at all in the usual commercial sense of the word, but a Committee, and not a very powerful one, since it had neither executive authority nor financial resources. It consisted of Hodgkin as Chairman and senior representatives (though not the Chairmen) of the Groups chiefly concerned with plastics—that is: Dyestuffs, Explosives, Fertilizers & Synthetic Products (Billingham), and Metals, who were concerned with moulding machinery.

It will be seen [wrote Hodgkin] that at the present stage it is not proposed that the Plastics Division shall have any direct executive power, although if matters develop favourably this may be advisable later on. But it is intended that it shall be from the first regarded as something definitely more than an advisory body, in the sense that its decision on any matter will have been arrived at by discussion between responsible members of all other groups concerned, and the latter should therefore normally acquiesce in such a decision, and have it carried out, without further consideration beyond that necessary to assure themselves that their interests in other directions are not likely to be adversely affected thereby.[22]

The statement of intent is laudable rather than convincing.

By the summer of 1933, then, ICI was at last committed, if 'committed' is not too strong a word, to the plastics industry through a 51 per cent interest in what Hodgkin called a 'small and struggling company' and a co-ordinating body which was little more than a discussion circle. Plastics Division and Croydon Mouldrite were linked through Hodgkin, Chairman of both, and through M. T. Sampson, a Delegate Director at Billingham. The whole enterprise was closely tied to Billingham, not only through dependence on it for materials but through research, under A. Caress and B. J. Wood, into phenol formaldehyde, urea formaldehyde, and 'Resin M'.

'Resin M', polymethylmethacrylate, was evolved through activity at more than one ICI Group. It was first made by Rowland Hill in Dyestuffs Group, but the synthetic method for making the monomer (methyl methacrylate) was a brilliant piece of work by J. W. C. Crawford at Ardeer.[23] His process, based on acetone, methanol, cyanide, and sulphuric acid, is still in use at the time of writing.

This major ICI invention was handled in a manner thoroughly characteristic of the times. It reached the market in 1935 in two forms, moulding powder ('Diakon', 'Kallodent') and transparent sheet ('Perspex'). There was clearly a demand, in the motor trade if nowhere else, for 'Perspex', but we have seen that ICI had been frightened off that market once already, in 1929, by fear of losing sales of soda ash to Pilkingtons (p. 341 above). At that time it was someone else's product ('Safetex' splinterless glass) which they refused to develop. Now their own product, 'Perspex', was at stake.

First, in November 1935, ICI made a ten-year agreement with Triplex by which Triplex were to be ICI's main customers for 'Perspex' sold as a substitute for safety glass and were promised 2½ per cent commission on other sales.[24] Triplex in return agreed not to make synthetic resins. With Pilkingtons, in July 1937, ICI agreed that if ICI's sales of flat transparent resin should reach 2,000 tons in four consecutive quarters Pilkingtons should have the right to call on ICI to form a joint company (51 per cent ICI, 49 per cent Pilkingtons) to sell the material, and the price to Pilkingtons for their 49 per cent interest should be related to the ICI profit being made.[25]

The Triplex agreement was modified during the war and the Pilkington agreement never came into effect, although production went beyond the specified limit in 1942. It was cancelled in 1948.[26] The two agreements, nevertheless, are important because, like many other agreements made in ICI's early years, they illustrate the strongly conservative cast of mind prevailing in ICI at the time, especially after the Depression. ICI were most unwilling to let any new development, however promising, upset existing customers. With such a mentality, which we shall come across again looking at synthetic fibres, anything like aggressive selling was impossible and the outlook for inventors, except those working along established lines of business, where the opportunity for innovation was apt to be small, was, to say the least of it, discouraging.

In pursuing other major developments, notably nitrogen, oil-from-coal, and dyestuffs, ICI was accustomed to work on a world scale. This had been the original intention in plastics, but the outcome was very different. There was a world plastics industry including familiar names —du Pont, IG, the Bakelite Corporation—with ramifications on both sides of the Atlantic; but ICI, by reason of their small and cautious investment, were not in this class, and the price of entry would be a much higher commitment of resources than the Board in the mid-thirties had a mind for.

ICI, with a bare majority holding in Croydon Mouldrite, remained in a small way of business in the plastics industry of Great Britain, particularly that part of it dealing in thermo-setting powders. It was not at all the class of business ICI were used to, being full of small competitors fiercely struggling to better themselves without regard for what Hodgkin called 'harmonious working'—that is, price-fixing. 'There is in fact', he said at the end of 1935, 'no organisation or community of interest whatever, in so far as moulding powders are concerned. Competition is keen, and at times unscrupulous. . . . Prices are cut as and when individual firms feel inclined to do so. The effect of this on Mouldrite's business since ICI bought control'—that is, in about eighteen months—'amounts to £24,000 per annum, or 38% on the paid-up capital.'[27]

The fundamental trouble, as Hodgkin was well aware, was that Croydon Mouldrite depended for its living on phenol formaldehyde powders, and they had long ago reached the point where patent protection had expired, the technique of manufacture was widely, if not always well, understood, little capital was required to come into the business, and in consequence phenol formaldehyde moulding powders were virtually bulk products from which only the largest and most efficient producers could expect a reasonable return on capital, and no one could expect a large one. Urea formaldehyde powders had not yet sunk so low, but 'it is certain', said Hodgkin, 'that the business . . . cannot continue for much longer to yield the present comparatively large margin of profit.'[28]

The way to prosperity from this situation was the same as in the dyestuffs business: continual research to find new products which, for a time, would earn monopoly profits. Croydon Mouldrite's research, before ICI came in, was almost non-existent. After ICI came in, time and money had to be spent in bringing products up to the level of those made by competitors—a process still incomplete at the end of 1935. 'While we have been doing this', said Hodgkin, 'our competitors have gone ahead and developed specialities which we shall have to develop later on.'[29] 'Developing specialities' would need money, but ICI's total budget for plastics research, including the work on existing products, was £19,350 in 1934, £19,500 in 1935, and £24,000 (projected) for 1936. 'A substantial research expenditure each year', said Hodgkin, and he did not mean the sums just quoted, 'must be a part of our Plastics programme if we really mean to make a success and to become the leading manufacturer instead of tailing along in the rear as we are now doing. This is fully realised by du Pont and the IG.'[30]

Hodgkin, by now a severely frustrated man, was in no doubt that the right thing for ICI to do was to buy out the other shareholders in Croydon Mouldrite and then use it as a basis for setting up a Plastics Group, 'since otherwise a properly concerted drive, conducted with the enthusiasm of a united organisation, will not be possible'. Then ICI, through Plastics Group, 'should be willing to face a period of financial loss in order to attain such outputs as will give us the lowest costs in the Industry', and at the same time 'adequate research must be conducted not only to cheapen and improve existing products but in order not to be left behind on new developments'.[31] About three weeks later he put the matter even more strongly: 'I am convinced that the present Mouldrite "organisation" and scale of operations cannot ever be an effective factor in the plastics trade; and that a continuance of present conditions will surely lead to disaster.'[32]

Hodgkin's urgency had some effect, though scarcely as much as he must have hoped. In April 1936 the General Purposes Committee agreed—yet again—that the plastics industry 'was one in which ICI should play a vigorous part (even if only on account of the important quantity of ICI raw materials consumed therein, apart from any future profits which should be made from the manufacture and sale of plastics themselves)'. Accordingly an offer was to be made for the Mouldrite capital not already owned by ICI, and 'on the completion of this purchase the whole of ICI's plastics effort [was] to be centralised, both in an executive and administrative sense, in Mouldrite Limited'. Plastics Division was to be abolished. 'The eventual intention', the Committee recorded, 'is to form a Plastics Group.'[33] The minority shareholders in Mouldrite Limited* were bought out in July 1936 for £24,275. Staff and the ownership of assets were transferred to

---

* 'Croydon' had been dropped from the name in April 1934.

Mouldrite from Fertilizers & Synthetic Products (Billingham). For the first time ICI had something approaching a centralized authority to direct its operations in the plastics industry, although Plastics Group was not formally constituted until March 1938.[34]

Between 1936 and 1938 the nature of ICI's plastics business began to alter radically. The importance of phenol formaldehyde resins declined until before the end of 1936 the question was being raised whether it was worth while to go on making them.[35] In fact they survived until 1967. 'Perspex', as demand built up from the aircraft industry under the stimulus of rearmament, began to look like ICI's first big independent success in plastics. The decline of thermo-setting powders and the rise of 'Perspex' mark, it may be thought, the decisive moment when plastics ceased to be looked on chiefly as a convenient outlet for raw materials and came to be seen as potential profit-makers on their own. At the same time, appropriately, research into plastics materials which might have a commercial application, of which there were many, was pressed forward under Caress.

The formation of Plastics Group did not solve the problems created by ICI's late and hesitant entry into the plastics industry. The vested interests of the established Groups were not speedily overridden, and demarcation disputes, conducted with careful courtesy, reached considerable heights of refinement and subtlety. Moreover, from 1935 onwards the most important plastic of all—polythene—was being developed in the Alkali Group, where it had first been made.

(ii) THERMO-PLASTIC ADVANCE: THE IMPACT OF POLYTHENE

In 1943 the thought occurred to Peter Allen, newly appointed Managing Director of the Plastics Group, that the situation in the chemical industry was rather like the situation in W. H. Perkin's day, nearly a century earlier, when the importance of a new range of raw materials—aromatic compounds—had been established but the great commercial successes were still mostly undiscovered and unpredictable. 'Chemically', he wrote in his diary on 15 July 1943, 'it [is] a good and exciting time to be alive like the early days of the dyes industry.'[36]

A little more than ten years before Allen wrote, in an ICI laboratory, a discovery had been made which may claim to rank among the major industrial discoveries of the twentieth century: polythene. The word 'discovery' is used advisedly. Polythene was not an invention, as nylon and 'Terylene' were in the sense of being a result of a premeditated attempt at producing a substance with valuable commercial properties. It was the unexpected upshot of a speculative experiment, in which several people were alert enough to see commercial possibilities.[37]

The discovery of polythene owed nothing to ICI's interest in the

349

plastics industry. It was made at Winnington and it came about through co-operation between two Groups which had no great liking for each other. One was the Dyestuffs Group, proud of their technical excellence but prickly about the depressed standing of dyestuffs in ICI. The other was the Alkali Group, heirs of Brunner, Mond, not less aware than Dyestuffs of their own particular brand of technical excellence but aware also of the central importance of the products of their Group for the survival of ICI in the conditions of the early thirties.

In Alkali Group's research laboratories, finished in 1928, there was a young team of physicists, physical chemists, and engineers, many of them recruited by F. A. Freeth. The Research Manager was H. E. Cocksedge, Freeth's brother-in-law and his collaborator on ammonium nitrate during the Great War. In 1930, after the Depression had set in, the management of Alkali Group was bold enough to assert a belief in the value of speculative research without obvious commercial application and to keep a limited programme going in spite of the very unsympathetic attitude which prevailed in some quarters at Millbank after the death of Alfred Mond. Freeth himself, as early as 1919, had made his views known: 'Very many of our own and other people's discoveries have been based upon work of what by no stretch of imagination could be called a practical character.'[38]

This kind of research at Winnington, in the late twenties and early thirties, was directed towards the effects of extreme physical conditions —very high temperature, very high vacuum, very high pressure. In high-pressure work the Winnington research staff had Billingham's experience to draw on, though they intended to go far beyond it, and they had also the advantage of Freeth's connections with the Thermodynamics Laboratory at the University of Leyden, and especially with Dr. A. Michels, who in the mid-twenties began a long association with ICI as a consultant on high-pressure techniques and apparatus.

By the spring of 1931 the first simple apparatus for high-pressure studies had been established at Winnington. William Rintoul, Freeth's colleague as joint Research Manager of ICI, invited Michels to address the Dyestuffs Group Research Committee. Besides members of Dyestuffs research staff and Rintoul himself, the Committee included three eminent organic chemists. One was Professor (Sir) Robert Robinson, later President of the Royal Society. Robinson was impressed by Michels, and being himself anxious to try a number of organic reactions under pressure, submitted, in the autumn of 1931, 'a memorandum in which a number of reactions suitable for high-pressure studies were suggested'.[39] The Committee strongly supported him, and suggested the Winnington laboratory, even though it did not belong to Dyestuffs Group, as the obviously suitable place to do the work. 'Co-operation with organic chemists', the Committee added, 'would be essential', which would mean, amongst other things, that Dyestuffs' interests would not be neglected.

Robinson's interest in high pressure was lucky for two of the young men at Winnington, J. C. Swallow and Michael Perrin. In January 1932 they proposed work on reactions under pressures ranging, perhaps, as high as 20,000 atmospheres, and they suggested that an organic chemist from Dyestuffs might move into Winnington. Their proposals were accepted and a programme of high-pressure research was put in hand which, as Perrin said in 1953, was 'on the lines of the Dyestuffs Division's scheme'.[40]

The chemists who worked on the project were R. O. Gibson and E. W. Fawcett. Gibson, like Perrin and Swallow, had worked under Michels in the Netherlands. He had also worked on high-pressure research at Billingham which had been killed by the slump. Fawcett, who had been with Gibson at Billingham, was the Dyestuffs nominee. The Research Engineer was W. R. D. Manning, who adapted, developed, and invented pressure vessels intended for work well beyond the range of pressure—a few hundred atmospheres—which was usual at Billingham. The work done by Fawcett and Gibson, chemically,

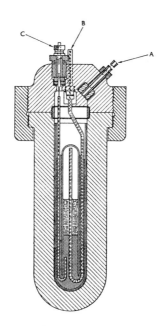

The gas-reaction vessel in which polythene was first observed.

Source: R. O. Gibson, 'The Discovery of Polythene', Royal Institute of Chemistry Lecture Series, no. 1, 1964, p. 16.

was almost entirely negative, but for the engineers it was more rewarding. Fifty or so reactions produced very little except some rather too exciting adventures with gas under pressure, but on that experience Manning and E. W. Colbeck, a metallurgist, were able to develop equipment which would handle gas at high pressure with reasonable safety, and by the spring of 1933 Fawcett and Gibson were ready to study half a dozen reactions between gases and liquids which Dyestuffs Group proposed to them.

One of the reactions proposed was between ethylene and benzaldehyde. Fawcett and Gibson decided to try this one first, because they had a cylinder of ethylene at hand but no carbon monoxide, which the other experiments in the programme required. On Friday 24 March 1933 they put 20cc. of benzaldehdye and ethylene in a pressure vessel (see diagram, p. 351) at 2,000 atmospheres and 170°C. and left it. The pressure held overnight, but during the weekend a leak developed, and when Fawcett and Gibson came in on Monday morning they found no pressure. The benzaldehyde had been blown out of the reaction tube and was mixed with the heating oil surrounding it. The two men took the apparatus apart, and Fawcett showed Gibson the end of the steel tube through which gas came into the reaction space. It looked as if it had been dipped in paraffin wax, and Gibson scribbled in his notebook: 'Waxy solid found in reaction tube.'[41]

This is the first record of the substance which later became known as 'polythene'. Neither Gibson nor Fawcett knew what it was nor how it had come to be formed, and it was far from being the result they expected from the experiment. Nevertheless they tried again. One attempt produced a little more waxy solid: another blew up. In April the minutes of the Dyestuffs Group Research Committee recorded:

The work on the reaction between ethylene and benzaldehyde at 2,000 a.t.m. has been abandoned. The first experiment gave a waxlike substance, probably a polymerized ethylene, but a repetition resulted in an explosion. . . .[42]

In spite of this decision, a little more work was done—enough to identify the waxy solid positively as an ethylene polymer of fairly high molecular weight, quite different from other ethylene polymers then known, and to establish some of its properties, although the total quantity made was about half a gram and the largest sample from a single experiment was 0·2 gm. Fawcett, Gibson thought, 'had a hunch' from the start that the new polymer 'might turn out to be important', but the experimental results were so inconclusive and the explosive possibilities so great that in July 1933 work on the reaction was given up in favour of other activities.

After that, until December 1935, the new polymer remained in the background. Dyestuffs Group, at the end of 1933, withdrew from sponsorship of work which had apparently proved unrewarding. Fawcett moved to work on molecular distillation. Perrin, who had been working in Amsterdam, came to Winnington and joined Gibson in Fawcett's place. The work they undertook, in place of the work sponsored by Dyestuffs, was on the mechanism of reactions under very high pressure, as against the chemical results of specific reactions. This represented a far-reaching change of outlook, coming much nearer to Perrin's sphere of interest than Gibson's. Gibson, by his own account, found the position uncomfortable. In October 1935 he was moved to another section of the Winnington research department and Perrin was put in charge of high-pressure work under Swallow.[43]

The most promising lines of investigation from the new point of view seemed at this time to point towards ethylene in reaction with water or with carbon monoxide. Although interest was now focused on the physical chemistry of reactions rather than on the organic compounds that might result from them, Fawcett's and Gibson's unfinished work of 1933 was brought out of obscurity and Perrin decided to have another look at it. One thing that had been established was that benzaldehyde played no part in the reaction and Perrin proposed to investigate the behaviour of ethylene alone under pressure. Unfortunately the rules of the laboratory forbade him to do so. After some of the more alarming results of the earlier experiments Cocksedge had banned any further tinkering with ethylene under pressure. The ban was still in force in 1935, though the reason for it was less cogent because brick cubicles had been built in the laboratory and high-pressure experiments could be remotely controlled. Moreover, Manning had greatly enlarged and improved the apparatus. Nevertheless Perrin thought it politic to go to work unofficially and out of hours.

About half-past six on the evening of 19 December 1935 Perrin, Manning, and L. Mills, a laboratory assistant, set out to repeat the experiment carried out by Fawcett and Gibson in 1933, except that they put no benzaldehyde in the 'bomb' (reaction vessel) but only commercial ethylene guaranteed to be at least 96 per cent pure. They got the pressure to 2,000 atmospheres and the temperature to 170 °C. but then, to Manning's surprise and dismay, because he thought a leak must have developed in his apparatus, the pressure dropped and could not be held at 2,000 atms. even when the bomb was repeatedly replenished: a process which they continued until all the gas was used up.

Perrin says that he was not surprised. The drop in pressure was what he was expecting as the ethylene polymerized, though he suspects that there may have been a leak as well. However, when the bomb was cooled and opened, 'it was with no great surprise, but considerable pleasure' that he found about 8 gms. of polymer inside it, though he considered the quantity disappointingly small. Manning seems to have been more emotional. 'You can imagine our excitement', he told an audience in 1964, 'when masses of snowy white polymer spilled out. . . . We had made at least 10 g. . . . something like a hundred times more than all the previous yields put together.'[44]

The key to the success of the experiment, though it was not discovered until later, lay in the impurity of the ethylene. It contained a trace of oxygen, which acted as a catalyst. If there had been less, nothing would have happened. If there had been more, the 'bomb' would probably have blown up, carrying with it, presumably, Michael Perrin's career in ICI.

As matters were, Perrin and his two colleagues had rediscovered polythene, and this time it stayed discovered. On the day after the first success, 20 December 1935, the experiment was repeated, this time officially, by E. G. Williams, J. G. Paton, and F. Bebbington. They too must have been lucky with their ethylene, for their experiment worked, whereas on several later occasions there was an explosion.[45] By February 1936 the first British patent specifications had been filed and by April a brand name—'Alketh', later 'Alkathene'—was on the register. Although the importance of oxygen in the reaction was not appreciated for several months, enough polymer was made in the first weeks of 1936, no doubt by sufficiently exciting methods, to provide samples for testing. By the end of January 1936 film had been cast from a hot solution on to plate glass, moulded material had been made, and outstanding dielectric properties were becoming apparent.[46]

During the remainder of 1936 and in 1937 work went ahead both on the chemistry of the reaction and on problems of chemical engineering. J. L. S. Steel, Commercial Director of Alkali Group, pointed out that for commercial purposes a continuous process of manufacture would have to be devised, for it was out of the question to dig batches of the material out of high-pressure vessels. With a continuous compressor, designed by Michels, working with mercury pistons, and with a stirred vessel with an internal motor, designed by Manning, a process was quickly developed.

By 15 June 1937 W. F. Lutyens, Chairman of Alkali Group, could lay down commercial policy to A. E. Hodgkin, Chairman of Mouldrite Limited through which ICI's plastics business was being carried on:

Manufacture will be undertaken at Winnington. The raw material will at any rate in the first instance be alcohol, although after the tonnage produced reaches some hundreds per annum it may be that we shall consider oil cracking as a source of ethylene. Alketh will only be produced as a powder and we shall not at Winnington work it up into any finished state with the possible exception of a crude sheet form.[47]

Lutyens's masterful tone was characteristic of the man and of his

Group. The 'gentlemen with a taste for chemistry', in their deceptively relaxed style, knew they were good, and their Chairman took every opportunity of telling the rest of the world, particularly the ICI world, how good they were. Within about eighteen months of Perrin's evening escapade, itself an episode fully in the Winnington tradition, they had produced a new plastic which was already within sight of being a saleable product. That was a notable achievement which Lutyens took care to bring prominently before the ICI Board. He had been discussing it with Gaskell and Coates the day before he wrote to Hodgkin, which was why he was able to write with such authority.

The Chairman of Alkali Group could bring great weight to bear. His Group contributed handsomely to the profits of ICI and it had inherited Brunner, Mond's prestige. Hodgkin's Group, although nominally in charge of ICI plastics, had as yet neither profits nor prestige. Therefore although polythene had nothing whatever to do with Alkali Group's established activities, that Group took a grip on commercial and technical development, the technical side of which was not finally relinquished until 1958. In any fight for control, Hodgkin was poorly placed.

A fight broke out as soon as the potentialities of polythene began to be recognized. 'We have a case here', Hodgkin wrote in April 1938 to A. Caress, Research Manager of Mouldrite, 'of a plastic (and it is nothing else) being developed by the Alkali Group which in many ways impinges on our field. . . . The proper attitude for the Alkali Group to take up in my opinion is that which the General Chemicals Group take . . . in regard to the manufacture of methyl methacrylate; that is, they manufacture the methacrylate and sell it to our Group who know how to employ it and who have the necessary contacts.' He went on to say that Alkali Group ought to manufacture polythene and then sell it 'to the existing Groups of ICI, e.g. Mouldrite, Dyestuffs, NCF [Paints] and Leathercloth for the purposes for which those groups are fitted to make use of it'.[48] 'Proper attitude' or not, it was never one which Alkali Group was willing to take up, and in pressing his point, as he continued to do up to the outbreak of war, Hodgkin was heavily outgunned. Caress was quite blunt about the power politics of the situation. 'I am naturally keen', he wrote, 'to get the production and development of all organic plastics into the Plastics Group. If we attempt to achieve this gradually by a continual tug-of-war with the other Groups, we are handicapped by the fact that, by reason of their size, they have men, money and facilities to draw on which we cannot have. . . .'[49]

Polythene, ICI's major discovery in the plastics field, was therefore taken through every stage from laboratory experiments to large-scale production and sale by Alkali Group and not by Plastics Group at all. The reasons were partly practical—Alkali Group had an organization far more highly developed in every respect, commercial as well as technical, than Plastics Group—and partly, probably far more importantly, psychological. When a transfer of polythene was recommended, in 1941, P. C. Allen wrote in his diary: '. . . if [the author of the recommendation] thinks we're going to hand all this across to the Plastics Group on a plate without protest, indeed without a hard fight, the little fellow is very much mistaken.' Somewhat more temperately, about five weeks later, Allen wrote: 'I take great exception to . . . the transfer of polythene sales control, sales policy, prices and profits away from the inventors as wrong in principle and bound to restrict similar inventions in future.'[50] The recommendation was dropped.

Until nearly the end of the war everyone concerned with polythene thought its importance lay almost entirely in its dielectric properties. In August 1936 W. H. H. Demuth, Development Director in Alkali Group, sent 20 gms. for testing to Dyestuffs Group. B. J. Habgood, who had recently been brought into the Group, by Cronshaw, from the cable industry, pointed out that polythene was like gutta percha but much better as an insulator for submarine cables. At about the same time it was recognized that no other material, natural or synthetic, could match polythene as an insulator at very high frequencies. Towards the end of 1938, at a joint meeting of the Research and Technical Executive Committees, Lutyens, Allen, and Swallow spoke of 'probable developments in the field of television, short-wave radio and high voltage power cable'.[51]

Beside the electrical uses of polythene the other uses seemed unimportant. At prices of 5s.–6s. (25–30p) a lb., realistic enough for the cable-makers, it was a material for the world of high technology but not for the mass market. By the mid-fifties packaging film and household goods, made by injection or extrusion moulding, would carry the world consumption of low-density polythene into six-figure tonnages, but in 1939 none of that had been thought of and in the United Kingdom, in any case, the required sources of cheap ethylene did not exist. The only use foreseen for moulded polythene—perhaps thirty tons a year—was shotgun cartridges, so when it was hinted to Hodgkin that his Group might take over moulding and casting, but not the electrical uses, his response, expressed in a letter to Caress, was to wonder 'whether . . . there is any advantage in our attempting to develop what amounts to the fag end of polythene while the Alkali Group are doing the rest'.[52]

On 1 September 1939 Alkali Group brought their first full-scale polythene plant into production at Wallerscote. It was supposed to be capable of 100 tons a year and more capacity was on the way, largely at the insistence of Peter Allen. Neither he nor anyone else, however, yet foresaw more than very moderate output. Allen, early in 1939, was talking in terms of 410 tons a year by 1942, nearly all taken up in electrical uses. That, he thought, would give a turnover of £138,000 and a very attractive profit—£57,000. These were small figures, but he urged 'that polythene . . . may begin to move very quickly' and he was anxious to anticipate demand so as to 'obtain profits or bargaining power . . . for as many years as possible of those conferred by the patent monopoly'.[53] On these grounds he gained sanction from Millbank for another 100 tons a year capacity—half what he hoped for—after the original 100-ton plant was in production.

Allen's case rested on the strength of ICI's patent position throughout the world, which was very great. In the USA, the most important market, ICI held a 'composition of matter' patent which protected polythene itself, regardless of the process by which it was made. This was important, because in Union Carbide, unknown to ICI, work was going on to develop a process entirely independent of ICI's, and it succeeded.

Thus by the outbreak of war polythene was launched, in a very small way, and by the beginning of 1940 Allen said it was 'out of the development stage and firmly established as an ordinary commercial proposition'.[54] Supply, during the war, did not catch up with demand

TABLE 29

POLYTHENE SALES: PRICES, ROYALTIES AND PROFITS, 1938–1948

	Sales tons	Price per lb.		Royalties[a]	Profit[b]	Loss
				£	£	£
1938						4,174
1939	10	5s.	25p			6,169
1940	105	5s.	25p			11,796
1941	177	5s.	25p			596
1942	557	5s.	25p		85,953	
1943	920	4s.	20p		105,530	
1944	1,274	3s. 9d.	19p		173,728	
1945	1,059	3s. 3d.	16p		40,927	
1946	1,314	3s. 3d.	16p	40,423	86,422	
1947	1,567	3s. 3d.	16p	71,509	68,522	
1948	1,514	3s. 3d.	16p	74,322		43,777[c]

a From licences of the American patents to du Pont and Union Carbide and Carbon.
b Profit including royalties after deducting Obsolescence and charge for central services.
c '. . . the cost of ethyl alcohol per pound of polythene which had been 6·25d. [2·6p] from 1941 to 1945 thereafter rose by steps to 16·7d. [7·4p] in April 1948.'

Source: W. F. Lutyens, 'Trading Results for the Year 1949', 1 November 1949, CP 'Polythene General'.

until 1943. Sales rose from 10 tons in 1939 to 1,274 in 1944, with an accompanying scramble, all the time, to get new plant sanctioned. In 1940 a 1,000-ton plant was put up at Winnington to meet demand from the Ministry of Aircraft Production for chlorinated polythene as 'dope' for aircraft canvas. No such use ever developed, but the extra capacity was not less welcome for that. By 1945 Alkali Group had plant capable of producing about 1,500 tons of polythene a year.[55]

During the war the entire output of polythene, apart from some sent across the Atlantic, went to the cable companies, on Government account, chiefly for electrical purposes, especially airborne radar. As soon as fixed ground stations became practicable a demand arose for mobile sets and particularly for sets in fighter aircraft, so that they could intercept bombers at night or through cloud. This demand posed formidable problems. The equipment had to be compact, robust, and weatherproof, and it had to work at very high voltages on ever higher frequencies representing, eventually, wavelengths under ten centimetres. No dielectric material, before polythene, could meet all these conditions. Development work, to meet the needs of Government research establishments, was put in hand under E. G. Williams.

Polythene, in the words of Sir Robert Watson-Watt, 'had a high dielectric strength, it had a very low loss factor even at centimetric wavelengths, it could fairly be described as moisture-repellent and it could be moulded in such a way that it supported aerial rods directly on water-tight vibration-proof joints backed by a surface on which moisture film did not remain conductive. And it permitted the construction of flexible very-high-frequency cables very convenient in use.'[56]

Radar sets with polythene insulation were in action in night fighters over England and in ships at the battle of Matapan during March 1941, and the pressure for greater production was intense. 'It maddens me', wrote Allen in his diary on 29 August 1941, 'that night fighter equipment should wait on our incompetence.'[57] Before the end of the war polythene-insulated radar in ships and aircraft had been used to defeat U-boats, to guide bombers in raids on Hamburg and Berlin, and in sinking the *Scharnhorst*.

The Germans, lacking polythene, were obliged to build much bulkier radar sets than the British. Yet the Germans only narrowly missed getting polythene from ICI. At the end of 1938, before the military uses of polythene had been thought of, at any rate within ICI, IG asked for information and Lawson suggested that in return for so important a disclosure ICI might get information on a whole range of products: polystyrene, 'Mipolam' and 'Igelite' (both based on polyvinyl chloride), 'Oppanol' (polyisobutylene), and the 'Buna' synthetic rubbers. The Research Committee agreed that information should only be exchanged for rights in some equivalent discovery, because the rights in polythene were 'too valuable to be sold for money alone, or for processes already purchaseable for money'. Soon afterwards the Groups Central Committee decided 'we are not at present ready to enter into any discussion with the IG on the matter', and it was dropped.[58] It was a perfectly ordinary commercial transaction, or rather non-transaction, and in ICI, certainly, it was regarded as nothing more, nor, probably, in IG, but there cannot have been many business propositions in history which have broken down with such advantage to Great Britain.

Lawson's proposal to exchange information on polythene for information on a large range of IG products indicates ICI's belief in the potential of polythene as well as their pre-war weakness in the general field of the plastics industry. Having gone into plastics late and timidly, and having spent sparingly on research ($£104,700$ in $5\frac{1}{2}$ years),[59] ICI in 1939 had only one product of real promise—'Perspex'—in quantity production. The turnover of Plastics Group was small, its profits were minute, its production facilities were scattered and inadequate, it was understaffed, and the Group Chairman's morale was low. He departed into the Army, no doubt with relief, at the outbreak of war.

The war brought a transformation. Demand for 'Perspex', as a glazing material for aircraft, shot up. At the same time demand rose rapidly for various forms of polyvinyl chloride, a plastics material not previously manufactured in Great Britain. Plastics Group, the neglected outlier, became the centre of activities which foreshadowed the coming shape of the chemical industry, particularly in their reliance on polymerization chemistry as a theoretical base and on heavy organic chemicals as a source of raw materials. Plastics Group's operations, too, required a degree of interdependence among ICI Groups which cut across the established pattern of organization. Alkali Group's self-contained polythene operation—the result, as we have seen, of a fortunate historical accident—became increasingly anachronistic.

In 1939 357 tons of 'Perspex' were produced: 4,700 in 1944. The turnover from 'Perspex' sales in 1940, at $£725,000$, was greater than the entire turnover of the Plastics Group ($£606,000$) in 1939, and the gross trading profits from the product show similar startling increases (Tables 30 and 31). The monomer, methyl methacrylate, was made by General Chemicals Group and passed to Plastics Group for polymerization and final processing, first at Billingham, then at Darwen, and later in an agency factory at Rawtenstall. Almost the entire output of polymethyl methacrylate went into aircraft as 'Perspex' but the moulding powders 'Diakon' and 'Kallodent' (used in dentures) demonstrated the versatility which is a leading characteristic of the plastic materials which were developing in the late thirties and early forties.

TABLE 30

'PERSPEX' PRODUCTION AND SALES, 1939–1949

	Production	Sales
	Tons	
1939	357	n.a.
1940	455	n.a.
1941	1,480	n.a.
1942	2,450	n.a.
1943	3,782	n.a.
1944	4,700	n.a.
1945	n.a.	1,952
1946	n.a.	4,353
1947	n.a.	4,525
1948 (est.)	n.a.	3,700
1949 (est.)	n.a.	4,200

Source: P. C. Allen, 'Rawtenstall Factory', 27 Aug. 1948, CP 'Government Contracts—Acquisition of factories'.

TABLE 31

PLASTICS GROUP TURNOVER AND PROFITS, 1937–1942

	'Perspex' Turnover	Gross Trading Profit	'Welvic' Turnover	Gross Trading Profit	Total Turnover	Gross Trading Profit
	£	£	£	£	£	£
1937	78,000	6,000			324,000	17,600
1938	189,000	19,500			399,000	17,200
1939	318,000	77,000			606,000	97,000
1940	725,000	293,000	24,000	3,000	1,118,000	315,000
1941	869,000	296,000	117,000	23,000	1,490,000	421,000
1942[a]	1,356,000	336,000	407,000	113,000	2,420,000	592,000

[a] All figures estimated

Other Products: Phenol formaldehyde and urea formaldehyde powders and resins
'Kallodent' and 'Diakon'
Nylon monofilament (1941–2)

Source: W. J. Worboys, 'Some Notes on the Plastics Group', 2 Feb. 1943, CR 922/252/252.

Polyvinyl chloride was a later starter. Its chief value in wartime, especially after the loss of rubber plantations in the Far East, was as sheathing material for electric and telephone cables, though it was also used in leathercloth and as a dope for aircraft fabric, as long as that was needed. It was developed in America and Germany before the war, and at the Leipzig Fair in 1937 and 1938 the attention of Plastics Group was attracted to products made from PVC by IG. Two—'Igelite' and 'Mipolam'—were proposed by Lawson as part of his package deal over polythene information. When war broke out development work was going on in Plastics Group and in Dyestuffs Group, but there was no production. Production began in 1940 with 'Welvic', a plasticized compound made by Plastics Group first at Billingham and later at Welwyn, from polymer imported from Carbide & Carbon Chemicals in the USA.

Vinyl chloride was manufactured from acetylene and hydrochloric acid, though it could also be made by a process using ethylene dichloride if that were available as a by-product of other manufactures. In either case, chlorine was at the root of the matter and the obvious Group to handle vinyl chloride manufacture was General Chemicals, who would then pass it on to Plastics Group to work up into a saleable form. Plant was put up at Runcorn for about 1,500 tons a year and also, after the disasters in Malaya, at Hillhouse, for 5,000 tons a year as a replacement for rubber in electrical uses. Hillhouse, in Lancashire, was chosen because on that site General Chemicals were already making poison gas and other products based on chlorine. The making of 'Welvic' was Plastics Group's responsibility, and the marketing also, though in wartime that was chiefly a matter of dealing with the appropriate official at the Ministry of Supply.

With polythene on the one hand and 'Perspex' and 'Welvic' on the other, and with the certainty of more plastics materials to come, to say nothing of closely related developments in nylon and other fibres, it was evident as early as the autumn of 1941 that ICI's organization to cope with plastics, existing and potential, would have to be rationalized and strengthened. In September 1941 Lutyens, as Development Director, commissioned an inquiry into the whole matter. It showed what a difficult and delicate task the reorganization of the plastics side would be, in the face of vested interests already established in the field by 'every Group of ICI', as the Report says, '(other, perhaps, than the Lime and Salt Groups)'.[60] For even where Groups could show no claim to research or manufacturing interests in plastics, as Alkali and Dyestuffs both could, they might still be interested in the use and/or sale of the products, as Metals, Leathercloth, Explosives, and Paints all were.

The reorganization of Plastics Group included a strong infusion of talent from Winnington. Caress, a Director since the foundation of the Group, remained, but the management as a whole was greatly changed and expanded. The new Chairman, W. J. Worboys, was an Australian Rhodes Scholar who had been Commercial Director at Billingham. The Managing Director was P. C. Allen and J. C. Swallow also joined the Board as Research Director. E. G. Williams and W. R. D. Manning joined the staff. 'I helped to put polythene on the map', wrote Allen in his diary late in 1942, '. . . now it is a great thriving process and it in turn has helped to make me.'[61] It was a sentiment which others might have echoed as so many of the developers of polythene moved into Plastics Group.

'The Plastics Industry', wrote W. J. Worboys early in 1943, surveying his new domain, 'is in its infancy . . . and there will almost certainly be big developments after the war. Both ICI and this Country are already behind in the race, and it is suggested that we should do everything we can to catch up now and not wait until the post-war race starts, when we would find ourselves even more behind.'[62] 'Big developments' in the plastics industry of 1943, and in many branches of it until much later, meant considerably less than 10,000 tons a year. In the case of poly-thene, however, a distant and imprecise prospect of much larger possibilities was beginning to appear. It arose, as these things so often do, from America and from the mass market, for as early as 1941 du Pont foresaw cheap polythene 'satisfactory for manufacturing films'—presumably for packaging—'which they believe will be the largest potential commercial market'.[63] By 1945 Alkali Group had come to much the same way of thinking—'very large peacetime outlets would exist if polythene were sold at a low price of 2/– [10p] per lb.—a price which would yet be profitable for a large plant using oil as a source of cheap ethylene.'[64]

So we are led back, as in the early forties ICI continually was, to the fundamental problem of post-war development—the choice of a source of heavy organic chemicals. Before we pursue the problem to its solution, however, we must examine the development of fibres: a development which, going on at the same time as the development of polythene and other plastics, was similarly wrenching the structure of ICI's business away from the patterns of the past.

REFERENCES

1 'Plastic Materials Industry'—Report for Executive Committee by Plastic Materials Committee, 8 Dec. 1927, CR 560/1/1.
2 Ibid. p. 3.
3 A. E. Hodgkin, 'ICI Plastics, General Review of the Position in December 1935', 30 Dec. 1935, GPCSP.

[4] H. J. Mitchell, 'Proposal for ICI to enter the field of Plastic Materials in the British Empire', 6 Oct. 1927, Annex 1, p. 2, CR 560/1/R.

[5] Ibid. p. 2.

[6] As (1), Appendix 1.

[7] Ibid. p. 2.

[8] As (4), p. 2.

[9] H. J. Mitchell, 'British Xylonite Co.', 13 Feb. 1928, CR 560/4/R1.

[10] L. J. Barley, 'Plastic Materials Industry', 13 Nov. 1928, CR 560/4/–.

[11] ECM 1258, 26 Feb. 1929.

[12] ECM 1481, 10 Nov. 1929.

[13] As (11).

[14] F. Walker to L. H. F. Sanderson, 13 Feb. 1932, CR 560/16/16.

[15] K. Gordon, 'Suggested ICI Programme for Plastics', 17 Aug. 1932, AD 'ICI (F & SP) Ltd —Development', DB 110.

[16] CACM 475, 25 July 1932; A. E. Hodgkin, 'Plastics. Note No. 1', 19 Dec. 1932, CR 147/5/110.

[17] F. Walker to Sir Walrond Sinclair, 6 Jan. 1933, CR 147/5/110.

[18] GPCM 484, 15 March 1933.

[19] As (16), refce. 2, p. 5.

[20] Ibid.

[21] A. E. Hodgkin, 'Plastics', 27 Feb. 1933, p. 2, CR 147/130/138.

[22] Ibid. p. 2.

[23] F. D. Miles, A History of Research in the Nobel Division of ICI (ICI Nobel Division privately printed, 1955), p. 132; private information.

[24] CACM 1203, 11 Nov. 1935.

[25] GPCMs 1285, 1 June 1937, and 1318, 28 July 1937.

[26] J. L. S. Steel, ICIBR, 20 May 1948.

[27] As (3), p. 5.

[28] Ibid. p. 6.

[29] Ibid. p. 36.

[30] Ibid. p. 37.

[31] Ibid. pp. 1–2.

[32] A. E. Hodgkin, 'PF and UF—Forecast of Future Possibilities', 22 Jan. 1936, p. 1, CR 147/41/41.

[33] GPCM 1079, 8 April 1936.

[34] ICIBM 5956, 9 March 1938.

[35] A. E. Hodgkin, 'PF Manufacture', 4 Jan. 1937, CR 147/52/77.

[36] P. C. Allen, History of Polythene—Extracts from P. C. Allen's diaries September 1939—November 1945, p. 112 (15 July 1942).

[37] The discovery of polythene is well documented by those who took part:
  (i) A. S. Irvine, The History of Polythene, draft, Technical Director's Department Winnington, 27 Jan. 1950, IF.PU/396. 624/01.
  (ii) M. W. Perrin, 'The Story of Polythene', Research 6, 1953, p. 111.
  (iii) R. O. Gibson, 'The Discovery of Polythene', Royal Institute of Chemistry Lecture Series, no. 1, 1964.
  (iv) W. R. D. Manning, 'High pressure in the Chemical Industry', The School Science Review 160, June 1965, p. 541.

[38] Quoted in (37 iii), p. 3.

[39] Minutes of the Committee quoted in (37 iii), p. 10.

[40] As (37 ii), p. 113.

[41] As (37 iii), p. 18.

[42] Ibid. p. 19.

[43] Ibid. p. 23.

[44] As (37 ii), p. 116; as (37 iv), p. 550.

[45] As (37 iii), p. 29.

[46] Ibid.

[47] W. F. Lutyens to A. E. Hodgkin, 'Alketh', 15 June 1937, CR 534/1/13.

[48] A. E. Hodgkin to A. Caress, 'Polythene', 6 April 1938, CR 534/1/13.

[49] A. Caress to A. E. Hodgkin, 'Polythene', 7 April 1938, CR 534/1/13.

[50] As (36), 4 Dec. 1941, 13 Jan. 1942.

[51] Minutes of a Joint Meeting of the Research and Technical Executive Committees on 6 Dec. 1938.

[52] A. E. Hodgkin to A. Caress, 'Polythene', 28 June 1939, CR 534/1/13.

[53] P. C. Allen, 'Polythene (Reasons in support of an extension beyond 70 tons per year capacity)', 18 Jan. 1939, pp. 10–11 and generally, CP 'Polythene—Capital Expenditure'.

[54] P. C. Allen, 'Polythene and Chlorinated Polythene', 21 Feb. 1940, CP 'Polythene General'.

[55] W. F. Lutyens, 'Trading Results for the year 1949', 1 Nov. 1949, CP 'Polythene General'.

[56] Sir Robert Watson-Watt quoted in (37 i).

[57] As (36), 29 Aug. 1941.

[58] D. R. Lawson to Groups Central Committee, 12 Jan. 1939, CR 534/1/13; Research Executive Committee Minute 4/39, 24 Jan. 1939; Groups Central Committee Minute 1711, 30 Jan. 1939.

[59] A. Caress, 'Three-Year Plan (1939–41)', n.d., Report No. 141 (R/M), Summary, CR 922/78/135R.

[60] W. F. Lutyens to Group Chairmen, 18 Sept. 1941, CR J922/22/227; E. D. Kamm, 'Reconstitution of Plastics Group', Nov. 1941, CR J922/22/227.

[61] As (36), 18 Dec. 1942.

[62] W. J. Worboys, 'Some Notes on the Plastics Group', 2 Feb. 1943, CR 922/252/252.

[63] E. J. Barnsley, 'Polythene Production in the United States', 21 Oct. 1941, CR 534/1/38.

[64] As (55), p. 8.

CHAPTER 20

Mid-Century Materials II: Fibres

(i) KEEPING OUT OF RAYON: COURTAULDS AVOIDED 1928–1939

PLASTICS AND FIBRES shared similar technology, but in ICI they were seen as presenting vastly different commercial problems. With plastics, people in ICI felt fairly happy. They represented business of a kind they understood and which, they considered, fell fairly within their field: business, that is to say, in supplying other manufacturers with materials for their own activities. Fibres were different altogether. The field was dominated by Courtaulds—rich, successful, supposedly technically very expert, and valued customers of ICI. Little was known in ICI about the technique of spinning and there were alleged to be arcane mysteries in the marketing of fibres to the textile industry. Moreover the consumer market—the fashion trade, even—lay not far away, and that was no place for a self-respecting chemical firm.

'Artificial silk'—viscose rayon*—became a commercial proposition about the beginning of the twentieth century, and by 1914 it was a very lucrative one. As time went on, fibres made by one industrial process or another established themselves throughout the textile industries, in competition or combination not with silk only but with wool, cotton, and other natural materials.

Textile fibres, artificial and natural, are based on molecules arranged in long chains. In nature, cellulose provides the molecular structure of cotton, flax, and other fibres of vegetable origin. Protein, from the silkworm or from other animals, is the long-chain basis of silk and wool. The structure of cellulose and protein, animal or vegetable, can be modified industrially to yield fibres not found in nature, and these were the basis of the original 'artificial silk', derived from cellulose. Later, means were found of building up long-chain structures not found in nature. This was the origin of the true synthetic fibres, nylon, 'Terylene' and the rest. The underlying chemistry was indistinguishable from the chemistry of plastics and indeed some of the plastics which were attracting attention in the thirties and forties could be produced in a fibrous form.

* The early documents quoted in this chapter all refer to 'artificial silk' which, with 'acetate silk', was the term commonly used in Great Britain before 'viscose rayon' and 'acetate rayon', or simply 'rayon', were accepted from the USA. The text follows the usage of the documents, reflecting the gradual change in terms.

The makers of artificial silk quickly became important customers of the heavy chemical industry for caustic soda and sulphuric acid. The explosives makers, well acquainted with cellulose, could easily go into artificial silk themselves. The Nobel-Dynamite Trust dabbled with it in the nineties.[1] Du Pont did much more. In 1920, allied with a French group, they established themselves as large-scale manufacturers of viscose and, after 1923, of the related packaging film, Cellophane. ICI kept clear. Why?

When ICI came into existence the undisputed leaders of the British artificial silk industry were Courtaulds Limited. The industry was booming. Courtaulds, with an immense American subsidiary, larger than themselves, had a total income considerably greater than ICI's.[2] Moreover Courtaulds bought caustic soda and other supplies from ICI but they would be perfectly well able to go into production themselves if ICI should seriously displease them. ICI, preoccupied with the development of Billingham and with problems of reorganization and rationalization, were understandably chary of plunging into artificial silk, no matter how attractive, on grounds of pure technology, it might seem to be.

Nevertheless, in the expansive days before the Depression ICI were tempted—once. In February 1928 H. J. Hopf of Basle wrote to H. J. Mitchell offering a licence for the British Empire under patents held by the Rhodiaseta (more commonly spelt 'Rhodiaceta') company of France, which belonged to the group with which du Pont were associated in the USA.[3] The patents governed a process for making cellulose acetate and cellulose-acetate fibre, a material which was being developed in Great Britain by British Celanese (of which McGowan had once been a Director) to the considerable discomfort of Courtaulds, who relied on viscose. 'I wish we could have a shot at it', wrote L. J. Barley, and no doubt others would have shared his opinion if they had known of the proposal, but it was not to be. 'I have spoken to most of the Executive Committee on the subject, and it seems quite impossible for us to break away from the policy of supplying raw materials to all these people [the makers of artificial silk], they in their turn keeping out of our field.'[4] Hopf was politely turned away, although 'if a Company becomes established . . . in the way you propose, we shall be delighted to take up the question of its requirements of basic raw materials'.[5]

Another approach soon came from Rhodiaceta, this time through Solvay. They were sure that some one would bring the Rhodiaceta process into Great Britain, they hoped it would be ICI, and they wanted to take an interest themselves. Henry Mond certainly, and probably other Directors, were attracted by the idea of a £2½m. company, in which the group that controlled Rhodiaceta would have a large holding, which should 'co-operate with the leading artificial silk manufacturers in this country in working the process'.[6]

Above: Physical Testing
Laboratory for rubber
products, Blackley Works, 1947.

Stoking a Lancashire Boiler
at Blackley Works, 1952.

'Mouldrite' Phenol Formaldehyde Works, Croydon, as it looked when production
ceased there on 31 December 1949 (page 344).

Wilton Works, 1956 (chapter 21). A general view looking east over the trunk road,
showing, left to right, No. 2 Olefine plant, 1, 2, and 3 Polythene plants,
Bain works, and power station.

Before the proposal could be carried to a conclusion, the ICI Directors who were dealing with Solvay in the matter decided 'that, in order to keep the position clear, it would be well for ICI in the first place to have a discussion on the whole matter with Messrs Courtaulds'.[7] Perhaps they were nervous. They had reason to be, for when Courtaulds were approached it turned out that they had already rejected a proposal from Rhodiaceta, offending Rhodiaceta as they did so; that they had a low opinion of the Rhodiaceta process and thought they could do better on their own; and that they had no mind to see ICI in artificial silk either directly or indirectly, on their own or in association with others.[8]

Samuel Courtauld wrote to Sir Alfred Mond on 4 June 1928, reporting his Board's reaction to a conversation between himself, Mond, and McGowan on 22 May. He congratulated Mond on his peerage, announced on the day he wrote, and got down to business. After dismissing Rhodiaceta's proposal scornfully, he went on to acknowledge that it was natural 'that you'—ICI—'as chemical manufacturers, should have felt drawn . . . in the direction of artificial silk'. Likewise Courtaulds, starting as textile manufacturers, had 'developed on the chemical side', and they had had suggestions 'from some of the largest chemical manufacturers on the continent'—IG? Or was he bluffing?— 'to extend our business in that direction'.

If we have not encouraged such suggestions [said Courtauld], this has been due to the fact that we felt that the ground was already covered by the ICI and we did not wish to encroach upon their field. Frankly, we do not see that any good interest would be served by each of us going out of our way to invade the other's special field of activity.

The proposal for a non-aggression pact was supported by a threat:

No doubt the consumption of artificial silk in this country will increase; the business will become highly competitive, the processes specialised, and subject to continual modifications. The ICI, if it remained outside, would doubtless do a very large business in supplying the chemicals required, but it can hardly be open to question that such business would be heavily jeopardised if the ICI were to enter the trade as a competitor. I hope you will give these questions consideration before finally tying yourselves up with the French group, and I shall be glad to have a further conversation with you, if you desire it.[9]

That was the end of the Rhodiaceta proposal. As soon as Courtauld's letter reached Mond, he wrote to France saying that ICI's Directors, 'for many reasons which have no bearing on the merits of the process',

367

N

had decided they could not take an interest in Rhodiaceta.*[10] In conveying the decision to Janssen of Solvay, a few days later, Mond was more explicit: 'As you know, Courtaulds are a very powerful firm, with whom it would not be in our interest to quarrel.'[11]

Instead of quarrelling, Mond would bargain. 'We have no doubt', he replied to Samuel Courtauld, 'that you would be prepared, in consideration of our abstention from your industry, to entertain a long term agreement for the purchase of your raw materials from us, and that you would refrain from entering the sphere of chemical manufacture. I suggest that we should consider an exchange of shares as a means of linking up our interests.'[12] The idea of a share exchange was accepted by Courtaulds in principle, or so Samuel Courtauld said, but they put it off indefinitely because 'their Company had no shares in their Treasury'—meaning unissued shares—'which prevented such an arrangement being carried into effect at the present time'.[13] In fact, the proposal probably had no attraction at all for Courtaulds. Their industry was booming, they had ample cash reserves, and there was no reason in the world why they should limit their freedom of action by a marriage with ICI. Thus the 1928 proposal for an ICI-Courtaulds link, the first in a long series, came to nothing, apparently with little regret and no hard feelings on either side.

Relations between ICI and Courtaulds, during the thirties, were not to be founded on anything so formal as an exchange of shares, but on a declaration of principle formulated on 17 June 1928, when Melchett, McGowan, H. J. Mitchell, Henry Mond, and J. G. Nicholson met Samuel Courtauld and Harry Johnson. 'Mr. Courtauld and Mr. Johnson', says the record of the meeting, 'agreed to the principle that Courtaulds should abstain from entering the field of Chemical manufacture, on the understanding that ICI should similarly abstain from entering the Silk industry.'[14] That meant, of course, that Courtaulds would take supplies from ICI instead of manufacturing for themselves. 'We are still of the opinion', wrote Samuel Courtauld ten days after the meeting with Melchett and McGowan, 'that you, as large chemical manufacturers, should be able to supply us with everything that we require as cheaply as we could produce it ourselves, whilst the advantage of our keeping out of the Heavy Chemical industry is obvious.'[15]

This agreement perfectly expressed ICI's commercial policy between the wars. It was founded on the conception of ICI as, above all, suppliers of materials to British industry, and it recognized a clear separation between the legitimate functions of ICI and their customers, giving each side a right to expect that the other would respect its

* This was the way Mond put it, although the proposal was for ICI to take an interest in a new company to be set up jointly with Rhodiaceta, not in Rhodiaceta itself.

interests. Informal, unenforceable at law, guaranteed by the good faith of the parties and underpinned by the realities of industrial power, actual and potential, the understanding with Courtaulds bore a strong resemblance to understandings between ICI and Levers (later Unilever), the Distillers and other valued customers who might, if provoked, become competitors. And of course the game could be played the other way round, as when ICI's Explosives Group from time to time demonstrated that they could perfectly well make glycerine if Lever Brothers should ask too high a price. Such crudities, however, were normally held in reserve.

(ii) GOING INTO NYLON: COURTAULDS EMBRACED 1938–1945
For eleven years after 1928 ICI and Courtaulds dealt with each other, amicably enough, as supplier and customer: nothing more. ICI, unlike du Pont, kept out of rayon and cellophane. No thought of chemical manufacture, it would seem, crossed the corporate mind of Courtaulds. The two firms neither competed nor co-operated. One bought what the other sold, that was all.

Not that in ICI they put artificial fibres out of consideration: quite the contrary. From 1935 onward the subject was pursued with increasing determination, but not along the cellulose path which had led Courtaulds to prosperity.

In 1935 Professor W. S. Astbury of Leeds and Professor A. C. Chibnall of London proposed to ICI the idea of preparing a synthetic fibre from vegetable proteins dissolved in urea.[16] The idea of using protein fibres had a history running back to 1875 and they were attracting a good deal of attention in the mid-thirties. An agreement was made between ICI and the two professors and their suggestion was passed for development to the research department at Ardeer, which already had some experience of fibres and of spinning. By 1939 a practicable process was beginning to be worked out for producing, from groundnut meal, a fibre which had many of the characteristics of one of the major natural fibres: wool. By this time the original theories of Astbury and Chibnall had been left far behind and the Ardeer chemists were in full control. They were nowhere near a commercial proposition, but the signs were hopeful. They called their new product 'Ardil'. Courtaulds, at the beginning of 1939, had not yet been told about it, but it was expected they soon would be.[17] They were themselves working on a fibre from milk protein ('Fibrolaine').

In January 1937, while 'Ardil' was being developed at Ardeer, Lammot du Pont wrote to H. J. Mitchell (McGowan being on his way to Australia) to say that for some years du Pont people had been working towards 'a new type of synthetic fiber useful for textile purposes and possibly other purposes', but they had not wanted to say anything about it to ICI until they had 'a satisfactory patent position'.[18] Lammot was long-winded and not very specific, mentioning only 'that the basis of the new fiber development is not cellulose but a material known as a polyamide . . . made by chemical synthesis', but he made it quite clear that in du Pont's view the invention was important and that they would want to deal with it, as they had a right to do under the Patents and Processes Agreement, as a 'major invention' for which special terms must be made. For the moment there was no hurry—no hurry at all—but du Pont wished ICI 'to be advised of the situation' so that there should be 'no misunderstanding' as to why they had not been advised before.

This was the first official news, in ICI, of the fibre which became known as nylon, though it probably came as no surprise to those who read American scientific journals. Mitchell replied non-committally and revealed that 'curiously enough' ICI too had a new fibre—'Ardil'—which they thought so well of that it 'had to be viewed as a major development which does not appear to come under our Agreement'.[19] Du Pont made no response to that overture, and matters remained in suspense for well over a year, except that Francis Walker, on a visit to the USA in the summer of 1937, was given a good deal more information about du Pont's new fibre, which he described as 'a hard opaque resin substance melting at 285 °C.'[20]

Serious bargaining started in May 1938 when W. R. Swint of du Pont's Foreign Relations Department wrote to Mitchell saying that du Pont were ready to give ICI information on the new development and inviting 'properly qualified representatives with the object of reaching some understanding sufficiently definite to permit of full disclosure and co-operation between our two companies'.[21] On 25 May 1938 C. S. Robinson, R. E. Slade, and L. J. Barley left for the USA.

The material which the ICI mission went to investigate was known at the time as '66 Polyamide', being the condensation polymer of adipic acid $(COOH(CH_2)_4COOH)$ and hexamethylenediamine $(NH_2(CH_2)_6 NH_2)$, each of which contains six carbon atoms per molecule.[22] It was described as tough, white, and translucent, and it could be produced as a fibre approximating in appearance to silk, or as bristle.[23] It had strength, elasticity and an attractive appearance—'essentially a high grade material . . . in some respects superior to real silk', and from the first du Pont expected a great future for it in stockings.[24] They promoted it with great confidence as a first-class material in its own right, necessarily rather expensive: never as a cheap substitute for natural materials. Nylon was never allowed to fall into the same marketing trap as margarine or 'artificial silk'.

Du Pont's proposal for the licensing of the '66 Polyamide' process was being discussed in ICI during July 1938.[25] They proposed to grant ICI an exclusive licence for the United Kingdom and the British Empire (CIL to have Canada) and to parcel out the rest of the world between themselves, their French and Italian rayon associates, and IG. Under the normal working of the Patents and Processes Agreement ICI would have been entitled to a non-exclusive licence in du Pont territories, and this substantially less favourable offer was not very graciously received. 'After all,' said C. S. Robinson, 'if Europe is to be parcelled out, I see no reason why the Germans should get most of it. "66" is in a sense a luxury article and in the Empire is unlikely to be purchased by count-less millions of coloured races having little or no purchasing power. Actually the white population of the Empire is nearly the same—rather less I think—as that of the new German Reich.'[26]

Moreover, as Wadsworth explained over the transatlantic telephone to Fin Sparre on 20 January 1939, the Government was urging British industry 'to maintain its export trade in the face of the unorthodox methods of competition of the Totalitarians'.[27] Could du Pont help in this economic warfare by granting ICI a licence which would enable them to export to Scandinavia—'five countries . . . known to wish to maintain their democratic character'? The political appeal, perhaps, had some limited effect. The terms as finally granted for an exclusive licence to ICI covered the British Empire (except Canada), Eire, Egypt, Palestine, Iran, Iraq, and British Mandated Territories, and a non-exclusive licence was granted for Norway, Sweden, Finland, Estonia, Latvia, Lithuania, Poland, and Portugal.[28]

Well before this point was reached, indeed as soon as there was any serious possibility of ICI taking up '66 Polyamide', the question of relations with Courtaulds arose, because any development of this kind would be likely to represent what C. S. Robinson, in November 1938, called 'a departure from our previous policy of abstention from the manufacture of textile fibres'.[29] Courtaulds, with du Pont's consent, would at the very least have to be informed, and it seems immediately to have been taken for granted that they would be offered a share in the venture. 'With Du Pont's consent', said Robinson, 'we can approach Courtaulds merely as potential manufacturers of polyamide; alterna-tively, we can endeavour to interest Courtaulds in the wider conception of a joint enterprise to develop a more or less new industry—the pro-duction of organic chemical fibres.'[30]

There were strong reasons, or at any rate reasons which in 1938 looked strong, for attempting to realize the 'wider conception'. Courtaulds' position, both in textiles and as customers of ICI, was as imposing as it had been in 1928 and those responsible for ICI's com-mercial policy, led by J. G. Nicholson, were just as anxious not to imperil the caustic soda contract. Moreover, a move into a market of which ICI had no experience was bound to be expensive and risky, but with Courtaulds' help both risks and expense might be kept down. Then, as soon as the new fibre came off du Pont's secret list, Courtaulds showed a lively interest and themselves got in touch with du Pont, rather to the embarrassment of ICI, where minds were still being made up.[31] Du Pont, knowing ICI's ignorance of the textile trades, were 'well disposed towards the idea of a joint ICI/Courtaulds venture'.[32]

By the time ICI and du Pont signed a definitive 'Nylon Agreement', on 31 July 1939, the Directors of ICI and Courtaulds had agreed to form a joint company and to work together on the manufacture and development of nylon yarn, and representatives of Courtaulds had gone with the ICI party which gathered technical information from du Pont in the spring of 1939.[33] The two businesses thus engaged themselves to enter married life together. Some twenty-five years later, after one of the noisiest corporate quarrels in the history of British business, the marriage ended in divorce. It is worth pausing to consider what kind of a partner, in 1939, ICI was so ready to take on.

Courtaulds Limited, in 1939, controlled the largest rayon business in the world. During the thirties it did not pay nearly so well as in the boom of the late twenties—1938 was a particularly bad year—and Courtaulds' total income, which had been greater than ICI's up to 1928, was consistently less afterwards. Even in the worst years, however, dividends were paid, though not always covered, and Courtaulds' financial policy was held in great respect in the appropriate circles.[34] As part of their policy, Courtaulds' Board invested heavily in gilt-edged and local government stocks. Their holdings stood at a book value of £19m. in 1934. After that they were raided to keep dividends up, but by the time the worst was over, in 1939, they still stood at £10m.

If anyone had suspected that this conservative financial policy, with its massive investments outside the business, might be an outward and visible sign of an inward technical sluggishness, he would have been right. After the brilliant shift from silk into viscose rayon before 1914, leading to the heady prosperity of the late twenties, Courtaulds, in the words of Professor Coleman, 'wholly failed to see the implications of chemistry and to set up a serious research effort', and as a result made no adequate provision for new products in the future.[35]

Courtaulds at this time thought of themselves as part of the textile industry, and in spite of Samuel Courtauld's remarks in 1928 about going into chemicals, they would have been ill-fitted to do so. The one chemist on their Board, in the late thirties, owed his position to long service, not to any special esteem in which his subject was held. The Board as a whole was made up, on the one hand, of self-made men with a commercial or technical background in rayon or textiles; on the other, of a group round the Chairman, of considerably higher social standing, whose experience lay in quite other fields. The Chairman himself, Samuel Courtauld (1876–1947), was aloof, troubled with an uneasy social conscience, suspicious of some of the methods of 'Big Business'. F. J. Rennell Rodd, later the 2nd Lord Rennell of Rodd, appointed in 1935, was a partner in Morgan, Grenfell and had been a diplomat. Courtauld's successor in the Chair (1946–62), (Sir) J. C. Hanbury-Williams (1892–1965), became a Director in 1931 after joining the company from a firm of Oriental merchants. He was the husband of a Byzantine princess; a Director of the Bank of England; a Gentleman Usher to the King.[36]

In ICI there was no doubt a deficiency of knowledge of the textile field, and some in Courtaulds, as in ICI also, were critical of the committee-ridden organization of ICI's central management, but in scientific attainment and applied technology ICI had no equals in British industry, and certainly not in Courtaulds. In an effort to make good their deficiency ICI were about to link themselves with a partner of impressive strength in one particular field, but with nothing like their own breadth of outlook beyond it. In 1939 the strength, particularly in specialized knowledge, seemed to be what mattered, but there might come a time when breadth of outlook would be important too.

ICI's agreement with du Pont[37] dealt with nylon in three main forms: multifilament yarns, staple fibres, and monofilaments—usually called monofils. Multifilament yarn, in which continuous filaments are twisted together, is a competitor of silk for stockings and other knitted goods. For 'staple fibre' the filament is cut into short lengths like cotton staple and it has the advantage that faulty filament can be more easily dealt with than in continuous yarn, so that quality control is easier. Single, much thicker, untwisted filament—monofil—is the material for bristle. As well as developing '66 Polyamide' itself du Pont had worked out methods of converting it into all these forms on a commercial scale. They had put together a development team of some 200 chemists and engineers who succeeded brilliantly, working very fast, under the leadership of Crawford H. Greenewalt, later President of du Pont. Du Pont's knowledge of factory processes, as is commonly the case with licensing agreements, was probably the most valuable part of the package offered to ICI.

As well as defining the territories for ICI's exclusive and non-exclusive licences, the agreement laid down terms for ICI to grant sub-licences to customers for the use and treatment of nylon in its various forms. With du Pont's consent, ICI might also grant sub-licences to nylon manufacturers, a provision essential if ICI and Courtaulds were going to set up a joint company and essential also for development overseas. Elaborate sliding scales of royalties may be summarized as follows:

Multifilament yarns and staple fibres, $7\frac{1}{2}\%$ of net selling price up to 1m. lb./year falling to 3% at more than 40m. lb./year.

Monofils, 10% of net selling price up to 200,000 lb./year falling to 3% at more than 10m. lb./year.

Other forms of 'synthetic linear superpolyamides', 5% of the net selling price 'of the polymer as such' up to 2m. lb./year falling to 2% at more than 20m. lb./year.

With the terms of this agreement settled in principle, ICI and Courtaulds agreed between themselves that the Dyestuffs Group of ICI should make nylon polymer which would be passed to the jointly-owned company for spinning into yarn and to Plastics Group of ICI for making into monofil. The first polymer plant, with a capacity of 200 tons a year, was to be put up at Huddersfield; the first spinning plant (360,000 lb./year of yarn) on a site belonging to Courtaulds next to their factory at Coventry; the first monofil plant (32 tons/year) at the Plastics Group factory at Welwyn Garden City.[38]

The company set up by ICI and Courtaulds was British Nylon Spinners Limited, incorporated on 1 January 1940 as a private company with capital of £300,000 subscribed equally by the two founders. On 5 January ICI granted BNS a manufacturer's licence under the terms of the Nylon Agreement and at the same time ICI, Courtaulds, and BNS entered into a three-cornered co-operation agreement. Two 'fundamental principles' were written into the text:

The first . . . is that both ICI and Courtaulds shall obtain the major proportion of any profits which may . . . arise as a result of the Company's [BNS's] operations by way of dividends . . . on the shares held by them. . . . The second . . . is that ICI and Courtaulds . . . shall not make any substantial profit from the supply of polyamide raw materials or services to the Company or from selling commissions on the sale of products manufactured by the Company.[39]

The Agreement went on to say that Courtaulds should supply factory services and ICI should supply nylon polymer at prices representing a return of 6 per cent on the capital employed for either purpose.[40]

BNS was set up with a Board of ten, drawn equally from Courtaulds and ICI, and the Chairman had no casting vote. Courtaulds' representatives included Samuel Courtauld himself and Hanbury-Williams, who was given to persistent but not always pertinent questioning and was described by one of his colleagues as 'a superb stuffed shirt'. The most active of the original Courtaulds Directors was P. S. Rendall, who had been on the Board of Courtaulds Limited since 1937 and was by training a yarn salesman.[41] The original BNS Directors from ICI were J. G. Nicholson, J. H. Wadsworth (replaced in 1944 by A. J. Quig, a far more forceful character), W. F. Lutyens and D. H. B. Wride, a Billingham chemist who was the rather surprising choice for BNS's first General Manager. Lutyens, until Quig came in, was the most active figure on the ICI side, and his relations with Rendall and with Courtaulds generally turned sour almost at once. He disliked almost everything about Coventry, he considered Courtaulds' representatives there insufferably provincial, and he thought Courtaulds' main Board controlled their nominees too closely, which hindered the erection of plant.[42] The Courtaulds side, for their part, suspected the all-ICI partnership between Lutyens and the General Manager, Wride.[43]

In BNS's first years war made anything like normal development impossible. Polymer was imported from the USA and the very small output of yarn, first from Coventry and later from another plant at Stowmarket in Suffolk as well, was all taken up by the Government, chiefly for parachute fabric and cords.[44] Towards the end of 1943 the pressure of war, in this as in other branches of ICI's activity, began to ease a little, so that post-war planning became possible. At once the strains in the ICI/Courtaulds relationship began to make themselves felt.

The underlying cause of strain was still the same as in 1928—each party's fear that the other would become a competitor. As 'Ardil' developed, the likelihood of competition increased, especially when Courtaulds, to Lutyens's intense irritation, showed themselves unenthusiastic about anything in the nature of a joint venture.[45] The two partners, therefore, had scarcely begun to manage BNS together before they started looking for ways of keeping their wider activities apart.

Between 1940 and 1943 negotiations went on, chiefly between J. G. Nicholson and Samuel Courtauld, with occasional references to McGowan, with the aim of turning the informal understanding of 1928 into a closely defined agreement, on similar lines to ICI's agreement with DCL, in which the parties' 'characteristic fields' and an 'intermediate field' between them would be delimited.[46] At the same time the idea of a link began to be aired again. It was suggested in 1941, and McGowan consulted the Government about it, that ICI and Courtaulds should each invest £5m. in the shares of the other, with an exchange of Directorships to match.[47] Nothing came of it all. Each party was torn between the desire to avoid competition and the determination not to be shut out of profitable possibilities. Courtaulds had just undergone, in the national interest, unanaesthetized amputation of their American subsidiary. It left them badly shocked but heavy with cash—£24m.—which sooner or later they would have to find a use for.[48] In ICI everyone could see that the old boundaries between textiles and chemicals were breaking down. In the summer of 1943 a party from ICI—Rowland Hill, Frank Osborne, W. Boon, and

Peter Allen—visited du Pont for the first time since the outbreak of war. They returned much impressed, and with enough knowledge to make them quite confident that ICI could, if necessary, go into fibres on their own. After that, even with £550,000 of sales to Courtaulds at risk, no one in ICI was willing to see ICI barred from the possibility of doing so.[49]

In November 1943 the Board of ICI and the Board of Courtaulds each recorded the failure to agree and each Board exchanged minutes with the other.

Courtaulds [says the ICI Board Minute] believe that, as the manufacture of the chemicals required in the production of synthetic textile filaments is probably the greater part of the process, they cannot agree to keep out of chemical manufacture. On the other hand, ICI recognize that production of textile filaments may only be a small proportion of the requirements of some plastics materials, and that it would be impossible to legislate for textile filaments separate from the rest of the production required by other trades. There is therefore the possibility that the two Companies may find themselves in competition either in the textile or chemical field, but they have agreed that should such a position arise they will both deal with the matter on a 'good neighbour' basis. Courtaulds have declared that the last thing they wish to do is to break their friendship with ICI, and ICI have equally expressed their desire to retain and foster good relations between the two Companies.[50]

In the calm of the upper atmosphere, where Samuel Courtauld, Lord McGowan and other Olympians floated on clouds of goodwill, talk of 'friendship' between the two companies no doubt sounded well enough. At the level of Coventry works, matters were otherwise. By December 1942 Lutyens was writing to Nicholson casting doubt on Courtaulds' 'technical efficiency and driving power' and complaining of their 'unhelpful behaviour in connection with BNS affairs in general'.[51] Throughout 1943 Courtaulds conducted a campaign, clean against the wishes of du Pont and ICI, to get nylon classified as a new kind of rayon, and Lutyens took it as evidence that Courtaulds were frightened of nylon as a competitor to viscose rather than keen to develop it in its own right.[52]

In 1943, also, the 'fundamental principles' of the Co-operation Agreement (p. 374 above)—that ICI and Courtaulds should take their profits by way of dividends from BNS, not from supplies and services to it—began to look less like foundation stones of co-operation than like rocks on which co-operation might be wrecked, because the 6 per cent return which ICI had agreed to accept for polymer manufacture was beginning to appear to them totally inadequate. In agreeing to it in the first place the ICI negotiators seem to have made a serious miscalculation. It would seem they assumed that the capital employed by

Courtaulds in providing services at Coventry, and the skill and effort which they put into marketing, would more or less balance the capital employed by ICI on the chemical side, particularly in polymer plant. In fact, by the summer of 1943 it was apparent that as soon as large-scale production became possible ICI would have to put so much capital into polymer that their share of the joint venture, including their holding in BNS, would be much larger than Courtaulds'. Then, whereas the return to be expected from BNS's operations might run as high as 25 per cent on the capital employed, ICI's return from making polymer would still be limited to 6 per cent, giving them a return on their total capital commitment much lower than Courtaulds' return on theirs.[53]

Courtaulds let it be known that for their part they would be quite happy to employ some of their spare cash at 6 per cent (much better than the yield on gilt-edged) in helping to finance polymer plant. That didn't help at all. It gave Coates and Nicholson a bad fright—the last thing they wanted was Courtaulds' intervention in polymer manufacture.[54] The knowledge that they were caught on the consequences of their own misjudgement did nothing to mollify feelings of injustice on ICI's side, and on the side of BNS (Courtaulds too, no doubt) there was an equally aggrieved feeling that Dyestuffs Division were reluctant to put a proper degree of effort behind the completion of polymer plant because the return on capital was so small.[55] On all three sides, the 6-per-cent clause imported lasting resentment into the nylon project, which was poisoned by it until a new agreement was arrived at in 1951.

The exchange of minutes in November 1943 was evidently intended, on both sides, to mark a major change of policy. By getting rid of the 1928 Agreement it opened the way for ICI to advance unhindered into the field of man-made fibres, a field seen to be technically inseparable, as the wording of the minute makes clear, from the other great field which was expanding rapidly at the same time—plastics. At the time, probably, the ICI Board had 'Ardil' chiefly in mind, but products more nearly akin to nylon and the newer plastics were well known to be on the way, and the declaration left ICI perfectly free, if so minded, to develop new man-made fibres for their own sole profit, without bringing BNS into the enterprise at all.

Plans for the peace-time development of BNS nevertheless went ahead, largely under the impetus of A. J. Quig (1892–1962) who, having joined the ICI Board in 1940, was exerting increasingly powerful influence in the general direction of radical change. He had made his career in the not over-refined school of the paint and varnish business, the nearest of all ICI activities to the retail market, and it may be that he felt more at home with man-made fibres—only one remove from the fashion trade—than those members of the Board who took it for granted

377

that the natural and only respectable function for ICI was to supply industrial raw materials.

Quig became a Director of BNS in 1944. He brought in, first as General Manager and then as Managing Director, F. C. Bagnall. Bagnall, who had read Modern Greats at Oxford (1928-31), had been on the sales side of ICI for a few years in the thirties but after that in management consultancy and farming, and with this background had been chosen by Quig as more likely to be acceptable to Courtaulds than a long-service ICI man. Quig himself worked easily with Rendall. Lutyens, though he remained on the BNS Board, was persuaded to defer to Quig.[56]

BNS, thus refurbished and reinforced, was launched during 1944 on a programme aimed at producing 10m. lb. of yarn a year—about ten times the capacity at Coventry and Stowmarket—for which capital expenditure of £6m.-£7m.* would be required, to be provided equally, as the Co-operation Agreement laid down, by ICI and Courtaulds. The spinning factory was at first intended to be at Banbury, Oxon., and a site was bought there in 1941, but in the war-time and post-war controlled economy nothing could be done without building licences and 'priority'—a familiar word of the day—in the supply of scarce materials. Both were in the gift of the Board of Trade, and they were refused unless the factory was built in a Development Area, which Banbury was not. Eventually a site was found at Pontypool, Mon., just across the hills from the mining valleys of South Wales. Building started in April 1945, but it was not until December 1950 that the plant came into full production.[57]

Spinning plant for BNS was the joint responsibility of ICI and Courtaulds. Polymer plant, to supply the spinners' raw material, was for ICI alone to provide. In 1944 there was no more than a small pilot plant on a cramped site at Huddersfield. As early as November 1943 it was suggested in Dyestuffs Group (later 'Division') that the smallest full-scale plant, to be economic and competitive with du Pont (there was a prospect of supplying polymer to CIL), should be of 5,000 tons/year capacity. This size fitted conveniently with BNS's plans. It was recommended to the ICI Board by the Technical Director in March 1944, and in April 1945 Dyestuffs Division put up a proposal for investing £2·8m. in a 5,000 tons/year nylon polymer plant at Billingham, using benzene as starting material and relying on processes and techniques developed by du Pont.[58]

Billingham is a long way from Pontypool. The site was chosen partly because plant could be put up there more quickly than elsewhere and there was room to expand; partly because factory services were

* £7·7m. by March 1945 (ICI BM 10,375, 22 March 1945).

available; partly because hydrogen, ammonia, methanol, and nitric acid could all be obtained in the existing Billingham Works. The expenditure proposal forecast a starting date at the beginning of 1948. It turned out to be over-optimistic. Design and building were hindered by the post-war shortage of engineering staff, and at the end of 1946, for reasons which will be examined in a later chapter, the flow of information from du Pont was suddenly cut off. The plant finally came into production, after a good deal of acrimony, in May 1949.

By the summer of 1945, when the war ended and something like ordinary commercial development became a practical possibility, ICI was committed to large-scale development of nylon in partnership with Courtaulds. Some £7m. of capital expenditure had been sanctioned, and beyond nylon other possibilities were visible in the expanding field of man-made fibres. In other circumstances they too might have passed into the joint control of the partnership. In fact, they did not, and we must examine the emergence of ICI as a competitor both of BNS and of Courtaulds.

(iii) FAILURE WITH 'ARDIL': COURTAULDS INVITED 1947–1952

The first candidate for joint development was 'Ardil'. Lutyens, and no doubt others in ICI, felt aggrieved when Courtaulds turned it down, though they had good reason for doing so. 'Ardil' was a competitor of wool, as nylon was of silk, but whereas nylon was stronger than silk, 'Ardil' was much weaker than wool, especially when wet, so that it could not be used on its own; but if it was mixed with wool the result could not be described as 'all wool', a description very important at the lower end of the trade. 'Ardil' could never be made pure white—the nearest was a pale cream—and although it dyed readily and felt attractively soft to the touch, these advantages were not enough to offset its faults. The faults were ameliorated but never got rid of, and in Dyestuffs Division, where they knew the textile trades well, there was never any confidence in 'Ardil'. One wit suggested that the only sensible way to make a fibre from peanuts was to feed the nuts to sheep and then shear them. 'Ardil' could never be attractive unless it was cheap, and to make it cheap a great deal would have to be sold—but where was the demand to come from?[59]

'Ardil', as a marketing proposition, was at the opposite end of the scale from nylon. Nylon was so attractive that it would sell at a high price until production built up sufficiently to justify a price reduction. 'Ardil' might have sold well enough if it had been cheap, but it was very difficult indeed to build sales while it was expensive. As late as 1955 'Ardil''s production and selling costs were 47d. (just under 20p) a lb., against 24d. (10p) a lb. for viscose staple fibre. Raw material costs were similar, so were manufacturing costs, and the difference arose entirely from differences in the scale of production.[60]

ICI, backing their own judgement against Courtaulds, and supported by enthusiastic noises from the Bradford Dyers' Association and other firms in wool and cotton, prepared for a major independent investment in 'Ardil', channelling it through Explosives Division at Ardeer where the development work was done, rather as polythene was developed at Winnington. At Ardeer they predicted a market of 10,000 tons a year, a figure representing 'by far the most economical' scale of production for a unit of plant. Contemporary estimates of the likely demand for nylon five years after the war ran little more than half as high. It is difficult to resist the suspicion that the marketing estimate may have been influenced by the plant designers' calculations, especially since ICI's sales representatives in the Northern Region were only about half as optimistic as the Scots.[61]

'Ardil' was very important to Explosives Division, which needed to diversify, and to south-west Scotland, which needed new industrial investment. Moreover, a good deal of Scottish national pride was invested in the project, too. The Division Board fought hard for it, in particular to get the plant sited at Dumfries rather than in north-east England. At Dumfries there had been an agency factory during the war and the site could be adapted. A proposal for capital expenditure of £2·1m., suggesting a return of over 7 per cent even when the plant was at 50 per cent capacity, was put up in July 1947.[62] Building began in 1949, parallel—it was hoped—with the building of plant by Unilever, at Bromborough on Merseyside, to supply groundnut meal as raw material.

Shortage of groundnuts, disputes with Unilever over processing charges, delay in completing the Bromborough plant, all hindered the start of production. When production did start, early in 1951, it was just in time for a severe depression in the textile trade. Of 316 firms who had shown interest in 'Ardil' in 1947, only 76 placed small trial orders in 1951. The plant was capable of 22m. lb./year, but in the five complete years during which it was running sales never rose above a yearly rate of 2·6m. and trading losses, from 1951 to 1957, came to £3·7m.[63] Paul Chambers, writing in 1952, thought 'Ardil' was several years too late: launched in time, while the price of wool was high, it might have made good profits.[64] The final close, in 1957, lies beyond the period of this history, but it seems clear that by then it had been amply demonstrated that the inherent faults of the product, lack of wet strength and bad colour, were never likely to be cured, and that the price could never be brought low enough to attract demand sufficient to support a profitable output. Courtaulds had been sceptical of 'Ardil' from the start, and they were right.

Protein fibres as a class made nobody's fortune. Others besides ICI took them up: others were expensively disappointed. The man-made fibres which were to dominate the emerging mid-century consumers' market were synthetic polymers of which nylon was the first major example. Alongside it, from 1944 onward, another polymer of first-class importance was being developed in ICI, although not in the first place an ICI discovery. This was 'Terylene'.

(iv) THE RISE OF 'TERYLENE': COURTAULDS EXCLUDED 1943–1951

In 1939 J. R. Whinfield and J. T. Dickson set out to develop new fibre-forming materials. Outside ICI, industrial research laboratories were still rather rare in Great Britain, but the Calico Printers Association Limited had a small one, and it was there that Whinfield and Dickson started their work. They chose to investigate the polyesters of glycol, a line which Carothers, the discoverer of nylon, had dropped when it seemed that polyesters were too susceptible to hydrolysis. The du Pont patents disclaimed polyesters, leaving a very strong position open if it should turn out that Carothers was wrong.

Whinfield and Dickson worked their way quite quickly through various esters until they came to that from terephthalic acid, a relatively rare substance used in small quantities as a dyestuffs intermediate. Here they found an exception to Carothers's generalizations on the instability of polyesters: in combination with ethylene glycol it produced a poly-ester with properties both useful and patentable.[65] As a reward for their intellectual courage in challenging Carothers, they gained an ex-ceptionally strong patent, which took full advantage of du Pont's disclaimer, and the material covered by it eventually came on the market under the brand 'Terylene'. It was an invention deliberately sought and found: not, like polythene, a discovery made as the result of careful and shrewd observation of an accident of technology.

'Terylene' (polyethylene terephthalate) has most of the properties of nylon, plus some of its own. It can be hot-drawn into thin, strong filaments of high elasticity. Its melting point is high. It is less affected by water than nylon, an advantage for some purposes but a drawback in dyeing. The fibres can be 'crimped' to give something like the feel of wool, which is more difficult with nylon. In short, if Carothers had carried his investigation of polyesters to the point which Whinfield and Dickson reached a few years later, it is conceivable that he would never have taken polyamides as far as nylon.

CPA brought 'Terylene' to the notice of the Ministry of Supply and a good deal of work was done on it at the DSIR's Chemical Research Laboratory. The Ministry, convinced (quite rightly) that the material could never be got into production in time to be used in the war, asked ICI's Plastics Group, in December 1943, to evaluate it, pointing out at the same time that CPA had patent rights over it. In February 1944 the Ministry, CPA, and ICI held a joint meeting, at which the Ministry withdrew from the project, leaving ICI and CPA to settle matters between themselves.[66]

A few 'small and somewhat impure samples' of 'Terylene', examined in Dyestuffs and Plastics Divisions in the early months of 1944, were enough to convince the experts of both Divisions that ICI ought not to let the product go, especially since some one else would certainly take it up. Apart from the intrinsic properties of the material, there was a good deal to be said for it from the technical point of view. Since it depended heavily on ethylene, and since cheap ethylene from a cracker was already a possibility, albeit rather a distant one (p. 395 below), it might well be cheaper to produce than nylon, which required benzene. The manufacturing process for the polymer looked as if it would be easier than for nylon, though development work on the production of terephthalic acid would be needed. For spinning the yarn the melt-spinning process developed by du Pont for nylon would be available.[67]

ICI's negotiators, under the general authority of Cronshaw, who knew CPA well, had no great difficulty in coming to terms. As early as July 1944 P. C. Allen reported that they were willing to grant a licence on terms broadly agreeable to ICI, and a detailed agreement was recommended to the Board about eighteen months later. It granted a twenty-year licence, world-wide and exclusive, at rates of royalty falling from $7\frac{1}{2}$ per cent to 3 per cent as sales rose from 1m. lb./year to more than 10m. lb./year. CPA were also granted the right to royalties at half these rates in 'a peripheral field of potential invention, which is defined and limited by the programme of research work which CPA had projected but owing to the War had been unable to undertake to any worth-while extent'.[68]

This last provision, in effect, left CPA with an interest in inventions which they might have made if they had not granted the 'Terylene' licence to ICI. It was devised as an alternative to forming a joint ICI/CPA company. This had been CPA's original suggestion, and it was not to ICI's liking at all. They had not disengaged from the 1928 agreement with Courtaulds with the idea of immediately tying themselves up with some one else. Nor indeed did CPA look strong enough. Their representative at one meeting told Allen that they could not go on alone because development might cost as much as £100,000. ICI were determined that any new development in man-made fibres which came their way should be under their own unfettered control. They made the point clear to Courtaulds in November 1943, and in private they were determined that 'Terylene' should be kept away from BNS. Once the CPA had agreed to forgo their own proposal for a joint company, the door was locked against BNS, for as Allen observed: 'Any suggestion that British Nylon Spinners might be the medium for the development of 'Terylene' would be unacceptable to CPA in view of the fact that ICI has already refused to consider a joint manufacturing company with CPA themselves.'[69]

This was not a position which either BNS or Courtaulds could be expected to accept passively. What it amounted to was that one of the shareholders in BNS was determined to develop a new fibre, closely similar to BNS's only product, nylon, without giving either BNS or the other shareholder, Courtaulds, any opportunity of sharing in the development themselves. 'Terylene' was not at first thought of as a direct competitor of nylon, being seen as a staple fibre to mix with worsted, but nevertheless the ICI representatives on the BNS Board were in a very delicate, not to say false, position. As Directors of BNS they were bound to fight single-mindedly for BNS's advantage, but as Directors of ICI they were committed to a policy framed—quite properly, from ICI's point of view—solely in the interest of ICI. It is hardly surprising that the Managing Director of BNS, at this period, was inclined to regard them as representatives of a hostile power.

Bagnall, the Managing Director, and Courtaulds' representatives on the BNS Board fought hard, on the line that ICI were debarred from using, outside the nylon field, information supplied by du Pont for operations within it. In particular, BNS contended, ICI had no right to order melt-spinning plant from Peter Brotherhood, who had built it for BNS from du Pont designs. Brotherhood had undertaken not to supply similar plant to any third party and BNS, taking their stand on that, forbade Brotherhood to supply the equipment to ICI Plastics Division, thus effectively blocking their programme for the development of 'Terylene'. 'Thus', wrote Lutyens angrily in February 1945, 'an impasse has been reached which, if allowed to remain, will result in the ridiculous situation where neither ICI, BNS, Courtaulds nor any third party in this country can use for work on filaments outside the nylon field either this particular machinery or, in fact, any machinery or process that is employed by BNS in the manufacture of nylon, whereas it is known that du Pont in the USA have adopted this practice.'[70]

The ridiculousness of the situation must have been as apparent to the Directors of Courtaulds and BNS as to Lutyens. Evidently they were trying to force ICI to share the 'Terylene' rights with them, for Lutyens also reported that one of Courtaulds' directors (unnamed) had 'hinted that, in his view, BNS might be able to agree to ICI's request in the matter of ordering machinery if the development of 'Terylene' was handed over to BNS'.

In the spring of 1945 ICI mounted an offensive against the 'impasse'. Lutyens, visiting the USA, obtained from du Pont a statement in writing 'that ICI are entitled to use nylon machinery and process knowledge

. . . for their work on synthetic fibres other than nylon'. Reluctantly, du Pont also agreed to the grant of a sub-licence, for substantially the same rights, to Courtaulds, so long as they did not employ nylon know-how in the field of rayon.[71] This concession, Lutyens thought, might help to soften Courtaulds' attitude in the 'Terylene' dispute, but in fact, when he got back, Courtaulds refused a settlement and suggested going to arbitration. E. A. Bingen, ICI's Solicitor, thereupon put a case to Counsel (H. U. Willink and John Brunyate). They returned opinions entirely favourable to ICI's view of the dispute over the use of du Pont's information. 'Du Pont having agreed,' they said, 'ICI are free to use for the melt spinning of fibres other than Nylon, Du Pont patents etc. which fall in part within and in part outside the Nylon field.'[72]

Armed with this opinion, John Rogers moved forward in his accustomed role as peacemaker. He gave a set of the papers to Samuel Courtauld and on 22 November 1945 had 'an interesting conversation' with Courtauld and then with Hanbury-Williams. Courtauld, who had been ill, said blandly 'he had not in mind any acute differences between his Company and ICI and thought that most of the difficulties . . . which had arisen in our BNS partnership had been more or less satisfactorily dealt with and solved'. Hanbury-Williams contributed to the general air of exalted detachment by expressing the view 'that the melt spinning difficulty should be kept, as he put it, "at a high level" '.[73]

The trouble was that whenever ICI/Courtaulds relations came down to earth, where in the long run they always had to be settled, they invariably became acrimonious. In this case Courtaulds agreed, rather tardily and unwillingly, to withdraw their opposition to the placing of orders for melt-spinning equipment with Peter Brotherhood,[74] but after that they put forward a claim to a sub-licence from du Pont, not only for themselves but for BNS also, on the same terms as those granted to ICI.

At this point even the 'high level' relationship snapped. Rogers told Hanbury-Williams, in a letter of 9 May 1946, that 'but for our intervention' du Pont would probably not have been willing to grant a licence to Courtaulds at all; that ICI would certainly not press du Pont to grant Courtaulds the same terms as ICI had ('you will appreciate that the relations between ICI and du Pont are on a very special basis'); and that he—Rogers—saw 'no particular reason why the activities of BNS should extend beyond the fields originally envisaged'. 'Our discussions', he said, '. . . have already been long drawn out and I am compelled to conclude that there is no prospect of an identity of views being reached on the outstanding points.' As a consequence, ICI would not expect BNS to raise their embargo on ICI orders to Peter Brotherhood, and ICI would go ahead on their own—'unfortunate . . . but in the circumstances . . . inevitable'.[75]

In fact by this time ICI's victory was for all practical purposes complete. The attempt to block their independent development of 'Terylene', by forbidding the use of du Pont process information, had failed. ICI were free to go ahead, as they had all along intended, without regard either for Courtaulds or for BNS. BNS, no doubt, were disgruntled, and with reason, but in nylon, after all, they had a proved success with a seller's market eager for all and more than all that they could make. 'Terylene', on the other hand, though full of promise, was as yet untried, and ICI, if they had somewhat roughly seized an opportunity, had also accepted a risk.

This victory was followed by another, enabling ICI at last to raise the rate of return from nylon polymer from the 6-per-cent basis which they had so ill-advisedly agreed to in the original Co-operation Agreement with Courtaulds. The victory was achieved in an unplanned, long-drawn-out, unexpected way, and it shows what surprising results may follow from a law-suit—on this occasion, the anti-trust suit *US v. Imperial Chemical Industries*.

In the autumn of 1946 the President of du Pont, Walter S. Carpenter, arrived in London, accompanied by Wendell Swint. What they were demanding, as part of the defence tactics in the anti-trust suit which we shall examine in chapter 24, was the cancellation of the 1939 Nylon Agreement with ICI. ICI had no option but to agree, and a new Licence Agreement was sealed in November 1946, to take effect at the end of the year. One consequence, and a serious one, was that all technical information from du Pont was cut off. Another consequence, with which we are immediately concerned, was that ICI were obliged to come to new terms both with BNS and with Courtaulds.

First, they gave each a present. Under the new agreement with du Pont, ICI were able, at some cost to themselves, to offer BNS licences in the nylon field at $2\frac{1}{2}$ per cent, which was less than they themselves were paying for licences which also covered fields other than nylon. At the same time they could offer Courtaulds the long-desired licence for melt-spinning outside the nylon field. On these terms Courtaulds agreed to 'take all steps in their power to secure that British Nylon Spinners . . . raise no objection to the conclusion by ICI of the new agreement [with du Pont]', and with both the BNS shareholders in agreement with each other—albeit for reasons which had nothing to do with BNS—there was no doubt what conclusion the Board of BNS would come to. 'It is believed', wrote Bingen, 'that the solution reached will clear some of the difficulties . . . experienced between ICI and Courtaulds in the past and lead to easier relations in the future.'[76]

Then ICI moved on to the 6-per-cent grievance. Their relations with Courtaulds and BNS were governed by the Co-operation Agreement drawn up in 1940. When a new Nylon Agreement was made between

ICI and du Pont, a new Co-operation Agreement must follow. Presented with this necessity, ICI were determined to turn it into an opportunity for getting the rate of return on nylon polymer increased, the more so since large-scale polymer plant was coming into production at Billingham, and BNS, thanks partly to low polymer prices based on the 6-per-cent rate of return, were becoming highly profitable. The prospect of an increased return on ICI's polymer capital was not particularly welcome either to the Managing Director of BNS, who was dissatisfied with Dyestuffs Division's performance at the polymer plant and wanted his polymer prices fixed and low, or to Courtaulds, who had indicated their willingness to join with ICI in polymer production on a 6-per-cent basis. Accordingly the negotiations were long and sometimes acrimonious, particularly when ICI sent BNS a debit note for £47,358 15s. 10d. representing interest at 6 per cent on the cost of plant under construction but not in production.[77] BNS took advice, refused to accept the note, and never paid. The affair as a whole gave almost unlimited scope for accountants' and lawyers' subtlety, and more delay was injected into the proceedings by the continuing hope, on the side of Courtaulds and BNS, that ICI might be persuaded to let BNS handle 'Terylene'.[78]

The issue was brought to a head by the rising demand for polymer. When BNS asked ICI to consider extending the capacity of the Billingham polymer plant by 20m. lb. a year, ICI were immediately placed in a strong bargaining position, since there was no alternative supplier. The final negotiations were carried on, for ICI, by Fleck. For BNS, the negotiator was P. S. Rendall, Vice-Chairman (the Chairman was an ICI Director, A. J. Quig). Courtaulds were represented by their new Technical Director, Alan Wilson. When the new Agreement was at last made, on 17 June 1951, it provided ICI with a return of 11 per cent on fixed and working capital, very carefully defined, employed in the production of nylon polymer.[79]

It is a curious reflection on the unpredictability of judicial processes, and perhaps on the quirkiness of historical causation generally, that an action started in the USA for the purpose of destroying restraints on American foreign trade should have had the effect of providing ICI with the opportunity of revising, in their own favour, financial provisions of an agreement under English law which they had long regretted entering into but from which, until the anti-trust suit caused the cancellation of the Nylon Agreement, they had seen no way of escape.

The decision to commit ICI to independent development of man-made fibres was taken, in the abstract, in November 1943, and it very soon took concrete shape with the unexpected offer of the rights to 'Terylene'. It was confirmed by the outcome of the warfare over melt-spinning plant between 1944 and 1946. By the time that was over, the eventual size of the new industry that was coming into being was beginning to be hazily apparent, and in the autumn of 1946 the far-reaching implications for ICI were being explored, particularly in two of the Divisions most closely concerned, Plastics and Dyestuffs. Two members of Dyestuffs Division, K. W. Palmer and R. Hill, in what they described as 'some personal musings', asked themselves, in November 1946, 'To what sort of size will the field of development for synthetic fibres in this country open out over the next twenty or thirty years?' They answered: 'A field that is tremendous in scope; a field that, measured in terms of capitalisation, might be as big as ICI and Courtaulds are together at the present time.'[80]

Official papers of about the same date were rather less expansive in tone, but there was no disposition to play down the possibilities of fibres. As in the related field of plastics, really large tonnages (by the standards of the day) began to be talked about, and not for one fibre only, but for several—nylon, 'Ardil', 'Terylene', the new 'Fibre A' (acrylonitrile) which du Pont were developing into 'Orlon' from a polymer, discovered in Dyestuffs Division, for which they had devised a solvent which enabled it to be spun by the dry-spinning process. On this basis, the possibilities of fibres were rated much higher than the possibilities of plastics. This turned out to be wrong, but at the time it seemed reasonable because polythene still looked much more like an expensive speciality for the electrical industry than like a cheap commodity for the mass consumer market. In the fibres field, on the other hand, it was fairly easy to calculate demand for clothing and thence to forecast a large market not merely for one man-made fibre but for several, and du Pont's booming nylon business in the USA (16,000 tons' output in 1946, with 40,000 authorized, and a prospective investment of $175,000,000)[81] lent substance to optimism.

All this would take time and the costs would be heavy. Development of 'Terylene', up to the end of 1950, cost about £5m. For expenditure on fixed assets, Allen in October 1946 expected to see an investment of about £7·75m. in nylon, £7m. in 'Terylene', and £5·45m. in 'Fibre A'. 'Fibre A' was left to du Pont, but expenditure on 'Ardil' came to £2m.–£3m. When the estimate for the first 'Terylene' plant finally came it was for £8·6m. rather than £7m. and, with working capital of £1·1m., came nearer £10m.[82]

The formal proposal for ICI's first 'Terylene' plant, of 5,000 tons, was put forward in September 1950. This proposal, comparatively small in itself, pointed towards massive investment, if it succeeded, in a new system of chemical industry. But it might not succeed. In a paper supporting the proposal Caress observed that even after seven years' development work there were still 'many gaps in our knowledge'.

McGowan, aware of the risks as well as the possibilities, presided at the decisive meeting. He was 76 and not at his best after lunch, but in the morning, Allen recalls, he was fully in control of the situation and asked all the right questions. The decision to invest was made.

The success of this decision, along with others which committed ICI to large-scale development in the emerging mid-century chemical industry, depended on plentiful raw materials from organic sources and on adequate manufacturing arrangements. Neither could be improvised from the resources of existing Divisions and each interacted with the other. We must take up again our examination of ICI's search for sources of heavy organic chemicals and the related decision to develop a very large multi-divisional manufacturing site at Wilton in the North Riding of Yorkshire.

REFERENCES

1 *ICI* I, pp. 177, 382–3.
2 D. C. Coleman, *Courtaulds*, 2 vols., Clarendon Press, Oxford, 1969, Vol. II, p. 255.
3 M. J. Hopf to H. J. Mitchell, 3 Feb. 1928, CR 123/2–5/–.
4 L. J. Barley, 'Note for Mr. Martin', 21 Feb. 1928, CR 123/2–5/–.
5 L. J. Barley to M. J. Hopf, 15 Feb. 1928, CR 123/2–5/–.
6 H. Mond to E. Janssen (Solvay), 23 Feb. 1928, CR 123/2–4/–; see also cable to Mitchell, 22 Feb. 1928, and 'Notes on informal meeting between Mr. Henry Mond and M. Desquin on 22 March 1928', CR 123/2–4/–.
7 'Notes of a Meeting . . . 3rd April 1928', CR 123/2–4/2.
8 Sir Frank Spickernell to H. J. Mitchell, 28 April 1928, CR 123/2–4/–; 'Interview with Mr. S. Courtauld, 22nd May 1928', CR 123/2–4/–; S. Courtauld to Sir A. Mond, 4 June 1928, CR 123/2–4/–.
9 As (8), 3rd refce., p. 3.
10 Sir A. Mond to M. Gillet, 5 June 1928, CR 123/2–4/4.
11 Sir A. Mond to E. Janssen, 8 June 1928, CR 123/2–4/5.
12 Sir A. Mond to S. Courtauld, 12 June 1928, CR 123/2–4/–.
13 'Interview with Courtaulds: 17th July 1928', CR 123/2–4/–.
14 Ibid.
15 S. Courtauld to ICI, 27 July 1928, CR 123/2–4/–.
16 'A Perspective of the Preparation of Protein Fibres', 29 June 1939, CR 700/33/97.
17 'Ardil Development. Notes of a Meeting . . . on 14th February 1939', DECP.
18 L. du Pont to H. J. Mitchell, 14 Jan. 1937, CP 'Du Pont General'.
19 H. J. Mitchell to L. du Pont, 5 Feb. 1937, CP 'Du Pont General'.
20 F. Walker, 'Du Pont's new synthetic Fibre', 9 June 1937, CP 'Du Pont General'.
21 Du Pont Committee Monthly Report for May 1938.
22 A. Caress, '66 Polyamide. Plastics Group Interest', 5 Oct. 1938, CR 302/20/20R.
23 'Nylon', June 1939, CR 302/20/20R.
24 Ibid.
25 MBM 90, 27 July 1938; '66 Polyamide Development—Du Pont Continental Arrangements', 28 July 1938, CR 302/20/20R.
26 C. S. Robinson to J. G. Nicholson, 25 Nov. 1938, CR 302/20/20; Overseas Executive Committee Minute 127/38, 5 Dec. 1938.
27 W. F. Lutyens, 20 Jan. 1939, CR 302/20/20.
28 W. F. Lutyens to Development Executive Committee, 22 Feb. 1939, CR 302/20/20.

[29] C. S. Robinson to J. G. Nicholson, 3 Nov. 1938, reporting telephone conversation with Du Pont, CR 302/20/20.

[30] C. S. Robinson, '66 Polyamide', 11 Nov. 1938, MBSP 15 Nov. 1938.

[31] Ibid.

[32] As (26), 1st refce.

[33] W. F. Lutyens, 'Nylon. Progress made during 1939', 5 Feb. 1940, DECP.

[34] Coleman II, p. 321.

[35] Ibid. p. 495.

[36] Ibid. pp. 224–6.

[37] 'Summary of Nylon Agreement dated 30th March 1939', 12 May 1939, CR 302/20/20.

[38] As (33).

[39] 'ICI, Courtaulds Limited and BNS, Co-Operation Agreement', 5 Jan. 1940, Art. VIII(1), CP 14/1/3.

[40] Ibid. Art. X, XI.

[41] Coleman II, p. 332.

[42] W. F. Lutyens, 'ICI/Courtaulds Relations', 2 Feb. 1945, CR 700/2/2.

[43] Private information.

[44] 'British Nylon Spinners—Formation and Progress of the Company: 1940–1947', 20 Sept. 1947, ICIBR.

[45] As (42).

[46] See: (i) W. F. Lutyens to Group Chairmen, 'Proposals for a written agreement with Courtaulds', 4 Oct. 1940, and Note by W. H. Coates, 9 Oct. 1940, CP 14/1/4.
 (ii) Numerous drafts in CP 14/1/4.
 (iii) MBM 713, 15 Jan. 1941; ICIBM 8,159, 13 Feb. 1941.

[47] J. G. Nicholson, 'ICI and Courtaulds', 9 Oct. 1941, MBSP; letters between S. Courtauld and J. G. Nicholson, 3 Feb., 1 April, 23 April 1942, CP 14/1/4.

[48] Samuel Courtauld to J. G. Nicholson, 3 Feb. 1942, CP 14/1/4.

[49] W. F. Lutyens to J. G. Nicholson, 'ICI and Courtaulds', 14 Dec. 1942, CP 14/1/4; information from Sir Peter Allen.

[50] ICIBM 9,645, 11 Nov. 1943.

[51] As (49).

[52] As (42).

[53] Coleman II, p. 332.

[54] As (42).

[55] Private information.

[56] Ibid.

[57] See: (i) As (44).
 (ii) Formal letters referring to increase of BNS capital, 23 Nov. 1944, ICIBR 4 Dec. 1944; 13 March 1945, CR 830/2/2.
 (iii) ICIBM 10,375, 22 March 1945.

[58] See: (i) Nylon Steering Committee Minutes 10 Nov. 1943, CR 830/92/143.
 (ii) ICIBR (Technical Director) March 1944.
 (iii) Dyestuffs Division Proposal A5/45, 19 April 1945, CP 14/1/3.

[59] J. Craik, ' "Ardil" . . . A Survey of the Position in Sept. 1952', CR 3/1/1; Fibres Division, 'Examination of the "Ardil" Project', 22 Aug. 1957, CR 16/3/1.

[60] 'The Cost of Production of "Ardil" Fibre', 9 Aug. 1955, CR 3/1/5.

[61] Explosives Division to Board of Trade, 21 Nov. 1947, CR 3/1/1; commercial discussion, 18 Aug. 1948, CR 700/1/9.

[62] 'Ardil', supporting memo to form A, 6/46, 8 July 1947, CR 3/1/1.

[63] As (59), 2nd refce.

[64] S. P. Chambers, 18 March 1952, CR 16/3/1.

[65] Hardie and Pratt, pp. 217–19; ICIBM 10,797, and Annex I, 13 Dec. 1945.

[66] ICIBM 10,014, 13 July 1944, and supporting papers.

[67] Ibid; as (65), 2nd refce.

[68] Ibid.

[69] As (66), 'Terylene', p. 3.

[70] W. F. Lutyens, 'ICI/Courtaulds Relations', 2 Feb. 1945, CR 700/2/2.

[71] 'Mr. Lutyens' Report on Visit to USA and Canada May–June 1945', CP 14/1/3.

[72] 'Case for the Opinion of Counsel, October 1945', CR 123/1/1; 'Questions submitted to Counsel and the Answers thereon', p. 3, 16 Oct. 1945, CR 123/1/1.

[73] John Rogers, 'Courtaulds and ICI', 27 Nov. 1945, CP 14/1/4.

[74] S. Courtauld to J. Rogers, 'Re British Nylon Spinners Limited', 23 Jan. 1946, CR 123/1/1.

[75] J. Rogers to J. Hanbury-Williams, 9 May 1946, CP 14/1/4.

[76] E. A. Bingen, 'Nylon', 4 Nov. 1946, p. 4 and attached letter, Hanbury-Williams to Bingen, 28 Oct. 1946, p. 2, ICIBR.

[77] 'BNS Board Meeting convened for 27 Feb. 1948', CR 830/14/14.

[78] A. J. Quig, 'The price of Nylon Polymer to BNS', 24 July 1950, CR 830/6/6.

[79] ICI/Courtaulds/BNS Agreement of 17 June 1951, clause 6, CR 830/45/45.

[80] K. W. Palmer and R. Hill, 'ICI and the development of synthetic fibres; some personal musings', 30 Nov. 1946, CR 700/5/5.

[81] R. Hill and K. W. Palmer, 'Memorandum for Executive Committee', 10 Oct. 1946, CR 700/5/5; W. F. Lutyens, 'Synthetic Fibres', 4 Oct. 1946, CP 14/1/3; Annex 1 to R. A. Lynex, 'Synthetic Fibres', 28 Oct. 1946, CP 14/1/3.

[82] 'Synthetic Fibres Development in ICI', 3 Oct. 1946, pp. 7, 11, CP 14/1/3; Plastics Division, 'Proposed "Terylene" Plant', 8 Sept. 1950, CR 700/10/19; 'Report to accompany Form A', 11 Sept. 1950, CR 700/10/19; verbal information.

CHAPTER 21

The Road to Wilton 1943–1951

By THE END OF 1947 the Board of ICI had sanctioned some £25m. of investment, by various Divisions, in a centre of chemical industry, at Wilton across the Tees from Billingham, of a kind formerly unknown in Great Britain.[1] £25m., to be invested by 1952, was a starting figure, and the capital invested rapidly grew beyond it as the various projects gathered way. The decision to invest on this scale at Wilton was the biggest decision of its kind taken since the collapse of the fertilizer industry in the early thirties, which had put an end to ICI's first design for prosperity. It marked, in the most emphatic way, a new start in a new direction, and it provided a base from which ICI could take off into unprecedented expansion in the chemical industry of the fifties and sixties.

The establishment of Wilton, like the establishment of Billingham many years before, was a massive declaration of faith in a new system of chemical industry. It represented the convergence of developments which we have been examining in earlier chapters, since it was intended to provide, on one site, the raw materials, the services, and the space which would be required for large-scale activity in the newest branches of ICI's business. The concentration of diverse yet interdependent functions on one site was the outcome of a new approach to problems of organizing production. At the same time a new approach was emerging to the national and international organization of the chemical industry as a whole. By the time Wilton began to come into production, in the early fifties, the world of the cartel-makers—of the founders of ICI and their generation—had passed away.

At the beginning of 1943 nobody in ICI could exactly foresee how the business would develop after the war. Nevertheless the general outline of things to come was taking shape, so much so that on 31 March the Development Executive Committee met, under W. F. Lutyens and Sir Holbrook Gaskell (Technical Director), 'to discuss what immediate action should be taken to determine where new manufacturing capacity should be placed for new and existing ICI chemical products'.[2] The meeting looked at several 'inter-related problems'. The main one was a proposal for a 4,000-ton unit for nylon polymer, which would require extensions to hydrogen and ammonia plant as well as plant for synthetic phenol and other organic chemicals. Greater capacity was also needed for electrolytic chlorine and for the ammonia-soda process, and there was a tentative proposal for 'a possible combined ammonia soda/muriate plant', intended for ammonium chloride fertilizer.

These projects were intricately interconnected, partly by their relationship to nylon and partly by dependence on salt as a raw material. It was proposed, therefore, to put them all on the same site, and that site a new one, since existing ICI sites were already too full to allow room for expansion. As time went by the list of projects was drastically revised, and the main one—the project for a large nylon polymer plant—eventually took shape on an existing site at Billingham South. The idea of a composite site, however, was never given up, and it is fair to regard the deliberations of this committee meeting, in 1943, as the genesis of Wilton.

The site that was needed would have to be near salt deposits and cheap fuel, with access to the sea for the import of materials and the shipping of finished goods. It would require water for cooling and process, and an outlet for effluent into a tidal river or the sea. Preferably it would be in a 'development area' so that the authorities would be disposed to grant the indispensable building licences and priorities for scarce materials.[3] Right from the start it was recognized that a site near the mouth of the River Tees would meet all the requirements listed above. It would also allow access by pipeline to Billingham for ammonia and hydrogen. The Post War Works Committee consulted the Divisions and by April 1944 representatives of both had picked on the Wilton Estate on the south side of the Tees between Middlesbrough and Redcar.

The property finally bought, late in 1945, is shown on the map opposite. It consisted altogether of some 3,500 acres and it included a nineteenth-century mansion, Wilton Castle. The block required for industrial development was 1,841 acres of flat agricultural land, sloping slightly towards the Tees, between Wilton village and the Middlesbrough–Redcar road. The remaining land, to the south of Wilton, partly farmland, partly woods, was chiefly intended as a site for housing. The owner was Colonel Lowther and the price agreed for the 1,841 acres was £143,500. The price for the whole property was £190,000.[4]

From the start it was recognized that from the point of view of the Divisions the deterrent to opening up a new site, rather than extending on an old one, was the cost of providing roads, railways, offices, laboratories, drainage and other services.[5] The opportunity which the expected post-war developments presented, of spreading these charges widely over different projects, seemed, at any rate at Millbank, to be too good to miss, but obviously the greater the spread, the better. Great reliance was therefore placed, to start with, on Dyestuffs

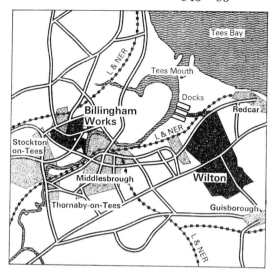

Figure 16. The Wilton Site

Source: Press handout, 10 Dec. 1945, CP 'Wilton Works'

Division. Not only had they the largest construction programme in ICI, but their existing sites would soon be full or overfull. Blackley in particular was unattractive for new developments, and it began to be suggested 'that eventually Blackley would cease to exist as a manufacturing site'.[6]

In Dyestuffs Division the idea of 'Site X' was at first received with enthusiasm. The Division was self-confident and ambitious, with great plans for a post-war dyestuffs output of 32,400 tons, to go into markets captured from the Germans, the Americans, and the Swiss. Moreover the Division regarded itself as the main centre in ICI of organic chemical industry, so that 'future manufactures of a large scale type such as nylon or synthetic rubber',[7] as R. S. Wright put the matter in November 1944, would fall naturally within its scope. There was consequently a great deal to be said for a new site, where plant for dyestuffs and intermediates would stand alongside plant manufacturing the primary inorganic raw materials, and where 'the development of new techniques' would gain impetus from 'the availability of cheap hydrogen and possibly ethylene or other manufactures based on imported raw materials such as oil'. The Dyestuffs' planners began to have visions of a new layout on the largest scale, like the great Bayer complex at Leverkusen. In time—say, 10–15 years—all the Division's earlier factories, at Huddersfield, Blackley, and elsewhere, could be given up.[8]

In Dyestuffs Division they were thinking in this expansive way throughout the early part of 1944, while the consultations about 'Site X'

393

were going on. Gaskell, perhaps because of wartime preoccupations, perhaps because of ill health, perhaps because of both, seems to have been slow to realize quite how large the Dyestuffs ideas were. When he did, in August 1944, he was, in his own words, 'rather horrified'. 'As far as I can make out', he wrote to Lutyens, 'the potential output . . . would be something like five times as much as the total output of all the existing works including the extensions . . . which it was originally intended should go to Huddersfield and Blackley.'[9]

Dyestuffs Division, represented usually by W. H. Demuth, the Chairman, and P. K. Standring, Technical Managing Director, thus came into collision with the Technical Director of ICI. Frustrated in their long-term ambitions, they began to argue that building on Site X would take too long for them to catch the market after the war, and they had better make extensions on their existing sites. They had a good deal of support at Millbank, where Cronshaw, lately the Division Chairman, had joined the ICI Board in November 1943, and Fleck thought it would be useless to try to force an unwilling Division on to the Wilton site, even though personally he wished to see them go there.[10] By the summer of 1945 it was beginning to look as if the whole Wilton project, Dyestuffs' support being largely withdrawn, might founder, and a Chairman's Conference at the end of July did nothing to resolve the dispute.

Where a Chairman's Conference failed, a discussion between Gaskell, Rogers—ever the conciliator—and Cronshaw seems to have succeeded. On 1 August 1945 Gaskell wrote to Demuth: 'Dr. Cronshaw when he sees you will be able to give you a much clearer idea of what is actually in our minds than would be obtained by reading any detailed notes of the somewhat discursive conversations . . . in the Board room on Wednesday last.'[11] A few days later he revealed a major change of plan:

. . . the whole Site X question has assumed a rather different aspect during the last week or two and it now looks as though we should have to concentrate first on a cracking plant and manufacture a certain number of products derived from the hydrocarbon gases from this plant and that consequently the direct Dyestuffs work will probably come rather later in the order of priority.[12]

Behind this announcement by Gaskell lay two fundamental decisions, recently taken, on the major technical question with which this Part of this volume opened: how best could ICI provide the heavy organic chemicals required for the new branches of the chemical industry which were in prospect? The first decision was that the best source of these materials would be petroleum. The second, which directly affected the future of the Wilton site, was that ICI should put up cracking plant for themselves rather than relying on the oil industry to do it for them.

These decisions were profoundly important to the development of ICI over the next quarter-century and more, and to the relations of ICI with other businesses, particularly the Distillers Company Limited and the Anglo-Iranian Oil Company Ltd., the predecessor of British Petroleum Ltd. We must see how they were arrived at.

(ii) ANGLO-IRANIAN, DISTILLERS, AND THE PROBLEMS OF PARTNERSHIP

Until some time in 1943, if not later, the view generally taken in ICI, on the subject of sources of heavy organic chemicals, was that petroleum was not of much interest. There was very little refinery capacity in Great Britain and in any case ICI's foreseeable demand, except perhaps for acetone (from propylene), hardly seemed likely to produce the economies of scale which naphtha-cracking offered as its chief advantage. 'For the present', wrote Lutyens in December 1941, 'there is no incentive for ICI to become interested in sources of ethylene other than alcohol, as the existing demand is small and prospective changes in the chemical industry are not likely to increase it materially.'[13] He was careful to point out that his remarks were 'not intended to represent a final view'.

By the summer of 1943 the view was in fact changing. Kenneth Gordon, having remarked that the development of organic chemicals in Great Britain had been held up by 'the unduly favourable position enjoyed by producers of fermentation alcohol'—that is, the Distillers—'and by the reluctance of ICI to compete', went on to point out that 'to enter into the complete range of aliphatic organic chemical compounds, it will be necessary to have access to cheap supplies of crude petroleum or natural gas'.[14] In February 1944 Lutyens decided that the whole subject ought to be comprehensively surveyed, not least because it seemed unlikely that the advantage conferred on the Distillers by the 'inconvenience allowance' would go on for much longer.[15]

In April 1944 R. F. Goldstein and S. H. Oakeshott of Dyestuffs Division reported on the importance of ethylene and propylene to ICI. It lay, they said, 'in the great reactivity of the unsaturated hydrocarbons and secondly their potential availability in enormous quantities at very low cost.' Here lay the crux. How 'enormous' would be the quantities that ICI would require? Would they be 'enormous' enough to ensure the 'very low cost'? Goldstein and Oakeshott could only make estimates within limits so wide as to be almost meaningless: 5,500 to 19,000 tons a year of ethylene, they thought, might be required, 'split between ethylene for polythene [2,200 to 8,000 tons] and its interpolymers [400 to 1,600 tons] and ethylene for ethylene oxide', and their estimate for propylene—an unspecified figure above 2,500 tons—was even vaguer. Nevertheless they were in no doubt that if ICI did not have access to ethylene and propylene at world competitive prices, 'this gap will be filled by competitors, such as the Oil Companies, British

Celanese or Carbide and Carbon, with loss to its current and potential interests'. They were in no doubt, either, that 'the only method of obtaining both these gases at world competitive prices is by the cracking of imported gas oil' and that 'in the event of ICI taking no definite action, the control of the aliphatic field will pass into the hands of others which may well prove a crippling handicap'.[16]

Others were more cautious. 'On the information immediately available within ICI,' wrote Ronald Holroyd and others in June 1944, referring particularly to ethylene, 'it is unsafe to make a clear-cut recommendation between the oil route and the molasses route since the cost of raw material is so important and so dependent on external factors in the arbitrary control of other companies.'[17] There were already negotiations in train, however, as Holroyd and his co-authors pointed out, which might be expected to clear the matter up, for in February 1943 Sir William Fraser, the Chairman of the Anglo-Iranian Oil Company, had approached J. G. Nicholson, whom he knew well (the two were 'Willie' and 'Jack' to each other), suggesting co-operation in the chemical industry between AIOC and ICI. On 17 February 1943 Fraser, with other representatives of AIOC, entertained Nicholson, Quig, Lutyens, and Gordon to dinner at the Dorchester. Fraser took the line of the national interest. 'Co-operation', he said, 'could only result in mutual benefit and greater speed, and Britain in the post-war period would require all the concentrated effort of loyal nationally minded combines such as ours.'[18] Nicholson agreed, saying that in 'the peculiar economic situation' of the post-war years 'it will only be the good fellowship and co-operative understanding of concerns such as ours that will enable us to get through this period', and Fraser rounded off these patriotic exchanges with the remark: 'I feel we made some progress . . . on lines . . . which we both feel might not only benefit our respective interests but do something for the Old Land.'[19]

The suggestion for co-operation was not new. We have already seen (p. 321 above) that AIOC had approached ICI more than once during the thirties. So had Standard Oil and there had been contact with Shell and with Trinidad Leaseholds.[20] There had never been any binding agreements, however, because the common ground between the oil industry and the chemical industry was so criss-crossed with prior commitments that it was hard to pick a way through them. It was especially hard for ICI. Nothing could be done, without du Pont's consent, which might imperil the security of information transmitted under the Patents and Processes Agreement, and du Pont were above all wary of leaks to the oil companies.

ICI had rebuffed AIOC before the war, but in 1943 their incentive to reach an agreement was much stronger. A joint committee of the two companies, with Quig as Chairman, pushed matters along so smartly that by May 1943 they had a draft basis of collaboration on paper. It provided for a joint company, equally financed by ICI and by AIOC, to manufacture 'intermediate products required for the chemical industry from raw materials available from AIOC'. In a passage strongly reminiscent of the BNS Agreement, it laid down that 'since the joint organisation is intermediate between activities of AIOC and activities of ICI it is fundamental to its whole success that both partners should aim at maximum profits to the Joint Company which will then be divided 50/50.'[21]

The activities proposed for the joint company at this stage pre-supposed a large chemical works alongside AIOC's refinery at Abadan on the Persian Gulf, as well as operations in Great Britain. From natural gas and waste refinery acid, ammonium sulphate fertilizer could be made. The ethylene in refinery gas could be converted to alcohol for convenient transport to the United Kingdom, where it would be converted back to ethylene: alternatively, 'special gas oil' or naphtha could be supplied for cracking in the United Kingdom to yield ethylene. With salt and electric power, caustic soda and chlorine could be made. The refinery would take (but not sell) caustic soda and some hydrochloric acid manufactured from the chlorine. The surplus chlorine could be used as such, or in the manufacture of bleaching powder or ethylene dichloride by combination with ethylene. Finally, propylene from refinery gases would provide a source of acetone for ICI.[22]

The success of the scheme rested on a balance of advantages. ICI and AIOC both hoped to profit from an exchange of technical knowledge and skill. Both hoped for an assured supply of cheap or cheapish raw materials. Neither expected the joint company to compete in any way with its own established activities. The balance was delicate, just as it was in the BNS arrangement between ICI and Courtaulds, and if it were to be upset the aggrieved party would not be likely to look very kindly on the joint company.

In November 1943 Heads of Agreement between ICI and AIOC provided for an even more ambitious scheme of co-operation. The joint company, Petroleum Chemical Developments Limited, was to 'promote co-operation . . . not only in the "petroleum chemical field" but also in all fields which both parties agree are of mutual interest . . . and where one or other of the parties has the necessary special materials or services available . . .'.[23] PCD itself was not to engage in manufacture. It was to investigate and recommend projects which would each be handed to an operating company formed for the purpose with Ordinary capital supplied by PCD. In time, therefore, PCD would become the holding company and directing centre for a joint ICI–AIOC group in petro-chemicals—no doubt a very large and powerful group indeed.

The ten voting Directors were to be drawn equally from ICI and AIOC. The ICI representatives included Lord McGowan, Sir (1944) John Nicholson, and A. J. Quig, and one of the AIOC nominees was Sir William Fraser. The Managing Director, without a vote, was Kenneth Gordon. He came to PCD from the direction of ICI's petrol enterprises, which by 1944 were fast losing whatever shreds of commercial credibility they had until then retained, so that even Gordon's fiery enthusiasm could not survive. He transferred it to petro-chemicals, along with his intolerance of delay and scorn for infirmity of purpose. If PCD needed technical brilliance, leadership, and drive, then the company probably had the best managing director in the business, but if patience, tact, and diplomatic subtlety were also important, the choice was scarcely a happy one.

PCD, so near birth, was never born. Briskness and optimism foundered in quagmires of delay. Du Pont, perhaps rather unwillingly, gave verbal consent to the agreement in November 1943 but did not confirm it in writing until a year later, and at the same time sent a copy of their letter to the State Department.[24] Solvay, through Baron Boël, protested loudly and at length. Boël disliked the whole proposal, considering that it would block ICI-Solvay exchange of information in the petro-chemical field after the war, and while Boël was still unsatisfied McGowan was unwilling to have PCD registered.[25]

Then the State Department intervened. 'As you know,' wrote B. F. Haley, Director of Economic Affairs, to Wendell Swint of du Pont, on 3 January 1945, 'it is the established policy of this Department to encourage and promote the international exchange of technological information on a non-restrictive basis. However, since the proposed arrangement which you have submitted'—that du Pont should pass technical information to PCD—'is to take place in the framework of the du Pont-ICI Patents and Processes Agreement, which is the subject of legal proceedings by this Government, this Department is unable to approve the proposal.'[26] Du Pont protested, but the Department remained firm, and du Pont were obliged to withdraw their consent to the disclosure to PCD of technical information transmitted to ICI.[27]

This was a heavy blow at the foundations of the whole edifice, and another was about to fall. Sir William Fraser was growing anxious—somewhat belatedly, it may be thought—about the terms of AIOC's concession in Iran, and whether the Iranian Government would accept the PCD agreement, with its proposal for a large chemical works at Abadan. Nobody in AIOC, it seems, had consulted the Iranian Government before the Heads of Agreement were initialled, or sought a competent legal opinion on them, and when Gordon was in Abadan in the winter of 1944-5, putting together a list of fifty-three new projects for PCD's attention, AIOC's men on the spot 'raised with Major

Gordon the question of whether the proposed activities of PCD would be permissible and within the limits of the concession held by AIOC in Iran'.[28] They raised the same question at home, naturally, and Fraser consulted Dr. V. R. Idelson, KC, an international lawyer of British nationality but Russian extraction, who had advised Fraser when the concession was negotiated in 1933. Idelson's opinion was anything but reassuring. As Fraser put it in a letter to 'Harry'—Lord McGowan—Idelson said that 'instead of having one tripartite agreement covering all PCD agreements, we should have two or more, of which one would relate solely to Persia'.[29] He made ominous remarks, too, about anti-trust dangers in the USA.

Sir William Fraser found all this very alarming. Co-operative arrangements of the PCD type, in 1945, were politically very unpopular, and, as he said at a meeting with ICI on 19 April 1945, he did not want 'to be tainted with the "cartel brush" ', especially as he had Government directors on the AIOC Board. Moreover, he had the Iranian Government to think about as well, and on top of everything else he was 'exceedingly apprehensive of the text of the PCD Agreement being asked for by the American authorities if once it got enmeshed with the du Pont/ICI relations'. He had not, he said, 'been able to give his attention to PCD as much as he felt he should', but what he wanted was 'an organisation . . . to emphasise the faith they had in a big development of first class national importance, but the activity would still be based on love and not on fear'.[30]

Sir William's anxiety, combined with Idelson's legal proposals, wrecked any chance there might still have been of setting PCD up as the centre of power in a group of manufacturing companies. Instead there appeared a plan for a company limited to technical service and research, leaving operations in Iran firmly under the control of AIOC. Gordon, who was barely on speaking terms with Sir William Fraser, relieved his feelings in a letter to his own Chairman, to John Rogers, who had taken Nicholson's place as chief negotiator with AIOC, and to Fleck:

The limited experience to date [he wrote bitterly] shows that an independent development company is nobody's friend. It is only welcome for so long as it is only a name. The moment it starts to formulate a practical scheme, it finds itself on territory in which some one considers he already has rights.

There were no objections to PCD schemes in Iran, until they were put down as practical projects during our recent visit. Then they became intrusions into AIOC's own business; the fact that the three months' visit was thus rendered futile did not appear illogical. I have even been restrained from circulating to my colleagues our report on the visit.

The sphere of the company having been whittled down to an organisation for carrying out research, which the ICI [—Gordon always wrote of 'the

ICI'—] could do much more quickly and efficiently itself, there seems to be every reason for carrying out this process to its logical conclusion and dispensing with the Development Company altogether.[31]

This complaint, imputing to AIOC jealousy of PCD, would have imported a new element of rancour into relations between ICI and AIOC. Rogers and Fleck were careful not to raise the issue. Gordon's vehemence did him no good and probably helped to destroy him. He was a man twice disappointed, having twice committed himself passionately to enterprises which seemed to promise great things, and then failed, and perhaps his experience with Tube Alloys was hardly less bitter. In 1948 he left ICI and stormed successively through Trinidad Leaseholds and Head Wrightson before he became Director-General of Ordnance Factories at the Ministry of Supply in 1952. In 1955 he died, at the age of 58.

In the spring of 1945 plenty of material existed for a quarrel between ICI and AIOC. Rogers and Fleck, on the ICI side, and Sir William Fraser in AIOC, were evidently determined that it should not break out, and under their emollient direction of affairs the proposal for a joint company remained alive, though failing, for another two years.

Nevertheless, when Gordon questioned the need for a joint company he may have been intemperate in his expression of opinion but he was broadly right in his judgement of affairs. If a joint company could neither have free access to information from du Pont—and the State Department's hard line was reaffirmed in the summer of 1945—nor control manufacturing companies in Iran, then the reasons for bringing it into existence were attenuated to the point of disappearance. It would be left with no function which either ICI or AIOC could not perfectly well carry out for themselves. From ICI's point of view the chief remaining attraction of a joint venture, apart from the provision of capital by AIOC, would be 'special fractions of oil for cracking, the supply of oil at advantageous prices, assistance in cracking-plant design and disposal of oil products'.[32] But there was no reason to suppose that these facilities might not be obtainable in the ordinary way of business, if not from AIOC, then from Shell.

On the other hand, ICI's need for a source of supply of petro-chemicals, whether they owned it themselves or shared the ownership with someone else, was rapidly becoming more urgent. In the Budget of 1945 the Chancellor, as he had long been expected to do, destroyed the fiscal advantages of making ethylene from alcohol. The 'inconvenience allowance' was withdrawn with effect from 1 January 1946 and at the same time import duty was remitted on hydrocarbon oils to be used in Great Britain for making chemicals. Both measures greatly strengthened the case for putting up cracking plant designed as part of the chemical industry—that is, to supply heavy organic chemicals, leaving motor fuel as a by-product—rather than as part of the oil industry, in which motor fuel would be the principal product. There was now, in fact, no question that ICI would need access to such a plant, and we have already seen (p. 394 above) that Wilton had been chosen as the most likely site for it as far back as the summer of 1945, though other possibilities—at Runcorn and at Heysham—were not for some time entirely ruled out.

The negotiations for joint operations with AIOC moved into their last phase with 'Project A', a proposal by ICI for large-scale cracking operations at Wilton. In the form discussed between April and September 1945, later much modified in detail but not in principle, the scheme called for 140,000 tons a year of specially selected oil from Abadan. It would be cracked to give the maximum yield of olefines—that is, for chemical manufacturers' materials, not for fuel—and it was expected to produce 18,000 tons of ethylene, 21,000 tons of propylene, and 15,000 tons of butylenes. The capital cost was estimated at £2·1m.[33]

The Budget of 1945 thus helped to force the issue of establishing cracking plant for producing chemical raw materials from oil. Inevitably it thereby forced another issue, too: the relations between ICI and DCL, for if DCL intended, as they did, to remain in the chemical industry, they would have to move from molasses towards oil. The problem was not sudden. It had been clearly in prospect for at least three years and recognizable for many years longer. After the Budget of 1945 it could be put off no longer: for DCL a crisis was at hand.

ICI's relations with DCL were governed by the agreement negotiated by Nicholson in 1938 (p. 324 above) which provided for the parties to keep out of each other's elaborately defined 'characteristic fields' and only to compete—but competition was not really of the essence of the matter—in an equally elaborately defined 'common field'. The agreement was in the restrictive, defensive mood of the thirties and earlier. It was never popular with Quig, Gordon, and others of the more aggressive spirits in ICI, for they considered it far too inhibiting to technical progress, and if it came to public notice in the political climate of 1945 it would certainly not be well regarded, especially with a Labour Government in power.

Cracking light oils (naphtha) for chemical raw materials would yield a wide range of products within DCL's 'characteristic field', and within the 'common field' a range even wider. Quig, in a paper drafted for him by L. Patrick, went so far as to say that 'all the more obvious products for exploitation—isopropanol and acetone, secondary butanol and MEK [methyl ethyl ketone] in particular—were in the DCL characteristic field and many others within the common field'. Of those in the common field the most important was ethylene, for if that were made from oil then ICI would no longer need ethyl alcohol for the purpose and would, moreover, be able to make alcohol in competition with DCL. Ethyl alcohol, in Quig's words, was 'the most characteristic DCL product of all',[34] and competition would be most unwelcome.

DCL would have to be told of ICI's proposed joint operations with AIOC and the possibility of conflict would have to be faced, but as early as July 1944 it was being suggested within ICI that, whatever DCL might do if the 1938 agreement broke down, 'the position may be regarded with equanimity'.[35] That remained the bedrock on which ICI's negotiators stood. The importance of heavy organic chemicals had become so great that they were not to be deterred from their chosen course of action even by the prospect of quarrelling with a customer worth, in 1943, £415,000, chiefly for soda ash and methanol. On the other side of the account, DCL supplied ICI with goods worth £996,000 a year, and they faced the ruin of that trade as oil displaced molasses. No wonder ICI could regard the position 'with equanimity'.

ICI's negotiations with AIOC were revealed to C. G. G. Hayman of DCL in November 1944. He in turn revealed, probably not greatly to the surprise of anyone in ICI, that DCL had an oil partner in mind too—Trinidad Leaseholds. However, said Hayman in February 1945, DCL were prepared to drop Trinidad if they were admitted to the ICI–AIOC schemes.[36] This proposal would avoid a breach with DCL and was not unwelcome to ICI's Board, who were at first inclined to favour a joint enterprise owned in equal shares by all three partners. But when at length, in September, a firm offer was made to DCL, after preliminary conversations conducted, for ICI, by Quig, the terms were much less generous. It was proposed that ICI and AIOC should own the cracking plant, DCL being firmly and explicitly shut out, and products from it would be sold to DCL, and to all the world besides, at commercial prices. DCL would thereby get supplies of isopropanol for conversion to acetone, which was important to them, and ICI would agree to leave the conversion of isopropanol in DCL's hands 'so long as the resulting acetone were sold more cheaply to ICI than ICI could manufacture it themselves'. It was hoped, said Quig—but can he have meant it?—'that sufficient inducement had been offered to DCL to meet their natural aspirations'.[37]

The inducement was nothing like sufficient. Before the offer was made, in conversation with Quig and Gordon, Hayman had indicated that DCL had found AIOC 'difficult to deal with', and later, writing to Quig, he remarked that the situation 'appeared likely to create less amicable relations' between DCL and ICI. He reminded Quig that DCL's business in aliphatic solvents had been developed mainly to meet

ICI's needs and that ICI had been such large buyers that they had been able 'to insist upon prices much below market level'. He had claimed the acetone market as DCL's 'legitimate sphere of operations' and had asked for a share in the cracking plant on the ground that DCL's demand for propylene would be complementary to ICI's demand for ethylene, so that the joint requirements 'should go a long way towards a balanced cracking programme', from which a jointly owned company would supply the gases to ICI and DCL 'on such a basis as would give a moderate return on the capital'.[38]

Quig was determined, notwithstanding the half-formed intention of the Board, to keep DCL out of the cracking plant. 'However modest [their] share might be,' he wrote to Gordon in November 1945, 'I am quite certain that Hayman would regard it as a pre-emptive right to all new developments.'[39] At meeting after meeting in the autumn of 1945 he pressed his views with his colleagues and with the representatives of DCL. Fleck, Rogers, and Coates all felt uneasily that ICI were going back on earlier conversations with DCL and on their own Board's decision, and Rogers, at one meeting, was prompted to inquire 'whether there would be any feeling of "dirty tricks" '. Quig maintained, in reply, that 'there could be not the slightest charge of ICI having gone back on a definite offer', but whether Hayman would have agreed with him is doubtful. Certainly Hayman resented ICI's preference for AIOC over DCL as a partner.

Nevertheless Quig had his way. The executive Directors, at the meeting on 4 December 1945 at which the doubters expressed their misgivings, came to four conclusions:

1. ICI have no occasion to fear a break with DCL.
2. ICI cannot accept DCL as partners in the cracking plant, though prepared to sell DCL products on best favoured terms for equal tonnage.
3. ICI will therefore go ahead with manufactures based on petroleum but lying within DCL characteristic field. If this is unacceptable to DCL, then the DCL/ICI Agreement must terminate.
4. The foregoing policy needs discussion with DCL at Board level.[40]

All that remained was to write the new policy into the minutes of the Board, which was done at some length on 20 December 1945, and communicate it to DCL. Rogers and Quig met Hayman and two of his colleagues on 30 January 1946. They expected Hayman to denounce the 1938 agreement, but he did no such thing, having evidently come to the conclusion that DCL would gain nothing from a fight with ICI and would do better to seek partners in petro-chemicals elsewhere. He simply asked for time for the Distillers to consider their position and went quietly away.[41] The agreement was not in fact ended until 1948, when at Quig's suggestion the ICI Board agreed to give notice to terminate. 'I am thinking', Quig explained, 'of the impact of the Monopoly (Inquiry & Control) Bill upon this and similar commercial arrangements, which seemed reasonable enough in pre-war days but may now be deemed to be against the public interest.'[42]

With the decision to keep DCL out of the cracking plant, and at the same time to go into their 'characteristic field', very much to their displeasure, a new hard line began to appear in ICI's relations with other businesses. It can hardly be accidental that the appearance of the new line of policy coincided almost exactly with the retirement, on 31 July 1945, of Sir John Nicholson, the architect of the 1938 DCL agreement. It was not only the methods of the thirties that were being challenged, but the traditions of almost a hundred years of regulated competition in the chemical industry. Some of the pressure, as Quig recognized, came from the movement of public opinion, but a strong push, as we have seen, came from within ICI itself.

If co-operation with DCL was unnecessary—indeed, undesirable— was there really anything to be said for co-operation with AIOC either? The question arose on 4 December 1945 when the Directors were considering their change of policy towards DCL, and Quig asked whether they would reconsider their policy towards AIOC, too. One somersault at a time was enough, and the matter was pushed aside as a digression. 'It was generally acknowledged that whatever the pros and cons, ICI were now irrevocably committed to the 50 : 50 basis with AIOC.'

From that time, nonetheless, ICI's policy towards AIOC settled to a line as hard and unwavering as the line of their policy towards DCL. ICI and AIOC set up a Joint Policy Panel, but as ICI's ideas developed, during 1946, the AIOC members became steadily more disenchanted with ICI's proposals for Project A. Fleck, who was in charge of them, was perfectly clear about what he wanted, and although his behaviour towards AIOC was more tactful than Gordon's, he was no less determined to have his way, being convinced that what he wanted was reasonable and right. Broadly speaking, the aim which he consistently pursued was to set up at Wilton plant for cracking light petroleum distillate, to be supplied by AIOC, for separation of the gaseous products, and for their primary chemical conversion into compounds needed in the chemical industry. The principal object of this plant, from ICI's point of view, was to supply heavy organic chemicals, especially ethylene, cheaply and in large quantities. Beyond that, in ICI they were determined that the products of the joint company should neither require information derived from du Pont through the Patents and Processes Agreement nor compete with products already manufactured by ICI.[43]

A firm proposal for Project A, within these limitations, was endorsed by Fleck and accepted by the Board in October 1946. It recommended the investment of £3·18m. in plant at Wilton designed for an output of 27,000 tons a year of ethylene. The quantity was not expected to be required until 1953, and the detailed figures of the proposal were worked out on estimates for 1950. 210,287 tons of light distillate, supplied by AIOC, would be cracked to yield 22,420 tons of ethylene and 23,550 tons of propylene, leaving as by-products 22,404 tons of fuel oil for the plant and 79,428 tons of petrol for AIOC.[44]

ICI intended to take all the ethylene at £40 a ton, which was about half the current cost of ethylene from Distillers' alcohol. They intended to use it as follows:

for polythene	5,790 tons
ethyl chloride	3,910
ethylene dichloride	870
ethylene oxide and ethylene glycol	11,850
	22,420

The propylene was to be pumped to Billingham, hydrated in existing plant, modified for the purpose, to wet isopropanol, and pumped back to Wilton for conversion to 2,480 tons of pure isopropanol for sale and 14,020 tons of acetone, chiefly for ICI. There was a difficulty here, because ICI's offer of isopropanol to DCL was still open, provided they could convert it to acetone more cheaply than at Wilton. DCL, however, had not said that they could, and in ICI it was taken for granted that they couldn't. 'It is assumed therefore', as the author of the expenditure proposal put it, 'that these difficulties are overcome.'

Gross realizations from the plant, for ethylene, isopropanol, acetone, and di-isobutylene (the value of by-product petrol was taken in as a credit against the cost of feedstock), were estimated at £1,485,175, against total annual expenditure (including obsolescence) of £1,242,622. The profit, at £242,553, represented a net annual return on £3,180,000 of 7·63 per cent.

When this proposal was discussed with AIOC at a meeting of the Joint Policy Panel the atmosphere, Gordon reported, was 'rather strained because of the feeling held by AIOC that ICI had reserved to themselves the most lucrative parts of the project'. What they chiefly meant was that ICI were now insisting on taking for themselves the entire manufacture of ethylene oxide and ethylene glycol, whereas AIOC had previously assumed that the joint company would be allowed to share in the trade. ICI, pointing to their trade of 2,000 tons a year in ethylene glycol and quoting Dyestuffs Division's manufacture of ethylene oxide, contended that both products fell within the range of their established interests. AIOC observed that the claim was based on 'prior commercial production of these products *by another route from non-petroleum sources* [WJR's italics]',[45] implying that they thought it came close to sharp practice.

ICI's determination to limit the activities of the proposed joint company finished the project for co-operation in petro-chemicals with AIOC. Sir William Fraser and his colleagues, in correspondence with John Rogers and at meetings with Fleck, tried to bring about a change of mind, but they failed. 'Once the door is open to the new joint company', Coates wrote to Fleck, 'to produce products already produced by ICI, or on which we have done serious work, it seems difficult to see the limits of the new company's encroachment on our interests.'[46] And encroachment ICI were in no circumstances prepared to tolerate.

In AIOC, other arguments were developing against the joint project. Whatever form it took, it would certainly be based at Wilton, because ICI had gone too far with development to wish to draw back and the cracking plant was essential to the viability of the whole project for a composite site for chemical plant. In AIOC, on the other hand, the wisdom of putting cracking plant at Wilton was beginning to look doubtful. Would it not be much better to put it alongside an oil refinery?

Broadly [wrote Sir William Fraser to John Rogers on 12 December 1946], the situation has undergone some change since we started to consider this matter in that the prospects of larger refining operations in this country are now very real. We are now aware that other oil companies are likely to be following a UK refining policy, so that apart from any projects that at present may be planned for the production of chemical products . . . we must expect competition from petroleum interests. It is therefore, I feel, of vital importance that any venture on which we embark should be started on a basis that will always be competitive with others, and I very much fear that the Wilton project does not answer to this description.

An operation to manufacture about 40,000 tons of chemicals from a raw feedstock of some 200,000 tons, leaving a large proportion in the form of petroleum products requiring finishing treatment can obviously be much more economically dealt with if such plants are located alongside a refinery.[47]

On 30 January 1947 three directors of AIOC explained their dissatisfaction with Project A, at length, to Fleck, Gordon, and Lutyens. Committed, on the one hand, to supplying ICI with cheap ethylene, and debarred, on the other, from profitable dealings in ethylene derivatives, the proposed joint company was an unattractive proposition to AIOC as it stood, and the limitations on the use of du Pont information made future developments problematical. Since no possibility of getting it altered emerged from the ICI side, the AIOC party turned it down. They did it with elaborate politeness, thanking Dr. Fleck for the work he had performed as Chairman of the Joint Policy Panel 'and for having contributed so greatly to the harmony of the meetings by presiding over them'.[48] Ritual letters, equally amicable, passed between Rogers and Fraser towards the end of February 1947, and the three-year-old project for ICI/AIOC co-operation, started with such patriotic enthusiasm and businesslike dash, peacefully died.[49]

At once the two partners rejected by ICI turned towards each other. In September 1947 DCL and AIOC announced a £5m. plan for the joint manufacture of chemicals in the United Kingdom, in close association with the refining operations of AIOC. ICI, it may be thought, had little to complain of, but Fleck was disgruntled. 'In view of the close nature of our recent discussions with the Anglo-Iranian,' he said, with a disarming air of injured surprise, 'I consider that that Company has not been as frank with us as we had reason to expect.'[50]

(iii) WILTON ESTABLISHED

In the early part of 1947, all partners cast aside, ICI stood on their own at Wilton. It was a site of a kind quite new to ICI, needing a new kind of organization, in which the necessary independence of Divisions to run their own factories would have to be reconciled with the degree of central control required to provide the cheap common services which were of the essence of the whole conception.

In May 1946 Sir Ewart Smith, recently returned from Government service to succeed Sir Holbrook Gaskell as ICI's Technical Director, presented a scheme to the Board. The central feature was the Wilton Council to co-ordinate activities on the site. Its members were representatives of the Divisions which had works there and its Chairman had the standing of the Chairman of a Division, with a General Manager under him. On the ICI Board a Group Director would be responsible for Wilton's affairs. The running of the individual works would be the job of the Divisions they belonged to.[51]

No one needed reminding that friction between the Divisions, or between the Divisions and the Wilton management, could very easily arise. A great deal would depend on the personality of the Group Director and even more on the personality of the Chairman of Wilton Council. Alexander Fleck was the Group Director. The Chairman was J. W. Armit, a Nobel man who had been Director General of Explosives and Chemical Supplies in the Ministry of Supply during the war, and afterwards briefly Chairman of Leathercloth Division. He took office at Wilton in May 1946. The first meeting of the Wilton Council was held in November.

On 13 November 1947 Fleck laid before the Board an estimate of the profit expected from capital which it was planned to invest in the first stage of development at Wilton, up to 1952.[52] The pattern was as follows:

	Capital	Return on Capital (Net)
	£m.	%
Billingham Division		
Olefine Plant	4·8	12·6
Ethylene Oxide and Glycol	2·3	12·6
Plastics Division		
Urea formaldehyde and Woodfilled Moulding Powder	1·1	12·4
Phenol formaldehyde Moulding Powder	1·1	14·5
'Perspex'	1·0	15·8
Dyestuffs Division		
'Lissapol' N	0·6	26·9
a-Naphthylamine	0·4	12·5
Alkali Division		
Polythene	1·5	10·0
General Chemicals Division		
Chlorine (2 units)	6·7	3·2
Wilton Works		
unproductive Capital	6·0	
Total	25·5	7·6
of which:		
Divisional Capital	11·8	
Wilton Works Capital	11·8a	
Working Capital	1·9	

a This figure represents the provision of services for the Divisions. It includes £6m. 'unproductive capital', i.e. capital invested in services not yet fully employed.

The scheme had changed radically from the scheme originally discussed in 1943, centring round Dyestuffs Division's projected nylon polymer plant, which in the event was built at Billingham. The only stable feature between the two schemes was the provision of a large chlorine plant. But although the details had altered, the principle remained. The roots of ICI's post-war development were being sunk in one composite site, which had been the intention from the start.

Wilton Works officially opened on 14 September 1949, when Lord McGowan planted a tree to mark the occasion. The cracking plant started up in July 1951. By 1956 over £57m. capital was employed at Wilton, including £15·5m. in a 'Terylene' plant. ICI was fairly launched into the new chemical industry, then in the full exuberance of expansion.

These matters are not our concern in this volume. We have examined, in this chapter, the converging pressures of technical development, working often in unexpected ways, which in the mid-forties combined to shatter the mould in which the chemical industry had for many years been set—certainly since the end of the Great War, and in many ways for much longer than that. We have seen the radical changes which were set in motion within ICI and in its relations with other businesses in Great Britain, particularly Courtaulds, DCL, and AIOC. We must turn now, to end this account of ICI's first quarter-century, to the most spectacular change of all: the enforced abrogation of the Patents and Processes Agreement with du Pont.

REFERENCES

[1] A. Fleck, 'Wilton Report for Board Meeting on 13th November 1947', 7 Nov. 1947, CP 'Wilton Works'.

[2] DECP 121, 19 April 1943.

[3] ICIBR (Technical Director), T 10/44, April 1944.

[4] ICIBR (Technical Director), T 15/45, 21 Sept. 1945; ICIBM 10,750, 8 Nov. 1945; ICIBR (Technical Director), T 17/45, 19 Nov. 1945.

[5] As (3).

[6] Ibid.

[7] R. S. Wright, 'Site X in relation to the post-war building requirements of the Dyestuffs Division', 10 Nov. 1944, PWWCP 74.

[8] Ibid.; PWWCP 31, 27 Jan. 1944.

[9] H. Gaskell to W. F. Lutyens, 12 Aug. 1944, CR 93/858/859.

[10] (i) As (7).
    (ii) 'Plants for Post-War Construction', PWWCP 75.
    (iii) PWWC, Minutes of 33rd Meeting, 16 Nov. 1944.
    (iv) H. Gaskell to PWWC, 12 Dec. 1944, CR 93/858/859.
    (v) C. J. T. Cronshaw, 'The Present Position in regard to Site X', 6 Jan. 1945, PWWCP 81.
    (vi) H. Gaskell, 'Site X', 23 July 1945, CP 'Wilton Works'.
    (vii) W. H. Demuth, 'Notes on Chairman's Conference 25 July 1945', CR 93/134/134. And other papers, chiefly in CP 'Wilton Works' and PWWCPs.

[11] H. Gaskell to W. H. Demuth, 1 Aug. 1945, CR 93/134/134.

[12] H. Gaskell to F. L. Clark, 10 Aug. 1945, CR 93/134/134.

[13] W. F. Lutyens to Group Chairmen, 2 Dec. 1941, CR J261/11/11.

[14] K. Gordon, 'Relations between Oil Industry and Chemical Industry', 7 July 1943, CR 261/11/16.

[15] W. F. Lutyens, 'Raw Materials for the future manufacture of Heavy Organic Chemicals', 19 Feb. 1944, CR 261/11/11.

[16] R. F. Goldstein and S. H. Oakeshott, 'The Importance of Ethylene and Propylene to ICI', 17 April 1944, CR 534/3/59.

[17] R. Holroyd, L. Patrick, J. Payman, F. H. Peakin, 'Raw Materials for Heavy Organics— Ethylene', 29 June 1944, CR 261/11/11.

[18] 'Notes of a Meeting with AIOC', 17 Feb. 1943, CR 59/8/587.

[19] Correspondence between J. G. Nicholson and Sir William Fraser, 18–19 Feb. 1943, Nicholson Papers.

[20] 'AIOC–ICI, 1st Meeting of Collaboration Committee', 11 March 1943, CR 59/8/587.

[21] AIOC–ICI, 'Draft Basis of Collaboration', 7 May 1943, CR 59/8/587.

[22] Ibid.

[23] 'Notes of Proposed Arrangement between ICI and AIOC', 10 Nov. 1943, Nicholson Papers; Minutes of Meeting held 16 Nov. 1943, Nicholson Papers.

[24] 'AIOC/ICI, Third-Party Agreements affected by PCD Agreement', 9 May 1944 (revised 26 May 1944), pp. 6–7, ICIBR; Wendell Swint to E. J. Barnsley, 1 Nov. 1944, Nicholson Papers.

25 ICIBR (Development Director) 8/45 and Annex; L. Patrick, file note, 'PCD and Solvay', 1 Jan. 1945, CR 39/31/31.
26 B. F. Haley, Dept. of State, to Wendell Swint, du Pont, 3 Jan. 1945, CR 39/15/16.
27 E. J. Barnsley to W. F. Lutyens, 26 Jan. 1945, CR 39/15/16.
28 K. Gordon and J. Ridsdale to W. F. Lutyens and others, 28 Feb. 1945, CR 39/14/14.
29 Sir William Fraser to Lord McGowan, 22 Feb. 1945, CR 39/14/14.
30 PCD—'Meeting at Britannic House', 19 April 1945, CP 'PCD Ltd'.
31 K. Gordon, 'Notes on the present position of PCD', 15 May 1945, CR 39/14/14.
32 K. Gordon at 1st meeting of the Petroleum Chemicals Management Committee, Nov. 1945, CR 39/26/28.
33 Paper for Discussion at Chairman's Conference, 10 Sept. 1945, CR 39/12/12.
34 A. J. Quig, 'Relations with Distillers Co. Ltd.' (by L. Patrick), 10 Dec. 1945, CP 'Distillers'.
35 'The Future of ICI/DCL Relations', 20 July 1944, CR 161/10/10.
36 As (34), p. 1.
37 Ibid. pp. 2–3.
38 K. Gordon, 'Petroleum Chemicals', 11 Sept. 1945, CR 45/3/3; C. G. Hayman to A. J. Quig, 21 Sept. 1945, CR 45/3/3.
39 A. J. Quig, 'Petroleum Chemicals', 6 Nov. 1945, CR 39/14/14.
40 As (34), Appendix B, 'Meeting of Executive Directors'.
41 'Meeting with Directors of DCL, 30 Jan. 1946', CP 'Distillers'.
42 ICIBM 12,233, 27 May 1948; A. J. Quig, 'Agreement with DCL', 25 May 1948, ICIBR.
43 A. Fleck, 'Petroleum Chemicals Development', 4 Oct. 1946, CP 'P C Developments'; Sir William Fraser to J. Rogers, 19 Nov. 1946, CR 39/12/12; minutes of Joint Policy Panel, 1 April 1946, CR 39/14/24; minutes of Petroleum Chemicals Management Committee Meeting no. 10, 9 Oct. 1946, CR 39/26/28, and many other papers.
44 As (43), 1st refce.
45 AIOC Paper, 'ICI/AIOC', 4 Nov. 1946, CR 39/12/12.
46 W. H. Coates to A. Fleck, 'Discussions with the Anglo-Iranian Oil Co.', 23 Sept. 1946, CP 'PCD Ltd.'.
47 Sir William Fraser to J. Rogers, 12 Dec. 1946, CR 39/14/14.
48 ICI/AIOC Joint Policy Panel minutes, 30 Jan. 1947, CR 39/14/24.
49 Correspondence between John Rogers and Sir William Fraser, 26 and 27 Feb. 1947, CR 39/14/14.
50 ICIBR (Group C Director), AA 73/47, 2 Sept. 1947.
51 ICIBM 11,064, 23 May 1946.
52 A. Fleck, 'Wilton Report for Board Meeting on 13th November 1947', 7 Nov. 1947, CP 'Wilton Works'.

PART V

FROM CARTELS TO COMPETITION
1939–1952

CHAPTER 22

The Cartel System under Strain

1939–1944

WE COME NOW TO THE CLIMAX of our story. From the late 1860s up to the outbreak of the Second World War competition in the national and international markets of the chemical industry was regulated by the kind of arrangements, often very elaborate, which go under the name of 'cartels'. Under the impact of war and technological change those arrangements began to fall apart and then from 1944 onward political pressure, originating in America, swept through them like a gale until the du Pont Agreement, so long the central pillar, if not the sacred ark, of ICI's commercial policy, was destroyed by a judgement under American anti-trust law. With it the system of regulated competition vanished, and the subsequent history of ICI, if it comes to be written, will be set against a very different background.

When war broke out the long-established cartel system in the world chemical industry was at its ultimate pitch of labyrinthine complexity, because it was essentially a means of dealing with the characteristic commercial problem of the depressed thirties: price competition in over-supplied markets. ICI's business was governed by 800 agreements or thereabouts,[1] and in the international field with which this chapter is chiefly concerned there were half a dozen or so agreements of cardinal importance.

Three cartels, the origins of which have been discussed in earlier chapters, regulated with a good deal of success world trade in nitrogen, in dyestuffs, and in the products of hydrogenation. The alkali business was governed by agreements between ICI and Solvay, the US Alkali Export Association (Alkasso), and the Russians, who for a handsome consideration kept their alkali off the export market. Underpinning the whole was the Patents and Processes Agreement with du Pont, of which it was said, in 1938, 'as ICI's obligations to du Pont are world wide and cover the majority of ICI products, it follows that all agreements or understandings, wherever contemplated, may have some bearing on the ICI/du Pont Agreement.'[2] It was renewed in 1939.[3]

War immediately disrupted a large part of this network, for apart from du Pont and, over a more limited field, Solvay, ICI's closest partner in the international chemical industry was IG. The two were linked in the three great cartels, of which they were the main architects and upholders, and in numerous other agreements. Apart from the advantage of regulated competition, the value of technical information flowing from Germany was freely recognized in ICI. It covered many subjects, but it was generally held to contribute more to the making of petrol and related products, under the terms of the International Hydrogenation Patents Agreement, than to anything else.[4] Petrol became steadily more important as the probability of war with Germany turned into certainty; and of course it worked the other way as well, in so far as information from ICI reached IG by way of the IHP patent pooling arrangement.

During the years immediately before the war new links had been forged between ICI and IG. In 1936 each took equal shares, along with Solvay, in the Aetsa Electrochemical Company, newly founded in Finland.[5] In 1938 ICI and IG jointly set up the Trafford Chemical Company to make dyestuffs in Great Britain, and IG carefully arranged to keep its own technical secrets although control of the company lay with ICI.[6] At the end of the same year ICI, IG, and Solvay signed long-term technical and commercial agreements for the regulation of the alkali trade.[7] There was constant consultation, bargaining, and co-operation over business in South America, and in 1939 ICI and Solvay made a commitment to IG to form a joint company in Argentina—Electroclor—which quickly became embarrassing and eventually led to unfounded suspicions of trading with the enemy.

All these elaborate arrangements, as well as the great cartels for nitrogen and dyestuffs, were swept away by the war. Co-operation with partners in neutral and friendly countries did not cease at once, but the German conquest of Europe practically abolished the categories 'neutral' and 'friendly' as far as nitrogen and dyestuffs on the Continent were concerned. Solvay itself became an enemy company, though dealings with Baron (later Count) René Boël, Solvay's representative outside occupied Europe, remained lawful, and the Solvay American Corporation remained free to manage large funds in the USA. The IHP Agreement remained in force because American and British oil companies were parties to it, but the decreasing importance of hydrogenation made it increasingly unlikely to survive in peacetime markets.

As well as destroying the great European cartels, the war threatened ICI's policy of responding to programmes of industrial development in foreign markets by establishing and developing overseas factories to replace the export trade. The difficulty of providing the necessary funds showed up in major matters such as finance for AE & I in South Africa, to match contributions from De Beers, and in comparatively trivial episodes such as an application to the Treasury, in 1942, for permission to send £50,000 to Brazil for the exploration of rock salt deposits which might form the basis of a local ammonia-soda plant. If such a plant were to be built in Brazil, ICI would much rather they did it than anyone else, and Coates, advised by Alkali Group, regarded £50,000—about the value of one month's sales of alkali products to Brazil—as 'a very moderate sum to spend as an insurance policy', but the Treasury refused to let it go.[8]

Difficulties of this sort were at first regarded as temporary annoyances inevitable in wartime, but by the end of 1943 it was recognized that they went deeper and would outlast the war, reflecting the general overstrain of the national resources as much as any weakness in ICI's own position. 'It is anticipated', wrote S. P. Leigh, Export Executive and Acting Overseas Manager (later Overseas Controller), on 1 December 1943, 'that the funds available for overseas manufacturing investment after the war will be strictly limited, both by the resources of the Company and by government regulation. Another limiting factor will be the shortage of technical manpower. . . .'[9] On both counts his forecast turned out to be accurate.

Export trade, as well as overseas investment policy, was distorted by the demands of war. At first it was seen, both in ICI and by the Government, as an offensive weapon of economic warfare. In Dyestuffs Group, as we have seen (chapter 15 (iii) (b)), a very large expansion scheme was put in hand, aimed at seizing German trade, especially in India. That scheme was curtailed in 1941, and long before then it had become apparent that the general plan of using exports offensively was unrealistic. British industry, including ICI, almost immediately came under such strain that there was very little to send abroad in the ordinary way of business, let alone for a great expansion.[10] As the war went on, every commercial consideration was sacrificed to the overpowering strategic necessities, until in reporting on exports for 1943 Leigh observed:

The primary limiting factors on the company's export trade during the year have been on the one hand, restriction of orders from all Allied countries to basic essential requirements and on the other, the reservation of supplies to meet wartime needs at home.

Such considerations as the general balance of payments and maintenance of contact with overseas customers have practically ceased to operate, except to a limited extent in the case of the neutral countries accessible to shipment.[11]

Wartime regulation of the export trade in the interests of the British war effort dislocated the peacetime system of regulating export trade in the interests of ICI's business. All arrangements with IG, including elaborate rules for the supply of explosives in South America and dyestuffs in India, were immediately cancelled, and at the same time the effect of the British blockade diverted demand towards those still able to meet it. If ICI could not deliver the goods, American manufacturers could, with the prospect that they might establish themselves in markets which ICI for so long had striven to protect.

In general, ICI tried to arrange for business in their traditional markets which they could not handle to be placed with du Pont, on the understanding—they hoped—that this would be a temporary arrangement inspired by common enthusiasm for the British cause. Soon after war broke out, when the policy of 'offensive exports' was in full vigour, Melchett wrote to Crane explaining that ICI, 'as a loyal and patriotic firm', would 'do everything in its power to obstruct the trade of the enemy in neutral countries . . . to displace such German goods as manage to penetrate the blockade, and to obtain their proper share of the supply of those goods for the benefit of their own country'. He went on to say that if ICI found themselves short of goods to supply the demand created by the absence of German exports, 'we may ask if you would adopt the magnanimous attitude of supplying these goods yourselves in the interim, leaving us the opportunity of resuming deliveries as soon as we are in a position to make them'.[12]

This astonishing request betrayed a complete misunderstanding of du Pont's position. It asked them, while still neutrals, to help in furthering the war aims of the British Government and at the same time to place themselves in danger from the anti-trust laws of their own country, and all this without taking any permanent competitive advantage of the situation for themselves. It was evidently a proposition which du Pont found highly embarrassing, and in reply Crane wrote:

We will naturally preserve a free hand and consult the best interests of the Du Pont Company. We will undoubtedly take on for our own account much foreign business, some of which was hardly obtainable by us previously. . . . There are areas in which our activities are logical, there are goods which it is logical for us to supply. On the other hand, there are areas and products which could be effectively served by us only for temporary shortages of goods in this great emergency. In all of these matters we want to be guided by sound business considerations as well as by a fair and sympathetic attitude towards yourselves.[13]

As the months and years of war went by, this matter and others which we shall notice later generated a good deal of irritation in ICI against du Pont. Like much else in their wartime relationship, the feeling over exports was a miniature model of the wartime relationship between Great Britain and the USA. The British side felt that they deserved special consideration, above and beyond strict contractual obligations, for their exertions in the common cause, especially before America came into the war, and there was a mounting sense of grievance that the Americans, indisputably the stronger power, were taking undue advantage of their strength and their immunity from attack.

As a consequence the du Pont Agreement came under strain. Some in ICI, and no doubt in du Pont also, began to question its continued value, as for instance E. J. Barnsley, ICI's senior representative in the United States, who in 1942 wrote a letter to Coates which was highly critical of American business in general ('all the talk was . . . a wonderful time for all, outside those doing the fighting'), of du Pont in particular ('Why not let us be wholly realistic . . . ? Du Pont are in business . . . for themselves only'), and of the Agreement ('Should not ICI . . . attempt to become entirely free . . . and push ahead as it sees fit?').[14]

This view, at that time, was not the view of the 'establishment', either in ICI or du Pont. It may not even have been Barnsley's normal view. He was a temperamental man who had suffered the loss of an arm, and he may have been writing in a passing mood of irritation. Nevertheless, there may have been those who would have agreed with him, as there certainly were a little later. He was swimming with the stream. Apart altogether from the effects of the war, technological change in the chemical industry was beginning to force a reconsideration of old alignments, particularly as they affected relations with Courtaulds, the Distillers, and the oil companies, and in spite of the fact that competition with the Distillers had been regulated by written agreement as recently as 1938. By the time the war ended a strong body of opinion in ICI was sufficiently self-confident to scrap agreements if they got in the way of desirable developments. No one, at that time, suggested sacrilege against du Pont, and perhaps no one would with McGowan still in the Chair, but we have seen that Quig was almost indecently eager, as soon as Nicholson retired, to get rid of the DCL Agreement, which was a sign that ICI was moving of its own accord, independently of outside pressure, away from the ancestral tradition of regulated competition.

The movement, nevertheless, was enormously accelerated, and in the end made decisive, by political pressure. It came from the USA, it originated in the Antitrust Division of the Department of Justice in the late thirties, and from 1944 onward was brought to bear on a cartel structure already weakened in the manner we have outlined. Its impact was devastating.

REFERENCES

[1] E. A. Bingen, 'Agreements Section', 5 Feb. 1946, CP 'Organisation-Agreements'.
[2] Remit to Du Pont Committee from Overseas Executive Committee, 17 June 1938, CR 93/49/431 R.

3 Copy of the 1939 Agreement in CR 595/7/46; 1929 Agreement is at Appendix IV.
4 L. Patrick, 'The International Hydrogenation Cartel', 10 Nov. 1943, CR 59/103/618.
5 Alkali/Chlorine Committee Minutes 7, 16, 80/36; 201/37; supporting papers.
6 USA, Dyestuffs and Organic Chemicals, Appendix II, 'History of Trafford Chemical Co. Ltd.'
7 CP 'ICI/IG/Solvay. Pre-War Agreement'.
8 Correspondence between Coates and S. D. Waley, 8 May and 22 May 1942, CP 'South America and Brazil, Alkali'; Overseas Executive Committee Minute 125/42, 26 May 1942.
9 S. P. Leigh, 'Overseas Manufacturing Investment', 1 Dec. 1943, CR 93/91/873.
10 J. H. Wadsworth, 'Expansion of ICI Export Trade', 13 Dec. 1939, Annex III to Commercial Executive Committee 259/39.
11 Overseas Controller (S. P. Leigh) to Overseas Director (J. H. Wadsworth), 'ICI Exports —1943', 4 April 1944, ICIBR (Overseas Director).
12 Melchett to Crane, 8 Sept. 1939, USA, Patents & Processes Agreement IVA, pp. 2–3.
13 Crane to Melchett, 25 Oct. 1939, USA, Patents & Processes Agreement IVA, p. 12.
14 E. J. Barnsley to W. H. Coates, 5 Aug. 1942, USA, Patents & Processes Agreement IV, Appendix E.

CHAPTER 23

The Anti-Trust Offensive 1939–1944

AMERICAN ANTI-TRUST LAW is heavily charged with elements of popular passion and deeply-held, irrational belief about the nature of big business and the sovereign virtue of free competition. In the words of a high American authority, it 'is not, in the conventional sense, a set of laws by which men may guide their conduct. It is rather a general, sometimes conflicting, statement of articles of faith and economic philosophy, which takes specific form as the courts and governmental agencies apply its generalities to the facts of individual cases in the economic and ideological setting of the time.'[1] It is therefore exceptionally unpredictable in operation, so that, as E. A. Bingen observed in 1944, '. . . the best business men and lawyers in the United States never know where they stand'.[2]

It also goes through periods of greater and lesser activity. In 1907, when an action was launched against du Pont, the anti-trust law was being enthusiastically enforced by Theodore Roosevelt's Administration, but in the twenties 'antitrust administration was not notably vigorous'. Then in the late thirties, in the day of F. D. Roosevelt's New Deal, Thurman Arnold became head of the Antitrust Division of the Department of Justice and propelled it into a period of activity which 'up to the present time [1960] has been without question the most exuberant in antitrust history'.[3]

McGowan and everyone else at the top of ICI were used to anti-trust law. For as long as the eldest of them could remember, long before ICI was founded, they had had to take account of it in any dealings across the Atlantic, especially with du Pont, but they had taken it for granted that in any proceedings that might be started the direct object of attack would be du Pont or some other American firm, not ICI. With Thurman Arnold in charge at the Antitrust Division, signs began to appear that this might no longer be a safe assumption, and when Richard Fort took over from G. W. White as President of ICI (New York) in 1938 he took legal advice about the Company's position under anti-trust law.[4]

He had cause for anxiety. ICI did no trade in the USA and the function of the New York office, as described by White himself, was mainly ambassadorial, meaning that the office was maintained for keeping in touch with du Pont and other American firms on matters of the very kind likely to arouse the deepest suspicions of Thurman

Arnold. White, writing in London to ICI's Overseas Committee, naturally with no thought that his words would ever come to the notice of the Antitrust Division, explained how the American business of the Nitrogen Cartel was handled through ICI(NY):

As a matter of fact, this particular item of ambassadorial work is at the moment [November 1938] being done by an individual. . . . There is no record kept of the transactions. There is no established link between the individuals and the companies concerned because the latter do not bind themselves to any action. The whole thing is in the form of a loose conversation which has to be interpreted and the interpretation transmitted by Transatlantic telephone to this [London] side. So far as the Nitrogen Cartel is concerned it works perfectly and as a consequence ICI profits to a considerable degree by it.[5]

In view of these remarks it is hardly surprising that ICI's first direct experience of anti-trust proceedings arose from the affairs of the Nitrogen Cartel. On 27 June 1939 representatives of the US Attorney-General appeared at the offices of ICI(NY) to serve subpoenas on four officials of ICI(NY), including Fort himself, and also on E. J. Barnsley, then Deputy Treasurer of ICI, who happened to be in New York and was therefore, from the Attorney-General's point of view, conveniently available. The summonses required those on whom they were served to appear with records before a Federal Grand Jury which was holding an investigation to determine 'whether ICI's relations with American nitrogenous fertilizer companies in the United States, as dealt with through the New York office, involved any violation of the US Anti-Trust legislation, in particular the Sherman Act'.[6]

The Attorney-General's men took ICI(NY) by surprise. Only M. G. Tate was there when they called, and they swept him before the Grand Jury before he had a chance to take advice, having first seized papers from the office and from Tate's flat. A summary of the Nitrogen Cartel Agreements 'prepared in London for the benefit of the Nitrogen Committee [of ICI]' contained the damaging passage: 'Discussions with the representatives of the American makers in London have . . . shown that there is a common point of view on many matters which will lead to mutual orderly marketing and maintenance of the price standard. Discussions . . . in New York . . . have led to orderly working of American imports and exports of Sulphate of Ammonia.'[7]

If anything more was needed to found a case against ICI(NY), Tate provided it. Hurried before the Grand Jury without legal advice, and 'brusquely handled', his admissions under cross-examination 'made it appear that he had been acting as an intermediary between the International Nitrogen Cartel and the American producers, that he had discussed with the American producers quotas and prices of

420

nitrogenous fertilizers, that he had had numerous conversations with London by means of Trans Atlantic telephone calls and that he had destroyed certain correspondence'. The evidence he gave, Bingen concluded, 'may well have created an impression that ICI had something to conceal'.

On the strength of statements extracted before the Grand Jury from Tate, Fort, and Johnson of ICI(NY), and from Barnsley, the Attorney-General began to press ICI to disclose not only copies of the Cartel Agreements, which would have been fairly harmless, but also all correspondence between London and New York on the subject of nitrogenous fertilizers. That would be very embarrassing to the American parties to the correspondence, the Barrett Company of Detroit, a subsidiary of Allied Chemical Corporation. They were already in trouble with the authorities over price discrimination and if, as a result of ICI's disclosures, they were indicted under the Sherman Act they would be, as Bingen put it, 'much upset', having doubtless, in his view, kept no such documents themselves.

There was really no help for it. If ICI refused disclosure the American authorities, through the State Department, the Foreign Office, and the English courts, could eventually force it and the refusal would 'produce the worst impression in the minds of the Federal Authorities'. On the other hand by disclosure ICI might hope to gain the authorities' goodwill, even if they lost Barrett's, and the case might be settled amicably. In August 1939, therefore, professing a desire to help the federal authorities, ICI sent F. C. O. Speyer to New York with the documents.

The result, for ICI, was as satisfactory as could be hoped. Thurman Arnold even told Speyer that 'in view of the co-operation received from ICI' he regretted that the Government had such a good case. The Grand Jury could, he said, indict all the members of the Cartel, including ICI London, but 'as a friendly gesture' he was prepared to ask them to indict only ICI(NY) and possibly its President, Fort ('he would be spared any indignities such as finger-printing etc'), and to keep the indictment 'sealed', thus avoiding publicity.[8]

The case was eventually settled in 1942, through the long-drawn procedures of American law, by a Consent Decree which did little more than register the existing wartime state of affairs. It went on the assumption that the Nitrogen Cartel was no longer functioning and forbade ICI(NY) to combine with others to fix fertilizer prices in the USA, to restrain shipments or sales to or from the USA, or to exchange unpublished information.[9]

The result of this particular case was mild enough, but the way it was conducted showed how broad a view the anti-trust authorities took of their duties. Clearly they would need very little encouragement to proceed not only against ICI(NY), within the American courts' jurisdiction, but also against ICI themselves, outside it. Moreover, as Melchett observed: 'There is a strong political atmosphere in the whole of these [nitrogen case] proceedings which somewhat clouds the legal position, which in this type of legislation is always obscure.'[10]

Between the launch of the nitrogen case in 1939 and the start of proceedings against the du Pont agreement at the beginning of 1944 there were numerous anti-trust actions in which ICI were more or less directly involved. They included a case, in 1941, aimed chiefly at IG through their American subsidiary, General Aniline and Film Corporation; another, in 1942, against eight firms accused of a dyestuffs conspiracy; and two more, in the same year, against du Pont and others for their activities in acrylic products and methyl methacrylate.[11] In September 1941 the New York offices were again raided, subpoenas were served requiring information on ICI activities since 1926 in dyestuffs, plastics, hydrogenation and explosives, and representatives of the authorities 'thoroughly scrutinized all files'.[12] 'We are getting acclimatized to all this spate of litigation', Bingen wrote in July 1942, 'and are now taking it in our stride.'[13]

The pressure behind these cases, in Lord Melchett's view, based on an interview with Thurman Arnold, came partly from an attempt to force industrial price levels down towards the level of agricultural prices by breaking 'any and every sort of arrangement' for the private regulation of competition. 'Of course', said Melchett, 'this is not law. It is a running enquiry under the threat of criminal prosecution into the day to day conduct of industry.'[14] No doubt, but its political attractiveness to F. D. Roosevelt's 'New Deal' Administration is obvious.

More pressure was built up by the outbreak of war. The Administration, convinced that supplies of strategic materials were being hindered by the sinister machinations of fascist cartels, was determined to unearth the ramifications of German industrial groups in the USA, and especially of IG, so as to break their hold on aluminium, magnesium, synthetic rubber and various drugs and medicines, and to make their patents and technical knowledge available for war preparations in the USA. Melchett himself was subpoenaed, while he was in the USA in 1940, to give evidence on the production of magnesium.[15]

Melchett's discussions with Arnold in 1940, even though the nitrogen case was going on, seem to have been amiable. Arnold and his chief, Leon Henderson, 'strongly advised me [Melchett] not to take these prosecutions too seriously. They are not intended as trials for punishment for offences against the law. They are rather enquiries to determine what the law ought to be in given cases when all the facts have been divulged. The criminal proceedings ensure that facts and figures will not be withheld. Further they ensure that the company will abide by the directions given. . . . It is all very American and very "New Deal".[16]

In 1940 Melchett could still take this detached and rather patronizing view because anti-trust proceedings and the political pressures behind them were still looked upon purely as matters of domestic American concern. It never entered Melchett's head, nor, probably, anyone else's in ICI, that the Americans might try to convert the rest of the world to the gospel of anti-trust. Yet as the war entered its later phases, with American power obviously the driving force for victory, this was precisely what the Americans began to do. They were utterly convinced, in this matter as in others, of the rightness of their motives, of the universal efficacy of their policies, and of their power to enforce them, if they pushed hard enough, throughout that part of the world dominated by the Western allies.

International cartels of the type favoured by ICI policy-makers in the thirties were particularly offensive to anti-trust doctrine, and they came under heavy attack in America during 1943. Early in 1944 Francis Biddle, Attorney-General of the United States, put the whole weight of his official position, and therefore presumably the weight of the American Administration, behind a statement of the American point of view and a declaration of policy.[17] 'At the end of the war', he told members of the Harvard Law School Alumni Association, '. . . there will be a striking contrast between the free enterprise system we value and a cartelised Europe.' The American system, he admitted, was not 'perfectly competitive', but he insisted on 'an incisive difference between a society which seeks to give to every man an opportunity to engage in business and to direct it in accordance with his own judgement and at his own risk, and a society which is ready to place in the hands of private groups or of public officials the authority to determine who may engage in business, the output quotas of those who are permitted to produce at all, and the market areas within which, and the terms upon which, each may sell.'

This was an orthodox statement, which might have been made at any time since the passing of the Sherman Act, about the superiority of American free enterprise to the 'rigid, stratified, hierarchical system of industrial control' which Europeans were said to prefer. In the past, however, it would probably have been added, implicitly or explicitly, that if that was the way Europeans liked it, that was their own affair, and in any case Americans had nothing to fear from such effete timidity. In 1944, however, Biddle took a stronger line. 'We recognize', he said, 'that we cannot and should not wish to impose a foreign way of life, even though it be our own, upon other people', but that was simply a prelude to an announcement that this was exactly what the Administration was proposing to do, in so far as they were most unwilling to see 'centralized channels of industrial control . . . maintained or recreated'. After the war Europe, starting with Germany, was to be cleansed of cartels.

So was the rest of the world. 'The division of world markets', said Biddle, 'cannot fail to affect us', and he went on to attack private agreements which had the effect, though it was not explicitly stated, of keeping foreign goods out of the United States in return for an 'understanding' that American producers would accept 'quotas and limitations' elsewhere in the world. 'This is an arrangement for a private tariff which has the advantage to the private parties that it need not conform to any law of Congress. It is also an arrangement to cut down American production. Beyond that, it is treaty making, allocating territories as dependent upon a particular foreign power for its [their?] economic life.'

The threat to the ICI/du Pont agreement was clear, and indeed by the time Biddle spoke legal action had been launched against it. Before we deal with that, however, it will be convenient to look at the prevailing British view of cartels and similar arrangements and, more particularly, at the views held within ICI.

In British official circles they seem to have been taken by surprise by the vigour of the American attack, and to have been looking for a lead. Just before Christmas 1943 Sidney Rogerson, head of ICI Publicity, met the Public Relations Officer of the Board of Trade, just back from the USA, 'where he found the feeling against cartels was so strong that it tended to prejudice cooperative working between British and American units'.[18] He had with him the PRO from the Ministry of Production, and both were looking for help. 'Neither of these two officers,' Rogerson told McGowan, 'occupying key positions and responsible at least in part for drafting the speeches of their Ministers, had any clear idea what a cartel was', and they were hoping that someone in ICI would tell them so that they could pass on the information to their Ministers and, presumably, to other officials, for they assured Rogerson 'that the same ignorance was widely spread in their offices'. If that was so, it is hardly surprising to find Biddle, in his address a few weeks later, observing with perceptible condescension: 'Possibly we may be forgiven for believing that the British have not fully seen what the cartels have done to them.'

However ill-informed the two PROs might be, neither they nor their departments were hostile to the idea of cartels, or to big business generally. Quite the contrary—they were anxious to see big business, led by ICI, defending itself not only against American opinion but also against what Rogerson called 'a small and active political minority' in Great Britain. The PRO from the Board of Trade would be willing, he said, 'to assist in putting the case for cartels before the public'.

424

In ICI a brief was already being prepared 'explaining exactly what a cartel was, why it was necessary, how it worked etc.', and a good deal of time and money were spent on examining and defending the case. Thus in the autumn of 1944 papers were commissioned from two economists, O. W. Roskill and J. H. Wilson, later Prime Minister, and at the beginning of 1947 a committee of the FBI, over which W. H. Coates presided, issued an elaborate report, 'Private International Industrial Agreements', which relied fairly heavily on the arguments put forward by the American Professor J. A. Schumpeter in *Capitalism, Socialism and Democracy*, then recently published.[19]

McGowan himself was an industrious propagandist, described by Coates in 1946 as 'the "plumed knight" in the cartel battle'. He was a life-long believer in regulated competition, his career had been founded on skill in negotiating market-sharing agreements, and he was perfectly frank about his opinions. In 1941, writing to Melchett, he said he did not accept 'the theory . . . that competition is essential to efficiency',[20] and later, when he felt that ICI was under attack, he repeatedly said the same thing publicly. In the House of Lords, on 5 July 1944, his succinct defence of international cartels presented the reasoning behind the commercial policy which ICI had pursued before the war and which McGowan, at least, had every intention that they should pursue afterwards:

I have elsewhere spoken and written of some of the benefits which can flow from international collaboration as a result of industrial agreements. I repeat that those benefits are numerous and substantial. The purpose of these agreements is, in the main, to regulate but not to abolish competition. Such agreements can lead to a more ordered organization of production and can check wasteful and excessive competition. They can help to stabilize prices at a reasonable level. . . . They can lead to a rapid improvement in technique and a reduction in costs, which in turn, with enlightened administration of industry, can provide the basis of lower prices to consumers. They can spread the benefits of inventions from one country to another by exchanging research results, by the cross-licensing of patents, and by the provision of the important 'know-how' in the working of those patents. They can provide a medium for the orderly expansion of world trade and can make a sub-stantial contribution . . . to the difficult problems of the post-war readjustment of production in countries greatly affected by the war. They can also assist in providing much greater stability of employment.[21]

McGowan was speaking in a debate on a White Paper (Cmd. 6527 of 1944) on Employment Policy. By the time he spoke the Government, with or without help from ICI, had made up its mind about cartels. Dealing with 'combines and . . . agreements, both national and international, by which manufacturers have sought to control prices and output, to divide markets and to fix conditions of sale', the White

425

Paper says (para. 54): 'Such agreements or combines do not necessarily operate against the public interest; but the power to do so is there. The Government will therefore seek power to inform themselves of the extent and effect of restrictive agreements, and of the activities of combines; and to take appropriate action to check practices which . . . work to the detriment of the country as a whole.'

This point of view and policy, formulated by the Coalition Government, was taken over by the Labour Government in 1945. It was difficult to refute the proposition that some trading agreements, though not all, might be against the public interest, and that consequently all such agreements ought to be open to public inquiry. ICI made no attempt to fight this eminently reasonable policy, but instead concentrated on making sure that such agreements as were in existence, and any that were proposed, would stand up to examination.[22]

The British Government's policy towards cartels fell a long way short of what the Americans were pressing for. In 1946 they were trying to get an international authority set up for regulating business practices. Their own principle, enshrined in anti-trust legislation, was that certain practices were bad and should be forbidden (unless, as in shipping and agriculture, American interests needed protection), and that was what they would have preferred. The British Government held out, however, and when the Monopolies Act* of 1948 was passed it was framed on the basis foreshadowed by the Coalition Government four years before: that although certain practices were by their nature suspect, circumstances might alter cases, and any practice that was complained of should be reported for investigation by the Monopolies Commission, set up under the Act, to determine whether or not it was against the public interest. The sentiment of the Act was unfavourable to ICI's long-established methods of regulating competition, but the Act did not forbid them and indeed recognized that under certain circumstances such methods might be justifiable.

The attack on the ICI/du Pont agreement, then, was part of an exceptionally vigorous anti-trust campaign, rooted in American home politics like similar campaigns in the past, but having in addition much more ambitious aims in the field of international trade. For ICI, it was an assault on business practices, perfectly legal outside the USA, which McGowan and others were prepared to defend as being not only in the interest of ICI but in the public interest as well. On this wider ground, the defence embodied the feeling, widespread in Great Britain, that the Americans had done well out of the war and that their post-war policy, in cartels as in other matters, was not at all friendly to British interests. 'America', wrote Lord Glenconner, a lay Director of ICI, in 1944, 'is

* The Monopolies and Restrictive Practices (Inquiry and Control) Act 1948.

strong, ready, confident of herself and on the job. We are not. Is it surprising that she feels she can sweep the board, and that Washington and New York will become the greatest economic, financial and political centres the world has ever seen?'[23] The moral he drew, and McGowan agreed with him, was: 'If . . . America takes the path of expansion and thinks to assign us only a minor role, we must look to ourselves. . . .'

REFERENCES

[1] Abe Fortas, Foreword to A. D. Neale, *The Antitrust Laws of the USA*, CUP, Cambridge, 1960, ch. 5.

[2] E. A. Bingen, 'Questions and Answers on the Anti-Trust Suit', 8 Jan. 1944, CR 595/67/67.

[3] Neale, p. 29.

[4] OECP, 10 Nov. 1938.

[5] OECP, 19 Nov. 1938.

[6] E. A. Bingen, 'ICI(NY) Federal Enquiry into alleged violations of the Anti-Trust Legislation', 11 July 1939, MBSP.

[7] Ibid. p. 2.

[8] F. C. O. Speyer, 'Proceedings for Indictment of ICI, New York and London', 31 Aug. 1939, OECP.

[9] Solicitor to Management Board, 'USA Enquiry into Fertilizer Industry', 27 May 1941, MBSP.

[10] Melchett to Management Board, 12 July 1939, MBSP.

[11] E. J. Barnsley to E. A. Bingen, 11 April 1941, FD 'USA 251, Administration 1941'; same to same, 15 May 1942, FD 'USA 26, Dyes Indictment'; same to same, 17 Sept. 1942 and Bingen to Barnsley, 22 Oct. 1942, FD 'USA 224, Moulding Powders 1942'.

[12] Overseas Executive Committee Minute 278/41, 25 Nov. 1941.

[13] Bingen to Barnsley, 16 July 1942, FD 'USA 26, Dyes Indictment'.

[14] Melchett, 'Reports on US Visit 1940', MP.

[15] Ibid.

[16] Ibid.

[17] F. Biddle, 'Cartels: An Approach to the Problem', an Address to be given at the Annual Dinner of the Harvard Law School Alumni Association, 23 Feb. 1944, CR 104/19/19.

[18] S. Rogerson to McGowan and others, 'Cartels and large Combines', 23 Dec. 1943, CR 104/19/19.

[19] Roskill's paper, MP 'Cartels, Correspondence'; Wilson's paper seems to be lost, but Quig commented: 'I know that he was not employed to put up a case for cartels and although he has attempted to put a brief case in favour he has obviously struggled hard to find his pro-Cartel arguments.' 26 Oct. 1944, CR 104/19/19; FBI Committee Report, CP 'FBI—Industrial Agreements Committee'.

[20] McGowan to Melchett, 4 April 1941, MP 'Post War Problems'; Coates, 'Visit to Canada and the USA', 14 Aug. 1946, ICIBR.

[21] Hansard, Lords, vol. 132, no. 67, col. 682 et seq.

[22] R. A. Lynex, 'International Cartel Policy', 22 May 1946, CR 104/4/4, covering memo of same title by E. A. Bingen.

[23] McGowan to W. A. Akers, 'Post-War Prospects', 7 Nov. 1944, CR 104/19/19, covering paper of same title by Glenconner, pp. 3–4.

CHAPTER 24

United States *v.* Imperial Chemical
Industries 1944–1952

THURMAN ARNOLD, addressing a sub-committee of the House of Representatives in 1943, spoke disapprovingly of an ICI/du Pont world-wide cartel, active since 1929.[1] There is little doubt that by that time his Division of the Department of Justice, in keeping with the militant zeal of the Administration, was preparing a comprehensive attack not merely on the ICI/du Pont alliance but on as much as the Division could get at of the network of agreements which for so long had governed the working of international trade in the chemical industry.

Arnold's speech was not the only portent. In June 1943 ICI and CIL were cited as co-conspirators in an anti-trust suit, over titanium, against du Pont and National Lead. By August it was known that proceedings were likely against Alkasso, the members of which, ever since its foundation in 1919, had considered themselves protected by the Webb-Pomerene Act. All these happenings convinced E. J. Barnsley, ICI's permanent wartime representative with ICI(NY), that a direct attack on the ICI/du Pont alliance must be imminent, and in fact it was.[2] It was launched in January 1944, and it was followed in March by the related suit against Alkasso, in which ICI was also a defendant.

On 6 January 1944 the Attorney-General of the United States filed a suit (Civil Action No. 24–13 in the United States District Court for the Southern District of New York) against ICI, ICI(NY), du Pont, the Remington Arms Company Inc., and certain individuals, of whom two were referred to, with austere republican virtue, as 'Harry Duncan McGowan, known as Lord McGowan' and 'Henry Mond, known as Lord Melchett'.*[3] The action was founded on Section 1 of the Sherman Act of 1890:

Every contract, combination in the form of a trust or otherwise, or conspiracy, in restraint of trade or commerce among the several States or with foreign nations, is hereby declared to be illegal. . . .

* Both were later dismissed from the suit.

The action being a civil one, not criminal, there is no indictment, but the plaintiff's case and the remedies sought are set out in the Complaint, an octavo volume of 94 pages plus 86 pages of exhibits. It is founded on thorough, if prejudiced, historical research (there was no shortage of documentary evidence) and it describes the relationship between du Pont and ICI and their predecessors from 1897 up to the date of the action. Since that has been a steady theme of both volumes of the present work it would be superfluous to go through the 194 paragraphs in detail. It is sufficient simply to say that the facts of a very complicated situation are ably set out and interpreted, of course from a point of view hostile to the defendants.

The attack was on a broad front, directed not only at the Patents and Processes Agreement itself and numerous other agreements, but at the joint ownership by ICI and du Pont of CIL in Canada, of the Duperial companies in Brazil and Argentina, and of CSAE in Chile.* ICI's and Remington's joint ownership of the Brazilian ammunition firm *Companhia Brasileira de Cartuchos* was also under attack, being linked to the main causes of action by du Pont's control of Remington.

The Attorney-General's essential allegation (para. 194) was that for many years ICI and du Pont had been linked in a 'combination and conspiracy . . . to restrain the foreign and domestic trade and commerce of the United States'. They had eliminated competition between themselves and with certain non-American companies (particularly IG) and as a consequence the export trade of each party had been curtailed and ICI had been prevented from manufacturing in the USA. The final sub-paragraph, a direct attack on ICI, is full of world-wide crusading zeal:

The combination and conspiracy herein alleged between ICI and du Pont has materially assisted ICI in formulating and carrying out its program for the world cartelization of the chemical industry. Because of du Pont's agreement not to compete with ICI in world markets as part of the combination and conspiracy herein alleged, ICI has been enabled to enter into numerous agreements with European manufacturers of chemical products pursuant to the terms of which such manufacturers have restricted exports to the United States and competition between du Pont, ICI and such manufacturers in world markets has been eliminated.

The Department of Justice launched the suit with a blast of hostile publicity. To ICI at home the most damaging items were insinuations in paragraphs 151 and 168 of the Complaint that arrangements made in South America before the war had been allowed to continue in wartime and had interfered with the British and American war effort.

---

* For pre-war South American arrangements, see chapter 12 (iii) above.

It was true that in Chile there had for many years been quota agreements with Dynamit Aktien Gesellschaft, IG's explosives subsidiary, and it was also true that very shortly before the war broke out ICI had undertaken in writing to allow IG a holding in an Argentine company, Electroclor, in which Duperial and La Celulosa, also an Argentine company, were the other shareholders.* In a wartime atmosphere it was unnecessary to do more than call attention to these arrangements for people to draw the worst conclusions. On the day the suit was filed the BBC went a long way towards encouraging them to do so: 'The Suit', a broadcast was reported to have said, '. . . charges both du Pont and ICI with instructing their Latin-American representatives to continue co-operation with the representatives of the German Corporation in Chile and Bolivia [presumably the reference was to CSAE only] after the outbreak of war.'[4]

McGowan's outraged response, directed at the Assistant Attorney-General of the US, Wendell Burge, rather than at the BBC, denied 'utterly and totally any suggestion that any action of ours during the war, and indeed before the war, was of any other character than designed to assist both the British and Allied Governments by every means within our power'. For the benefit of ICI's own employees a highly emotional letter was sent to Divisional Chairmen pointing out that McGowan's second son had been dangerously wounded, that one son-in-law had been killed and another taken prisoner, and that 'the whole male families' of McGowan and Melchett were on active service.[5]

Neither of the matters complained of had been hidden from the British and American authorities and they had been ended as soon as possible. In Chile the quota arrangements with foreign explosives manufacturers had always been an object of suspicion, with reason, to the Chilean Government, and the forced wartime abandonment of them was greeted with relief by the local management, as we saw in chapter 12, for purely local reasons.[6] But in ending the arrangement letters passed, on ICI's behalf, between New York and Chile, which displayed a degree of consideration for German feelings embarrassing in wartime. One such, written by J. W. Squirrell of ICI(NY), was quoted in the Complaint. In Chile, Petrie of CSAE wrote to DAG's representative regretting 'exceedingly' the end of a 'long and harmonious association'.[7] Even during a war, politeness is not treasonable, and the agreement had in fact been ended as quickly as its terms would allow: that is, on 31 December 1939.

The Electroclor entanglement was less easily broken. It had certainly been unwise to enter into a binding agreement with a German company in a foreign country, likely to be neutral, when war was already virtually

---

* For details of CSAE and Electroclor, see pp. 219, 228 above.

certain, and why the decision was taken remains a mystery. The result was to put Duperial Argentina in an impossible legal position, since one of the two shareholders—ICI—was required by English law to do everything possible to procure a breach by Duperial of a contract perfectly sound in Argentine law. With the consent of the British Embassy in Buenos Aires a settlement was arrived at, involving payments to Anilinas Alemanas, IG's Argentine subsidiary, but after America went to war fresh complications arose and the eventual result, in 1946, was the payment of 253,866·66 pesos, plus 48,934·40 pesos interest, as damages to Anilinas Alemanas for breach of contract.[8] Again, there was no willing co-operation with the enemy, but the situation, given a little foresight in 1939, need never have arisen.

The aspersions on ICI's patriotism, though wounding at the time, were by way of being a temporary embarrassment and there is no evidence that they did ICI's reputation any lasting harm, but they did display the temper of the Department of Justice and of Roosevelt's Administration as a whole. That was shown also, with much more solid menace, in the unprecedented way in which the anti-trust suit was designed to range audaciously beyond the jurisdiction of the American court.

The legal standing of the Patents and Processes Agreement, obviously, could be called in question before the American court and since ICI, through ICI(NY), maintained an office in New York, there were reasonable grounds for making ICI a defendant. In the matter of the jointly owned companies, however, the conduct complained of was outside the USA, which might have been thought to put ICI, as an English company, out of reach. Yet the American authorities, relying on an American statute, sought to bring ICI within American jurisdiction for acts of commercial policy blameless under English law. Their action was part of the new policy of exporting anti-trust doctrine. It marked a long advance in the powers attributed to the American courts by the American Department of Justice, which had never formerly taken quite so lofty a view.[9]

As soon as the Department of Justice had launched their attack on the ICI/du Pont alliance they moved on another pre-war arrangement for regulating international competition in the chemical industry. This was the 'Alkasso Agreement', under which for many years ICI and the members of the US Alkali Export Association had shared a joint figure of export business in fixed proportions. In March 1944 ICI and ICI (NY) found themselves, alongside the American parties to the agreement, defendants in the suit *US v. US Alkali Export Association*.

This suit, like *US v. ICI*, represented an extension of anti-trust practice.[10] Section 2 of the Webb-Pomerene Act 1918, apparently on the principle, not unknown in other countries besides the USA, of letting business men do to foreigners things they would not be permitted

431

to do at home, had permitted American exporters to make agreements in foreign trade which in the home trade would have violated the Sherman Act. There were safeguards for free competition within the USA and for the rights of exporters who did not wish to join any agreement, but for the better part of twenty-five years the members of Alkasso were under the impression that their activities were perfectly legal, even laudable, and that the wrath of the Antitrust Division would never be visited upon them. They were now to find out how wrong they had always been, and so were ICI. It was open to ICI, in both these cases, to move to quash them for want of jurisdiction. In the Alkasso case they did, and failed. In *US v. ICI* no motion was made and ICI prepared to defend themselves.

To avoid interference with war work, ICI sought support from the American War Department and Navy Department to get proceedings stayed. The Department of Justice was unwilling and the matter was carried to the British Ambassador and by him to the President. That, Bingen thought, annoyed the Departments, but eventually a formal agreement between the War Department and the Department of Justice provided for postponement of trial until after the end of the war, though no one seemed sure whether that meant the end of war in Europe or in the Far East as well.[11] Matters in the Alkasso case, meanwhile, moved at a stately pace towards a trial.

During the spring and summer of 1945 the pause in legal proceedings coincided with the relaxation of wartime pressure following the defeat of Germany. Coates, Lutyens, Bingen, Patrick, and others crossed the Atlantic, then and in the following two years. It was the last opportunity, under the old conditions of partnership, for an examination of the whole relationship between ICI and du Pont which was coming under such heavy attack. How did matters stand?

ICI and du Pont had wide areas of interest in common, otherwise there would have been no basis for the long alliance. The American lawyers, however, in presenting their case, emphasized the community of interest almost to the extent of making the two businesses sound like identical twins. The reality, recognized both at Millbank and in Wilmington, was that there were deep differences in scope, organization, outlook, and policy, and these differences produced strains in the alliance independently of anything the Department of Justice might do. What rankled most on the ICI side, amongst those well-qualified to judge, was an uneasy feeling that the achievements of ICI's research, in spite of polythene and 'Perspex', did not stand up well in comparison with results from du Pont.

This uneasiness in ICI-du Pont relations was exacerbated by the uneasiness general in Anglo-American relations late in the war—a reluctant recognition of American power, wealth, and technical ability, all much enhanced by the effects of the war itself. 'Well,' as Peter Allen put it privately, 'as long as we are poor relations and take their money, we can do nothing about it but grin and bear it; but after the war we shall have to go out after the American chemical industry or we'll become just a little appendix to the great colossus. That's why what ingenuity and ability we can command in ICI, and God knows there's none too much of it, must be organised in the wisest way.'[12] Barnsley at the same time, as we have already remarked (p. 417 above), was expressing the same kind of irritation with American business in general and with du Pont in particular. Such tension in the wartime air, arising from so many causes, did not make for easy relations, and in ICI there came to be a disposition very readily to take offence.

When the irritation in ICI came down from general causes to particular cases, a good deal of it centred on the exchange of information under the Patents and Processes Agreement. In ICI there was a wide-spread feeling that du Pont's negotiators, as a matter of policy, went out of their way to belittle the value of information offered to them by ICI. Of one of them, Wendell Swint, P. C. Allen observed: 'his technique is to advance deep into our territory, then retreat a quarter of the way back, yelling that he has been robbed.' The object was to set up a strong bargaining position in negotiations for licences or at the five-yearly valuation. 'It is unfortunate', wrote Slade in 1942, referring to 'Perspex', 'that du Pont so often fail to appreciate the importance of processes which we send them. Their neglect . . . affects the balance of our credit at the quinquennial valuation.'[13]

There was no question in anyone's mind, in ICI, that in polythene ICI had a major invention comparable with nylon, and it was on that basis that polythene was offered to du Pont in 1939. Du Pont's reception of the offer, cool and leisurely, ended after about a year in a claim to have developed, independently of ICI, a low-cost production process. In September 1941 McGowan, writing personally to W. S. Carpenter, President of du Pont, in terms drafted by P. C. Allen, offered ICI's knowledge of polythene to du Pont, for war purposes, with a settlement after the war.[14] A party from du Pont came over in November 1941, to see what ICI had to offer, and the upshot infuriated Lutyens, Allen, Swallow, and others of the group responsible for polythene in ICI. Du Pont, they felt, had been working behind ICI's back with the object of bringing down the rate of royalty which they would have to pay on any licence they might take out, or in other words that they had been pursuing exactly the kind of policy complained of, in a different con-text, by Slade. 'As I feared,' wrote Allen of du Pont's response to ICI's licence proposals, 'it is grasping and unreasonable. They claim, as I expected, that they have done much on the process side and that we only invented the product and a half-baked way of making it.'[15]

The strength of feeling in ICI, conveyed, presumably in diplomatic language, by letters from Lutyens and Barnsley, evidently took du Pont by surprise. Swint, writing to Barnsley, soothingly attributed 'the misunderstanding which has arisen' to the wartime necessity of negotiating by letter instead of face to face. On the same day W. S. Carpenter wrote to McGowan, also in conciliatory terms. Neither Swint nor Carpenter, however, gave anything away, and both held to du Pont's claim for their own process. 'We do not think', Swint concluded, '. . . in the midst of present uncertainties it is feasible to draw up by correspondence a detailed commercial agreement for polythene', and in fact the matter was left until Coates and Lutyens went to the USA in the summer of 1945.[16]

The ICI/du Pont alliance, then, like the alliance between Great Britain and the USA, had its internal stresses, and in the lesser alliance as in the greater some ran deep while others may have been passing wartime annoyances. To the matters already mentioned there could be added a lingering feeling, in ICI, that du Pont might have granted more generous terms in the Nylon Agreement of 1939; and deep differences between the two partners over the proper way of running their joint Duperial companies in South America, which had hardly appeared before war broke out but began to become uncomfortable as time went on.

By the end of the war none of the causes of irritation we have been examining had got to such a pitch as to threaten the premature break-up of the alliance, certainly as long as the older generation, represented by McGowan on the one side and Lammot du Pont on the other, remained in charge. Perhaps they never would have done. Perhaps they represented no more than the inevitable friction in the working of elaborate machinery. Nevertheless younger men on both sides of the Atlantic were beginning to question the continuing value of the ancestral wisdom, and it is at least open to question whether the alliance would have continued in its old form if no anti-trust proceedings had been brought and the Patents and Processes Agreement had come up for renewal, as its terms provided, on 30 June 1949.

Against this background of ICI/du Pont relations the legal proceedings in *US v. ICI* slowly gathered way or, as E. A. Bingen preferred to put it, pursued their weary course, and at the same time, slightly faster, the Department of Justice pressed the Alkasso suit towards a trial. In both suits the Department had a strong case and was not disposed to settle by Consent Decree on any terms likely to be suggested by the defendants.

US v. Alkasso came to trial in March 1947. Judge Bright died in 1948, but his successor, Judge Kaufman, found for the plaintiff on all the facts and issues involved, and in his Opinion held that the Webb Act had been passed to enable smaller producers to form co-operative selling agencies to compete with large foreign firms, but not to permit export associations to make restrictive arrangements with foreign competitors. 'Viewing the Webb Act in the light of contemporaneous interpretation of the anti-trust laws, considering the import of the Act when read as a whole, and giving careful attention to the entire legislative history of its passage,' wrote Judge Kaufman, setting an important precedent as he did so, 'the conclusion is irresistible that the Webb-Pomerene Act affords no right to export associations to engage on a world-wide scale in practices so antithetical to the American philosophy of free competition.'[17]

None of the defendants appealed. Judgment was given in January 1951, ordering the cancellation of the Alkasso Agreements, granting an injunction against revival, and forbidding ICI to refuse to sell alkali in the USA. The Agreements had in any case long ceased to operate and Bingen considered the judgment 'innocuous'. The Department of Justice, in the Alkasso case, had succeeded in its aim of getting rid of the only acknowledged cartel in the chemical industry in which American firms had ever considered themselves legally entitled to take part. After some thirty years of this misguided conduct, Judge Kaufman had firmly pointed out to them, and to ICI as well, how wrong and un-American it had been.

When *US v. Alkasso* came to trial in 1947, *US v. ICI* was still three years away from court. The case was much the more complex of the two, with a great weight of documentation. When the trial finally opened both sides put in, as exhibits, some of the papers they had examined. The plaintiff's 1,436 exhibits take up thirteen volumes and the defendants' 2,264, seventeen. From this luxuriant verbal forest it is possible to carve out two main issues, or groups of issues. One centres round the Patents and Processes Agreement, which provided for cross-licensing of patents between ICI and du Pont and for the exchange of technical information against payment. These arrangements, the plaintiff contended, were a disguise for a market-sharing agreement, since the licences granted by each party to the other were exclusive within designated territories, so that ICI and du Pont were prevented from competing with each other. The other main aspect of the case was an attack on the joint companies in Canada and Latin America. These companies were all outside the USA but they all operated, said the plaintiff, in restraint of US foreign trade.

In the matter of the Patents and Processes Agreement, as is evident from the narrative set down in this volume, it was difficult to deny the substance of the plaintiff's allegation, though attempts were made. Indeed in January 1945 Bingen, in a memorandum for internal circulation, went so far as to say 'the Patents and Processes Agreement is in fact what it sets out to be and . . . there is no commercial arrangement or understanding for a division of markets between ICI and du Pont which can be inferred from the Agreement or which rests on any un-written understanding or arrangement outside the Agreement.'[18] Later, however, he showed little confidence in the defence which ICI were advised to put forward, which was that the US market was closed to them by tariffs and other measures beyond their control and that they had exercised their independent judgement in arriving at the decision not to try to do business in the USA with products covered by the Patents and Processes Agreement. Bingen did not think it likely that such an argument would convince the courts who, he said, in cases based on the Sherman Act, did not 'approach wide issues very realis-tically' and were 'inclined to base decisions on a strict construction of the Act without regard to background considerations'.[19]

Du Pont's legal advisers, from the start, seem to have had no real hope of defending the Agreement. Like their predecessors faced with anti-trust proceedings before 1914 (Vol. I, pp. 199, 212), their instinct was to anticipate the findings of the Court and remove causes of offence. In 1906 and again in 1913 du Pont had cancelled agreements without waiting for an injunction, and in 1946 it became evident that their minds were moving the same way again. When Walter Carpenter, President of du Pont, was in London that autumn, Swint, who was with him, proposed cancelling the Nylon Agreement of 1939 in favour of 'a new form of Licence Agreement . . . more in keeping with present tendencies in the United States and, therefore, more likely to be acceptable to the Department of Justice'.[20] Swint, again like his pre-decessors thirty-odd years earlier, was in a hurry. With a view to a Consent Decree du Pont 'were anxious', in Bingen's words, 'that their house should be put in order before any discussions started with the Department of Justice'. Accordingly, a new Nylon Licence Agreement was sealed by ICI early in November 1946, to take effect at the end of the year.[21]

Under the new agreement du Pont assigned the nylon patents, at prevailing royalty rates, to ICI and ICI gained the right to use the patents, royalty-free, outside the nylon field, as for instance in the melt-spinning of 'Terylene'. Against that, and this was the point intended to conciliate the Department of Justice, ICI would lose the right, which they had under the old agreement, to new nylon inventions made by du Pont and to technical information from du Pont. After 31 December 1946 du Pont would no longer co-operate with ICI in the nylon field and the two businesses would be in competition, which was exactly the result which the Department of Justice would hope for, over the entire field of the chemical industry, from *US v. ICI*.

This early success of the Department of Justice had results, as we saw in chapter 20, far beyond the jurisdiction of the American courts and far beyond the causes of action in the suit. It obliged ICI to alter their whole relationship—much to their own advantage—with Courtaulds and BNS.

With the 1939 Nylon Agreement out of the way negotiations for a Consent Decree went on throughout 1947, but the Department of Justice showed no willingness to compromise. Whatever might happen to the Patents and Processes Agreement, ICI and du Pont were hoping to keep the joint companies in being, but in spite of an attempt to 'educate the Department'—Bingen's phrase—'on the joint company problem and to dissuade them from pressing for divestiture', they continued to do exactly that, and at the same time to insist on the dismantling of the exclusive licence arrangements which were the foundation of the Patents and Processes Agreement. Evidently the Department meant to get the negotiations broken off, and 'in view of the ideological attitude of the Department and their constantly increasing demands' there seemed to be no alternative, so far as Bingen could see in July 1947, to fighting the case.[22]

In May 1948 a strong delegation from du Pont, headed by the new President, Greenewalt, met McGowan and other ICI Directors in London. What was said at the discussions is not on record, but evidently it amounted to an announcement that du Pont intended to cancel the Patents and Processes Agreement, as they had cancelled the Nylon Agreement, without waiting for the legal proceedings to go any further. The idea had been in the air for some time and ICI had opposed it, on the ground that cancellation could do no good and might merely be taken by the Department of Justice as an admission of guilt. Du Pont, however, were under heavy pressure at home. Apart altogether from the ICI suit, two other anti-trust actions in which they were involved had recently been tried, and others were to come, so that they were a principal object of the Department of Justice's attentions. ICI felt obliged to defer to them, and the Patents and Processes Agreement, which du Pont, at six months' notice, could have cancelled unilaterally on 30 June 1949, was cancelled twelve months earlier—on 30 June 1948—'by mutual consent of the parties'.

The cancelled agreement was replaced by one which broke off all technical co-operation between ICI and du Pont, which did away with the long-established system of cross-licensing, and which abolished exclusive licences held by du Pont under ICI patents, so that in the former du Pont 'exclusive territories' (USA and Central America) ICI became free to offer licences to du Pont's competitors, even though ICI's exclusive licences from du Pont, as they stood at the making of the agreement, were not overthrown. The new agreement, as this clause indicates, was intended to placate the Department of Justice, and in return it was evidently hoped that the Department might be persuaded not to attack the Nylon Agreement of November 1946 and the Polythene Agreement of January 1947, neither of which was affected by the cancellation of the Patents and Processes Agreement.

Thus the main support of the ICI–du Pont alliance was unceremoniously knocked away. For twenty years an alliance with du Pont had been the centrepiece of ICI's foreign policy; for thirty years before that, of Nobels', and the cancelled agreement expressed the ideas by which, for nearly three generations, the chemical industry of the world had regulated its affairs. No clearer indication could have been given of the passing of an age. The manner of the cancellation was not at all to ICI's liking. Bingen, in a document explaining it to the Board, was outspoken about the political muscle behind the judicial hammer which struck the blow:

While ICI do not accept the contention of the Department of Justice that the Patents and Processes Agreement is in any sense unlawful under the Sherman Act and a decision on its legality or otherwise would require a court ruling, nevertheless the evident desire of the United States Authorities that this important agreement should not be continued, coupled with the trend of recent judicial decisions of the United States Courts since the agreement was originally made, led the parties to the view that in the circumstances the best course would be for them to terminate this particular agreement of their own volition.[23]

After the destruction of the Patents and Processes Agreement the rest of the elaborate linkage between ICI and du Pont fell rapidly apart. An attempt was made, for the sake of appearances, to preserve the joint-company operations in South America, but the partners were critical of each other's methods of management and had no real desire to remain in wedlock. McGowan, visiting South America in 1949, came back critical of du Pont's policy, and by 1952 Bingen was sure that du Pont would want to cut loose from ICI even if they were not compelled to, and that was what ICI wanted, too:

My [Bingen's] own feeling—which I believe is shared by the Board—is that, on balance, a split of the joint interests in South America would be advantageous to ICI rather than the reverse. . . . Duperial Argentina and Brazil and their operation under joint ownership have always presented difficult problems . . . particularly . . . because of the divergent approaches of ICI and du Pont towards problems such as the introduction of local capital into the companies, the return on capital which ought to be obtained from money invested in South America and the degree of autonomy which ought to be conceded to local managements. . . .[24]

This was not at all the view taken in ICI of the joint ownership of CIL. Du Pont and ICI had their differences over that, particularly in the degree of autonomy to be allowed to the Canadian management, and matters might get worse with the breaking of links 'down the line' after the scrapping of the Patents and Processes Agreement. But the jointly-owned Canadian company—McGowan's first achievement in the large-scale merger field—was so much a part of the structure of ICI, especially for the older generation, that its disappearance was almost inconceivable. 'CIL's business', as Bingen put it, 'is so well established, its management is largely autonomous, the territory in which it operates is so accessible both to ICI and du Pont, that the same problems [as in South America] do not really present themselves here.' It was a much older investment, he was saying, in a much less foreign environment. Moreover it was a very profitable investment, and Bingen knew how much was owing to du Pont. 'ICI's investment in CIL,' he said, 'which has been due in no small measure to du Pont's share of the management and du Pont's inventions, has shown a greater rate of development and greater profitability than any of ICI's other manufacturing investments overseas.'[25] ICI would be very unwilling indeed to see the joint ownership of CIL broken up. But broken up it was to be, when the Department of Justice insisted.

The comprehensive upheaval in the summer of 1948 brought to an end, for practical purposes, the close partnership of fifty years between du Pont on the one hand and ICI and their predecessors on the other. The exchange of technical information across the Atlantic stopped at once and the former partners began to move into competition with each other. From 1948 onwards du Pont set up sales agencies in ICI's 'exclusive territories' and in 1950 ICI bought a 70 per cent interest in an American dyestuffs company—Arnold, Hoffman & Co. Inc., of Providence, Rhode Island—as a base for manufacturing operations in the USA.

The cutting-off of technical information was much more serious for ICI, in the short run, than the somewhat tentative beginnings of competition. The technical provisions of the Patents and Processes Agreement were treated with unmerited scepticism by the American legal authorities, who professed to regard them simply as a cloak for illegal market-sharing, but they were much more than that, and cancellation was most unwelcome to ICI. The effect it might have had already been seen when, after the cancellation of the Nylon Agreement of 1939, information on the manufacture of nylon polymer was cut off from the beginning of 1947 (although a little more got through during that year only). Dyestuffs Division's polymer plant was then at a critical stage of development, which was made more difficult and slower by the cessation of information, with the incidental result of souring still further the relations between Dyestuffs Division and BNS.

The decisions of 1948 put the ICI–du Pont alliance in its coffin, but at the Department of Justice they were determined to screw down the lid. They had long memories and they did not forget that du Pont had voluntarily scrapped agreements in the past, only to replace them with others no less reprehensible. This time they meant to have a Decree, and being confident of their case they were not prepared to settle on any terms short of complete surrender. That was more than ICI and du Pont were prepared to accept, so during the summer of 1948, even as the new ICI–du Pont arrangements came into effect, the Department of Justice moved steadily towards a trial. In September 1948 came the surprising, rather disturbing, spectacle of the Department's agents, acting on the order of an American court, making a thorough search of ICI's files in London. Although some documents were privileged, the investigators found, as no doubt they expected, that over many years ICI had been far less guarded in their paperwork than du Pont. Commenting, later, on declarations made by Fin Sparre during the negotiations leading up to the 1929 Patents and Processes Agreement (chapter 3 (iii)), the trial judge observed:

It would be difficult to conceive a more explicit acknowledgement of the existence of a commercial understanding between du Pont and ICI. . . . It is significant that these damaging utterances—explicit and unqualified reports of statements by a responsible du Pont official—are to be found only in the files of ICI.[26]

There was plenty more of the same sort, which was natural, since ICI had never done anything illegal under English law and no one, until the launch of *US v. ICI*, had supposed that they would be brought within the reach of the Department of Justice.

The suit was tried before Judge Sylvester T. Ryan from 3 April to 30 June 1950. Verbal evidence filling 4,000 pages or more was called, including testimony by Sir William Coates and others seeking to establish that 'tariffs, freights and other economic circumstances' had determined ICI's policy of not exporting to the USA. That may have been true enough—it is certainly true that a prosperous British export trade in alkali was killed by the Dingley Tariff of 1897, and ICI would not have wanted to repeat that experience—but it did not meet the Department of Justice's central allegation that for many years ICI and du Pont had kept up a network of agreements which, under cover of patent-licensing arrangements, had been designed to get rid of competition between them.

'I have had a pleasant time with this case,' said Judge Ryan towards the end of the trial, in the amiable conversational style of much of the proceedings, 'and with the light that God gives me I will try to do justice to it.' Before the trial began he had announced that he owed it to himself and his family to take two months' vacation—not having had a vacation for three years—during the summer, and it was not until 28 September 1951 that he filed his Opinion, delivered in 207 pages of mordant lucidity.

Discounting the verbal testimony, the Judge based his findings almost entirely on the documents, remarking of the Patents and Processes Agreements: 'We find these latter . . . , allegedly motivated solely by technological considerations, fit with miraculous neatness into the general pattern of territorial arrangements. We are unable to accept the proposition that this happy harmony was the result of sheer coincidence.'[27] In this spirit of scepticism, expressed again and again in phrases neatly barbed, he found for the Department of Justice on every issue, saying that 'the defendants entered into a conspiracy to divide among themselves the territories of the world and that the agreements considered herein—principally the Patents and Processes Agreements and the Joint Company Agreements—were all parts of the conspiracy, devices intended to carry out that purpose.'

Ryan delivered his Final Judgment on 30 July 1952, 8½ years after the filing of the suit. Even then, 'final' though it was said to be, it was repeatedly modified up to 21 April 1953, and work required to put into effect those parts of it which related to CIL was not finished until the spring of 1954.

The terms of the Decree required to give effect to the Judgment were finally settled, according to American practice, at '"shirt-sleeve" conferences' (Bingen's phrase) between Judge and counsel for both sides in Judge Ryan's chambers on 19, 20, 26, and 27 June 1952. Bingen, who attended, said they were 'of an informal and friendly character throughout'.[28]

So they may have been, but the Decree was none the less thoroughgoing for that. The sweeping nature of the Department of Justice's demands, all of which, with minor exceptions, were met, has already been indicated and a detailed analysis of the Decree, which is on public record, is not necessary. Broadly speaking, it confirmed the cancellation of the Patents and Processes Agreement and destroyed the exclusive rights which ICI still held; it ordered revision of the Nylon Agreement and the Polythene Agreement, both in the direction of getting rid of exclusive rights; it directed ICI to grant any licences that might be applied for in the USA under patents covering 'common chemical products' (a phrase which might be held to exclude 'Terylene') which were in production on 30 June 1950; and it directed ICI and du Pont to arrange the dissolution of their joint ownership of CIL and the two Duperials. The only demand of any consequence which the Decree did not grant was dissolution of the joint ownership of CSAE (ICI 42½%: du Pont 57½%) and CBC (50/50, Remington/ICI). These were allowed to stand, but measures were taken against any device to regulate competition between the shareholders, who were forbidden, for instance, to sell materials to CSAE on a quota basis or to divide business or profits on export sales of sporting ammunition to CBC, as ICI and Remington had been accustomed to do. In case any of the defendants should feel tempted to evade the provisions of the Decree, the Court retained for five years jurisdiction and the right of access to records.

There was talk of an appeal, but the justice of Judge Ryan's findings, as a matter of law, was hardly open to challenge. High-handed no doubt the American authorities were, in asserting jurisdiction over a foreign corporation, but their case was strong and the whole conduct of the proceedings was fair. That for many years an elaborate system of regulated competition had existed was something which no one in ICI, qualified to judge, could with a good conscience deny. It was perfectly legal under English law and there were sound, reputable reasons of policy behind it, but neither of these points constituted a defence against allegations founded on the Sherman Act. 'We deem irrelevant', said Judge Ryan in his Opinion, 'any enquiry into whether the arrangements between the parties actually injured the public interest, or whether the public benefited thereby.' No doubt, but we may be permitted to take a wider view.

The du Pont alliance, it has been contended throughout this volume and the last, must be set against the general background of an industry in which regulated competition was the rule, not the exception. In particular it must be set against the circumstances of the chemical industry between 1926 and 1939, the first years of ICI and the last years of the old system of world-wide cartels.

During these years, especially early on, it was generally held that the world leader of the chemical industry, technically and in industrial power, was IG, and in du Pont and ICI they took the view that if they were not to go under, some form of joint defence was necessary. Since IG's supremacy depended largely on powerful and well-directed research, then for ICI and du Pont one obvious line of joint action lay in pooling technical information and patents, thereby cutting out expensive duplication and helping to bring each party more quickly to level terms with IG.

This is not meant as a denial of the market-sharing aspect of the successive Patents and Processes Agreements, which certainly existed. It is an assertion of a point which Judge Ryan found hard to accept: that technical co-operation between ICI and du Pont was genuine, valuable, and had a purpose of its own. At times it hindered some of ICI's operations, especially where oil interests were concerned. Towards the end, they may have felt in du Pont that they were putting more in than they were getting out. These points may be admitted, but

442

they confirm rather than weaken the general assertion of the independent importance of the exchange of technical information.

As to market-sharing, it has to be borne in mind that this was a long tradition in important branches of the chemical industry, notably explosives and alkali, and it was based on a recognition, generally unspoken, that no individual firm could do much to expand the market or support it if it were contracting. If a firm's resources were to be used to the best advantage, therefore, and particularly if violent swings between over-capacity and under-capacity were to be avoided, some measure of co-operation with potential competitors was an obvious policy, and at least it seems to have had the effect of preventing the violent cyclical surges which were characteristic of some other industries, notably shipbuilding.

In the thirties this line of argument was reinforced by the world-wide failure of demand. With the market in so many products, especially nitrogen and dyestuffs, collapsing beneath them, what were manufacturers to do? Cut each other's throats, perhaps, until huge plants had to go for scrap, bringing disaster to the working community that depended on them. Or should they work out some measure of orderly control of productive capacity, even at the risk of being abused, then and since, as cartel-mongers and heartless rationalizers?

Even with cartels in operation, ICI lost some 10 per cent of their capital in the early thirties and men were dismissed on a scale never known before either in the time of ICI or their predecessors. To press for unrestricted competition in times like those would have been to press for industrial ruin and social suicide. It was surely a public benefit rather than a public injury, to use Judge Ryan's terms, that American competition, as well as German, was kept away from the British market just as in the USA and Germany British competition was similarly regulated.

By the time the war ended, everything was altered, not because of the war itself, but because inventions in the chemical industry were beginning to open up, for the first time in the period we have been dealing with in these volumes, a prospect of that apparently limitless advance towards synthetic materials which was to be a cause for world-wide optimism in the fifties and for world-wide gloom in the seventies. In these circumstances, it may be that the public interest demanded uninhibited competition as a condition of growth. Certainly opinion at the time thought it did.

At this point the Department of Justice struck. It struck, and the old order, already much weakened, collapsed before it. Neither in du Pont nor in ICI, perhaps, was there much real regret. The prospect before the chemical industry was so enticing, and demand so buoyant, that the elaborate regulation of output and competition seemed a time-

443

wasting reminder of old, unhappy, far-off times, gone with the thirties never to return. The allegedly short-sighted, self-interested timidity of cartel-makers became part of conventional wisdom and the industry surged forward, the head and front of economic growth, that supreme objective of industrial and national policy.

REFERENCES

1 E. J. Barnsley to Lord McGowan, 24 Aug. 1943, FD 'USA 300, Administration—General 1943'.
2 Ibid. Webb-Pomerene Act: see *ICI* I, p. 344.
3 Civil Action No. 24–13, Complaint paras. 6, 7.
4 6 Jan. 1944, CR 595/67/67.
5 Statement by McGowan and letter to Division Chairmen, both 7 Jan. 1944, CR 595/67/67.
6 CSAE Sales Report, 14 Sept. 1939, USA, CSAE V, p. 17.
7 Petrie to Lore, 29 Dec. 1939, USA, CSAE V, p. 31.
8 USA, DA VIII, p. 80 and generally.
9 Neale, p. 327.
10 Neale, p. 303.
11 E. A. Bingen, 11 April, 8 May, 5 June 1944, CR 595/67/67; E. A. Bingen, 'Anti-Trust Legislation', 5 Jan. 1945, ICIBR.
12 P. C. Allen, *Diary*, 30 May 1942.
13 R. E. Slade to M. G. Tate, 7 July 1942, CR 595/7/19.
14 Lord McGowan to W. S. Carpenter, 17 Sept. 1941, P. C. Allen, *Diary*.
15 P. C. Allen, *Diary*, 5 June 1942.
16 W. R. Swint to E. J. Barnsley, 'Polythene', 24 Sept. 1942; E. J. Barnsley to W. F. Lutyens, 25 Sept. 1942; W. S. Carpenter to Lord McGowan, 24 Sept. 1942. All CR 534/1/38.
17 E. A. Bingen, 'American Litigation', 24 Aug. 1949, ICIBR; Judge Kaufman's Opinion, Neale, p. 303.
18 E. A. Bingen, 25 Jan. 1945, CR 595/2/2.
19 E. A. Bingen, 23 July 1945, ICIBR.
20 The Solicitor (E. A. Bingen), 'Nylon', 4 Nov. 1946, ICIBR; ICIBM 11,348, 14 Nov. 1946.
21 As (20), refce. 1, p. 4.
22 E. A. Bingen, 21 July 1947, ICIBR.
23 The Solicitor to the Board, 'Agreements with E. I. du Pont de Nemours & Company . . .', 8 July 1948, p. 2, ICIBR.
24 E. A. Bingen, 'South American Joint Companies', 4 Jan. 1952, ICIBR.
25 E. A. Bingen, 30 April 1952, Armstrong Papers, 3/1/6.
26 Sylvester J. Ryan DJ, 'Opinion' in *US v. ICI*, quoted by E. A. Bingen, 'United States of America v. Imperial Chemical Industries . . .', 5 Oct. 1951, pp. 3–4, ICIBR.
27 Ibid.
28 E. A. Bingen, 'American Litigation (Interim Report)', 7 July 1952, ICIBR.

CHAPTER 25

ICI at Mid-Century

AT THE TIME OF ICI's twenty-third Annual General Meeting, on 8 June 1950, four of the fifteen full-time Directors had been appointed before the war. Seven had come on to the Board during the war and four had been appointed in 1945 or later. The Board of 1950 was very different in composition and outlook from the Board of only seven or eight years earlier, and it faced a rapidly changing chemical industry in a rapidly changing world. Yet of the two men who could reasonably claim to have founded ICI, one was still there, still active, still Chairman.

On the day of the mid-century Meeting McGowan was five days past his 76th birthday. He spent a great deal of his time, in the last years of his career, travelling round the outposts of ICI. He moved in semi-regal splendour, meeting Heads of State on terms not far short of equality. When he went to see South American Presidents it was a pity, no doubt, that they spoke no English, but the British Ambassador would be at hand to interpret for Lord McGowan.[1] The damage done to his authority by the rebellion of the Board in 1937–8 had been repaired, if not forgotten, and although McGowan's dictatorial powers had never, in form, been restored, there were few who would withstand the old despot in person.

He had every intention of nominating his successor and of remaining in office until he died. He had made both points clear to Melchett in 1941 (p. 308 above) and there is no reason to think he changed his mind later. At the end of 1950, however, his three-year service agreement would run out and the decision to reappoint him—or not—lay with the Board. The general opinion was that he should retire, but no one wanted to tell him so. In February 1950 he went abroad, as once before he had gone abroad when the Board was restive, and the Directors took their chance. John Rogers, the senior Deputy Chairman, wrote to him in South Africa.

McGowan did not take the Board's decision well, and is said to have reproached Rogers for his part in it. The Directors, however, more resolute than their predecessors in 1938, held to their course. On the last day of 1950 Lord McGowan ceased to be Chairman of ICI, and became Honorary President until he died, on 13 July 1961.

The Directors had provided for the succession by suggesting Sir Frederick Bain, one of the Deputy Chairmen. He was remarkable rather for amiability than force, but perhaps that was what they wanted after McGowan. He had come into ICI at the merger, as a Director of the United Alkali Company, and he reached the ICI Board in 1940 after being Chairman of the General Chemicals Group. During the war he was on Government service for three years and he became a Deputy Chairman in 1945. In 1950 he was sixty-one and had recently served for two years as President of the Federation of British Industries. On the evening of 15 November 1950 Bain was seized with a fit of coughing at a public dinner. He left the room, slipped and fell downstairs, being unable to save himself because he lacked one arm, lost in the Great War. He was badly hurt, the true nature of his injuries was not diagnosed in time, and on 23 November he died.

With five weeks to go before McGowan retired, there was no recognized successor and no time for lengthy consultations to find one. John Rogers, four years younger than McGowan, agreed to take the Chair for a couple of years to give time for making a choice between two or three possible candidates for a long-term appointment. The contest for the succession to McGowan was thus put off, but not abandoned.

TABLE 32

EMPLOYMENT OF CAPITAL IN ICI MANUFACTURING ACTIVITIES
IN THE UNITED KINGDOM, 1927–1952

	1927	1932	1937	1942	1947	1952
	per cent (rounded)					
Heavy Chemicals	50	37	42	35	34	32
Explosives	23	13	15	15	12	9
Fertilizers, hydrogenation &c.	18	31	20	16	13	24
Dyestuffs &c.	4	7	7	12	14	17
Metals	4	8	10	14	15	12
Paints, lacquers, leathercloth	1	4	5	7	6	3
Plastics					4	3
	100	100	100	100	100	100

Source: see Appendix II, Table 2.

In the mid-century ICI, over which John Rogers was so unexpectedly called upon to preside, the balance of activities was setting decisively into the pattern dictated by the changes which had been coming over the chemical industry since the thirties. Table 32 opposite shows how the relative importance, within ICI, of heavy chemicals and explosives had been dropping ever since the merger. In 1927 they accounted for nearly three-quarters of the manufacturing capital then employed by ICI in the United Kingdom. Twenty-five years later the capital employed in heavy chemicals had dropped to about one-third of a much larger total and the capital in explosives, badly affected by the decline in mining at home and the existence of local manufacture in the largest markets abroad, had dropped below one-tenth.

ICI's false start at Billingham shows up in the figures for fertilizers and hydrogenation, and the true change of direction begins to show, between 1937 and 1942, in rising investment under the heading 'Dyestuffs &c.', which blankets some of the newer and more promising organically-based activities, including weed-killers, insecticides, synthetic detergents, pharmaceuticals, and nylon polymer. Investment in plastics begins to show up towards the end of the period, though not yet on a large scale. Outside the boundaries of the chemical industry altogether, metals show an independent vigour of their own.

At the root of capital investment in the chemical industry lies that intricate web of mental, psychological, and physical processes covered by the flag of convenience 'Research and Development'. ICI's research policy during the war (p. 303 above) had all been directed towards greater activity, and by 1945 expenditure was already launched on the steeply rising slope which it has followed ever since. This policy was seen not only as good business but as public duty. Scientific discovery, material progress, and economic growth were in those days highly regarded, and the prevailing sentiment was expressed in 1946 by Herbert Morrison:

. . . least of all nations can Great Britain afford to neglect whatever benefits the scientist can confer upon her. If we are to maintain our position in the world and restore and improve our standard of living, we have no alternative but to strive for that scientific achievement without which our trade will wither, our Colonial Empire will remain undeveloped and our lives and freedom will be at the mercy of a potential aggressor.[2]

'The principle of decentralisation', said R. M. Winter, writing of ICI's research organization in 1948, 'has been carried to almost extreme lengths.' Winter was Research Controller in succession to Slade, who had retired at the end of 1945. On the Board was the first ICI Director specifically charged with the supervision of research: Sir Wallace Akers. A bachelor, rather reserved on first acquaintance, he was nevertheless easily approachable, and to his juniors warm and

447

friendly, with something of a donnish touch, and an engaging talker. He surprised everyone by getting married in the year of his retirement, 1953. The research staff at Head Office was described by Winter as 'a very small administrative unit'.[3]

Decentralized research meant Divisional research, an ICI tradition fiercely maintained in the Divisions themselves. At the centre, nevertheless, there was a hankering, displayed in Winter's choice of phrase ('decentralisation . . . carried to almost extreme lengths'), after more positive central direction, or at any rate a desire for research on behalf of ICI as a whole as well as the Divisions separately. That desire had been met by the purchase, before the war (p. 92 above), of a site for laboratories where scientists might engage in fundamental research 'freed from the urgencies and distractions of applied research'.[4]

In the Divisions it was suspected that an ivory tower of great luxury was being prepared for a privileged *élite*. Nevertheless, as soon as possible after the war the plan was carried forward, though not on the site orginally intended, and a research station was founded which in 1955 took the name 'Akers Research Laboratories'. By 1949, in the grounds of The Frythe near Welwyn, twenty-seven graduates and twenty-one other scientific staff were at work in mycology, plant physiology, organic chemistry, physical chemistry, inorganic chemistry, and physics. The will of the centre had prevailed, for the time being, over the misgivings of the circumference.

Instrument research and research into industrial hygiene was also centrally directed, but the great bulk of ICI's research and development went on, as it always had done, in the Divisions. The output, in terms of new products discovered and invented, and of manufacturing processes developed, has been discussed in earlier chapters and will be discussed further below. As the policy of expansion gathered way, in the early post-war years, the problem of relating output to expenditure naturally arose.

Figures opposite (Table 33) show how research money was spent in ICI, and how much, in 1938 and in 1952. The increase, even allowing for inflation, is striking. It struck, among others, one of the lay Directors, Lord Weir. At the end of a Board-room discussion of new projects he was heard to inquire not how research could be expanded but whether anyone knew how it could be stopped.[5] Weir's point was not that ICI should withdraw from research, but that some method should be devised for appraising the commercial value of projects undertaken and of cutting off those which showed no likelihood of a profitable outcome. He thus raised the central difficulty of controlling applied scientific research, which is to foster a lively spirit of inquiry without letting costs go unquestioned. With more and more being spent, from year to year, it was a matter of rapidly rising importance.

448

TABLE 33

ICI GROUPS—RESEARCH AND DEVELOPMENT EXPENDITURE

	1938	1952	Increase per cent
	£000	£000	
Group A			
Alkali	97·8	239·9	
General Chemicals	109·2	467·5	
Lime	9·7	39·8	
Salt	—	12·8	
Total	216·7	760·0	251
Group B			
Dyestuffs	185·3	1,108·0	
Pharmaceuticals	— a	397·7	
Total	185·3	1,505·7	713
Group C			
Billingham	163·9	946·7	
Central Agricultural Control	40·4	189·8	
Total	204·3	1,136·5	456
Group D			
Metals	20·4	315·9	1,448
Group E			
Nobel (formerly Explosives)	89·4	681·7	622
Group F			
Leathercloth	4·7	108·2	
Paints	20·3	284·1	
Plastics	15·4	505·6	
Total	40·4	897·9	3,183
ICI Total b	782·6 c	6,043·3 d	672

a Presumably included with Dyestuffs.

b Includes items for Wilton, Akers Research Laboratories, ICI Technical Department and Central Engineering Research, ICI Medical Department and Industrial Hygiene Research Laboratory, and ICI Head Office Research + Fees, Expenses and Publications.

c Unadjusted figure. See Note to Table 4 in Appendix II.

d As well as b above, includes £528,400 for Fibres.

Source: 'Some Salient Features of ICI Research', Appendix XII, circulated to Chairman and Directors by J. Ferguson, 21 April 1960, CR 2/7/3.

TABLE 34

NUMBER OF TECHNICAL OFFICERS ENGAGED ON
ICI RESEARCH IN ICI GROUPS

	1938	1952	Increase per cent
Group A			
Alkali	42	43	
General Chemicals	86	95	
Lime	2	8	
Salt	1	3	
Total	131	149	14
Group B			
Dyestuffs	145	352	
Pharmaceuticals[a]	—	—	
Total	145	352	143
Group C			
Billingham	78(1939)	135	
Central Agricultural Control	15	30	
Total	93	165	77
Group D			
Metals	20	49	145
Group E			
Nobel (formerly Explosives)	60	111	85
Group F			
Leathercloth	7	21	
Paints	13	39	
Plastics	12	84	
Total	32	144	350
ICI Total	481	1,034[b]	115

[a] Included with Dyestuffs.
[b] Includes Technical Officers engaged on research in Fibres, Industrial Hygiene Research Laboratory, Central Instrument Section, and Akers Research Laboratories.

Source: Records in ICI Research and Development Department.

The results of ICI's research and development could be looked at under three main headings:
(a) improvement in the efficiency of existing processes and in the quality of existing products;
(b) the discovery of new products;

(c) the development of new manufacturing processes for existing products, for new products discovered in ICI, or for products, such as nylon and 'Terylene', discovered elsewhere but over which ICI held rights.

Results under these headings were fairly easy to identify. What was far more difficult was to show whether the research leading to them had paid for itself and left a handsome margin over costs—in other words, that it had been profitable.[6]

During the fifties work was going on, along lines suggested by A. M. Roberts and (Sir) Ronald Holroyd, Akers's successor in 1953 as Research Director, to measure the results of research into existing processes and products. By 1969 J. D. Rose, the Research Director of the day, claimed accuracy within 10 per cent above or below, 'which is pretty good', and went on to say that this kind of research had been shown to pay for itself in about eighteen months, 'and this of course represents a very handsome return on the outlay in anybody's language'.[7] He was speaking of sums representing more than half the money which ICI was then spending on research and development.

Under headings (b) and (c)—exploratory research for new products and development of processes—the main difficulty arises. It is extremely hard to isolate the costs of bringing a particular product to a marketable form and then to relate them to profits earned. J. Ferguson, surveying in 1960 the results of ICI research between 1927 and 1958, listed under the heading 'discovery of . . . new and valuable products' the following:[8]

Methyl methacrylate	'Mysoline'
Polythene	'Antrycide'
Benzene hexachloride	'Fluothane'
Hormone weedkillers	'Monastral' Blue and 'Alcian' dyes
'Paludrine'	'Procion' dyes

Under the heading 'the working out of a number of processes' he listed:*

Methyl methacrylate	Nylon
Polythene	'Terylene'
Petrol	Titanium
Polyvinyl chloride	Carbonylation products

Concentrating on products rather than processes, and pointing out that his list was not exhaustive, Ferguson produced the figures tabulated below (Table 35), showing trading profits from the more important products, discovered in ICI laboratories, which by the time

*Most of the products and processes listed by Ferguson are mentioned in the text (see index). For brief notes on those which are not ('Antrycide', 'Fluothane', 'Procion' dyes, Titanium and Carbonylation products), see Appendix I.

TABLE 35

PRODUCTS DISCOVERED FROM ICI RESEARCH, 1927-1958—
TOTAL NET TRADING PROFITS EARNED IN PERIOD FROM START OF
MANUFACTURE TO 1958

	Date of Discovery	Date of first Production	Total net Trading Profit
			(£m.)
Polymethyl methacrylate	1931	1936	18·74
Polythene	1933	1938	34·13
Phthalocyanine pigments	1934	1938	0·64
'Velan'	1936	1937	0·55
'Nonox' EX, EXN, EXP	1938	1938	0·73
'Sulphamezathine'	1940	1943	1·56
'Methoxone' and 'Agroxone'	1940	1946	1·05
Benzene hexachloride	1942	1945	1·17
'Paludrine'	1942	1947	3·70
'Alcian' Blue, Yellow, Green	1947	1947	2·17
'Mysoline'	1952	1952	1·06
Total			65·50[a]

[a] Includes Divisional profits and profits arising within ICI from the transfer of materials between Divisions.
For details of products listed, see Appendix I.

Source: 'Some Salient Features of ICI Research', Table XV, circulated to Chairman and Directors by J. Ferguson, 21 April 1960, CR 2/7/3.

he wrote had been 'manufactured for a sufficient time to earn profits'. That limitation, he pointed out, cut out 'Procion' dyes and 'Fluothane'. The total net trading profits he gives as £65·50m., and by adding an estimate of £7m. for profits from 'minor dyestuffs discoveries' he arrived at a total of £72·5m. net trading profits from new products discovered through ICI research from 1927 to 1958.

Over the period Ferguson was dealing with ICI laid out £16·8m. on 'exploratory work, including screening tests, aimed at the discovery of new products'—that is, on exploratory research narrowly defined. The corresponding figure for research and development as a whole is £105·1m., but to get an idea of the return on that expenditure it would be necessary to take account of expenditure on existing products and processes and of royalty income from processes developed in ICI, especially the polythene process. Rose, nine years later, emphasized the indispensability, the complexity, and the high cost of the development work which must follow any scientific discovery before it can go into

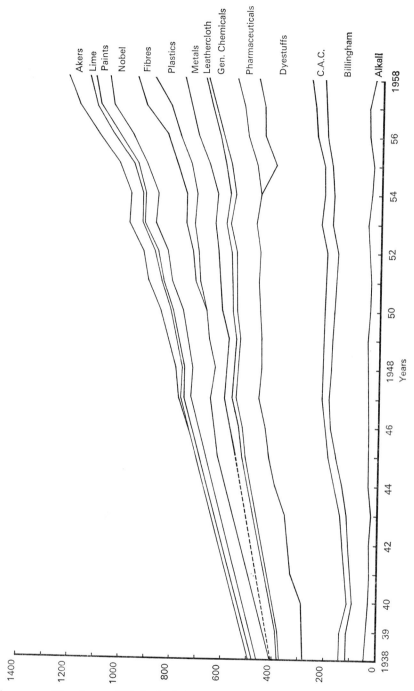

Figure 19. Numbers of Technical Officers in Research Departments of ICI
Divisions, 1938–1958

Source: ICI Research & Development Department

453

production, and he drew attention to the very attractive royalty income that may follow while patent rights last. To relate the profits earned from a product directly to the costs of research and development, however, would be a gross over-simplification. Rose concluded, and Ferguson would no doubt agree, that expenditure on long-term speculative research 'is the area in which it is difficult to relate outlay to return, and I have given up trying to do it'.[9]

What stands out from Ferguson's figures, as he pointed out, is that two products—'Perspex' and polythene—show up as vastly more profitable than any others in the list. That, perhaps, indicates the truth about research expenditure—that the effort is great, the successes are few, and the really big successes are very rare indeed. Any product which reaches the market from research is almost certainly the survivor of a great deal of abortive work involving, probably, the testing of thousands of possibilities. Before the discovery of the pest-control properties of benzene hexachloride, marketed as 'Gammexane' (p. 456 below), some 4,000 compounds had been tested, and after the discovery, up to 1959, they tested 30,000 more without finding any other outstanding insecticide.[10]

There is no standard method of making discoveries. Methodical investigation over a wide field, scientific reasoning within a comparatively narrow one, intelligent observation of the unexpected, chance and intuition have all produced outstanding successes in ICI and elsewhere, and any research organization must allow them free play. For that reason, if for no other, research cannot be minutely controlled financially or operationally. Scientists have a good deal in common with artists, and neither lend themselves readily to tidy administration. Research is an act of faith, and without faith it is most unlikely that there will be any works.

Research was particularly important to ICI's advance into the two promising and closely related fields of plant-protection and pharmaceuticals. That advance was hindered by a problem which seems continually to have baffled ICI's powers of organization: how to develop new products which did not fit easily into the field of interest of an existing Division or, worse, fell within the fields of two or more. The newly emerging branches of the chemical industry bristled with problems of this nature, and in plant-protection and pharmaceuticals they were unusually complicated.

The basic organic chemistry of all the products was closely related to that of dyestuffs, and Dyestuffs Division claimed general rights of proprietorship over the whole area, but their claim could be disputed by General Chemicals and to some extent by Explosives. Moreover, for anything in the nature of heavy engineering, they were dependent on help from Billingham. Central Agricultural Control had an obvious interest in the marketing of plant-protection products, but they had applications also in industry and public health. Pharmaceutical products might be either for human or animal ailments, and on the veterinary side Agricultural interests would come into play again, although Imperial Chemicals (Pharmaceuticals) Limited was a dependency of Dyestuffs Division.

The business in weedkillers and pesticides was bedevilled by the 1937 agreement between ICI and Cooper, McDougall & Robertston, which kept ICI out of the trade in protective 'dips' for farm animals and vested the marketing of plant-protection products in a jointly-owned company. Manufacture, however, was the business of ICI Divisions and of factories taken over by ICI under the agreement, which were run by Dyestuffs Division. Manufacturing profit was restricted under the agreement (as it was also under the BNS agreement) and the Divisions resented dealing with Plant Protection Ltd., the joint company, which found itself the unloved child of jarring parents. PPL's Directors considered their interests neglected by ICI Divisions, and relations were embittered. 'The Directors of PPL', wrote W. M. Inman, ICI Sales Controller, in August, 1945, 'have apparently given no thought to ICI's major interests . . . but have persistently tried to further a monopoly position, direct or indirect, for their own company.'

Even in this atmosphere, PPL's progress was striking, being greatly

TABLE 36

PLANT PROTECTION LIMITED: SALES AND PROFITS, 1938–1949

	Total Sales	Net Profits before Tax
	£	£
1938	227,708	16,065
1939	284,989	16,744
1940	390,611	40,903
1941	502,383	57,402
1942	689,904	97,998
1943	752,441	66,977
1944	751,392	2,688[a]
1945	920,283	32,613
1946	1,256,891	128,099
1947	1,761,784	261,175
1948	2,025,463	248,135
1949	2,100,522	87,556

[a] '. . . we were caught with difficult stocks which had to be liquidated and this . . . checked our profits for two years.'

Source: S. W. Cheveley (Managing Director, PPL), 'Plant Protection Limited', January 1950, Appendix 1, Quig Papers 12/2/1.

helped by wartime and post-war conditions which provided a sellers' market both at home and abroad, where a great deal of PPL's business lay. Staff rose from 200 in 1938 to 690 in 1949,[11] sales rocketed and profits, rather jerkily, rose with them (see Table 36 above).

Fleck negotiated a new agreement with Cooper, McDougall & Robertson in 1946 which, amongst other provisions, transferred to PPL the factories formerly run by Dyestuffs Division and arranged for a large increase in PPL's capital, which remained jointly held.[12] The new agreement was out of date almost as soon as it was made, because important new products were coming forward in ICI from which Coopers, under this agreement and others, gained benefits which, from an ICI point of view, seemed disproportionate to the services which Coopers could provide. Moreover, as long as ICI left the 'dip' trade to Coopers they lacked a sufficient base to build up, independently, a selling organization for animal medicines, which was a hindrance to the development of their pharmaceutical business. Over the whole field there was an overlap between ICI and Coopers in research, development, and manufacturing of pest-control products and animal remedies which ICI was becoming increasingly dissatisfied with. In 1948 three ICI Directors—Fleck, Quig, and Steel—suggested that ICI should take over Coopers.[13] Nothing came of the proposal and the two partners pass, still uneasily yoked together, beyond the pages of this History.

In the late forties PPL were relying heavily on sales, in various forms, of the insecticide 'Gammexane' and the selective weedkiller 'Methoxone':[14]

PPL Sales of:	1947 £	1948 £	1949 £
'Gammexane'	353,000	315,000	227,000
'Methoxone'	423,000	642,000	802,000
Other insecticides and fungicides	367,000	486,000	419,000
All other Products	621,000	605,000	669,000
Total[a]	1,764,000	2,048,000	2,117,000

a The difference between these figures and those quoted in Table 36 (page 455) is nowhere explained.

Both 'Gammexane' and 'Methoxone' were major ICI inventions, developed in the early forties. In their early history, both were unfortunate.

'Gammexane' (gamma benzene hexachloride) was developed in 1942, on the basis of research running back to 1934, to replace supplies of derris, a casualty of the Japanese conquest of Malaya. It was powerful and versatile. It killed locusts, grasshoppers, and many insects in the soil, including wireworm, and it could be used as a seed dressing.

On farm animals it killed lice, ticks, mites, and sheep blowfly, and since it killed ticks which arsenic would not touch it had a wide application in 'dips'—but ICI had left dips to Coopers. It killed malarial mosquitoes, human lice, bedbugs, houseflies, clothes moths. In granaries it protected the stored grain.[15] In short it was an insecticide so deadly in its effect and so universal in its applications that it might have been expected, like DDT, to have a major effect on the health and food supplies of the world, especially the poorer regions of it.

The discovery of 'Gammexane' was first publicly discussed by Slade in his Hurter Memorial Lecture at Liverpool in March 1945, and he quoted figures to show that it was far more deadly than DDT to houseflies, yellow-fever mosquitoes, and locusts.[16] By that time, however, 'Gammexane's' commercial prospects had been seriously damaged, partly by secrecy imposed by the Government but largely by the organizational complexity of ICI itself and the lack of any co-ordinating body to make sure of exploring and exploiting all the possible uses of the product.

'Gammexane' was discovered and developed by co-operation between the Research Department of General Chemicals Division and the Hawthorndale Biological Laboratories at Jealott's Hill. Jealott's Hill came under Central Agricultural Control. Naturally, therefore, the agricultural possibilities of 'Gammexane'—undeniably important—were investigated and others, including possibilities in public health, were left aside. When Geigy brought DDT into the United Kingdom in the latter part of 1942 they were able to attract the undivided attention of Government departments to its powers against lice and mosquitoes. 'As a consequence', says an ICI paper of November 1945, '. . . jungle trials were arranged by the Services before ICI had called attention to the possibilities of 666 ['Gammexane'], and when in 1943 sporadic approaches were made by General Chemicals Division and Development Department . . . and quite independently by the Research Department . . . it was clear that DDT had secured a lead in the Services which it would be virtually impossible for 666 to challenge.'[17] The largest single market left for 'Gammexane' was in the dip trade, and that was reserved for Coopers.[18]

It appears that 'Gammexane' fell between at least four of ICI's numerous stools—General Chemicals, Agriculture, Development Department, Research Department—none of which could reasonably be expected to take the responsibility of co-ordinating the efforts of the rest. The point was taken. In March 1945 the Development Director set up a Pest Control Panel. It was too late to be of much help to 'Gammexane', but 'Methoxone' was on the way.

'Methoxone' (4 chloro-2-methyl phenoxy acetic acid) was described in 1945 as 'an entirely new weedkiller'.[19] It neither sterilized the soil, as sodium chlorate and sodium arsenite did, so that no vegetation could survive, nor did it scorch the leaves of broad-leaved plants, like an acid. Instead, acting by processes which in 1945 were not fully understood, it was absorbed into the structure of some plants, but not others, causing drastic and gruesome distortion and slow death. Of the plants which it killed, as distinct from merely damaging, some were among the most serious competitors with cereal crops or with grass, but it would do no harm to the cereals or the grasses themselves.

'Methoxone' was developed from a discovery made at Jealott's Hill in 1940. Another compound suitable for a selective weedkiller was discovered at the same time—'Chloroxone' (2-4-dichloro-phenoxy acetic acid). Trials by General Chemicals' Research Department in 1942-3 showed 'Methoxone' to be the more active of the two, and 'Chloroxone' was set aside. Since work on weedkillers was 'largely merged with the investigation of certain military projects of the highest degree of secrecy',[20] the Government forbade the filing of patents abroad and ICI were prevented from putting anything like normal commercial development in hand until after the end of the war.

Meanwhile, in the USA, the action of 'Chloroxone' had been described in a magazine article under a different name, there had been a great deal of publicity, patents had been taken out, and the product had been launched. The patent situation was obscure, and in ICI there were suspicions that information might have leaked through official channels, but however all that might be it was clear that 'Methoxone' had lost its chance in the USA because, as F. C. O. Speyer wrote in 1947, 'the cost of "Chloroxone" and its application in agriculture throughout North America is so low that there is no possibility of "Methoxone" being able to compete unless it can be shown that "Methoxone" is far more effective or alternatively unless the cost of producing "Methoxone" can be reduced.'[21] Even under these somewhat depressing circumstances, 'Methoxone', as the sales figures already quoted show, became big business for PPL.

Pharmaceuticals, the chemistry and technology of which are closely related to those which produced the plant-protection chemicals 'Chloroxone' and 'Methoxone', illustrate yet again the uneasy fit between ICI's organization, technical advance, and commercial policy. The early development of pharmaceuticals, reasonably enough, had been placed with Dyestuffs Group in the thirties, and there was no lack of talent, as wartime work showed (p. 286 above). 'Paludrine', very widely used against malaria, and 'Sulphamezathine', at one time the sulphonamide most generally used in the United Kingdom, were both ICI inventions from the Dyestuffs Group. Nevertheless, in spite of rapidly rising turnover between 1942 and 1951, and very considerable, though fluctuating, profits, ICI pharmaceuticals were still, in the words of P. K. Standring, the Group Director responsible for them, 'a losing business . . . in the doldrums' at the end of our period.[22]

With the growth of the pharmaceuticals business, the question of separating it from Dyestuffs Division became more and more insistent, but it was not one to which the Board of ICI addressed itself with alacrity. Imperial Chemicals (Pharmaceuticals) Ltd. was set up as a selling company, mainly for tax reasons, in 1942, and Quig and Lutyens indicated an intention to transform it, at some indefinite future date, into a manufacturing company with its own plant.[23] The difficulty of disentangling the existing pharmaceutical plant from dyestuffs plant, however, was considerable; there were tax advantages in not doing so,[24] and it may be supposed that the Board of Dyestuffs Division showed no very great eagerness to be deprived of what might in time turn out a thriving source of business. The ICI Board set up Pharmaceuticals Division in 1944,[25] but the decision had so little practical effect that in 1947, when P. A. Smith became Chairman of IC(P), he was surprised and rather alarmed to find himself a Division Chairman—'the experimental evidence so far', he wrote to Cronshaw, 'leads me to believe that I should be rather out of my class'.[26] Throughout the period we are concerned with, Pharmaceuticals Division remained in a state of ill-defined but thorough dependency on Dyestuffs Division, which was strongly represented on the Pharmaceuticals Board and controlled the research laboratories and the manufacturing plant on which the pharmaceuticals business depended.

The pharmaceutical industry at large, during the forties and early fifties, was undergoing the profoundest upheaval in its history. One after another, synthetic products were transforming the whole basis of pharmacy and of medical practice. At the same time DDT was getting rid of mosquitoes and other disease-carrying insects. Disease was becoming preventable, curable, or controllable on a scale unimaginable ten or fifteen years earlier, and the effects, particularly on the balance between population and resources, were rapid and immense.

Seen against this background, ICI's approach to the pharmaceutical industry looks curiously half-hearted, and there is evidence that the procrastination in separating Pharmaceuticals Division from Dyestuffs Division was bad for morale. 'At present', wrote P. A. Smith in 1954, in what appears to be a studiously moderate tone, 'doubt often exists as to where the responsibilities of the two Divisions lie. The staff of both agree that a closer definition is desirable, and that in some departments a change of control would greatly improve efficiency and enthusiasm.' P. K. Standring, supporting Smith, went further. He wanted Pharmaceuticals divorced entirely from Dyestuffs, and said: '. . . what is really needed is good and well-informed leadership plus a close binding together of the three essential parts of the business, i.e. research,

459

production and selling—with research in all its aspects the concern of one man.'[27]

ICI might have advanced more rapidly in pharmaceuticals if they had taken over an existing business, as they originally intended to do. Various approaches were made, but they all fell through,[28] and eventually a policy developed, strongly supported by Cronshaw, of depending entirely on ICI's own resources, 'on the consideration that the best way thoroughly to master a new field of endeavour is to learn by one's own efforts at all stages'.[29] Accordingly the Division deliberately avoided taking licences or buying information.

This was magnificent, but was it wise? It ran clean contrary to accepted practice in the pharmaceutical industry, where advance was so rapid and multifarious that no single manufacturer, especially a new one, could hope to keep up with developments and build a large and varied range, to support a healthy turnover, without access to other people's knowledge. Soon after the war it was decided, on Cronshaw's advice, that ICI should work out a process for the deep culture of penicillin instead of buying information from the USA,[30] and ICI became probably the only manufacturers in the world who were trying to make penicillin from their own knowledge alone. The price of penicillin fell from 10s. 6d. (52½p) for 0·5 m u in 1947 to 3s. 6d. (17½p) in February 1950 and to 1s. 9d. (about 9p) in April 1954. ICI

TABLE 37

ICI PHARMACEUTICALS: SALES, RESEARCH EXPENDITURE, PROFITS, 1942–1953

	Sales Turnover	Research Expenditure	Trading Profit[a]	Loss
	£000	£000	£000	£000
1942	247	79		
1943	773	125	135	
1944	1,487	193	376	
1945	1,339	191	66	
1946	1,436	237		17
1947	1,729	313		95
1948	1,713	339		84
1949	2,494	368	143	
1950	3,566	372	607	
1951	5,076	420	1,156	
1952	4,270	493		89
1953	4,006	549		275

[a] After charging research, central services, depreciation.

Source: P. A. Smith, 'The Relation between the Dyestuffs Division and the Pharmaceuticals Division', 14 May 1954, ChaP 48/1/1.

could not follow it down with any profit, but their UK competitors, Glaxo and Distillers, were said to be 'reasonably satisfied'.[31] Glaxo and Distillers had both gone to the USA for information.

By the early fifties ICI was more firmly established at the 'heavy' end of the newly developing chemical industry—in plastics, fibres, petro-chemicals—than in the 'fine chemicals' end which we have just been examining. This was not due to technical incompetence. It arose partly from accidents of war, hindering normal development, but far more from defects in the organization of ICI itself. Indeed, it might almost have been deduced from the original balance of activities in ICI at the time of the merger, with its emphasis on heavy chemicals and its neglect of the 'fine' side of the industry. We have seen that ICI's organization, designed originally with the heavy chemical industry in mind, had never, up to the fifties, been really successfully adapted to further the needs of fine chemicals, even after their importance had been grasped.

The weakness of the organization, right through the period covered in this volume, was that no really satisfactory way was ever found of co-ordinating activities in the chemical industry unprovided for in the original structure of the business. That was designed in 1926 for the kind of work thought to be in prospect for the thirties. When that prediction turned out wrong no attempt seems to have been made to appraise the work of the second half of the century and to redesign the organization accordingly. Instead, the framework of 1926, patched and cobbled, was forced with increasing difficulty to serve purposes un-imagined when it was set up.

The size of ICI and the weight of established interests in the larger Divisions, directly represented as they were on the Board, made rapid decision impossible on the creation of new Divisions, the amalgamation of old and proud ones, and the transfer away from some Divisions of activities over which they had presided from the start, in which they took pride, and which they had run as profitable enterprises. Over matters of this kind a great deal of emotion could be generated, and decisions on the grouping and control of activities within ICI which, on purely rational grounds, might have been taken, and perhaps should have been taken, at any time from the mid-thirties onward, were in fact delayed until the mid-fifties, the sixties and even, in some cases, until 1971.

Problems arising from the growth and complexity of ICI were ob-scured in the two or three years after the war by the immediate prob-lems of the day. A great deal required to be done, both in maintenance and in expansion, but every proposal of any importance had to be steered not only through ICI's internal machinery—that was com-paratively easy—but through an all-enveloping web of Government regulations and past the obstacles raised by scarcity of materials, especially steel, and of staff, especially engineers.[32] In these circumstances it was enough if the Divisions' plans could be carried through, in something like the form desired, with something like the urgency dictated by the long wartime stop on ICI's natural development. The problems of the future could wait.

But not for ever. In the late forties, as the worst of the post-war stringencies eased and the view ahead began to clear, it became apparent that capital proposals were coming forward from the Divisions at a rate which over the coming years would need far more money than ICI could provide from retained profits and more even, perhaps, than could readily be raised on the market. The Divisions, quite properly, were considering their own interests rather than ICI's total resources or borrowing power, and it was becoming a matter of urgency for the main Board to face the problem of the central planning of capital expenditure.[33] They could hardly do that without asking fundamental questions about size, growth, and objectives; without asking themselves, in short, what ICI's business really ought to be.

In December 1949 ten executive Directors—all the executive Directors, that is, except the Chairman and Deputy Chairmen—formed a Capital Programme Committee under the Finance Director, S. P. Chambers, who, it seems clear, had forced the issue of capital planning over a report presented by the Technical Director Sir Ewart Smith, a couple of months earlier, setting out the likely demand for capital, from Divisions in the United Kingdom only, over the five years immediately ahead. The schemes the Divisions intended to put forward, he said, taken with those already sanctioned, would require, over five years, about £20m. of new money per annum and that figure took account only of expenditure on fixed assets, not of purchase of ex-Government factories or of land, houses, or rolling stock; nor did it allow anything for working capital or for finance for associated companies at home or abroad, and the overseas associated companies were hungry for capital.[34] 'Probably', said Chambers to the Committee, 'not more than an average of £8m. per annum could be provided out of the Company's own cash resources. . . . If, therefore, such a construction programme is to proceed, very large sums . . . will have to be obtained from outside sources. . . . Questions that arise for early consideration, therefore, are: ought the Company to contemplate large expansion, and if so, in what directions and upon what broad conditions?'[35]

Chambers thus brought the Committee straightaway to confront the widest aspects of their problem. Rough notes in his own hand, apparently taken at or soon after the meeting, amplify the record in the minutes. They suggest that the Committee viewed the Divisions' estimates of their capital requirements sceptically. 'We do not know',

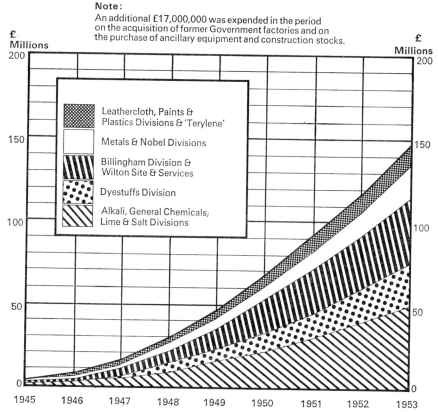

Note:
An additional £17,000,000 was expended in the period on the acquisition of former Government factories and on the purchase of ancillary equipment and construction stocks.

Legend:
- Leathercloth, Paints & Plastics Divisions & 'Terylene'
- Metals & Nobel Divisions
- Billingham Division & Wilton Site & Services
- Dyestuffs Division
- Alkali, General Chemicals; Lime & Salt Divisions

Figure 20. Cumulative Capital Expenditure, 1945–1953 (Working Capital is not included)

Source: ICI Annual Report 1953

Chambers scribbled, 'what total capital to be raised will be because we do not know . . . the likely demands from Divisions for capital. Divisions do not know whether projects should be put forward because they do not know what the policy of ICI is or whether the capital will be forthcoming.' After recording this succinct and comprehensive condemnation of ICI's techniques of planning and communication, Chambers noted that the figures supplied by the Divisions were for various reasons 'inadequate'.[36]

Assuming that the inadequacies could be made good, the Board would be able to to see how much would be required if they went on with everything. 'This total is likely to be too large because of the strain upon high grade manpower as well as upon extent to which market will let us have the money.' Choices would have to be made, and so a policy for making choices would have to be worked out. 'Major question',

463

R

Chambers's notes conclude, 'will be whether we ought to let ICI go into so many different fields that the problem of control by a single Board of Directors becomes unpractical and inefficient. Should activities be hived off for this reason and should fresh activities in new fields be refused for this reason?'

During the early part of 1950 the estimates of capital requirements were revised to get rid of 'inadequacies' and provide the Committee with a firm foundation for their deliberations.[37] The new figures included estimates for the associated companies at home and abroad as well as for United Kingdom Divisions, and for the five years 1950–4 they came to about £185m. Of this, somewhat less than £8m. was for overseas companies, and although Chambers commented 'This cannot be comprehensive', it suggests that in the early fifties ICI's commitment to expansion overseas was expected to be very small in comparision with expansion at home. With the details of the expenditure proposed, whether at home or abroad, we need not concern ourselves, since many projects, as was to be expected, were modified or given up during the five years which the estimates covered. What is important to us is the Committee's reaction to the figures prepared for them.

In August 1948 ICI had gone to the market for £20m., raised by an issue of 10,093,023 Ordinary shares at 40s. 6d. (£2·02½). In the summer of 1950 they raised another £20m. by the issue of 4 per cent Unsecured Loan Stock, privately placed and repayable between 1957 and 1960. ICI in the past had only very rarely gone to the market for capital, and the members of the Capital Programme Committee, meeting in the short interval between these two large operations, were nervous of trying to repeat the performance too often. They were now presented with a provisional programme which, Chambers said, would probably require £30m. a year for five years, and £20m. a year would have to come from the market.[38]

The Committee's first comment was that unless ICI's design and construction staff could be increased there would be no possibility of spending as much as £30m. a year, because the utmost ICI's existing engineering resources would be capable of would represent expenditure of about £21m. a year. Secondly, they thought it would be 'unwise to assume' that as much as £20m. could be raised on the market every year. 'Each issue of equity capital', they said, 'cuts out some of the market capacity . . . and the figures indicate that ICI would need to ask for more cash from the public than could be provided. The shortage of cash to take up equity capital is world-wide.'

The Committee's third conclusion was 'that in a few years' time ICI's capital will be doubled'. As a consequence, 'the Committee felt that they should closely consider the wisdom or otherwise of dividing the Company into two or more independent units, either by products or by basic industries in line with proper industrial grouping.' Their fear was 'that perhaps one Board might under future conditions be unable to give sufficiently close and knowledgeable attention to all schemes (especially those of the unusual or "off-pattern" kind) that result from the growing multiplicity of interests throughout the world comprising the Company's affairs.'

'If we aren't careful,' the members of the Capital Programme Committee were saying, 'ICI will grow too big to be manageable.' An unusual state of mind, one might think, for the Directors of a successful business, but then they were in an industry which by about 1950 was in a highly unusual state. New possibilities, as we have seen in previous chapters, were opening up with bewildering rapidity. New industrial links were forming, especially with petroleum in the direction of raw materials and with fibres in the direction of finished products. At the same time, as general industrial growth throughout the world surged forward, there was a lively and growing demand for the staple products of the old heavy chemical industry, especially alkali in various forms, chlorine, and nitrogen. The implications of all this hopeful turbulence, for ICI, were discussed by the members of the Capital Programme Committee in a report presented to the Board early in 1951.[39]

'The Company's capital has grown substantially since the Company's formation in December 1926, and the present prospects are that it will grow at a greater rate in the next few years. There is need for a more conscious policy in relation to future capital commitments if the Company is to avoid adding to the range of its miscellaneous activities without adding to its inherent strength and if it is to have an adequate defence against external criticism.' These opening sentences of the Report set out the two interwoven strands of thought which run through it, and indeed through much of ICI's policy-making, especially in the years after the war: on the one hand, concern for ICI's 'inherent strength' as a profitable business, privately owned; on the other, regard for ICI's standing in public life, particularly as a monopolist of certain essential industrial materials.

ICI's position as a monopolist of alkali, chlorine, nitrogen, and industrial explosives—the main commodities concerned—arose directly from the purposes behind the original merger, in which, as we saw in Volume I, there was a large element of public policy. The intention was to get a British chemical group powerful enough to compete internationally, and it succeeded, but at the cost of entrenching monopolies at home. In handling these monopolies, the members of the Capital Programme Committee maintained, the Board of ICI had always recognized their responsibilities to the country as a whole as well as to their stockholders. 'For this reason, when decisions to increase capacity have been taken, the wider responsibilities to maintain adequate supplies to British industry at reasonable prices have been as much in mind as the maintenance of a fair return on the stockholders' capital.'[40]

However conscious they might be of the purity of the Board's motives and the rectitude of their conduct, the members of the Capital Programme Committee were uneasily aware that since 1926 there had been, as they put it, 'a development of the public attitude towards the ownership and management of industry . . . particularly in those instances where the policy adopted has resulted in monopoly'. The passing of the Monopolies and Restrictive Practices (Inquiry and Control) Act, 1948, without substantial opposition, showed that public opinion generally, and not only on the Left, had come to regard price policy and control of the volume of output as matters of public concern.

As far as the older bulk products were concerned—alkali, chlorine, nitrogen—the members of the Capital Programme Committee would not be sorry to see ICI's monopoly broken, and indeed they were very ready to break it themselves, provided that could be done without depriving British industry of supplies or transgressing agreements, such as the Solvay Agreement, which preserved trade secrets. Their reason was that demand was rising very fast—in some cases at a rate of 3 per cent compound a year—and any attempt to keep up ICI's traditional policy of supplying the country's entire requirements, as well as providing for the export trade, would divert capital expenditure away from projects coming forward from research and from newer developments generally. The older activities of the business were very hungry for capital and as long ago as 1947 Lutyens, as Development Director, had thought too much attention was being paid to them.[41]

The path of the reluctant monopolist, however, was not an easy one. If commercial advantage had been the only consideration, as it would have been with a smaller business less central to the needs of the national economy, then it would have been easy enough to get rid of monopolies no longer wanted. But ICI, having long ago accepted the responsibility for supplying the country's needs of certain essential materials, could no more throw the responsibility over without notice, when it no longer suited them, than a nationalized corporation could. Any deficiencies that might arise would have to be made good from other sources. We therefore have the somewhat surprising spectacle of ICI Directors debating among themselves, in the pages of the Capital Programme Committee's Report and elsewhere, how other producers might be encouraged to enter fields hitherto dominated or entirely occupied by ICI.[42]

Competing for capital with the monopoly products, over the years to come, there would be, the Capital Programme Committee hopefully recorded, 'a steady stream of profitable projects . . . resulting from the expenditure . . . of about £5¼m. a year on research and development'.[43]

Certainly it would be a major disappointment if such projects were not forthcoming, after the encouragement which had been given to the Divisions, since the war, to increase their research effort by something between one-third and a half, but there would nevertheless be inconveniences. 'If', said the Committee, 'all the sound projects were adopted . . . the Company's activities would tend to expand in an ever-growing circle. Some of the fields in which the Company operates are widely separated from others.' Difficulties of management would grow, and besides that there would be a danger of ICI 'reaching a position in which its size and the importance of its role in British industry might be regarded as a matter of public concern whatever kind of Government was in power.'

The Committee's deliberations thus led them again and again to a point at which they were bound to consider whether ICI was becoming too large. Whether they looked at the raising of capital, or at problems of management, or at the development of public opinion and Government policy, the same question always arose: should ICI be split up?

'To any independent observer,' said the Committee, 'ICI's activities cover many different fields . . . and as a unit of industrial organisation the Company does not appear to be very logical.' The reason was that over a period of seventy or eighty years very different motives had operated, from time to time, for bringing various activities under common ownership: not least, although the Committee did not mention it, McGowan's ambitions in the motor trade which had caused him, about the end of the Great War, to extend Nobel Industries' metal interests.

It was easier, however, to take things on than to shed them. Staff became members of ICI Pension Funds, and if they were ambitious they could see a career in ICI as a whole which was much more attractive than a career restricted to one part of it. Moreover suppliers and customers might be upset if the ICI link were broken. 'Practical considerations of this kind', the Committee concluded, 'outweigh any advantages which might be gained by attempting to shed small units which do not logically fit particularly well into the ICI organization.'[44]

But what if ICI were to be split into two or more units 'so large that each would rank as an industrial concern of first-class size and importance'? Quite different considerations would apply, but the problem would arise of where to make the cut, or cuts. The interdependence of ICI's activities in the chemical industry, as we have seen in previous chapters, was close and becoming closer, and although the separation of Alkali Division was looked at, the idea was rejected.[45] Explosives Division, presumably, might have been a possibility, though its appeal to independent investors might not have been great. There was one obvious candidate: Metals Division.

The origins of Metals Division ran back to the politics of the explosives industry of the eighteen-nineties, when Nobels had bought their way into the Birmingham non-ferrous metal industry to give them a weapon against the ammunition trade of Kynochs, their competitors.[46] Towards the end of the Great War McGowan, at the head of Nobel Industries, in which the old Nobel business and Kynochs were both merged, had diverted resources out of explosives, for which he foresaw a dim future, into metals, for which, largely because of motor cars, he saw a bright one.[47] After the ICI merger the metals side grew both by takeover and by the extension of its activities, and it became the biggest business in the non-ferrous metals industry of the United Kingdom, though in none of its fields a monopolist. By 1951 Metals Division, with a net profit of £3·2m., showed a return of just under 12 per cent on capital employed of £27·3m.[48]

This was very satisfactory, and a flattering comment on McGowan's business acumen thirty years earlier, but Metals Division had never fitted easily into a business concerned with the chemical industry. The technology was different, the labour relations were different, the whole atmosphere in which the business was carried on was different, and Metals had always been run more or less as an independent enterprise. Then why not turn it loose entirely? The proposal had been made before the war, but at that time it would have been difficult to make the balance sheet attractive and the idea had been dropped.[49]

By the time the Capital Programme Committee came to look at the suggestion, the pre-war financial obstacles had disappeared, and the objections which could be raised against shedding one of ICI's small units did not apply. Metals Division could be floated off, from the start, as one of the largest British businesses, and there seemed to be no reason why either staff or payroll should suffer. Prospects for the staff might even improve.[50] Looking at their original problem, the raising of fresh capital, the members of the Capital Programme Committee thought the shedding of Metals Division might make matters easier both for ICI and for Metals. Chambers, thinking of the large institutional investors, said they would be much more likely to put up money for two companies rather than one, and his colleagues agreed with him. 'The mere size of ICI', as one of their reports puts it, 'becomes a disadvantage when investment managers have instructions not to put more than a certain proportion of their resources into one investment. These investors would welcome the separate high class investment which segregation of a large unit would afford.'[51]

When the members of the CPC looked at the size of ICI, they refused to admit that it was ever likely to become unmanageably large. The furthest they would go was to suggest that greater freedom of action might be a good thing for the Metals Board and that ICI's Board would be left free 'to concentrate on the remaining field which in any event might be expanding in other directions with the introduction of new products'.

Politically, nevertheless, they thought the splitting of ICI might be wise. 'Political action to reduce the power of the Company', they said, 'will become more difficult to avoid as the Company grows.' Perhaps if ICI were seen to be 'a very large company specializing almost entirely in the manufacture of chemical products' it might be 'less open to attack than one with activities extending into other fields as well'.[52]

In the early nineteen-fifties, when the period covered by this History closes, as in the early eighteen-seventies, when it opened, a rising gale of technological change was sweeping through the world's chemical industry. Under its force ICI was growing so fast and so variously as to oblige the Board to look deeply and critically at the course along which they were being hurried. They were prompted, first, by Chambers's warning that if growth went on at this rate they might find funds impossible to raise, but from that rather limited starting-point they launched on an inquest into the roots of ICI's being.

They found themselves compelled to examine the basic assumptions which had brought ICI into existence, to ask themselves whether they could run a business as big as ICI was becoming, and to balance ICI's interest as a private enterprise against its duty as a major public institution. They recommended—and in 1962 the recommendation was carried out by the formation of Imperial Metal Industries Ltd.—that a large part of ICI should be prised away from the main body.

All this suggests that in about 1950 ICI stood at a critical point, and that the Board knew it. Behind, in the quarter of a century covered by this volume, lay the establishment of the Company, the false start at Billingham, the shocks of the thirties, the age of the great cartels, the Second World War. Ahead lay expansion on a scale still not fully grasped, although it was already apparent that great risks would go hand-in-hand with great opportunities.

At the end of 1950 Lord McGowan retired. In 1951 the naphtha-cracking plant at Wilton started up. In 1952 Judge Ryan found for the United States and against ICI. This cluster of mid-century events may be said to mark, as clearly as any events might be expected to do, the end of the old ways and the beginning of the new. As well as that, they illustrate interwoven strands of narrative—personal, corporate, technological, politico-legal—which we have traced through nearly a hundred years of the history of one of the great enterprises of the modern world. With them, fittingly, the opening phase of the history of ICI may be said to come to an end.

REFERENCES

1 Lord McGowan, 'Brazil', 12 April 1949, ICIBR.
2 White Paper on Scientific Man Power, Cmd. 6824 of 1946.
3 R. M. Winter, 'Research in ICI', 24 July 1948, CR 31/34/34.
4 M. T. Sampson and R. M. Winter, 'The Butterwick Research Laboratories', 12 April 1949, ICIBR (Research Director).
5 Private information.
6 J. D. Rose, 'Research for Survival', *Proc. Roy. Soc.* A 312, 309–316 (1969), p. 309.
7 Ibid. p. 310.
8 'Some Salient Features of ICI Research', circulated to the Chairman and Directors by J. Ferguson, 21 April 1960, CR 2/7/3.
9 As (6), p. 310.
10 J. Ferguson, 'Discoveries of new Products by ICI—some Case Histories', 20 Aug. 1959, ChaP 10/1/1.
11 W. M. Inman, 'ICI/CMR/PPL Agreement', 20 Aug. 1945, CR 53/69/69; S. W. Cheveley, 'Plant Protection Limited', Jan. 1950, Quig Papers 12/2/1.
12 A. Fleck, 'Agreement with Cooper McDougall & Robertson Ltd. and Plant Protection Ltd.', 8 Nov. 1946, CR 53/69/69.
13 A. Fleck, A. J. Quig, J. L. S. Steel, 'Relations between ICI and Cooper McDougall & Robertson Ltd.', 23 Nov. 1948, CR 53/69/108.
14 Plant Protection Limited, Financial Statement for the year ended 30 Sept. 1949, CR 255/5/18.
15 N. A. McKenna, 'Gammexane', 7 Jan. 1947, CR 93/19/72.
16 General Chemicals Division Report GC/D/54 (2nd Draft), 5 Feb. 1945, CR 53/20/20.
17 N. A. McKenna, 'The development of "Gammexane" and "Methoxone"', 22 Nov. 1945, CR 53/20/20.
18 As (12), p. 4.
19 W. G. Templeman and E. Holmes, 'Methoxone: the new selective Weedkiller', n.d., CR 53/36/36.
20 A. O. Ball and D. J. Branscombe, 'Selective Weedkillers—Patent Situation in USA', 2 Sept. 1946, p. 2, CR 53/38/38.
21 F. C. O. Speyer to Sir W. Akers, 'Methoxone and Chloroxone in Canada and USA', 12 May 1947, CR 53/36/36.
22 P. K. Standring, 'IC(P) Ltd. Reorganisation Proposals', 17 May 1954, ChaP 48/1/1.
23 A. J. Quig, W. F. Lutyens to Management Board, 'ICI and the Pharmaceutical Industry', 12 Jan. 1942, CR 296/105/105.
24 A. J. Quig to P. A. Smith (not sent), 25 Feb. 1946, CR 110/14/14.
25 ICIBM 9,700.
26 P. A. Smith to C. J. T. Cronshaw, 23 Oct. 1947, Cronshaw Papers 46/1/1.
27 P. A. Smith, 'The Relation between the Dyestuffs Division and the Pharmaceuticals Division', 14 May 1954, p. 11, appended to (22); as (22), para. 6.
28 As (27), 1st refce., p. 1; as (23).
29 H. W. Thompson, N. W. Cusa, R. G. Hoare, A. R. Steele, 'Draft Report on the Future of the Pharmaceuticals Division', March 1958, ChaP 48/1/1.
30 C. J. T. Cronshaw, 'Penicillin', 23 April 1946, Cronshaw Papers 46/6/1.
31 As (27), 1st refce., p. 6.
32 ICIBRs (Technical Director), 1945–8.
33 ICIBRs (Financial Director), esp. 15 Aug. 1949; ICIBRs (Technical Director), esp. 21 Oct. 1949; ICIBM 12,948; numerous other papers.
34 ICIBM 13,058; note by R. A. Lynex, 13 Dec. 1949, CR 93/5/281; ICIBR (Technical Director) T20/49, 21 Oct. 1949; ICIBM 12,997.
35 CPCM 1, 20 Dec. 1949.
36 'CPC 1st Meeting . . . Rough Notes', ChaP 5/36/36.
37 A. Lyons, W. Robson, 'ICI Capital Programme', 19 April 1950, ChaP 5/36/36.
38 CPCM, 25 April 1950.
39 CPC Report to the Board, 22 Feb. 1951, CR 93/5/85R.
40 Ibid. p. 6.
41 W. F. Lutyens, 'Priorities', 18 July 1947, CR 93/2/41.

[42] 'ICI Capital Programme—Summary of ICI Board Discussion on 10th May 1951', pp. 1, 4, CR 93/5/36; ICIBM 14,434 (Chlorine Manufacturing Policy); 'The Future of the Salt Division', 27 June 1952, CR 93/5/235; 'Ammonia/Methanol Expansion Policy', 16 Oct. 1952, CR 93/5/234; 'Alkali Division: Future Policy', 7 Sept. 1952, CR 93/5/237.

[43] As (39), p. 4 (2(v)).

[44] Ibid. p. 9.

[45] CPC, 'Alkali Division: Future Policy', 7 Nov. 1952, CR 93/5/237.

[46] *ICI* I, p. 147.

[47] *ICI* I, pp. 382–4.

[48] CPC, 'Report on the Desirability of segregating the Company's Metal Interests', 26 March 1952, Appendix 2, CR 93/5/249.

[49] As (39), Appendix 1, p. 1.

[50] Ibid., Appendix 1 and generally; as (48), Appendix 2 and generally.

[51] As (39), p. 9.

[52] As (48), p. 1.

CHAPTER 26

The End of the Beginning

PRIVATE CAPITALISM may not have a very long future ahead of it; nor, indeed, the whole structure of industrial society, if the more enthusiastic prophets of doom are right. In either case the history recorded in the two volumes of this work—that is, the history of the forerunners of ICI from 1870 to 1926 and of ICI from 1926 to 1952—may quite soon be of no more immediate interest than the history, say, of the sixteenth-century Muscovy Company. Moreover, ICI's first quarter-century, which is all that this volume covers, may turn out to be a considerable proportion of ICI's entire history, if by the beginning of the twenty-first century, or earlier, ICI and the other great corporations of Western business have been swept, unregretted, like the *noblesse* of the *ancien régime*, into the dustbin of Revolution.

These are possibilities, but no more, and probably not the most likely. Leaving them on one side, there is no reason inherent either in the economic system at large or in the nature of ICI itself why ICI, along with other large corporations, should not last for centuries. Indeed the probability is that it—and they—will, for the survival power of large and wealthy organizations, given moderately competent management, is great. In that case the history of ICI between 1926 and 1952 is the merest beginning of ICI's story, and a beginning, moreover, which nearly went disastrously wrong.

Assuming, then, that what we have been looking at is the beginning of something that is likely to last a long time, and that we are looking at it at short range, without much in the way of hindsight, what is it that we see emerging in these first twenty-five years or so?

ICI has always been large. Size, indeed, was of the essence of its conception, for it was formed to balance, from the British side, the great German and American businesses which already existed in the chemical industry. There was no question of development in the conventional manner, like Lever Brothers, Courtaulds, the Dutch margarine firms, du Pont and many others, from a family business of moderate size to a very large group of enterprises. From the first ICI was Big Business, and on an international, not merely British, scale.

ICI, then, has always been a large corporation, during a period when large corporations, 'privately' organized or organized by the State, have been increasingly coming to dominate the economic and social landscape. On the 'private' side of the line these very large corporations, in Great Britain and elsewhere, are still organized as limited liability companies. By reason of their size and their importance in the economy, however, they are very different in nature from the businesses for which the limited company was originally devised as a convenient form of organization, and that difference has been observable in ICI from the outset.

First, it has never really been appropriate to speak of ICI as a 'private' undertaking, if by that it is meant that the business could be run in the interest of a narrow group of private owners. The shareholders, although legally the owners, would at all times have been more exactly described as 'investors', looking for income and capital growth, but not for a part in decision-making, and there have been no private individuals or families, as there have been in some American corporations, with holdings great enough to give a decisive voice on policy. Behind McGowan's original plan for the merger, moreover, there was a large element of public policy, making the Board conscious from the start that in all their major measures they would have to take the public interest into account, even, if necessary, ahead of the interest of the shareholders. They were frequently not free, as du Pont's management usually were, to consider simply the return on capital employed or the purely commercial aspects of a situation.

During the period we have been concerned with in this volume, there was a continual dialogue between ICI and Government, sometimes with Government looking for action from ICI, as in the case of the dyestuffs industry right at the beginning, and sometimes with ICI seeking help from Government, as in the protection of home-produced petrol and in the development of agricultural policy. During the war the interlocking between ICI and Government was very close indeed, and immediately afterwards, in the changing chemical industry of the day, there arose the question whether ICI should continue to guarantee supplies of the older bulk products, still essential to British industry but less attractive to ICI. The relationship, here, between public duty and commercial policy was very delicate indeed, and there was no obviously right course to pursue, because too dedicated a concern for past guarantees might blight hopes of future development.

Since there were no owning families, ICI was professionally managed from the first. To this generalization the two Monds are partial exceptions, but partial only. They owed their directorships to family connections, but they were able, full-time executives and it is impossible to believe that McGowan or the other Directors would have tolerated them if they had been incompetent. Management selection, unsystematic in the early days and, under Freeth, sometimes eccentric, was nevertheless always aimed at discovering ability regardless of

473

family interest or other considerations. The attempts of McGowan—no less—and Mitchell to establish hereditary rights were fiercely fought off.

It was impossible to make a great private fortune from a management position in ICI. Directors and higher managers were generously paid—some might have said McGowan was extravagantly paid—but the assets of the business did not belong to them, and it is from rights of property in capital assets that the millions are made. Managers, as a rule, were not men of means, and it might well be that a senior ICI manager's grocer or garage owner would be a far wealthier man than he.

Nor would an ICI manager's position give him a direct interest in the profits, which belong to the shareholders, not to the management. Main Board full-time Directors, until 1941, were paid commission, it is true, and in McGowan's case at least it was very substantial, but in principle they and all the rest of the management were paid by salaries not directly related to the results of the business from year to year, though naturally what was good for ICI would in time be good for ICI's managers; and the reverse would hold true also, as it did, very sharply, in 1931. In the way in which their income was derived, ICI managers were far closer to the men on the payroll, on the one hand, and to civil servants, on the other, than to independent business men.

It is necessary to insist at some length on the position of Directors and managers in ICI—and, indeed, in other large corporations—because there is a widespread tendency to confuse Directors and managers with owners, to assume that the interests of both are identical, and to assume, further, that their combined interest is inevitably and at all times opposed to the interest of 'the workers'. This confusion is not peculiar to the critics of capitalism. It often invades the minds of the managers themselves. The reasons for confusion are largely historical. They derive from the way in which the ownership and management of business enterprises have developed over the last century or so—over, in fact, the period covered by the two volumes of the present work. They are well illustrated in the history of ICI and its forerunners.

In 1870 it was rare to find a business in Great Britain, apart from the larger railways, which was not family property. Nobel's Explosives Co. Ltd. was an exception to this rule from the first, being financed by a group of investors unrelated to each other, and professionally managed, but in heavy chemicals the Leblanc businesses were family firms and Brunner, Mond, too, was owned by the founders and their families. Limited liability did not at first make much difference. Even when a business became a public company, as Brunner, Mond did in 1881, the original owners usually kept control, and the management would rest with them, their relations by blood and marriage, and such outsiders as they might choose to admit.

With rapidly growing businesses, however, ownership and management would begin to separate, and the tendency was observable both within and outside the chemical industry before the turn of the century. Owning families, finding it difficult to provide all the money needed for expansion, would be obliged to dilute their voting power—though the process might take a very long time—and a fast-growing business would also outrun family resources in another way, by needing more managers than even a large cousinhood could supply. Sooner or later—sooner in the case of Brunner, Mond; very much later in the case of du Pont or Solvay—the representatives of the original owners would have to take in strangers. Moreover, business ability and inclination run more plentifully in some families, such as the Monds, than in others, such as the Brunners, and again, if managers could not or would not be found within the family circle, they would have to be sought outside.

In Brunner, Mond at the date of the merger, the Monds were still strong as owners and as managers, and owning families were prominent in the management of the United Alkali Company, too. In Nobel Industries, on the contrary, there had never been owner-managers on the Nobel side and from Kynochs the Chamberlains had been displaced, while British Dyestuffs Corporation, Herbert Levinstein having been ejected, was professionally run throughout. As a condition of forming ICI, the owning families of the founding firms—Monds, Brunners, Gaskells, Muspratts—exchanged their shares for shares in ICI and their voting power became negligible. Those who chose to remain in active management, notably the two Monds and Holbrook Gaskell, held their places by professional ability, not as owners.

ICI, that is to say, like Unilever after it, was never anything but a professionally managed corporation. Once that position has been reached, it is by no means axiomatic that the interests of owners and managers will be identical. For owners, particularly absentee owners, asset-stripping may have strong attractions: not necessarily so for professional managers, who do not stand to gain personally but, with their careers in mind, have to look to the continuing goodwill of staff, workpeople, and customers. Hence, as we saw in chapter 25, the ICI Board in the early fifties rejected the idea of shedding any part of ICI which was not large enough, as Metals Division was, to provide satisfaction for all concerned. Moreover, there may be proposals where the public interest should take priority over considerations of maximum profitability. The professional manager does not stand to suffer loss, whereas the owner does, so that the professional manager is likely to find his judgement less clouded by self-interest in arriving at a decision.

What begins to appear, from all that we have so far considered, is that as an organization for managing large resources and indispensable national assets the large corporation, of the size and style of ICI, stands midway between a company in private business, properly so-called, and a public corporation of the kind that has been established, since ICI was founded, for running nationalized industries. ICI has always been free of formal Government control, on the one hand, but against that it has been far too large to be manipulated by any conceivable alliance of shareholders. This position, no doubt, raises questions of irresponsible power and public accountability, which we shall return to. To any mind less than fully convinced of the omnicompetence and far-sighted benevolence of the State, it has also certain advantages.

First, although ICI's position throughout the period covered by this volume obliged the Board to take the public interest into account, yet the Board were not obliged to accept the official view of what the public interest was, and could on occasion follow their own line. Thus in 1947 ICI refused to design and build gaseous diffusion plant for atomic energy, considering that it would overstrain their engineering resources and distract them from their proper work in the chemical industry. Lord Portal of Hungerford expressed 'grievous disappointment'.[1] To take a line of this sort, particularly under a Labour Government not at all averse to the idea of nationalization, must have needed considerable strength of purpose. In the event, there is no doubt that ICI's judgement was sound, not only in their own interest but in the public interest too, if the public interest lay in the development of the post-war chemical industry; but it must be asked whether, if the Board had been running a nationalized corporation, they would have been allowed to go their own way. At the very least, there would presumably have been a prolonged inter-departmental struggle.

A corporation organized independently of the Government is not only free of direct Government control of its operations. It is also free of direct Government control—or support—in its labour relations. A wage dispute, therefore, is less likely than a wage dispute in a nationalized corporation to bring trade unions into open conflict with the lawful authority of the State—a point of some constitutional importance which was less obvious when the main nationalized corporations were being set up than it has become since.

Relations with the British Government, for a corporation heavily engaged in international trade and investment, are not the only matter in issue. Relations with Governments overseas have also to be considered, particularly where local business is undertaken in partnership with local interests. In activities of this sort, especially in ex-Imperial territories, it is at least reasonable to assume that a nationalized corporation, owned by the British Government, might find itself less welcome, or more unwelcome, than a corporation, like ICI, independent of direct Government control. In ICI, when there was a possibility of nationalization in 1949, it was pointed out that difficulties might arise overseas, for a nationalized corporation, in employing nationals of other countries in senior positions, in relations with local shareholders, and in dealings with customers.[2] With the passage of time, none of these arguments have lost any of their force, and the importance of international trade and investment is even greater.

But in corporations as large as ICI, it may be said, the advantages of independence are outweighed by the risks, for here are the makings of over-mighty subjects. Should so great a concentration of resources as there has been in ICI since its foundation be at the disposal of a management which is autonomous and self-perpetuating? Should such power as those resources confer lie in any hands but those of the State?

Power is at the heart of the matter of this book, because the primary purpose of setting up ICI was to concentrate power: power, that is, in world markets, especially the markets of the British Empire. The Empire needed defending, in 1926, against the industrial and trading force of Germany and the USA just as, within a few years, it began to need defending against the military force of Japan, Germany, and Italy. The British Government was not directly responsible for forming ICI, that not being the fashion of the day, but the Government made its wishes known in the matter,[3] and it is legitimate to see the ICI merger as a stroke of Imperial policy.

From that point of view, it was successful. The British chemical industry, in spite of the gigantic misjudgement at Billingham, was immensely strengthened, so that it survived the Depression, contributed handsomely to the war strength of the nation, and faced the nineteen-fifties fit for the immense development and expansion of those years and later. Overseas, the position in traditional British markets was stabilized and held, the worst effects of the Depression were mitigated in the cartel arrangements of the thirties, and the rising chemical industries of South Africa, Canada, Australasia and, later, India and other countries were primed with capital and technical knowledge from Great Britain rather than elsewhere. At the same time a very large and expanding export trade was carried on. None of this would have been possible, it seems safe to say, without the concentrated power of the merger.

This is a record of British success, in the harsh politics of world trade, during a period chiefly remarkable for the feebleness of British policies and for the disintegration of Great Britain's world power. Success, however, may not be its own justification, and the question remains whether power capable of action on such a scale is too great to be left in private hands, particularly since it can be turned inward on the economy and society of Great Britain as well as outward on the world at large. In the late nineteen-forties, as we saw in chapter 25, the Board of ICI were apprehensive that some Government, not necessarily of

477

the Left, should decide that ICI's power was too great, and the question is not one that has gone to sleep with the passage of time.

Power is the essential material of politics. It is legitimate, therefore, to look at the organization of ICI in political terms. When ICI was set up, and for many years afterwards, there was no suggestion that democratic processes had any place in the running of business—no suggestion, that is, from any quarter that McGowan and his colleagues were likely to take any notice of. Accordingly the organization of ICI followed the normal structure of a limited liability company, becoming thereby an autocracy accountable only to the nominal owners, the shareholders: an autocracy, moreover, with powers limited only by the law and the Articles of Association, and protected by a veil of secrecy in which the only compulsory gaps were those required by the company law of the day—gaps which were not large enough to reveal £18m. goodwill in the original £65m. capitalization.

The limited company, as we have already remarked, had a form of organization designed, at a time when the rights of ownership were not seriously questioned, for enterprises much smaller than ICI, in which owners were frequently managers, and when 'business' was considered to be no business of the State. For ICI, a very large undertaking closely identified with public policy, it provided a Board which for practical purposes was self-selecting and self-perpetuating, and from the authority of the Board every appointment in ICI, as well as all policy, ultimately flowed. The 'lay Directors' could at times exercise decisive power, and it was generally recognized that they represented, in some sense not closely defined, the public interest; but their selection and appointment lay not with any public authority but with the Board itself, and in McGowan's time very largely with McGowan.

This autocratic organization carried with it, like other autocracies, its own internal politics which limited in practice an authority in theory absolute. In ICI, as in all large businesses, there was an unresolved conflict between necessary central control and equally necessary freedom of action in the Groups at home and in the overseas companies. The self-interest of the Groups (later, Divisions) themselves might stand in the way of positive action, though it could also provide a powerful drive, as it did in Alkali Group's development of polythene. The sheer size of the undertaking made delegation, consultation, and committee work imperative, and McGowan's attempt to push the Managing Director's powers to the uttermost, in the thirties, at last drove the Board into rebellion. But as long as the Board was not divided, there was no constitutional check on the central will, which within broad limits was autonomous and had at its disposal a powerful, if sometimes slow-acting, engine of decision.

Just how dictatorial the ICI central management could be, when unhindered either by internal tension or external countervailing power, is shown by the conduct of labour relations in the twenties and thirties. Mond's policy, which became ICI's, was enlightened in content and benevolent in motive. It was also thoroughly repugnant, in some of its most important aspects, to trade union sentiment. The unions at the time were weak, and the whole programme was carried through with no more than a polite nod in their direction, in the hope that it would detach the men's loyalty from them and fix it firmly to ICI.

In foreign policy, the conduct of ICI's affairs in Chile, between the wars, shows how a situation overseas might be exploited where there was neither a powerful Government on the spot nor powerful local shareholders. ICI and du Pont, but ICI especially, used the joint Chile company as a pawn in their wider game, and no one in Chile could stand up to them. How different matters might be where local interests were adequately represented was shown by events in Argentina, Brazil, and the British Dominions.

These two examples suggest that in ICI, as in other powerful organizations, there has been a tendency to push strength as far as it will go. The corrective is countervailing power on the other side, and the narrative set down in this History suggests, in ICI, sufficient political skill and sense of public responsibility to respond to movements of power and opinion in the world at large and to adjust policy continually to the mood of the times. Post-war labour policy in Great Britain, for instance, was consciously shaped to fit the new strength and temper of working-class opinion; and commercial policy generally after the war was framed with an eye to public opinion and Government policy towards monopolies and restrictive practices, though at the same time the Board was prepared to stand against pressure in what it considered mistaken directions.

In all these matters opinion outside ICI was reflected within: it was not altogether a case of unwelcome change being forced on a reluctant Board. Indeed the executive Directors, acting as the Capital Pro-gramme Committee, put themselves through prolonged self-questioning in which they showed themselves prepared to contemplate very drastic measures of change, taking the broadest view of ICI's responsibilities to investors, employees, customers, and the nation at large.

During the whole period of this History, indeed, the ICI Board could never fairly be accused of taking a narrow view of their responsibilities as they saw them: that is, from the point of view of convinced supporters of private capitalism. In considering all the major projects of the period —fertilizers, oil-from-coal, plastics, fibres, and the supply of heavy chemicals to industry—we have seen that the Board had always a strong regard for the public interest as well as for private profit, and they saw no reason to suppose that the two need be incompatible: rather the

479

reverse, for they would argue that the public interest required a proper regard for private profit, without which the system would collapse.

Destructive criticism is a popular spectator sport: never more so than when it is directed at capitalism and the larger capitalist enterprises. It is easy to show, and it has been shown in these pages, that ICI during the first quarter-century of its history had an organization ill-adapted to the developing requirements of the chemical industry; that one enormous error of planning nearly destroyed the merger within a few years of its foundation; that the constitution of the company placed great power in private hands and that it was sometimes used, in labour relations and in the affair of the Chile company, in a way that does not commend itself, forty years later, to opinion self-labelled 'enlightened'. The political thrust behind complaints of this sort is powerful, particularly when it is linked with public suspicion of the profit motive and of the activities of large, multi-national corporations.

Mond and McGowan's fundamental service to the nation, in 1926, was to act on the consequences of the unpalatable recognition that the British chemical industry needed drastic overhaul to give it any serious hope of sharing world markets, or indeed the home market, with the Germans and the Americans. In taking this view of their own industry—in which, after all, their own firms were reasonably prosperous—they showed far clearer vision than most of their contemporaries in the 'basic industries'—shipbuilding, coal, iron-and-steel, the textile trades—who were far too much inclined to blame their state of chronic depression on unprincipled foreign governments, on cheap foreign labour, on the bloody-mindedness of their own labour, or on anything rather than the deficiencies in organization, marketing, and technology of their own industries.

The construction and consolidation of the ICI merger, much the greatest merger that had ever been attempted in British business, was a major feat of leadership. The obstacles to success in a merger are not so much financial as practical and, above all, psychological. Fears must be calmed; old corporate rivalries put away; a sense of common purpose created in their place. The process can take years, and in IG Farbenindustrie, evidently, it was not complete when the Allies dismembered the business after the Second World War.[4] In ICI Mond and McGowan addressed themselves straightaway, with great energy and skill, to easing the internal strains and building a business in which diversity would be a source of strength, not weakness. The process was not complete by 1939, but nevertheless, when after only three or four years ICI was put to the severest test that any business in recent times has had to face—the economic crisis of 1930-1—a remarkable degree of underlying unity had already been achieved.

The unified business thus built up achieved its primary purpose, as we have already observed in this chapter, in establishing the re-organized British chemical industry in world markets and in encouraging the expansion of the chemical industry in various territories overseas, especially the British Dominions. In the disordered world of the thirties, ICI was strong enough to defend its own interests, which were also British interests, far more powerfully than the constituent firms, acting separately, would have been able to do. It is difficult to believe, for example, that Brunner, Mond could have negotiated the Nitrogen Cartel or British Dyestuffs Corporation the cartel in the international dyestuffs trade. As a consequence of ICI's strength, in spite of the disasters at Billingham, major sources of employment were preserved, and later expanded, for British workmen. People in two depressed areas, north-west England and Teesside, had particularly good reason to be glad of the presence of ICI Groups, and they were well aware of it.

The technological base of the British chemical industry was pre-served, revived (in Dyestuffs Group), and expanded in ICI, during our period, by the application of resources to research, development, and production on a scale never possible before in Great Britain. Moreover, ICI's bargaining strength supported the negotiation of agreements for the exchange of technical knowledge with du Pont and IG at a time when ICI's own research effort was in its earliest stages. Hindered by depression and then by war, it was being greatly expanded towards the end of our period, but the full results had hardly begun to show. Even so, a number of inventions of first-class importance, including poly-thene, 'Perspex', various dyestuffs, sulphonamides and antimalarials, were made in ICI laboratories from the earliest days onward, and during

TABLE 38

INVESTMENT OF £100·00 IN ICI ORDINARY STOCK IN 1926,
AND HELD CONTINUOUSLY UNTIL MAY 1973:

A. Including scrip issues but excluding rights issues, the investor would have come to hold
£418·00 nominal.
B. At the closing price on Friday 4 May 1973—276p—the value of the holding would be
£1,154·00, equal to 5·3 per cent growth a year since the original purchase.
C. Income in 1972–3 would be
£58·52 gross
£35·84 net after tax at standard rate of 38·75 per cent.
D. The cumulative *net* income, 1926–7 to 1972–3, would be £591·00.

Source: ICI Treasurer's Department.

the war the development of 'Terylene' began. Moreover, from 1944 onward ICI money was being spent freely on the encouragement of general scientific research in the universities.

In the broadest terms, what the founders of ICI set going in 1926, and what was built up over the succeeding twenty-five years, was an organization for increasing the general wealth by the profitable provision of industrial supplies. A small proportion of the increase was paid out to the shareholders in the enterprise, though hardly on such a scale as to make any individual shareholder very rich, either by means of income or capital growth. £100·00 (see Table 38 above) invested in ICI Ordinary stock at the formation of the company would have produced stock worth rather more than £1,000 at market prices in the spring of 1973: a respectable rate of growth, but hardly one to transform a small investment into a great fortune. Cumulative income over the same period would have amounted to £591.

Far the greater part of the wealth created by ICI in its first quarter-century flowed, by many channels, to society at large. It flowed by way of reinvestment in ICI itself, for greater productive capacity; by way of wages and salaries paid out to employees; by way of payments to suppliers of goods and services; and by way of goods produced, for other industries to use, from ICI materials. By 1952 the greatest growth was still to come, but foundations had been put down for a business which was at all times a national enterprise and upon which, to some considerable degree, the growth of the national welfare had already come to depend.

REFERENCES

[1] Lord Portal to J. Rogers, 13 Feb. 1947, and associated correspondence and papers, SCP; see also Gowing, *Independence and Deterrence*, ch. 17 (draft).
[2] *ICI and its Overseas Interests considered in relation to the Question of Nationalisation*, July 1949, CR 93/267/267.
[3] *ICI* I, pp. 422, 451, 453.
[4] Karl Winnacker, *Challenging Years. My Life in Chemistry*, Sidgwick & Jackson, 1972, p. 65.

Technical Notes

by F. G. Lamont

INTRODUCTION

EVEN IN WRITING an essentially non-technical history such as this, it is impossible to avoid using some technical terms. The following notes have been written in as non-technical language as possible to assist those without any scientific background but inevitably some technical detail has had to be included which may be of interest to those with some little acquaintance with chemistry. They are not intended as scientific descriptions or definitions.

Two main characteristics which distinguish chemical compounds from mixtures are, firstly, that the properties of compounds are quite markedly different from those of their components—thus the union of one atom of sodium, an extremely reactive metal, with one of chlorine, a very corrosive and poisonous gas, forms a molecule of common salt. Secondly the elements, in forming compounds, combine in rigidly fixed proportions which are simple multiples of basic 'equivalents', characteristic of each element. From such considerations was deduced the atomic theory of matter and chemical reaction.

Early chemistry was mainly concerned with the transformation of naturally occurring substances by reaction with others, e.g. the smelting of metals and the production of lime and alkali, but long before the physical existence of atoms had been demonstrated, chemists, armed with the atomic theory, had been able to show that, not merely were the basic units (molecules) of a compound built up of a precise number of particular atoms but that these were arranged in a definite structure so that the same assemblage of atoms in the same precise proportions could in certain cases form more than one distinct compound if arranged in different structures. Once this was recognized chemistry naturally turned to synthesis or the building up of desired structures; either from the elements or, more often, from intermediate compounds.

The atoms of some elements are capable of forming only a single linkage, or 'bond', with a single adjacent atom but others can form two or more such bonds and thus build up elaborate molecular structures. Carbon is outstanding in this; its atoms exhibit a 'valency' of four, i.e. can form four bonds, and also show a unique propensity to link up with each other, to form chains, rings, or even three-dimensional patterns. This is the basis of organic chemistry. These arrangements are represented in chemical formulae thus:

C–C–C–C–C–C–C
(chain)

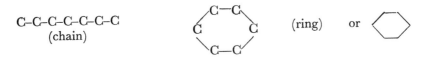

(ring)   or

Other elements (e.g. hydrogen, oxygen, chlorine, etc.) may be attached to or incorporated into these structures and innumerable elaborations of these basic structures are possible to produce compounds with desired properties of strength, elasticity, solubility or rigidity, etc.

A further very important property of carbon is the ability to form double, or even triple, bonds (represented thus, C=C and C≡C) between two adjacent carbon atoms as in ethylene and acetylene respectively. Contrary, perhaps, to expectation these multiple bonds involve internal 'strain' in the molecule and are therefore much more easily opened than a single bond. Compounds containing them are very reactive.

Modern chemical industry, the emergence of which is described in this book, is mainly 'organic' and based on the manipulation of simple 'building blocks', mostly compounds of carbon and hydrogen, to build up complex molecular structures, but a synthesis which can be demonstrated in the laboratory may be impracticable industrially until formidable problems in chemical engineering have been solved, particularly those which require or generate extremes of heat or pressure. Those who may wish to pursue these matters further may refer to:

John Read. *A direct entry to organic chemistry.* 1948.
A. Atkinson. *Modern organic chemistry.* Stanley Thomas Ltd., 1973.
F. S. Taylor. *A History of Industrial Chemistry.* Heinemann, 1957.
F. W. Gibbs. *Organic Chemistry Today.* Penguin, 1961.
A. Findlay. *Chemistry in the Service of Man.* 8th ed. Longmans, Green, 1957.
J. Knight. (rev. G. Bruce Macalpine). *Teach yourself chemistry.* E.U.P., 1971.
R. W. Thomas and P. Farago. *Industrial Chemistry.* Heinemann Educational Books Ltd., 1973.
B. G. Reuben and M. L. Burstall. *The Chemical Economy: A guide to the technology and economics of the chemical industry.* Longman, 1973.

ACETIC ACID   A colourless pungent liquid, the essential basis of vinegar, is an important intermediate for solvents, etc. Prepared originally by fermentation, it is now produced by oxidation of alcohol or synthetically.

ACETONE   A colourless, volatile liquid with a characteristic odour important as a solvent in paint, lacquer, and cordite manufacture, etc., and as an intermediate for the synthesis of methacrylates and many other organic compounds. Most commonly met as nail varnish solvent or remover.

ACETYLENE   A highly reactive hydrocarbon gas, the molecules of which are composed of two atoms of hydrogen attached to two atoms of carbon which are linked by a triple bond (represented thus: HC≡CH). Because the triple bond involves great intra-molecular stress acetylene is extremely reactive and indeed pure acetylene gas alone can be detonated with great violence. Acetylene is readily produced by the careful addition of water to calcium

APPENDIX I

carbide as in the old bicycle lamp. For a time it was used as illuminating gas and the oxy-acetylene flame is now widely employed for metal-cutting and welding. It is a convenient 'building block' for many organic syntheses.

ACID   The common inorganic acids (e.g. sulphuric, nitric, hydrochloric) are strongly corrosive, dissolve metals and neutralize alkalis (q.v.) or alkaline earths forming salts (q.v.) The simpler organic acids are soluble in water, the solutions being quite strongly acid. With increasing complexity the acidity and solubility decrease, the acids becoming more and more bland, oily, and eventually solid. The characteristic organic acid group is the carboxyl group —CO.OH, which imparts acidic properties to the molecule.

ACRYLIC   Derived from acrolein $CH_3=CH.CHO$ which in turn is derived from glycerine by the removal of two molecules of water. The presence of a double bond between the two carbon atoms permits the polymerization (q.v.) of acrylic compounds which are industrially important particularly in the case of methyl methacrylate ('Perspex') and acrylic fibres.

'ALCIAN'   The ICI trade mark for a series of dyes, some derived from phthalocyanine pigments (q.v.) and used mainly for cotton-fabric printing.

ALCOHOLS   Organic liquids containing the hydroxyl group —OH attached to a carbon atom, other than one forming part of an aromatic (q.v.) ring, e.g. methanol (methyl or wood alcohol), CH_3OH; ethanol (ethyl alcohol), C_2H_5OH, etc. Those containing long carbon chains, e.g. octanol, $C_8H_{17}OH$, are conventionally known as 'higher alcohols'. Alcohols have the property of combining with acids to form esters (q.v.) somewhat analogous to the neutralization of acids by alkalis, though the esters are very different in character from salts.

ALIPHATIC   Organic compounds of which the molecular structure is essentially based on carbon atoms linked together in linear chains are said to be 'aliphatic' (v. aromatic, petrochemical). The term is derived from the Greek for fat since this chemistry was first concerned with natural fats and particularly the 'fatty acids' dervied therefrom (v. hydrolysis).

ALKALI   (From the arabic al-qaliy, to roast in a pan. O.E.D.) Originally applied to the soluble products extracted from the ashes of vegetable or marine plants—hence potash. Their essential property is the ability to neutralize acids. Widely used throughout industry.

ALKYD   See under GLYPTAL.

ALLOY   A solid solution of two or more metals. Though a mixture and not a chemical compound, the presence of even a small amount of one component may so alter the crystal structure of the other as to produce great differences in the properties of the alloy from those of the components.

AMIDE   A compound containing the amide grouping —CO.NH$_2$ which may be obtained by the removal of the elements of water (H_2O) from the ammonium salt of an organic acid. Industrially the most important example is nylon (q.v.)

485

AMINES   May be regarded as 'organic ammonias' in which one, two, or three of the hydrogen atoms in ammonia (NH_3) are replaced ('substituted') by an organic group. The amines combine directly with organic or inorganic acids to form salts, as does ammonia, and have many industrial applications. Compounds with two amino groups are known as 'diamines', e.g. hexamethylene diamine, one of the two basic constituents of nylon.

AMMONIUM SULPHATE $(NH_4)_2SO_4$. The salt produced by the neutralization of sulphuric acid with two equivalents of ammonia. Its main importance is as a fertilizer either alone or in admixture. Originally (and still) produced from gasworks or coke-oven ammonia, it is now mostly based on ammonia synthesized directly from the elements nitrogen and hydrogen.

'ANTRYCIDE'   The ICI trade mark for a synthetic drug for the treatment of trypanosomiasis, popularly referred to as sleeping sickness.

AROMATIC   The term applied to that division of organic chemistry dealing with compounds having a structure based on closed chains or 'rings' of carbon atoms, particularly those derived from benzene (C_6H_6); naphthalene —a double ring ($C_{10}H_8$); and anthracene ($C_{14}H_{10}$)—a triple ring. So called because the earliest work in this field was concerned with natural substances which were 'aromatic' in the ordinary sense of the word.

BENZENE   A colourless volatile liquid of characteristic smell, the simplest aromatic (q.v.) compound. Originally obtained by carbonization of coal, it is now also produced from petroleum. The molecule consists of six carbon atoms linked in a ring structure with only a single hydrogen atom attached to each. It is quite reactive, forming a starting point for many organic syntheses (notably nylon) apart from its original applications as a solvent and fuel.

BENZENE HEXACHLORIDE   A powerful insecticide powder discovered by ICI in 1942, and almost simultaneously in France.

'BUNA'   The trade mark adopted by the IG for its synthetic rubbers based on butadiene (q.v.). 'Buna' S was a 'co-polymer' (v. polymer) of butadiene and styrene (q.v.) to which corresponded the American GRS synthetic rubber produced during the 1939-45 war.

BUTADIENE   A gaseous hydrocarbon which forms the major component of most synthetic rubbers. It is the most important 'di-olefine' (v. olefines), the molecule consisting of a four-carbon chain with two double linkages (bonds) between adjacent pairs of carbon atoms thus: $H_2C=CH—CH=CH_2$. It is therefore very reactive and readily polymerizable.

BUTYLENE   A gaseous hydrocarbon with one double bond and four carbon atoms in the molecular chain which may be either straight (normal) or branched (iso). Iso-butylene is readily polymerized and is important in the synthetic rubber field.

CARBONIZATION   The breaking up, by heating strongly, of complex organic substances into simpler compounds, generally volatile, leaving carbonaceous residues; e.g. the production of town gas, tar, and coke from coal.

CARBONYLATION   A reaction involving the addition of carbon monoxide, and frequently also hydrogen, to an unsaturated (q.v.) compound, as in the production of higher alcohols (q.v.).

CATALYST   A substance which, sometimes even in minute concentration, accelerates and/or directs a chemical reaction but is not consumed in it. Catalysts are therefore most important industrially and frequently permit use of less drastic conditions, such as lower temperatures and/or pressures, e.g. in synthetic ammonia, ammonia oxidation, polythene, oil-cracking processes, etc.

CONDENSATION   In physics, the transformation of a vapour into a liquid. In organic chemistry, the union, or combination, of two molecules by the elimination of the elements of water.

CRACKING   The treatment of petroleum fractions at high temperatures, whereby the molecular carbon chains are ruptured to yield simpler, more volatile products. Originally developed to increase the yield of petrol (gasoline) from heavy oils by heat alone, cracking now generally employs catalysts (q.v.) ('cat cracking'), and has become the mainstay of the petro-chemical industry, producing ethylene (q.v.), propylene (q.v. under olefines) butylene, butadiene (qq.v.), etc. from light oils described as naphtha (q.v.). The term is also sometimes applied to other similar reactions such as the splitting of ethylene dichloride to vinyl chloride and hydrochloric acid.

DYES   Intensely coloured organic compounds which can be firmly 'fixed' to various fibres, plastics and other materials to impart colour. They must resist fading by sunlight and removal by washing or dry cleaning; so must be insoluble in their final state, though in order to apply them it may be necessary to impart temporary solubility or render them so strongly adherent to the material as to resist removal by water or organic solvents. Thus the class known as vat dyes, insoluble as dyes, can be 'reduced' (so to speak, de-oxidized) to a soluble form which adheres to the fibre. On removal from the vat the soluble form is oxidized (often by the air alone) back to the insoluble dye which is now trapped in the fibre. The intense colouration is due to the presence in the dye molecule of certain atomic groupings known as chromo-phores, quite few in number (e.g. the 'azo' group —N=N—). The actual colour or shade is determined or modified by the rest (often the greater part) of the molecule which also determines the chemical behaviour, light and wash fastness, etc.

The advent of cellulose acetate and, later, polyester fibres which are virtually impenetrable by water necessitated new techniques. Means were ultimately found of dispersing the insoluble dye in such a finely divided form that it was adsorbed strongly by the fibre.

The search for reactive dyes, i.e. dyes which are fixed to the fibre (cotton) by reaction with the cellulose, bore fruit in early 1953 when ICI developed the 'Procion' (q.v.) range, first marketed in 1956.

ESTERS   The simpler esters are organic liquids, frequently of agreeable odour, and many are important as solvents and plasticizers. The more complex esters may be oily (e.g. olive or palm oils), waxy, or solid. The poly-ester of ethylene glycol with terephthalic acid is the basis of polyester fibre and film (v. glycol, phthalic, polymerization). Esters are formed by the union, with the elimination of water, of a molecule of an alcohol (q.v.) with one of an acid (q.v.), inorganic or, more generally, organic, somewhat analogous to the neutralization of an alkali by an acid, except that the esterification reaction is slow and reversible, and the esters, though neutral, are not salt-like in character.

ETHYLENE   A hydrocarbon gas, an important raw material for the manufacture of many products: e.g. polyethylene and polyvinyl chloride (v. olefines).

ETHYLENE GLYCOL   A colourless liquid, less viscous than glycerine which it somewhat resembles. Used as an anti-freeze and, most importantly, in the production of polyester fibre ('Terylene') and film. Contains two alcoholic (OH) groups in the molecule (v. glycol).

ETHYLENE OXIDE   The oxide of ethylene (q.v.), a very volatile and highly reactive liquid of great importance in many organic syntheses.

FINE CHEMICALS   Chemicals manufactured on a relatively small scale and generally by batch processes (v. heavy chemicals).

'FLUOTHANE'   The ICI trade mark for Halothane B.P., a derivative of ethane C_2H_6 containing one atom of chlorine, one of bromine, and three of fluorine in the molecule thus

$$F-C-CH\begin{matrix} F \\ F \end{matrix}\begin{matrix} Cl \\ Br \end{matrix}$$

Outstandingly successful as an anaesthetic (inhalant).

FUSEL OIL   A mixture of alcohols of intermediate boiling point (largely amyl alcohol) formed as by-products of alcoholic fermentation and separated during the distillation of spirits. Before the development of other (synthetic) solvents amyl alcohol was important as the basis for amyl acetate, one of the most useful solvents for nitrocellulose in the production of nitrocellulose lacquers and finishes.

GLYCERINE   A colourless heavy viscous liquid, a main constituent of the esters composing fats and vegetable oils. It is used as a 'humectant' or moisture retaining agent and as a softener for 'Cellophane'. Nitrated, it yields nitro-glycerine. The molecule contains three carbon atoms to each of which is attached an alcoholic (OH) group.

GLYCOLS   Are 'di-hydric' alcohols, i.e. compounds containing two alcoholic (hydroxyl) groups, e.g. ethylene glycol ($HO—CH_2—CH_2—OH$) used as anti-freeze and one of the components of 'Terylene' polyester fibre.

GLYPTAL   A contraction of glyceryl phthalate, designates a series of synthetic resins based on the polymerized ester of glycerine and phthalic acid or anhydride but modified by the addition during polymerization (q.v.) of linseed and/or other natural drying oils, etc. Used as the basis of synthetic paints and finishes. Also known as alkyd resins.

HEAVY CHEMICALS   The term indicates chemicals produced in large tonnages, usually now by continuous processes, as distinct from 'fine chemicals' (q.v.) produced on a small scale, frequently by batch processes.

HIGHER ALCOHOLS   See ALCOHOLS above.

HYDRATION   The chemical addition of water or its elements, to a molecule; or the addition of water in the formation of a crystal structure.

HYDROCARBONS   Compounds composed only of carbon and hydrogen, in which, therefore, all the carbon atoms are linked to each other, forming rings or chains, straight or branched.

HYDROGENATION   Chemical processes involving the addition of hydrogen to organic compounds, generally with the help of catalysts, and often under pressure.

HYDROLYSIS   The splitting of a compound by reaction with water, as liquid or vapour, frequently with the assistance of a catalyst; e.g. the splitting of natural fats and vegetable oils into glycerine and fatty acids.

INORGANIC   Originally implied chemistry and chemicals of mineral nature, i.e. not derived from or connected with living organisms (v. organic); includes a few simple carbon compounds (such as carbonates) and all elements other than carbon, and their compounds.

ISO-BUTYLENE   See under BUTYLENE.

ISOMERS   Chemical compounds composed of the same elements in exactly the same proportions, but arranged in different molecular structures and hence having different properties.

ISO-PROPANOL   An alcohol (q.v.) with a three-carbon branched chain thus:

$$\begin{matrix} CH_3 \\ \end{matrix} \!\! \diagdown \!\! \diagup \!\! CH\ OH$$
$$CH_3$$

obtained by the hydration (q.v.) of propylene (v. olefines); with many applications in solvent and other organic syntheses. Readily oxidized to acetone (q.v.).

ISOTOPES   Atoms with the same nuclear charge and hence the same number of orbital electrons similarly arranged and therefore indistinguishable chemically, but differing in the number of neutrons in the nucleus and hence differing slightly in mass (weight). Most elements occur as mixtures of two or more isotopes which explains why their atomic weights cannot be expressed as whole numbers. Thus chlorine, atomic weight 35·457, is a mixture of two isotopes, Cl 35 and Cl 37; uranium of U235 and U238.

METHANOL   Methyl or wood alcohol, a liquid boiling at 65 °C., was originally obtained by the distillation of wood tar, but is now made synthetically by the addition of hydrogen to carbon monoxide under high pressure and with the assistance of a catalyst. It is used as a solvent and in some countries as an anti-freeze. It is also an important intermediate or 'building block' for many organic syntheses.

'METHOXONE'   The trade mark for a growth-regulating plant hormone, discovered by ICI in 1940. Used as a selective weedkiller.

METHYL METHACRYLATE   A clear water-white liquid, the ester (q.v.) of methylacrylic acid and methanol, is produced from acetone, hydrocyanic (prussic) acid, and methanol. It can be polymerized (v. polymerization) with the aid of catalysts (q.v.) to a glass-clear highly transparent solid which is the basis of 'Perspex' and other such acrylic sheets.

MOLASSES   The syrupy liquors remaining from the successive crystallizations in the extraction and refining of cane or beet sugar. The most important in the chemical industry is that from the first stage known as 'blackstrap' molasses, which is very impure, unfit for human consumption but used for sweetening cattle foods and as a cheap source of fermentable sugar, e.g. for the production of alcohol, citric acid, acetone, and other products.

MONOMERS   The basic chemical units (molecules) which, linked together in chains or other molecular structures, form the polymeric substance of plastics, rubbers, and fibres (both natural and synthetic), etc.

'MYSOLINE'   The trade mark for an improved drug for the control of epilepsy, developed by ICI to avoid the undesirable side-effects of phenobarbitone previously used.

NAPHTHA   Originally applied to light tar oils, the term naphtha is now applied to very volatile liquid light petroleum fractions. These naphthas were originally used mainly as dry-cleaning media and as solvents for rubber, varnish, gums, etc. but are now the main raw material for petro-chemical processes.

'NONOX'   The ICI trade mark for a series of organic anti-oxidants used in the compounding and vulcanizing of rubber.

NYLON   The first fully synthetic fibre, invented by W. H. Carothers of du Pont, is the 'poly-amide' (v. amide) derived from the salt of adipic acid $(HO.OC.CH_2.CH_2.CH_2.CH_2.CO.OH)$ and   hexa-methylene   di-amine (v. amine) and hence known as nylon 66 (v. under polymerization) as both components contain six-carbon chains. These two building blocks were both synthesized from phenol or benzene. Other possible starting points were furfural and butadiene. A closely similar polymer was later developed in Germany, based on 'caprolactam', the anhydride of 'amino-caproic acid', $H_2N.CH_2$—CH_2—CH_2—CH_2—CH_2—$CO.OH$. With an amine (q.v.) group at one end and an acidic (q.v.) group at the other this can form long molecular chains in a manner exactly similar to adipic acid and hexa-methylene di-amine. This variety is known as nylon 6 (v. polymerization).

OCTANE NUMBER A measure, based on comparison with iso-octane (v. paraffins) at 100, of the suitability of a fuel for use in high-compression internal combustion engines.

OIL FROM COAL PROCESS This process may be regarded as a combination of cracking and hydrogenation and was originally developed by Bergius for brown coal (which contains considerably more hydrogen than bituminous coal) and achieved some small degree of success. The cracking (q.v.) of the coal structure results in a wide range of 'unsaturated' hydrocarbons which were simultaneously hydrogenated. Originally the powdered coal was mixed with creosote (from tar distillation) to facilitate injection into the continuous process but later it was found preferable to treat creosote alone which avoided trouble from the abrasive action of the ash in the coal. Propane and butane (v. paraffins) could be obtained as by-products.

OLEFINES Straight or branched chain hydrocarbons with one or more double linkages (bonds) between two adjacent carbon atoms and consequently readily reactive, e.g. ethylene $CH_2=CH_2$, propylene $CH_3-CH=CH_2$, butadiene (a di-olefine) $CH_2=CH-CH=CH_2$ etc. The name derived from ethylene which was early found to combine with chlorine giving an oily liquid (ethylene dichloride) and hence known as 'olefiant gas'.

ORGANIC Lemery in 1675 proposed this classification to cover those compounds occurring in the animal or vegetable kingdoms. Following the recognition that they all contained carbon and were generally more complex and markedly different in character from mineral substances, it was supposed that they could only be formed in a living organism under the influence of a 'life-force'—Berzelius' 'vis vitalis'—but this theory was exploded in 1828 by Wohler's synthesis of urea (q.v.)—a typical animal secretion. It is now recognized that every organic compound is (at least potentially) synthesizable from its elements, once its constitution is established, and that the distinctive character of so-called organic compounds is due to the unique propensity of carbon atoms to link up with each other in chains, rings, etc., thereby building complex molecular structures. The term 'organic' is therefore retained for convenience to cover the chemistry of all but a very few of the simplest compounds of carbon (e.g. carbides and carbonates).

'PALUDRINE' The ICI trade mark for Proguanil Hydrochloride B.P. This outstanding anti-malarial drug discovered by ICI in 1944 was the result of an intensive research effort to find drugs superior to the German 'Atebrin' (IG) or Mepacrine B.P. It was widely used in the 1939–45 war, and without such serious side effects.

PARAFFINS (from *parum affinis*—little affinity) are straight, normal, or branched iso chain hydrocarbons containing the maximum number of hydrogen atoms attached to the carbons, which latter, therefore, can only be connected together by single linkages. The molecules are therefore very stable and relatively unreactive. The paraffin series ranges from methane CH_4, the simplest, through ethane C_2H_6, propane C_3H_8, on to octane C_8H_{18} and beyond to kerosene, lubricating oils, and paraffin wax.

'PERSPEX' The ICI trade mark for glass-clear sheet polymer based on methyl methacrylate (q.v.). Though an ester (q.v.), it is not a polyester in the sense that the polymerization (q.v.) does not depend on building up a chain of ester linkages as in 'Terylene' but on the double bond in the acrylate structure.

PETROCHEMICALS A loose term, indicating chemicals produced from raw materials derived from petroleum. It would be better to refer rather to petrochemical *processes* since all so-called petrochemicals can be produced from other sources of carbon and hydrogen.

PHENOL Commonly called 'carbolic acid', a colourless crystalline solid of characteristic odour, is a most important intermediate or building block for many chemical syntheses, notably nylon. Chemically it is benzene with an OH (hydroxyl) group attached to the benzene molecule in place of a hydrogen atom, viz.

PHTHALIC ANHYDRIDE A white crystalline solid, important for the production of plasticizers, resins (v. glyptal), and as an intermediate in dyestuffs syntheses. Chemically it is the anhydride

of ortho-phthalic acid

It is manufactured by the catalytic

oxidation of naphthalene or xylene.

PHTHALOCYANINES A group of pigments and dyes based on the combination of four molecules of 'phthalonitrile' with a metallic atom, notably copper (fast blue lake dyestuffs). The phthalocyanines are extremely stable pigments. It proved very difficult to find means of conferring the temporary solubility necessary to convert them into dyes without sacrificing their fastness but ultimately the 'Alcian' dyestuffs, used for textile printing, were derived from them.

PIGMENTS Essentially intensely coloured insoluble substances, in the past mainly mineral or inorganic but now often organic, which are dispersed in a medium to coat and colour a surface, or in bulk to colour a solid. Some take the form of 'lakes', i.e. dyes precipitated on a carrier such as alumina.

PLASTICS Organic polymers (q.v.) which at some stage are suitable for forming to desired shapes by moulding, pressing, extrusion or some such process, generally assisted by heating. Classified in two main groups: (a) *Thermo-setting*, which harden permanently under the influence of heat, e.g. phenol-formaldehyde ('Bakelite'), urea-formaldehyde, melamine, etc.;

(b) *Thermo-plastic*, which soften under the influence of heat and can therefore be re-worked or processed further after forming, e.g. polyethylene, polyvinylchloride, polystyrene, polymethylmethacrylate ('Perspex'), celluloid and cellulose acetate.

POLYESTER A polymer (q.v.) in which the chain is built up by ester (q.v.) linkages between the basic units of the monomer (q.v.). The most important is that obtained from ethylene glycol and terephthalic acid which is the basis of polyester fibre ('Terylene') and similar fibres (v. under polymerization). Similarly synthetic resins derived from glycerine and phthalic anhydride form the basis of most paints (v. glyptal).

POLYMER A compound, the structure of which is, or may be regarded as, built up by the chemical linking together of a series of relatively simple molecular units (monomers) to form larger, and in most cases very large, molecules. The linking up is generally mainly in the form of linear chains. If much cross linking occurs between chains the polymers become infusible, but if not they are usually softened by heat (v. plastics). A co-polymer is one built up by similar linkage of two (or more) dissimilar monomers.

POLYMERIZATION The process of converting monomers (q.v.) to polymers (q.v.), which may involve different types of reaction in different cases. When only a single component is involved, polymerization generally depends on the presence in the monomer of one or more double linkages between pairs of adjacent carbon atoms. Such double bonds are reactive: one of the two links is readily broken and can then link up with a similar broken link in another molecule of the monomer, and so on, and thus build up a chain, carbon to carbon. E.g. ethylene polymerizes to polythene:

$$H_2C=CH_2 \longrightarrow \ldots -CH_2-CH_2-CH_2-CH_2-CH_2-CH_2- \ldots$$

and vinyl chloride to polyvinyl chloride (PVC).

$$CH_2=CH.Cl \longrightarrow \ldots -CH_2-CH-CH_2-CH-CH_2-CH- \ldots$$
$$ \underset{Cl}{|} \underset{Cl}{|} \underset{Cl}{|}$$

The polymers of propylene, butadiene, styrene, and the acrylics are of this type.

The second type of polymerization involves reactions between two different 'groups' to link up the constituent molecules. Thus the production of nylon (66) involves the union of two constituents, adipic acid ($HO.OC.CH_2.CH_2.CH_2.CH_2.CO.OH$)—a di-acid containing a chain of six carbon atoms. with an acidic (q.v.) group (—$CO.OH$) at each end, and hexamethylene diamine also a six-carbon chain compound but with an amine (q.v.) (—$CH_2.NH_2$) group at each end. When solutions of these (in methanol) are mixed 'nylon salt' is immediately formed and precipitated.

$$H_2N-(CH_2)_6-NH_3-O.OC-(CH_2)_4-CO.OH$$

By the crudest non-technical analogy this might be regarded as a sort of 'hook-and-eye' reaction: one of the two constituents (adipic acid and hexamethylene diamine) being regarded as having hooks at each end and the other eyes, so that the nylon salt has a hook at one end and an eye at the other. When the nylon salt is heated to about 280°C. further salt linkage ('hook-and-eye') takes place, building up extremely long chains. The salt linkage itself is not adequately stable but water is split off at each ester link converting it to the stable amide (q.v.) group, so the final polymer can be represented thus:

$$\left[\quad -NH-(CH_2)_6-NH-CO-(CH_2)_4-CO- \quad \right] \quad \text{N times}$$

Polyester is produced in a similar manner from ethylene glycol (q.v.) $HO.CH_2-CH_2OH$ and terephthalic acid (q.v.) $HO.CO-(C_6H_4)-CO.OH$, the polymer being represented:

$$\left[\quad -O.CH_2-CH_2O.CO-C_6H_4-CO- \quad \right] \quad \text{N times}$$

In this particular case the ester linkage is quite stable.

POLYTHENE (A contraction for polyethylene.) A tough translucent plastic produced by the polymerization (q.v.) of ethylene (q.v.), which can be extruded and drawn into film or tube or moulded into complex shapes. Has outstanding properties as an electrical insulator, especially at very high frequencies.

POLYVINYL CHLORIDE A thermoplastic material widely used for insulating electric cables, for flooring tiles and numerous other domestic and industrial applications; the polymer of vinyl chloride which is formed by the addition of hydrogen chloride ('hydrochloric acid gas') to acetylene (v. polymerization),

$$HC{\equiv}CH + HCl \quad \underline{\hspace{2cm}} \quad Cl.CH{=}CH_2$$
acetylene $\qquad\qquad\qquad$ vinyl chloride

or by the cracking (q.v.) of ethylene dichloride.

'PROCION' The ICI trade mark for 'reactive' dyes for cotton which are very simply applied from aqueous solution since the dye is fixed to the fibre by actual chemical reaction with the cellulose. Prior to this, dyeing of cotton depended on fixation of the dye to the fibre by purely physical means—adsorption, solution in the fibre substance, or mere entanglement in the fibre (as in vat dyeing). Only in the case of wool or silk was there any chemical attachment, in that case salt formation.

PYROLYSIS Literally 'treatment by fire'; refers to reactions effected simply by heating strongly. This would include the carbonization of coal; cracking (q.v.) is an extreme case.

RAYON The generic name for artificial or 'man-made' fibres based on cellulose. Chardonnet first produced rayon commercially by spinning solution of nitrocellulose which was subsequently de-nitrated to give a strong (but relatively expensive) fibre of regenerated cellulose. This was followed by cupramonnium fibre from cellulose dissolved in cupramonnium (strong ammoniacal copper oxide solution). Most rayon is now produced from *viscose* or cellulose xanthate, prepared by the reaction of carbon disulphide with cellulose impregnated with caustic soda ('soda-cellulose') and spinning the solution into dilute sulphuric acid. Acetate rayon is produced from cellulose acetate.

SALTS Compounds formed by the neutralization of an alkali by an acid, inorganic or organic, with the elimination of water. So called because of the resemblance to common salt of the typically crystalline inorganic salts known to early chemists.

STAPLE Originally rayon was produced in lustrous silk-like continuous filaments, hence described as 'artificial', or 'art' silk. Since then the main output has been in fibre cut to lengths to resemble cotton or wool, the term staple, borrowed from these natural fibres, referring to the length of the cut fibres.

STYRENE A colourless organic liquid; may be regarded as a compound of ethylene (v. olefines) attached to benzene. Due to the ethylene group it may be polymerized (v. polymerization) to polystyrene thus:

$$C_6H_5CH=CH_2 \longrightarrow \left[\begin{array}{ccc} -CH-CH_2-CH-CH_2-CH-CH_2- \\ | \qquad\qquad | \qquad\qquad | \\ C_6H_5 \qquad\quad C_6H_5 \qquad\quad C_6H_5 \end{array} \right] N \text{ times}$$

'SULPHAMEZATHINE' The trade mark for an improved sulphonamide drug developed by ICI. Following the recognition that the activity of the original sulphonamide drug, prontosil rubrum—an azo dye—was due to sulphanilamide formed by the breakdown of the dye in the body, efforts were made to increase the effectiveness of sulphanilamide and reduce its toxicity by addition to the molecule of other groups. Guanadine was early adopted and sulphaguanadine is still in use. Amino pryidine was then tried, followed by sulphadiazine which contains a pyrimidine ring (v. 'Paludrine'). 'Sulphamezathine', though prepared by a quite different synthetic route, also contains the pyrimidine ring but with two methyl (CH_3) groups attached thereto, giving still further improvement.

Sulphanilamide   Sulphaguanidine

'Sulphamezathine'

S

SULPHONATION   The addition of the sulphonic acid group —SO_2OH to organic, and particularly aromatic, compounds. This group is strongly acidic in character, forming salts etc., and is often introduced to confer water solubility.

SUPERPHOSPHATE   The early artificial fertilizer prepared by the action of sulphuric acid on phosphate rock which consists essentially of insoluble tri-calcium phosphate mixed with various impurities. Most of the phosphate is converted to the soluble mono-calcium phosphate, the rest of the calcium remaining in the product as gypsum (hydrated calcium sulphate) along with the other impurities. A typical analysis may show from 14 per cent to 20 per cent of soluble phosphate (expressed as P_2O_5). If phosphoric acid is employed instead of sulphuric, the formation of gypsum is avoided and 'triple super' is obtained, containing up to 50 per cent available phosphate (as P_2O_5).

SYNTHESIS   The building up of a desired compound from its elements or simpler groups.

TEREPHTHALIC ACID   One of the two components for the production of 'Terylene', is a solid 'di-acid' containing two acid (q.v.) groups in the molecule attached to a six-carbon benzene ring, thus

(cf. phthalic anhydride, q.v.)   HO.CO ⟨‾‾‾⟩ CO.OH

'TERYLENE'   The ICI brand of polyester fibre produced by the polymerization of ethylene glycol terephthalate, and melt-spinning the polymer (v. polyester, polymerization).

TITANIUM   A light metal, intermediate in density between aluminium and steel, with a high strength-to-density ratio, and resistant to chemical attack. First produced commercially by the Kroll process, involving the treatment of titanium tetrachloride with metallic magnesium, for which ICI later substituted sodium.

UNSATURATION   The carbon atom has the capability of combining with four hydrogen or certain other atoms and is therefore termed 'quadrivalent'. Usually in the formation of the characteristic chains of carbon atoms only one of these 'valencies' (links) is involved between each two adjacent carbon atoms, the remaining valencies being occupied by hydrogen or other atoms. The compound is then said to be 'saturated'. In certain cases, however, two or even three links or valencies may connect an adjacent pair of carbon atoms, with a corresponding reduction in the number of other attached atoms. The compound is then described as 'unsaturated'. An unsaturated compound is more reactive than a saturated one.

UREA $H_2N.CO.NH_2$. A most important organic intermediate, for example in thermo-setting resins (v. plastics), and rapidly increasing in importance as a high-strength nitrogenous fertilizer, formed by the reaction of two molecules of ammonia with one of carbon dioxide, a molecule of water being eliminated. The process is carried out under pressure and involves serious problems of corrosion and other difficulties.

'VELAN'   The ICI trade mark for a synthetic waterproofing agent, resistant to washing.

ICI Statistics 1926–1952

THESE STATISTICS (like those quoted in Appendix IV of Volume I) should be taken as indications of trend rather than exact statements of quantity. For that reason it has not been thought necessary, even if it were possible, to seek an exact reconciliation between figures quoted in the text, from various sources, with figures in this Appendix.

For the compilation of Tables 1 and 2 the historian is indebted to ICI Treasurer's Department.

TABLE I

ICI GROUP SALES, FINANCES, EMPLOYEES

Year	Sales	Capital Employed	Profit	Depre- ciation	Return on Capital Employed	5 year Moving Average	Employees (UK)
	£m.	£m.	£m.	£m.	%	%	'000
1927	26·9	72·8	4·9	0·9	7	—	47
1928	31·7	85·4	6·1	0·9	7	—	53
1929	35·0	98·9	5·9	1·2	6	—	57
1930	31·4	102·5	5·0	0·8	5	—	42
1931	32·1	99·1	3·9	1·3	4	6	37
1932	33·5	97·6	5·9	1·3	6	6	36
1933	37·3	97·9	6·6	1·4	7	6	42
1934	39·4	96·6	6·8	1·4	7	6	49
1935	42·7	96·1	7·0	1·4	7	6	50
1936	43·6	91·6	7·5	1·4	8	7	52
1937	54·1	94·8	8·4	2·0	9	8	57
1938	52·8	96·6	7·9	2·1	8	8	N/A
1939	62·1	97·1	11·4	2·7	12	9	75
1940	78·5	98·7	14·2	3·1	14	10	N/A
1941	94·4	98·5	15·2	3·0	15	12	N/A
1942	103·0	99·2	16·8	3·6	17	13	N/A
1943	111·4	100·1	13·3	3·6	13	14	N/A
1944	113·1	100·1	11·6	4·5	12	14	N/A
1945	103·5	104·7	8·3	4·8	8	13	N/A
1946	116·5	111·3	14·6	5·7	13	13	100
1947	136·7	115·2	14·4	7·3	13	12	N/A
1948	163·9	143·1	20·1	8·3	14	12	N/A
1949	174·6	147·6	16·0	8·2	11	12	120
1950	220·8	285·6	31·4	8·7	11	12	124
1951	262·7	307·4	41·0	8·9	13	12	131
1952	276·3	342·3	30·7	10·1	9	12	127

Notes:

(1) Pre-1938 profits as published have been adjusted to be consistent with profits of later years.

(2) Capital employed includes goodwill.

(3) The falls in capital employed between 1930 and 1931, and between 1931 and 1932, were mainly due to the writing down of physical assets following general reassessments. In addition, in 1931 The Magadi Soda Co. Ltd. was not consolidated as it had temporarily ceased to be a subsidiary because the Preference dividend was in arrears.

(4) The fall in capital employed between 1935 and 1936 was mainly due to the writing down of physical assets (principally the petrol and fertilizer plants at Billingham).

(5) Of the increase in capital employed between 1949 and 1950, £97m. was due to the revaluation of physical assets at 1 January 1950, and £20m. arose from the issue of a new loan stock.

(6) Profit is after depreciation but before taxation and fixed loan interest. In 1944–9 the following amounts have been charged against profits above, although at that time they were treated as appropriations of profit. The obsolescence and depreciation reserve formed by these appropriations has been deducted from fixed assets in calculating capital employed. These appropriations have been included under depreciation.

1944	£1·0m.	1947	£3·0m.
1945	£1·5m.	1948	£3·0m.
1946	£1·5m.	1949	£1·5m.

(7) The Profit figures include the following amounts of Investment Income:

Year	Investment Income	Year	Investment Income
	£m.		£m.
1927	1·2	1940	1·4
1928	1·1	1941	1·3
1929	1·7	1942	1·2
1930	1·5	1943	1·2
1931	1·1	1944	1·2
1932	1·2	1945	1·0
1933	1·2	1946	1·5
1934	1·4	1947	1·6
1935	1·4	1948	1·6
1936	1·5	1949	1·6
1937	1·7	1950	2·6
1938	1·3	1951	2·7
1939	1·5	1952	2·8

(8) Employees:

1927–1937	Main Home companies only.
1939 and 1946	Estimated.
1949–1952	Actual.
Other years	Not available.

TABLE 2

MANUFACTURING ACTIVITIES IN THE UNITED KINGDOM

These figures, compiled under the headings of ICI Divisions as they existed in 1970, are intended to illustrate the varying balance of activities within ICI during the period covered by the History. Because of changes in organization and accounting principles they can only be regarded as approximations.

Divisions (1970)	Year	External Sales	Capital Employed (Approx. figures)	Trading Profit (Approx. figures)	Return on Capital Employed
		£m.	£m.	£m.	%
Mond	1927	10·7	25·1	2·14	9
Heavy Chemicals	1932	8·5	24·9	2·74	11
including Alkali,	1937	14·0	31·3	2·31	7
General Chemicals, Lime	1942	23·7	26·9	5·68	21
and Salt Groups	1947	26·9	29·1	4·14	14
	1952	50·0	89·2	9·23	10
Metals	1927	3·8	2·0	·22	11
	1932	4·7	5·4	·40	7
	1937	10·1	7·3	·89	12
	1942	26·2	10·5	3·26	31
	1947	25·3	12·4	3·73	30
	1952	42·7	33·6	3·76	11
Paints	1927	1·0	·7	·20	29
(incl. Leathercloth)	1932	1·5	2·4	·30	13
	1937	3·1	3·3	·55	17
	1942	3·5	3·5	·54	15
	1947	6·7	5·1	·92	18
	1952	12·7	10·0	·41	—
Dyestuffs	1927	1·7	2·1	·05	2
(including fine	1932	2·7	4·5	·37	8
chemicals)	1937	4·5	5·2	·53	10
	1942	8·3	9·4	1·56	17
	1947	13·4	12·2	2·15	18
	1952	18·8	46·2	1·85	4
Nobel	1927	4·2	11·3	·78	7
Explosives	1932	3·4	9·1	·84	9
	1937	5·3	10·8	1·35	13
	1942	11·6	12·2	1·78	15
	1947	7·0	10·9	·97	9
	1952	15·1	23·7	1·68	7
Agricultural	1927	1·2	8·8	·10	1
including oil-from-coal	1932	5·7	21·0	·50	2
and other activities	1937	9·1	14·5	2·63	18
of Billingham Group	1942	18·0	12·5	3·37	27
and its predecessors	1947	29·2	11·4	3·03	27
	1952	59·6	65·5	8·00	12

TABLE 2—*continued*

Divisions (1970)	Year	External Sales	Capital Employed (Approx. figures)	Trading Profit (Approx. figures)	Return on Capital Employed
Plastics	1927	—	—	—	—
	1932	—	—	—	—
	1937	·3	·3	·02	7
	1942	2·0	·3	·34	—
	1947	6·2	3·4	1·11	33
	1952	10·4	10·0	·34	3
Pharmaceuticals	1927	—	—	—	—
	1932	—	—	—	—
	1937	—	—	—	—
	1942	·2	—	—	—
	1947	·8	·5	·03	6
	1952	2·7	1·4	·05	—

TABLE 3

CAPITAL EXPENDITURE

(a) Annual Capital Expenditure in the United Kingdom 1927–36

	£m.		£m.
1927	5·1	1932	0·9
1928	9·5	1933	1·2
1929	10·9	1934	4·4
1930	4·1	1935	2·8
1931	1·0	1936	2·4

'... in some cases the figures given do not agree with those shown in the ICI Annual Reports [because] the Report figures were in some cases provisional.'

Source: E. G. Minto to W. H. Coates, 3 Jan. 1944, CP 'Post-War General Memoranda'.

TABLE 3—continued

(b) Capital Expenditure in UK Divisions (formerly Groups), 1937–52

Group/Division	1937	1938	1939	1940	1941	1942	1943	1944	1945	1946	1947	1948	1949	1950	1951	1952	Total
	£000	£000	£000	£000	£000	£000	£000	£000	£000	£000	£000	£000	£000	£000	£000	£000	£000
Alkali			579	479	533	510	488	298	403	825	1,305	1,696	2,383	3,341	3,960	4,166	20,966
Billingham			456	635	312	199	196	160	237	420	759	1,879	3,991	5,099	4,396	3,501	22,240
Dyestuffs			180	790	1,627	1,554	1,180	364	309	855	1,468	3,283	2,501	3,376	4,291	4,749	26,527
Explosives/Nobel			243	140	78	110	100	74	52	251	455	509	757	1,463	1,378	1,297	6,907
General Chemicals			197	633	544	345	192	183	148	310	855	1,907	3,054	3,187	3,312	4,937	19,804
Leathercloth			26	10	6	4	4	4	12	33	128	227	264	282	188	243	1,431
Lime			94	48	33	37	43	15	87	137	340	502	358	141	154	305	2,294
Metals			614	622	585	350	286	173	194	341	565	670	975	1,813	2,027	2,222	11,437
Paints			34	33	11	30	30	34	34	125	238	241	254	248	281	344	1,937
Plastics			47	73	86	72	123	144	139	304	619	1,186	1,104	542	900	1,179	6,518
Salt			17	39	13	64	50	37	37	53	124	189	352	232	245	174	1,626
Central Agricultural Control								2	5	51	82	73	70	70	73	67	493
Wilton Works										729	1,405	1,927	1,653	1,320	2,411	1,204	10,649
Terylene Council										2	28	147	44	33	154	400	808
Total	3,121	3,033	2,487	3,502	3,828	3,275	2,692	1,488	1,657	4,436	8,371	14,436	17,760	21,147	23,770	24,788	133,637

Notes:

(1) These figures cover construction for the eleven Divisions, Central Agricultural Control, Wilton Works and Terylene Council. They do not cover working capital, special maintenance, or purchase of Government assets.

(2) In 1947 £3,166,000 was used to buy Government assets; in 1948, £2,775,000.

(3) Capital expenditure on behalf of the Government is not included.

(4) Total figures only are available for 1937 and 1938.

Source: Treasurer's Department Files 44/2/1–2.

TABLE 4

ANNUAL RESEARCH AND DEVELOPMENT BY ICI
IN THE UNITED KINGDOM SINCE 1927

Year	Research & Development Expenditure excluding Tech. Service (£000)	Year	Research & Development Expenditure excluding Tech. Service (£000)
1927 1928 1929	1,169* †	1940 1941 1942	829 948 1,119
1930	561* †	1943	1,325
1931	528* †	1944	1,877
1932	372†	1945	2,411
1933	419†	1946	3,022
1934	483†	1947	3,371
1935	591†	1948	4,090
1936	708†	1949	4,506
1937	818†	1950	4,830
1938	848	1951	5,177
1939	784	1952	6,044

* Includes Research Expenditure of 'Oil from Coal' at Billingham originally charged to Capital.

† Estimated by proportion from Research and Development + Technical Service + Process Control.

Note: Figures before 1946 have been adjusted by adding an estimate for Library and Patents costs to make them comparable with later figures.

Source: 'Some Salient Features of ICI Research', Appendix III, 21 April 1960. Prepared for ICI Board by ICI Research and Development Dept. CR 2/7/3.

ICI Directors, Secretaries, Treasurers, and Solicitors, 1926–1952

ICI DIRECTORS 1926–1952

Full-time Directors marked F

Name		From	To
Sir Alfred Mond (Created Lord Melchett 1928)	F	7.12.26	27.12.30 (Died)
Sir Harry McGowan (Created Baron 1937)	F	7.12.26	31.12.50
Harold J. Mitchell (Ceased to be full-time 31.7.38)	F	7.12.26	12.6.41
Henry Mond (Lord Melchett from 27.12.30)	F	7.12.26	24.7.47
John G. Nicholson (Knighted 1944)	F	7.12.26	31.7.45
Colonel George P. Pollitt (Ceased to be full-time 31.5.34)	F	7.12.26	31.12.45
Benjamin E. Todhunter	F	7.12.26	31.12.44
Lord Ashfield		7.12.26	4.11.48 (Died)
Sir John Brunner		7.12.26	13.10.27
George C. Clayton (Knighted 1933)		7.12.26	31.3.42
Sir Max Muspratt		7.12.26	20.4.34 (Died)
Lord Reading		7.12.26	30.12.35 (Died)
Sir Josiah C. Stamp (Created Baron 1938)		7.12.26	14.6.28
John Rogers	F	8.12.26	30.6.53
Lord Colwyn		14.7.27	22.5.39
Ernest J. Solvay		8.12.27	4.10.40
Lord Weir		14.6.28	18.6.53
Lord Birkenhead		8.11.28	30.9.30 (Died)
James H. Wadsworth	F	10.1.29	29.3.49
William H. Coates (Knighted 1947)	F	11.7.29	30.9.50
Holbrook Gaskell (Knighted 1942)	F	10.5.34	31.12.46
Harold O. Smith	F	9.1.36	31.12.51
Sir John Anderson		12.5.38 (first appointment)	1.11.38

Name		From	To
Peter F. Bennett		12.5.38	3.11.51
(Knighted 1941; created Baron 1953)			
Sir Andrew Rae Duncan		20.9.39	8.1.40
		(first appointment)	
Frederick W. Bain	F	11.4.40	23.11.50 (Died)
(Knighted 1945)			
Alexander J. Quig	F	11.4.40	31.12.56
Wallace A. Akers	F	9.1.41	30.4.53
(Knighted 1946)			
William F. Lutyens	F	9.1.41	30.4.53
Lord Glenconner		11.6.42	31.3.67
Cecil J. T. Cronshaw	F	11.11.43	31.12.52
Digby R. Lawson	F	11.11.43	7.12.47 (Died)
Alexander Fleck	F	8.6.44	29.2.60
(Knighted 1955; created Baron 1961)			
Lord Linlithgow		8.6.44	2.2.45
Arthur J. G. Smout	F	8.6.44	28.2.53
(Knighted 1946)			
F. Ewart Smith	F	27.9.45	31.3.59
(Knighted 1945)			
J. Lincoln Steel	F	27.9.45	24.3.60
(Knighted 1965)			
Sir Andrew Rae Duncan		27.9.45	30.3.52 (Died)
		(second appointment)	
S. Paul Chambers	F	10.7.47	29.2.68
(Knighted 1965)			
Valentine St. J. Killery	F	18.12.47	5.8.49 (Died)
Sir John Anderson		1.1.48	4.1.58 (Died)
(Created Viscount Waverley 1952)		(second appointment)	
Walter J. Worboys	F	27.5.48	31.10.59
(Knighted 1958)			
Peter C. Allen	F	26.7.51	29.3.71
(Knighted 1967)			
Eric A. Bingen	F	26.7.51	31.3.63
(Knighted 1966)			
Robert C. Todhunter	F	26.7.51	31.3.65
Alec T. S. Zealley	F	26.7.51	31.1.55
(Knighted 1957)			
John L. Armstrong	F	25.9.52	29.2.56
Percival K. Standring	F	25.9.52	31.3.57
Richard A. Banks	F	13.11.52	31.3.64
Ronald Holroyd	F	13.11.52	31.3.67
(Knighted 1963)			
Charles R. Prichard	F	13.11.52	30.6.60
James Taylor	F	13.11.52	31.8.64
(Knighted 1966)			
David J. Robarts		11.12.52	31.3.72

ICI SECRETARIES 1926–1952

Name	From	To
James H. Wadsworth	8.12.26	10.1.29
Philip C. Dickens	10.1.29	14.11.29
John E. James	14.11.29	31.12.44
Richard A. Lynex	1.1.45	31.3.62

ICI TREASURERS 1926–1952

Name	From	To
William H. Coates	8.12.26	14.11.29
Philip C. Dickens	14.11.29	1.11.45
John L. Armstrong	1.11.45	23.10.52
John H. Cotton	23.10.52	31.3.60

ICI SOLICITORS 1926–1952

Name	From	To
William Morris	1.1.27	31.12.38
Edgar C. G. Clarke (jointly with Bingen)	1.1.39	30.11.40
Eric A. Bingen	1.1.39	11.10.51
John W. Ridsdale	11.10.51	30.9.65

Text of the du Pont
Agreement, 1929

ARTICLES OF AGREEMENT made as of the 1st day of July, 1929, between IMPERIAL CHEMICAL INDUSTRIES, LIMITED, a corporation organized under the laws of Great Britain (hereinafter called "I.C.I."), party of the first part, and E. I. du PONT de NEMOURS & COMPANY, a corporation organized under the laws of Delaware, United States of America (hereinafter called "duPont"), party of the second part,

WITNESSETH:

WHEREAS, both I.C.I. and duPont are engaged in the development, manufacture and sale of a broad line of chemicals and chemical products, both in their respective home countries and in other countries, and maintain research and development organizations for the purpose of expanding their present activities as well as developing new industries; and

WHEREAS, each of the parties hereto desires the right to acquire licenses in respect of the patented and secret inventions of the other party, upon and subject to the conditions hereinafter set forth;

NOW, THEREFORE, in consideration of the premises and of the covenants herein contained, the parties have agreed as follows:

I. EXCHANGE OF INFORMATION.

(a) Each of the parties shall disclose to the other as soon as practicable, or in any event within nine months from the date of this agreement, or from the date of filing application for letters patent covering patented inventions, or from the time any secret invention becomes commercially established, information in respect of all patented or secret inventions now or hereafter during the life of this agreement owned or controlled by it, relating to the products hereinafter specified, sufficient to enable the other party to determine whether it desires to negotiate for licenses covering any or all of such inventions.

(b) Each of the parties agrees, whenever and so often as requested by the other, to furnish copies of all claims, specifications, applications and patents in respect of any such patented invention, and copies of all writings setting forth any such secret invention, and such further information as the other party shall request in respect of inventions relating to the products specified herein.

(c) Each of the parties shall forthwith appoint one or more competent, trustworthy and experienced persons in its employ for the purpose of receiving from the other party the information required to be disclosed under the foregoing provisions, and shall notify the other party of such appointment. Whenever and so often as the other party shall request, and at the expense of such other party, each party shall supply experienced chemists, engineers, foremen and other experts to assist such other party in investigating or testing any invention disclosed as aforesaid, or in applying or using any invention covering which license may have been granted to it hereunder, provided, however, that the party called upon for such technical assistance may arrange to furnish same at such time and in such manner as will not materially impede or interfere with its own activities and operations.

(d) An invention shall be deemed to be controlled, within the meaning of this agreement, whenever either party shall be able to grant to the other a license covering such invention within any territory or territories in which the other party may be entitled to demand exclusive or non-exclusive licenses under the terms hereof.

(e) Governmental objection or prohibition shall be a valid plea on the part of either of the parties to decline to reveal or to convey any rights under an invention which, but for such objection or prohibition, would come within the operation of this agreement.

II. RIGHTS TO ACQUIRE LICENSES.

(a) I.C.I. shall, upon request, grant to duPont the sole and exclusive license to make, use and employ, within the countries of North America and Central America, exclusive of Canada, Newfoundland and British possessions, but otherwise inclusive of the West Indies, and within all present and future colonies and possessions of the United States of America, any and all patented and secret inventions now or hereafter, during the life of this agreement, owned or controlled by I.C.I., relating to the products hereinafter specified, and to sell within said territories any and all of said products containing such inventions. (As referred to above, Central America shall be deemed to comprise the region between North and South America, extending from about N. latitude 7° to N. latitude 18°, that is from Colombia to Mexico, between the Caribbean Sea and the Pacific Ocean; and the West Indies shall be deemed to comprise those groups of islands lying off the southeast coast of North America and extending from near the coast of Venezuela northward to the latitude of North Carolina.)

(b) duPont shall, upon request, grant to I.C.I. the sole and exclusive license to make, use and employ, within the countries of the British Empire, inclusive of Egypt but exclusive of Canada and Newfoundland, any and all patented and secret inventions, now or hereafter, during the life of this agreement, owned or controlled by duPont, relating to the products hereinafter specified, and to sell within said territories any and all of said products containing such inventions.

(c) Each of the parties shall, upon request, grant to the other a non-exclusive license to make, use and employ, within any and all countries, other than Canada and Newfoundland, not within the exclusive territories specified above, any and all patented or secret inventions, now or hereafter, during the life of this agreement owned or controlled by the licensor, relating to such of the products hereinafter specified as are now manufactured by both parties, and to sell within said territories any and all of said products containing such inventions.

(d) Countries and territories not within the exclusive license territory of either party as defined above, but which may now or hereafter be administered under mandate by the British Empire or by the United States of America, or which may become a part of either sovereignty by proper authority, shall be considered as part of the British Empire or of the United States, respectively, so long as so administered; but whenever the respective sovereign power no longer exercises full political control over or administers any such country or territory, it shall be considered as non-exclusive territory under sub-paragraph (c) above.

(e) It is recognized that each of the parties may have established an internal trade in or export trade to a country or countries within the territory which under this agreement is designated as the exclusive license territory of the other party, and that in any such instance the other party as licensee may not be in a position to utilize the license or licenses granted to it with respect to such country or countries for the time being. It is, therefore, understood and agreed that in granting exclusive license or licenses covering any territory in which the licensor may have an established business, either internal or export, the licensor may, nevertheless, continue and fully enjoy the benefits of its operations therein until given reasonable notice by the licensee that the latter is in a position to utilize adequately its license in such territory.

(f) Licenses granted as aforesaid under any patented invention shall remain in effect to the end of the term for which such letters patent shall be granted or extended in the countries covered thereby, and licenses granted as aforesaid under any secret invention shall remain in effect so long as such invention shall remain secret, or, in event letters patent are subsequently obtained covering such invention, to the end of the term for which such letters patent shall be granted or extended in the countries covered thereby.

(g) Licenses granted as aforesaid shall be subject to adequate and justifiable compensation to be agreed upon by separate negotiations, but it is understood that such compensation will be determined under broad principles giving recognition to the mutual benefits secured or to be secured hereunder, without requiring detailed accounting or an involved system of compensation.

III. PRODUCTS.

The exchange of information provided in Section I, and the rights to acquire licenses granted in Section II, shall apply to all inventions relating to the following products and industries, subject to the exceptions set forth below:

(a) Explosives, other than military powders.

(b) Compounds of cellulose and its derivatives, including nitrocellulose compounds such as plastics and film, but excluding rayon, cellophane, explosives, and products covered under sub-paragraph (c) below; provided, however, that the activities of Societa Italia Celluloid and Societa Anomina Mazzucchelli (in which duPont has substantial stock interests), in this industry within the exclusive license territory of I.C.I., will continue until such time as may be mutually agreed upon between the parties hereto.

(c) Coated textile products, including components of those covered under sub-paragraph (b) hereof; provided, however, that with respect to inventions relating to such products the countries of Germany, Italy and France, including colonies and possessions thereof, shall be considered as the exclusive license territory of duPont, subject to application of sub-paragraph (e) of Section II to the present activities of I.C.I. in said territories.

(d) Paints, varnishes and lacquers, including the cellulose finishes known as Duco and Belco, and similar chemical finishes, and inclusive of synthetic resins and colloiding agents for use in paints, varnishes and lacquers, and plastics derived from cellulose; provided, however, that with respect to inventions relating to such products the countries of Germany, Italy and France, including colonies and possessions thereof, shall be considered as the exclusive license territory of duPont, subject to application of sub-paragraph (e) of Section II to the present activities of I.C.I. in said territories.

(e) Pigments, lakes and colors.

(f) Acids, both organic and inorganic, for both the heavy chemical industry and special industries.

(g) Chemicals of the general heavy chemical industry, excluding products of the general alkali industry.

(h) Dyestuffs, their intermediates, and other organic chemicals, including rubber chemicals; provided, however, that—

(1) while it is recognized that India, as coming within the British Empire, is the exclusive license territory of I.C.I., nevertheless, as an exception, it is agreed that, owing to the exceptional conditions obtaining in and the circumstances appertaining to that market, the dyestuffs activities of duPont in India may continue until such time as may be mutually agreed upon by the Presidents of the two companies, and that in determining the amount of compensation to be paid by I.C.I. under such licenses, due consideration shall be given to the extent of duPont's activities which are thereby terminated in such territory;

(2) the provisions of this agreement shall not apply to tetra-ethyl lead, but licenses with respect to said product may be the subject of separate negotiation;

(3) while inventions relating to dyestuffs and their intermediates are included in this agreement and subject to the provisions hereof, it is mutually agreed that, due to the exceptional conditions of said industry and the tentative negotiations during recent years with I. G. Farbenindustrie A. G., either party shall be free at any time to enter into separate agreement or arrangement with the latter company covering said industry. Upon the execution of such agreement or the entering into of such arrangement, this agreement insofar as it relates to the dyestuffs industry shall cease and terminate; provided, that all licenses theretofore granted under the terms of this agreement shall continue during the period for which granted, but all exclusive licenses so granted by each shall automatically become non-exclusive. Each

of the parties agrees, however, that in negotiating or upon entering into such an agreement or arrangement with said I. G. Farbenindustrie A. G., it shall use its best efforts to extend same to include the other party hereto.

(4) While inventions relating to dyestuffs and their intermediates are included in this agreement and subject to the provisions hereof, it is recognized that existing agreements and arrangements with other parties may prevent a full and mutual exchange of information and licenses relating to particular products of this industry; and it is therefore agreed that neither party shall be obliged to disclose information or to grant licenses under inventions relating to products of this industry, whenever in its opinion a full and reciprocal disclosure or grant of licenses relating to such products by the other party may be in conflict with existing agreements or relations of the latter.

(i) Synthetic ammonia, synthetic alcohol, and other products and by-products of the fixed nitrogen industry.

(j) Fertilizers.

(k) Synthetic products from the hydrogenation of coal and oil.

(l) Insecticides, fungicides and disinfectants.

(m) Alcohols manufactured by either synthetic or fermentation processes, other than synthetic alochol as covered in sub-paragraph (i) above.

The application of the rights granted hereunder relating to the products specified above, shall be subject to the terms of all existing relations and agreements between the parties hereto and between either or both of the parties hereto and other parties, as provided in Section X hereof. For purposes of reference only, a list of such agreements is attached hereto, marked "Schedule A";* it being understood, however, that such list is not intended to be all-inclusive.

IV. ELECTION TO ACCEPT LICENSE.

Whenever the party owning or controlling an invention relating to the products specified herein shall decide that it is advisable to utilize such invention or to exploit any product containing same within the territory which under this agreement is designated as the exclusive license territory of the other, it shall serve upon the other party a notice in writing setting forth the terms and conditions upon which the other party may obtain such exclusive license thereunder. The other party shall elect within a reasonable time after receipt of said notice whether it accepts such license upon the terms and conditions set forth in said notice, or upon such other terms and conditions as the parties may agree upon; but if the parties shall fail so to agree within a reasonable time, such license shall be deemed to have been rejected, and the party owning or controlling the invention shall be free to use same or to exploit the products containing same, and/or to license others so to use or exploit such invention or products, within such territory; provided, however, that no such license shall be granted to others upon terms and conditions

*Not reproduced.

more favorable than those offered to and rejected by the other party hereto, without giving to the latter a reasonable opportunity to accept such license upon such other terms.

V. NON-EXCLUSIVE LICENSES TO OTHER PARTIES.

Each party agrees that it will not sell, convey, or grant licenses or any other interest in or under any patent or invention relating to the products specified, to any other person whomsoever, covering any territory in which a non-exclusive license under such patent or invention has been or may be obtained by the other party under this agreement, without first advising the other party of its intention to make such grant or conveyance.

VI. CO-OPERATION IN SECURING NEW LICENSES.

Each of the parties agrees that if, during the continuance of this agreement, it shall obtain, acquire or possess a right in or license under any patented or secret invention relating to the products specified herein, which right or license is so limited that it can make no grant or license to the other party upon the terms and conditions herein set forth, it shall use its best efforts to assist such other party to obtain or acquire a right in or under such invention upon the terms and conditions herein set forth; but neither party shall be under any obligation to purchase or pay for any right or license for the benefit of the other.

VII. AID IN PROTECTING LICENSES.

(a) Each of the parties agrees to execute and deliver all such instruments in writing as may be necessary or proper for the purpose of further assuring and confirming any license granted pursuant to this agreement, or for the purpose of enabling such grants to be filed or recorded in any public office, and further to do whatever may be reasonably necessary to carry out the intent of this agreement.

(b) Should it appear at any time that any of the inventions covering which license has been granted to either party is the proper subject for letters patent in any territory for which rights have been so granted, the licensor will in conjunction with the first and true inventor, upon the request and at the expense of the licensee, apply for and use its best efforts to obtain the grant of letters patent or similar protection in respect of any of such inventions in such of said territories as the licensee may require, unless the party disclosing such invention demands that it be kept secret.

(c) Neither party shall be bound to defend any letters patent under which any license shall have been granted hereunder, but each of the parties agrees, whenever and so often as requested by the other party, but at the expense of such other party, to assist to the fullest possible extent in defending or protecting any such letters patent.

(d) Each party shall pay all fees and expenses for the maintenance of any patents in any territory in which the exclusive right shall have been granted to such party, and each party shall pay one-half of the fees and expenses for the maintenance of any patents in any territory in which joint rights exist under said patents in accordance with this agreement. Maintenance herein shall be deemed to include only payments of official fees, taxes and incidental expenses, but shall not include expenses of litigation.

VIII. DUTY NOT TO IMPAIR RIGHTS OF OTHER PARTY.

Each of the parties agrees not to make or consent to any disclosure or to do or consent to any other act that shall impair or depreciate the value of any license granted by it in pursuance of this agreement, or that shall impair or depreciate the value of the right, title and interest retained by the other party in any such patented or secret invention, and to take all reasonable care to prevent any such disclosure or act, but shall not, in the absence of bad faith or gross negligence, be liable in damages therefor.

IX. SUB-LICENSES.

Each of the parties to whom any license shall have been granted as herein provided may grant, within the limitations of such license, sub-licenses in respect thereof to any or all of its respective subsidiary companies; but every such sub-license shall be subject to all the terms and conditions contained in the grant of the license so sub-licensed and shall also contain terms, conditions, and obligations requiring such sub-licensee to do such acts as may be necessary or proper to enable the party granting such sub-license to observe all the terms and conditions and to perform all the obligations on its part contained in the grant of the license so sub-licensed. No sub-license in respect of any such license shall be granted by any sub-licensee, nor by either of the parties hereto, except as hereinbefore provided, without the consent in writing first obtained from the original licensor.

X. EFFECT OF EXISTING AGREEMENTS.

It is understood that both parties have established business relations through stock ownership in affiliated corporations and under agreements with other companies relating to the products specified herein, and each of the parties expressly recognizes that the provisions of this agreement are subordinate and subject to all such existing relations or agreements wherever it may conflict therewith. Each of the parties agrees, however, that in negotiating for the renewal of any of such relations or agreements which may expire during the existence of this agreement, it shall endeavor to effect such renewals on such basis or terms as will harmonize as fully as possible with the provisions of this agreement.

XI. ARBITRATION.

Should any difference or dispute arise between the parties hereto touching this agreement, or any clause, matter, or thing relating thereto, or as to the rights, duties, or liabilities of either of the parties hereto, the same shall be referred to the President for the time being of E. I. duPont de Nemours & Company and the President for the time being of Imperial Chemical Industries, Limited, who shall arbitrate, and their award shall be final. Should they not agree, they shall appoint an umpire, whose award shall be final, and the following provisions shall apply: If the question or matter to be decided is brought forward by I.C.I., the umpire shall be European; if, on the contrary, the question or matter to be decided is brought forward by duPont, the umpire shall be an American. Should the Presidents disagree as to the appointment of an umpire, then the umpire if a European, is to be appointed by the President of the Incorporated Law Society of England, and if an American, to be appointed by the President of the Association of the Bar of the City of New York.

XII PARTIES IN INTEREST.

(a) The benefits and obligations of this agreement shall inure to and be binding upon the parties hereto, and their respective legal representatives and successors, but shall not be assignable by either party without the consent in writing first obtained from the other party.

(b) The terms and provisions of this agreement shall apply to inventions owned or controlled by the respective subsidiary companies of each of the parties hereto, and each of said parties undertakes and assumes, for and on behalf of its subsidiary companies, all the duties and obligations of this agreement relating to such inventions.

(c) As used throughout this agreement, the term "subsidiary company" shall be deemed to mean any corporation in which either party owns or controls a majority of the outstanding voting stock, and any corporation similarly owned or controlled by any subsidiary or subsidiaries.

XIII. TERMINATION.

This contract shall continue in effect for a period of ten years from the date first hereinabove written.

IN WITNESS WHEREOF, E. I. du PONT du NEMOURS & COMPANY has caused its corporate seal to be hereunto affixed and this agreement to be signed in its corporate name by its President and Secretary, and IMPERIAL CHEMICAL INDUSTRIES, LIMITED, has caused its common seal to be hereunto affixed in the presence of and this agreement to be signed by one of its Directors and its Secretary at the City of London, England, as of the day and year first above written.

E. I. duPONT de NEMOURS & COMPANY

Attest   M. D. Fisher   By   I. du Pont

Assistant Secretary   President

IMPERIAL CHEMICAL INDUSTRIES, LIMITED

Attest   R. A. Lynex   By   H. McGowan

Assistant Secretary   President and Director

Source Material

PUBLISHED SOURCES

BOOKS

(The place of publication is London unless otherwise stated.)

Ashworth, W. *Contracts and Finance*. (HMSO and Longmans, 1953)
Association of British Chemical Manufacturers. *Report on the Chemical Industry, 1949*. (ABCM, Sept. 1949)
Badische Anilin & Soda-Fabrik AG. *In the Realm of Chemistry*. (Econ. Verlag, Düsseldorf/Vienna, 1965)
Barnett, Corelli. *The Collapse of British Power*. (Eyre Methuen, 1972)
Bäumler, E. *Ein Jahrhundert Chemie*. (Econ. Verlag, Düsseldorf, 1968)
Birkenhead, F. W. *The Prof. in Two Worlds*. (Collins, 1961)
Bolitho, H. *Alfred Mond, First Lord Melchett*. (Secker, 1933)
The British Economy Key Statistics 1900–1966. (Times Newspapers, n.d.)
Bruce, M. *The Coming of the Welfare State*. (Batsford, 1961)
Bullock, A. *The Life and Times of Ernest Bevin*. (Heinemann, Vol. 1 1960, Vol. 2 1967)
Cartwright, A. P. *The Dynamite Company*. (Purnell, Cape Town and Johannesburg, 1964)
Central Statistical Office. *Annual Abstracts of Statistics*. (HMSO)
——. *Statistical Digest of the War*. (HMSO and Longmans, 1951)
Chandler, A. D. (Jr.) *Strategy and Structure*. (MIT Press, 1962; Paperback Edition 1969)
Chandler, A. D. (Jr.) and Salsbury, S. *Pierre S. du Pont and the Making of the Modern Corporation*. (Harper & Row, New York, 1971)
Clark, R. W. *The Birth of the Bomb*. (Phoenix House, 1961)
——. *Sir Edward Appleton*. (Pergamon, Oxford, 1971)
Churchill, Sir Winston. *The Second World War, Vol. 1*. (Cassell, 1948)
Coleman, D. C. *Courtaulds*. 2 vols. (Clarendon Press, Oxford, 1969)
Concise Dictionary of National Biography. (Clarendon Press, Oxford, 1958)
Couzens, E. G., and Yarsley, V. E. *Plastics in the Modern World*. (Penguin Books, 1968)
Dutton, W. S. *Du Pont—One Hundred and Forty Years*. (Scribner, New York, 1942)
Evely, R., and Little, I. M. D. *Concentration in British Industry*. (Cambridge University Press, Cambridge, 1960)
Flechtner, H. J. *Carl Duisberg*. (Econ. Verlag, Düsseldorf, 1959)
Galbraith, J. K. *The New Industrial State*. (2nd edn. Hamish Hamilton, 1972)

Garnsey, G. *Holding Companies and their Published Accounts.* (Gee & Co., 1936)

Gowing, Margaret. *Britain and Atomic Energy 1939–1945.* (Macmillan, 1964)

Haber, L. F. *The Chemical Industry During the Nineteenth Century.* (Clarendon Press, Oxford, 1969)

——. *The Chemical Industry 1900–1930.* (Clarendon Press, Oxford, 1971)

Hardie, D. W. F., and Pratt, J. Davidson. *A History of the Modern British Chemical Industry.* (Pergamon Press, Oxford, 1966)

Hargreaves, E. L., and Gowing, M. M. *Civil Industry & Trade.* (HMSO & Longmans, 1952)

Haynes, W. *The American Chemical Industry.* 5 vols. (Van Nostrand, New York, 1948)

Hempel, E. H. *The Economics of Chemical Industries.* (Chapman and Hall, 1939)

Hexner, E. *International Cartels.* (Pitman, 1946)

Hobsbawm, E. J. *Industry and Empire.* (Penguin Books, 1969)

Hornby, W. *Factories and Plant.* (HMSO and Longmans, 1958)

Houghton, D. H. *The South African Economy.* (Oxford University Press, 1964 [3rd edn. 1973])

Jones, R., and Marriott, O. *Anatomy of a Merger.* (Cape, 1970)

Kaufman, M. *The First Century of Plastics.* (The Plastics Institute, 1963)

Keeble, Sir Frederick. *Fertilizers and Food Production.* (Oxford University Press, 1932)

League of Nations Statistical Year Book 1937–38. (Geneva, 1938)

Mason, E. S. (ed.) *The Corporation in Modern Society.* (Oxford University Press, 1960)

Medlicott, W. N. *Contemporary England 1914–1964.* (Longmans, 1967)

Mond, Sir Alfred. *Industry and Politics.* (Macmillan, 1927)

Moody's Manual of Investments. (New York, 1930)

Mowat, C. L. *Britain between the Wars 1918–1940.* (Methuen, 1968)

Neale, A. D. *The Antitrust Laws of the USA.* (Cambridge University Press, Cambridge, 1960)

Nichols, B. *Cry Havoc.* (Cape, 1933)

Norval, A. J. *A Quarter of a Century of Industrial Progress in South Africa.* (Juta, Cape Town, 1962)

Pelling, H. *A History of British Trade Unionism.* (2nd edn., Macmillan, 1972)

Pollard, S. *The Development of the British Economy 1914–1950.* (Edward Arnold, 1962)

Postan, M. M. *British War Production.* (HMSO and Longmans, 1952)

Postan, M. M., Hay, D., and Scott, J. D. *Design and Development of Weapons.* (HMSO and Longmans, 1964)

Reader, W. J. *Architect of Air Power, the Life of the First Viscount Weir.* (Collins, 1968)

Redfarn, C. A. *A Guide to Plastics.* (Iliffe, 1951)

Rees, G. *St. Michael.* (Weidenfeld & Nicolson, 1969)

Richardson, H. W. *Economic Recovery in Britain 1932–39.* (Weidenfeld & Nicolson, 1967)

Sasuly, R. *I.G. Farben.* (Boni & Gaer, New York, 1947)

Schumpeter, J. A. *Capitalism, Socialism and Democracy.* (4th edn., 1954; Unwin University Books, Geo. Allen & Unwin Ltd, 1970)

Scott, J. D., and Hughes, R. *Administration of War Production.* (HMSO & Longmans Green & Co., 1955)
Sloan, A. P. (Jr.) *My Years with General Motors.* (Sidgwick & Jackson, 1965; Pan Books Limited, 1967)
Statesman's Year Book. (Macmillan, 1934, 1940)
Taylor, A. J. P. *English History 1914–1945.* (Clarendon Press, Oxford, 1965)
ter Meer, F. *Die IG Farben.* (Econ. Verlag, Düsseldorf, 1953)
Threlfall, R. E. *The Story of 100 years of Phosphorus Making 1851–1951* (Albright & Wilson Ltd., Oldbury, 1951)
Turner, G. *Business in Britain.* (Eyre & Spottiswoode, 1969)
Walker, E. A. *A History of Southern Africa.* (Longmans, USA, 1957)
Warrington, C. J. S., and Nicholls, R. V. V. *A History of Chemistry in Canada.* (Chemical Institute of Canada, Toronto, 1949)
Who's Who and *Who Was Who.* (A & C Black, various dates)
Wilson, C. H. *History of Unilever.* 2 vols. (Cassell, 1954)
Winnacker, K. *Challenging Years. My Life in Chemistry.* (Sidgwick & Jackson, 1972)
World Economic Survey. (League of Nations, Geneva, 1934–5)

ARTICLES, LECTURES

Anon. 'A Century of Progress', *Contact; the Magazine of Canadian Industries Limited,* vol. 23, no. 6, June 1954.
Baddiley, J. 'The Dyestuffs Industry—Post-War Developments', *Journal of the Society of Dyers & Colourists,* 55, 1939, p. 236.
Biddle, F. 'Cartels: An Approach to the Problem'. Address to be given at the Annual Dinner of the Harvard Law School Alumni Association, 23 Feb. 1944.
Cook, J. W. 'Scientific Research', *The Times,* 21 April 1951.
Gibson, R. O. 'The Discovery of Polythene', *Royal Institute of Chemistry Lecture Series,* no. 1, 1964.
Gordon, K. 'The Development of Coal Hydrogenation by Imperial Chemical Industries, Ltd.', *Journal of the Institute of Fuel,* 9, Dec. 1935, p. 69.
Green, A. G. 'Landmarks in the Evolution of the Dyestuffs Industry during the past Half-Century', *Journal of the Society of Dyers & Colourists,* Jubilee issue, 1934, p. 49.
——. 'The Renaissance of the British Dyestuffs Industry', *Journal of the Society of Dyers & Colourists,* 46, Oct. 1930, p. 341.
Hackspill, L. 'Industrial Progress Due to the Use of High Pressures', *Chimie et Industrie,* 53, June 1945, pp. 387–92.
Holroyd, R. 'Hydrogenation', *Petroleum Technology,* 1936, p. 230.
——. 'Synthesis of Simple Nitrogenous Products'. Presidential Address to the Society of Chemical Industry Annual General Meeting in Dublin, *Chemistry & Industry,* 20 Aug. 1966, p. 1430.
Jensen, W. G. 'The Importance of Energy in the First & Second World Wars', *The Historical Journal,* XI, 3 (1968), p. 538.
Levinstein H. 'The Impact of Plastics on Industry—The Need for National Planning', *Chemistry & Industry,* 58, no. 9, 4 March 1939, p. 189.

Manning, W. R. D. 'High Pressure in the Chemical Industry', *The School Science Review*, 160, June 1965, p. 541.

Mitchell, H. J. 'Relationship between Headquarters' Departments and the Operating Groups of Imperial Chemical Industries, Ltd.' Paper to Seventh International Management Congress, Washington D.C., 1938, Administration Section.

Morton, Sir James. 'Dyes & Textiles in Britain', as delivered before the British Association, 1930.

——. 'History of the Development of Fast Dyeing & Dyes', *Journal of the Royal Society of Arts*, 29, April 1929, pp. 544–68.

Paine, C. 'Obituary Notice: C. J. T. Cronshaw', *Journal of the Society of Dyers & Colourists*, 77, April 1961, p. 163.

Perrin, M. W. 'The Story of Polythene', *Research*, 6, 1953, p. 111.

Robinson, C. S. 'Kenneth Bingham Quinan, 1878–1948', *The Chemical Engineer*, Nov. 1966, CE290.

Rose, J. D. 'Research for Survival', *Proceedings of the Royal Society*, A312, 1969, pp. 309–16.

——. 'Thirty-five Years of Organic Chemistry in the North West', Levinstein Memorial Lecture, *Chemistry & Industry*, 28 Nov. 1970.

——. 'Pollution in Perspective', *Chemistry & Industry*, 1971, p. 266.

Scott, W. D. 'The Chemical Raw Material Potential & its Relation to the Trend of Synthetic Resin Development', *Chemistry & Industry*, 29 July 1944, p. 274.

Slade, R. E. 'The Gamma Isomer of Hexachlorocyclohexane (Gammexane)— An Insecticide of Outstanding Properties'—Hurter Memorial Lecture, given in Liverpool, 8 March 1945, *Chemistry & Industry*, 13 Oct. 1945, p. 314.

Slade, R. E., Templeman, W. G., and Sexton, W. A. 'Plant Growth Substances as Selective Weed-killers', *Nature*, vol. CLV, 28 April 1945, p. 497.

Strevens, J. L., and Cross, A. C. 'Motor Spirit from Coal', *Nature*, 141, 7 May 1938, p. 812.

Swallow, J. C. 'Twenty-five Years of Polyolefines', *Chemistry & Industry*, 31 Oct. 1959, p. 1367.

Willson, D. R. 'Atomic Energy in the Pre-Harwell Era', *New Scientist*, 15 Aug. 1957, p. 31.

PARLIAMENTARY PAPERS

Government Papers

Statement Relating to Defence, 1935. Cmd.4827.

Statement Relating to Defence, 1936. Cmd.5107.

Employment Policy, 1944. Cmd.6527.

Removal of certain Excise Restrictions on Distillation and of Allowances on Industrial Alcohol and Exported Spirits, including as Appendix the Report of Industrial Alcohol Committee, 1945, Cmd.6622.

Proposals for Consideration by an International Conference on Trade and Employment, 1945. Cmd.6709.

Statements Relating to the Atomic Bomb, 1945. H.M. Treasury, HMSO.
Scientific Manpower, 1946. Cmd.6824.
Capital Investment in 1948, 1947. Cmd.7268.
Reports of the following:
Dyestuffs Industry Development Committee, 1930. Cmd.3658.
Royal Commission on the Private Manufacture of and Trading in Arms
(1935–6), 1936. Cmd.5292.
Committee of Imperial Defence Sub-Committee on Oil from Coal, 1938.
Cmd.5665.
Sir William Beveridge on Social Insurance and Allied Services, 1942.
Cmd.6404.
Committee on Hydrocarbon Oil Duties, 1945. Cmd.6615.
Committee on Coal Derivatives, 1960. Cmnd.1120.

Acts of Parliament
British Hydrocarbon Oils Production Act, 1934.
Monopolies and Restrictive Practices (Inquiry and Control) Act, 1948.

Hansard
Commons 1932–33, Vols. 279 and 280. Discussions on hydrogenation and oil
production from coal, June and July 1933.
Commons 1933–34, Vols. 285 and 286. British Hydrocarbon Oils Production
Bill.
Commons 1933–34, Vols. 284 and 286. Dyestuffs (Import Regulation) Bill.
Commons 1943–44, Vol. 397. Cartels and Monopolies, 24 Feb. 1944.
Lords 1943–44, Vol. 132. Employment Policy, 5 July 1944.
Lords 1944–45, Vol. 135. Cartels, 21 March 1945.
Commons 1944–45, Vol. 410. Ways & Means, Financial Statement, 24
April 1945.
Commons 1944–45, Vol. 411. Consolidated Fund debate, Cartels &
Monopolies, 13 June 1945.
Commons 1947–48, Vol. 449. Monopoly (Inquiry & Control) Bill, 22 April
1948.

UNITED NATIONS

Report of First Session of Preparatory Committee of United Nations Conference on Trade and Employment with Annexure 11, USA Draft
Charter, Oct. 1946.

PRIVATELY PRINTED AND PAMPHLETS

A National Policy for Industry, Nov. 1942. Statement by a group of industrialists.
(Mears & Caldwell)
*Addresses delivered on the occasion of the First Meeting of the Commercial and Tech-
nical Committees of Imperial Chemical Industries Limited (Chemical Group).*
(ICI, 30 Nov. 1926)
An Outline of the Chemistry and Technology of the Dyestuffs Industry. (ICI Dyestuffs
Division, 1953 and 1968)

Centenary of the Alkali Industry 1823–1923. (United Alkali Co., Widnes, 1923)
Fifty Years of Progress: The Story of the Castner-Kellner Alkali Company, 1895–1945. (ICI 1947)
Gavin, Sir William. *Jealott's Hill—A Record of Twenty-Five Years.* (ICI, July 1953)
Imperial Chemical Industries Limited—A Short Account of the Activities of the Company. (ICI, February 1929)
I.C.I. and its Overseas Interests Considered in Relation to the Question of Nationalisation. (Issued by ICI Private Industry Committee, July 1949)
I.C.I. The Parent Company and its Interests in the United Kingdom. (Issued by ICI Private Industry Committee, September 1950)
Imperial Chemical Industries of Australia & New Zealand Limited. A short account of the activities of the Company, its Subsidiary Companies and Associated Interests. (ICIANZ, n.d., about Jan. 1939)
Imperial Chemical Industries Record. (ICI 1927)
Labour Believes in Britain. (Labour Party National Executive, April 1949)
Landmarks of the Plastics Industry. (ICI Plastics Division, 1962)
Miles, F. D. *A History of Research in the Nobel Division of I.C.I.* (ICI Nobel Division, 1955)
National Situation—Report of Proceedings of Conference of Employers, 7th March 1946. (Issued jointly by British Employers' Confederation and Federation of British Industries)
Nobel Division. (Revised to January 1st, 1964, ICI)
Parke, V. E. *Billingham—The First Ten Years.* (ICI Billingham Division, 1957)
Private International Industrial Agreements. (Federation of British Industries Report, 1 January 1947)
Rationalisation of German Industry. (National Industrial Conference Board Inc., New York, 1931)
Sundour Golden Jubilee. (Speeches given at a Dinner held at Carlisle on 15 March 1956, Shenval Press)
Synthetic Organic Chemicals Manufactured by Carbide & Carbon Chemicals Corporation. (Carbide & Chemicals Corporation, New York, August 1934)
The Chemical Industry 1949. (Association of British Chemical Manufacturers Report)
The Launching of a New Synthetic Fibre: ICI 'Terylene' Polyester Fibre, A Historical Survey. (ICI June 1954)
The Manufacture of Agricultural Fertilizers at Billingham. (Synthetic Ammonia & Nitrates Ltd., May 1929)
The Possible Nationalisation of ICI. (Issued by ICI Private Industry Committee, 1949)
The Products of the Dyestuffs Division. (ICI April 1957)

UNPUBLISHED SOURCES
Air Ministry papers from Public Record Office
Cabinet papers from Public Record Office
Papers from archives of Solvay et Cie, Brussels
Trials of War Criminals from National Lending Library for Science and Technology

Unpublished Histories and collections of material

Anon. *African Explosives & Industries Ltd. History of the Ten Years 1927–1936.* (2 Feb. 1940)

Allen, P. C. *History of Polythene—Extracts from P. C. Allen's diaries September 1939—November 1945.*

Barncastle, H. J. *British Australian Lead Manufacturers Pty Ltd. History of the Company.* (Typescript, 3 volumes, 30 Sept. 1955)

Blainey, Geoffrey. *History of ICIANZ.* (Draft, Nov. 1959)

Bradley, F. *History of ICI's War Effort.* (Collected materials and some draft text)

Bunbury, H. M. *History of Imperial Chemical Industries Ltd 1926–1939.*

Bunbury, H. M., and Hodgkin, A. E. Material collected for Bunbury's History (see *ICI* I, p. 528).

Carter, G. F. *ICI's War Record.* (Annotated card index of documents)

Irvine, A. S. *The History of Polythene.* (Draft. Technical Director's Department, Winnington, 27 Jan. 1950)

Mumford, L. S. *Imperial Chemical Industries Ltd. A Concise History.* (Draft, 1964)

Osborne, W. F. *A History of the Early Development of 'Terylene' Polyester Fibre by Imperial Chemical Industries Ltd. November 1943—March 1951.* (Typescript, March 1960)

Penny, V. *Notes and Working Papers for ICI Plastics Division History 1923–1945.*

Stevens, F. P. *History of ICI 1939–1945.* (Typescript, 19 Feb. 1963)

The following are from ICI Head Office Archives:

ICI Directors' papers

Sir William Coates's collection of papers which covers almost the whole period and includes many papers duplicated elsewhere has been heavily drawn upon.

The papers of Henry Mond (Lord Melchett from 27.12.30) have also been heavily drawn upon. They are referred to as Melchett papers (MP). There is no collection of papers of the first Lord Melchett.

Other collections of Directors' papers (some small, some large) have been used to a varying extent and also collected papers of some members of ICI staff.

ICI Committee Minutes & Papers Mainly Used

Commercial Committee (Chemical Group)
Technical Committee (Chemical Group)
Executive Committee
Finance Committee
Fertilizer/Agricultural Committees
Research Council
Chairman's Conference Notes
Commercial Conference
Meetings of London Directors of ICIANZ (at first also called London Committee of ICIANZ or Australian Committee of ICI)
Administrative Committee
Board of Management

General Purposes Committee
Central Administration Committee
A very few papers for/notes of Chairman's Conferences
Sales Executive/Sales Committee
London Committee of Foreign Merchant Companies Group/Overseas Sales
Dept. London Committee
General Purposes Committee Research Committee
Commercial Committee
Alkali/Chlorine Committee
Special Meetings of Directors
Finance Executive Committee
Management Board
Commercial Executive Committee
Groups Central Committee
Personnel Executive Committee
Overseas Executive Committee
Research Executive Committee
Technical Executive Committee miscellaneous minutes
Nitrogen Committee
Du Pont Committee
Development Executive Committee
ICI/IG Committee
Government Contracts Committee
Engineering Supplies Committee/Government Relations Committee
Agricultural Policy Committee (subsequently called Agricultural Committee)
Secret War Committees (Main & Group)
Post War Committee
Main (Blacker) Bombard Committee/Special Arms Committee
Education Committee
Post War Works Committee/Construction Priority & Licences Committee
Some papers for/minutes of various Directors' Conferences
Notes of Quarterly Meetings of ICI Directors & Division Chairmen
Raw Materials for Heavy Organics Panel
Agricultural Division/Central Agricultural Control
Billingham Division Petroleum Chemicals Management Committee
Engineering Committee
AIOC/ICI Joint Policy Panel
Notes of Meetings of Certain Executive Directors
Monopoly Legislation Committee
Private Industry Committee
ICI/Solvay Committee
Capital Programme Committee

ICI Annual Reports and Proceedings of General Meetings
ICI Board Minutes and Reports to ICI Board
ICI Central Registry papers and letters
ICI Companies Information Book—maintained by Secretary's Department

ICI Departmental Annual Reports, 1927, 1928, 1929 (printed bound volumes)

ICI Foreign Department/Overseas Department papers

ICI Legal Department papers—especially documents and commentaries prepared for the defence to the action *USA v. ICI et al.* In them papers which have since been destroyed are quoted from and discussed. They include the Complaint (Civil Action No. 24—13) and Final Judgement (30 July 1952). Some collections are in sub-heads, which are given under Abbreviations, page xiii.

ICI Magazine, January 1928 onwards

ICI Organization Book—maintained by Secretary's Department

ICI Personnel Department papers

ICI Published Accounts

ICI Research and Development Department papers

ICI Secretary's Confidential papers

ICI Treasurer's Department papers

The Finance Company of Great Britain & America Ltd. Minute Books 1928-1933

Trafford Chemical Company—Share Register and Minute Book, 1938-1940

Papers have been kindly lent by the following:

Allen, Sir Peter	Steel, Sir Lincoln
Bristowe, W. S.	Weir, Lord (and Churchill College,
Hodgkin, A. E.	Cambridge)
Perrin, Sir Michael	Williams, L. H.
Sanderson, L. H. F.	Witt, F.
Smith, Sir Ewart	

People who have kindly supplied information:

Allen, Sir Peter	Chandler, A. D., Jr.
Arnold, Mrs. L.	Cheveley, S. W.
Bagnall, F. C.	Clapham, Sir Michael
Barker, Professor T. C.	Cockram, C.
Barley, L. J.	Coleman, Professor D. C.
Batty, J. K.	Cook, C. B.
Bayliss, G. M.	Cotton, J. H.
Beeby, G. H.	Cousin, J. D.
Beeching, Lord	Cox, A. W. J.
Bingen, Sir Eric	Craven, F. J. P.
Blewett, W. V.	Cross, D.
Bristowe, W. S.	Fleck, Lord
Brown, R. B.	Forbes, Sir Archibald
Bunbury, H. M.	Ford, H.
Butchart, H. D.	Gibb, R. M.
Caress, A.	Goodfellow, R. M.
Chambers, Sir Paul	Gowing, Professor Margaret

Haber, L. F.
Hinton, Lord
Hodgkin, A. E.
Hodgkin, E.
Holroyd, Sir Ronald
Hughes, J.
Irvine, A. S.
Irvine, F. M.
Kamm, E. D.
Kearton, Lord
Lamond, J.
Langford, E. J.
Leigh, S. P.
Lutyens, W. F.
Lynex, R. A.
Manning, R. E.
Manning, W. R. D.
Marsh, D.
Melchett, Lord
Munro, L. A.
McGowan, W. J.
Nicholson, D. J. S.
Nories, W. M. C.
Paine, C.
Patrick, L.
Payman, J.
Pearce, G. O.

Pennock, R. W.
Perrin, Sir Michael
Preston, R. W. D.
Prichard, C. R.
Robinson, C. S.
Robinson, J.
Robinson, Professor Sir Robert
Rogers, J.
Rose, J. D.
Rideal, Professor Sir Eric
Sanderson, L. H. F.
Smith, Sir Ewart
Smith, H.
Springford, Miss W. M.
Steel, Sir Lincoln
Storey, W. R.
Taylor, A. W.
Taylor, Sir James
Todhunter, R. C.
Vaughan, E.
Whetmore, S. A. H.
Williams, L. H.
Witt, F.
Woods, A. G.
Woods, R.
Wride, D. H. B.

Index

The heading 'Imperial Chemical Industries Ltd.' does not appear in this index. Instead, this title is assumed in all headings dealing with the aspects of the Company's organization and activities covered in the text.

T

heavy chemicals and industry: based on ammonia-soda process and ammonia synthesis, 183; compared with fine chemicals, 183; compared with organic chemicals, 129–30; controlled by BM and UAC, 15–17; development, decline and resurgence, 5, 129–30, 183, 447, 465; founded on family firms, 474–5; high pressure technology, 162; suppliers to plastics industry, 342; suppliers to the textile industry, 366–8; tradition of regulated competition, 6, 368

heavy chemicals and industry (Australia), 209, 212

heavy chemicals and industry (Canada), 214

heavy chemicals and industry (Germany), 33

heavy chemicals and industry (USA), 36

heavy chemicals (organic), *see* heavy organic chemicals

heavy organic chemicals and industry: AIOC/ICI production at Wilton, 404; Barley's views, 319; early neglect and late development, 4–5, 319–25, 329; heavy chemicals compared, 129–30; manufacturing policy, 325–7, 394; non-competitive policy, 325–7; as raw materials, 319–27, 394–5; source materials, 84, 319–25, 394–7, 401, 403, 406–7

heavy organic chemicals and industry (Germany), 130

heavy organic chemicals and industry (USA), 36–7, 129–30, 319, 325

heavy water, 289–90

'Heliogen' Blue B, *see* dyestuffs and industry (Germany)

Henderson, Leon, 422

Heysham agency factory, 266, 275

high pressure technology, 98, 128, 162, 168; *see also* ammonia synthesis; hydrogenation

higher alcohols: question of manufacture, 324–5, 327

Hill, Rowland, 346, 375–6, 387

Hillhouse Works, *see* General Chemical Group

Hirsch, Alfredo, 224–5, 228

Hodgkin, Adrian Eliot: Mouldrite chairman, 354; Plastics Division chairman, 345–6; proposals for plastics, 340, 344–8; views IHP Agreement, 170; views on N fertilizers, 99–101; views on oil-from-coal, 172–3, 177–8; views on polythene, 355–6

Holford Works, *see* Metal Group

holidays, 300, 302

Holliday, Major L. B., 185

Holliday, L. B. & Co., 185, 192, 194

Holroyd, Sir Ronald, FRS, 396, 451

Home Controller of Sales Organization, 235

Honorary President, *see* President

Hopf, H. J., 366

household goods: market for polythene, 356

Howard, F. A., 168

Huddersfield agency factory, 263, 274

Huddersfield Works, *see* British Dyestuffs Corporation Ltd.; Dyestuffs Group

Hudson, Rt. Hon. R. S., 284–5

Humphrey, H. A., 44, 82

Hurst, Sir Alfred, KBE, CB, 173

hydrocarbon gases, *see* oil refinery gases

hydrocarbon oils, *see* oil; *see also* oil-from-coal; oil fuels

Hydrocarbon Oils Production Act 1934, *see* British Hydrocarbon Oils Production Act 1934

hydrocarbons, 169, 319–27, 395–6

hydrochloric acid, 16, 361

hydrochloric acid (Australia), 209

hydrochloric acid (Iran), 397

hydrogen: production at Billingham, 128, 158–9, 263, 275, 392: at Cardington, 275: at Weston-super-Mare, 275; as a raw material, 128, 162, 379, 391–3; redundant plant used, 128, 171, 180; war production, 263, 274

hydrogenation: agency factory at Heysham, 266, 275; and ammonia synthesis, 10, 128, 162; and Patents and Processes Agreement 1929, 166, 168–9; and US anti-trust litigation, 422; high pressure technology, 128, 162; IHP Agreement, 167–70

Hydrogenation, Cabinet Sub-Committee on Coal, *see* Cabinet Sub-Committee on Coal Hydrogenation

Hydrogenation Cartel, 170, 174, 195, 266, 413–14

hydrogenation of coal, *see* oil-from-coal

hydrogenation of crude oil, *see* oil cracking

hydrogenation of gas oils, *see* oil cracking

hydrogenation petrol, *see* oil fuels; *see also* oil-from-coal

hydrogenation products, *see* oil cracking; *see also* oil fuels; oil-from-coal

T

SA, *see Sociedad Anonima*

SA & N, *see* Synthetic Ammonia & Nitrates Ltd.

SAI, *see* Scottish Agricultural Industries Ltd.

SAIC, *see Sociedad Anonima Industrial e Comercial*

S-IG, *see* Standard-IG

SPC, *see* Solvay Process Company Inc.

Safetex Safety Glass Ltd., 341

safety fuse (Australia), 207–8, 212

safety fuse (New Zealand), 207

St. Clair du Rhône, *see* Compagnie Francaise des Produits Chimiques et Matières Colorantes de St. Clair du Rhône

St. Denis, *see* Société des Matières Colorantes et Produits Chimiques de Saint Denis

salaries: Administrative Committee's powers, 140; chemists, 78, 84; cuts due to the Depression, 119–20; Chairman, 134, 246; directors, 23, 474; disparity among Groups, 77–80; disparity with wages, 299; McGowan, 133–4, 246, 308; Managing Director, 133, 246; President, 133; Salaries Committee, 246, 268

Salaries Committee, 246, 268

sales: agents and agencies, 199–200, 204, 226, 255, 281; central control, 26, 137, 200–1, 235; direct sales to farmers advocated, 106–7; government business in war-time, 256–7, 279; Home Controller of Sales Organization, 235; Nicholson's views, 137, 235–6; no competition with customers, 6, 323, 342, 346; no direct sales to the public, 335; organization dictated by market-sharing, 235; by overseas subsidiaries, 200–2; Sales Committee, 235; selling prices controlled by McGowan, 235; 'venture trading', 200, 226; war-time figures, 256–7, 279–80; *see also* exports; government contracts

Sales Committee, 235

sales offices overseas, *see* subsidiary and associated companies (overseas)

Sales Organization, Home Controller of, *see* Home Controller of Sales organization

sales policy, *see* commercial policy

sales quotas, *see* markets and market-sharing

salt, 324, 392, 397

salt (Brazil), 415

salt (Iran), 397

Salt Division, 298

Salt Group, 269

Salt Union Ltd., The, 269

Sampson, Col. M. T., 98–9, 346

Sanderson, L. H. F., 239

Sandoz SA, 191

Scheme M. RAF Expansion, *see* aircraft and industry

Schmitz, Hermann, 40, 42, 147, 149–50, 188

Schnitzler, Dr. Georg von, 44–5, 191–2

scholarships, *see* ICI Scholarships

schools, 70–6

Schumpeter, Prof. J. A.: *Capitalism, Socialism and Democracy*, 425

Scientific Advisory Committee, 290–1

scientific staff, *see* technical staff

Scottish Agricultural Industries Ltd., 108–9

Scottish Dyes Ltd., 185–6

seal of the company, 8

Second World War, 254, 257, 281, 286, 358, 413

Secret War Committee, 288, 290

Secretary: Administrative Committee member, 140; Central Administration Committee member, 142; Dickens, 73; Groups Central Committee member, 269; James, 267; responsibilities, 24; Wadsworth first Secretary, 24

segregation, *see* organization

selection of staff, *see* management and staff

selling agents, *see* agents and agencies

selling organization, *see* sales

selling prices, *see* prices

service agreements, *see* agreements

service departments, *see* Head Office

shadow factories, *see* agency factories; *see also* armaments industry

Share Ownership Scheme, *see* Employees' Share Investment Scheme

shareholders, 23, 118, 468, 473, 478

shareholdings, *see* investments

shares: Deferred shares at merger, 19–20, 238; effect of Depression, 117–18; Employees' Share Investment Scheme, 62–3; exchange with Courtaulds proposed, 368, 375; exchange with founding families, 475; exchange with IG proposed, 40–3; exchange with merger companies, 7, 18–20, 127; 4% Unsecured Loan Stock 1950, 464; McGowan's share-dealings questioned, 241–3, 246; A. Melchett's support of depressed

Africa), 9, 199, 204–6; *see also* African Explosives & Chemical Industries Ltd.; African Explosives & Industries Ltd.

subsidiary and associated companies (South America), 38, 199–200, 219–31

'Sulphamezathine', 458

sulphate of ammonia, *see* ammonium sulphate

Sulphate of Ammonia Federation, *see* British Sulphate of Ammonia Federation

sulphonamides, 281, 317, 458

sulphur and sulphur products (Argentina), 224

sulphuric acid, 16, 366

sulphuric acid (Canada), 214

sulphuric acid (Iran), 397

Sun Engraving Co. Ltd., 287

superphosphates, 100, 108

superphosphates (Australia), 208–10

superphosphates (Canada), 214

superphosphates (South Africa), 205

Supply, Ministry of, *see* Ministry of Supply

Swallow, J. C., 351, 353, 362

Swint, W. P., 39, 227, 370, 385, 433–4

Swinton, 1st Earl (Sir Philip Cunliffe-Lister), 21

Swiss IG, 45, 185, 187–93

synthetic ammonia, *see* ammonia and ammonia synthesis

Synthetic Ammonia & Nitrates Ltd., 41

synthetic detergents, 6, 447

synthetic detergents (Germany), 33

synthetic fibres: acrylonitrile, 387; Barley's views, 366; cellulose, 89, 365–6; cellulose acetate, 366; chemical nature, 365; competition or co-operation with Courtaulds, 164, 365–9, 371–9; decision on independent development, 386–7; early development, 286, 317; 'Fibre A', 387; 'Fibrolaine', 369; from groundnuts, 369, 380; H. Melchett's interest, 366; from polyamides, 370–2; from polyesters, 381; from protein, 365, 369, 381; '66 Polyamide' (nylon), 370; post-war potentialities, 387; raw materials, 366–9, 379–82; *see also* individual fibres, e.g. 'Terylene', nylon, 'Ardil'

synthetic fibres (France), 366

synthetic fibres (Germany), 33

synthetic fibres (USA): 'Orlon', 387

synthetic gasoline, *see* oil-from-coal; *see also* oil fuels

synthetic nitrogen, *see* nitrogenous fertilizers and industry

synthetic petrol, *see* oil-from-coal; *see also* fuels

synthetic polymers, *see* synthetic fibres

synthetic resins, 84, 318, 329, 335, 339, 346; *see also* plastics and industry

synthetic resins (Germany), 33, 332

synthetic rubber, 332, 334–6, 393

synthetic rubber (Germany), 32–3, 332, 334–5, 358

synthetic rubber (USA), 36, 334

Synthite Ltd., 340

TGWU, *see* Transport and General Workers' Union

TNT, *see* trinitrotoluene

TUC, *see* Trades Union Congress

tanks: explosives for anti-tank weapons, 276

tanks, fuel, *see* fuel tanks

tar, *see* coal tar and creosote

Tar Residuals Ltd., 180

tariffs: British Hydrocarbon Oils Production Act 1934, 129; Dyestuffs (Import Regulations) Act 1921, 45, 184, 186, 192, 329; Imperial Preference, 9, 130, 216, 218; import duty on oil, 171, 177, 323, 400; 'inconvenience allowance' on industrial alcohol, 322–3, 395, 400; protection for ammonium sulphate, 149: for dyestuffs, 45, 184, 186, 192, 216, 329, 331: for methanol, 218: for industrial alcohol, 323, 395, 400: for oil-from-coal, 129, 171, 178–9, 264, 323

tariffs (Chile), 220

tariffs (USA), 436, 440

Tate, M. G., 420–1

taxation: and establishment of overseas companies, 200; and liquidation of NIL, 237; excess profits tax (EPT), 280; expertise of Coates and Chambers, 310; on oil-from-coal profits, 180; on overseas profits, 200; on profits on sales of shares, 237; on wartime profits, 257, 280–1

'tear gas', 262

Technical Committee, 25–7

Technical Department, 86

Technical Director, 303, 307, 407

Technical Executive Committee, 268